T0342282

DEFECTS IN SOLIDS

DEFECTS IN SOLIDS

Richard J. D. Tilley
Emeritus Professor, University of Cardiff, Wales

WILEY

A JOHN WILEY & SONS, INC., PUBLICATION

Library of Congress Cataloging-in-Publication Data:

Tilley, R. J. D.
  Defects in solids/Richard J. D. Tilley.
     p. cm.
  Includes indexes.
  ISBN 978-0-470-07794-8 (cloth)
  1. Solids—Defects. 2. Solids—Electric properties. 3. Solids—Magnetic properties.
4. Solids—Optical properties. I. Title.
  QC176.8.D44T55 2008
  620.1′1—dc22   2008001362

10 9 8 7 6 5 4 3 2 1

To Professor F. S. Stone
who engendered a unique research group spirit
and my contemporaries in the University of Bristol,
who augmented it.

■■■■■■ CONTENTS

■ PREFACE

During the course of the last century, it was realized that many properties of solids are controlled not so much by the chemical composition or the chemical bonds linking the constituent atoms in the crystal but by faults or defects in the structure. Over the course of time the subject has, if anything, increased in importance. Indeed, there is no aspect of the physics and chemistry of solids that is not decisively influenced by the defects that occur in the material under consideration. The whole of the modern silicon-based computer industry is founded upon the introduction of precise amounts of specific impurities into extremely pure crystals. Solid-state lasers function because of the activity of impurity atoms. Battery science, solid oxide fuel cells, hydrogen storage, displays, all rest upon an understanding of defects in the solid matrix.

As the way in which defects modify the properties of a solid has been gradually understood, the concept of a defect has undergone considerable evolution. The earliest and perhaps simplest concept of a defect in a solid was that of a wrong atom, or impurity, in place of a normal atom in a crystal—a so-called point defect. Not long after the recognition of point defects, the concept of linear defects, dislocations, was invoked to explain a number of diverse features including the mechanical properties of metals and the growth of crystals. In recent years it has become apparent that defect interactions, aggregation, or clustering is of vital importance.

The advance in understanding of defects has been made hand in hand with the spectacular development of sophisticated experimental techniques. The initial breakthrough was in X-ray diffraction—a technique that still remains as the foundation of most studies. A further change came with the development of transmission electron microscopy that was able, for the first time, to produce images of defects at an atomic scale of resolution. Since then, advances in computing techniques, together with the availability of powerful graphics, have thrown quite new light on the defect structure of materials.

Besides the multiplicity of defects that can be envisaged, there is also a wide range of solid phases within which such defects can reside. The differences between an alloy, a metallic sulfide, a crystalline fluoride, a silicate glass, or an amorphous polymer are significant. Moreover, developments in crystal growth and the production of nanoparticles have changed the perspective of earlier studies, which were usually made on polycrystalline solids, sometimes with uncertain degrees of impurity present.

All of these changes have meant that the view of defects in solids has changed considerably over the last 20 years or so. This book is aimed at presenting an overview of this information. However, the topic encompasses a huge subject area, and selection has been inevitable. Moreover, courses about defects in solids, whether to undergraduate or postgraduate students in chemistry, physics, geology, materials science, or engineering, are usually constrained to a relatively small part of the curriculum. Nevertheless, the material included in this book has been chosen so that not only basics are covered but also aspects of recent research where exciting frontiers lie. Unfortunately, the influence of defects upon mechanical properties is mostly excluded, and the important area of surfaces and surface defects is only mentioned in passing. Similarly, the area of studies prefixed by nano- has been bypassed. This is because the literature here is growing at an enormous rate, and time was not available to sift through this mountain of data. Although these omissions are regrettable, they leave scope for future volumes in this series. Finally, it must be mentioned that most emphasis has been placed upon principles, leaving little space for description of important experimental details. This is a pity, but within the constraints of time and space, became inevitable. Again, a future volume could correct this shortcoming.

The first four chapters introduce basic concepts that are developed to build up a framework for understanding defect chemistry and physics. Thereafter, chapters focus rather more on properties related to applications. Chapter 5 describes diffusion in solids; Chapter 6, ionic conductivity; Chapters 7 and 8 the important topics of electronic conductivity, both intrinsic (Chapter 7) and extrinsic (Chapter 8). The final chapter gives a selected account of magnetic and optical defects.

To assist in understanding, each chapter has been prefaced with three "introductory questions" that focus the reader upon some of the important points to be raised in the following text. These are answered at the end of each chapter. In addition, end-of-chapter questions aid understanding of the preceding material. The first set of these consists of multiple choice questions—a "quick quiz," to test knowledge of terms and principles. This is followed by a number of more traditional problems and calculations to build skills and understanding in more depth. In addition, supplementary material covering the fundamentals of relevant topics such as crystallography and band theory are included so that, in the first instance, a reader will not have to look elsewhere for this information.

Each chapter contains a short list of additional sources that expand or give a different perspective on the material in the preceding chapter. Most of these are books or original scientific literature, and only a few web sources are listed. The Internet provides a data bank of considerable power but has two drawbacks: It is easy to become swamped by detail (often a simple search will throw up a million or more allegedly relevant pages), and much of the information located has an ephemeral nature. Rather than list a large number of sites, the reader is encouraged to use a search engine and keywords such as "magnetic defects," which will open an alternative perspective on the subject to that presented in this book.

I have been particularly helped in the compilation of this book by family, friends, and colleagues. Professor R. B. King first suggested the project, and Drs. E. E. M. Tyler, G. J. Tilley, and R. D. Tilley made suggestions about contents and scope. Professor F. S. Stone offered encouragement and kindly offered to read sections in draft form, resulting in invaluable advice and comment that added substantially to the clarity and balance. Mr. A. Coughlin was constantly encouraging and gave assistance when mathematical discussions of topics such as polynomials or random walks became opaque and provided valued help into the proofs. The staff of the Trevithick Library, University of Cardiff, have been helpful at all times and continually located obscure references. The staff at Wiley, gave constant advice, help, and encouragement. Dr. John Hutchison, University of Oxford, provided stunning micrographs for which I am greatly indebted. Finally, my gratitude to my wife Anne cannot be understated. Her tolerance of my neglect and her continued encouragement has allowed this project to reach a conclusion.

Comments and queries will be gratefully received and can be sent to tilleyrj@cardiff.ac.uk or rjdtilley@yahoo.co.uk.

R. J. D. TILLEY

CHAPTER 1

Point Defects

What is a point defect?
What is the "effective charge" on a defect?
What is an antisite defect?

1.1 INTRODUCTION

Defects play an important part in both the chemical and physical behavior of solids, and much of modern science and technology centers upon the exploitation or suppression of the properties that defects confer upon a solid. Batteries, fuel cells, displays, data storage, and computer memories all directly utilize, or have evolved from, an understanding and manipulation of defects in inorganic materials. This technology has been developed over some 80–100 years and started with the simplest concepts. However, as the effect of defects upon the properties of the solid gradually became appreciated, the concept of a defect has undergone considerable evolution. The simplest notion of a defect in a solid was the idea of a mistake such as a missing atom or an impurity in place of a normal atom. These structurally simple defects are called point defects. Not long after the recognition of point defects, the concept of more complex structural defects, such as linear defects termed dislocations, was invoked to explain the mechanical properties of metals. In the same period it became apparent that planar defects, including surfaces and grain boundaries, and volume defects such as rods, tubes, or precipitates, have important roles to play in influencing the physical and chemical properties of a solid.

Defects can thus be arranged in a dimensional hierarchy (Fig. 1.1*a*–1.1*d*):

1. Zero-dimensional defects—point defects
2. One-dimensional (linear) defects—dislocations
3. Two-dimensional (planar) defects—external and internal surfaces
4. Three-dimensional (volume) defects—point defect clusters, voids, precipitates.

This and the following chapter are concerned with point defects.

Defects in Solids, by Richard J. D. Tilley
Copyright © 2008 John Wiley & Sons, Inc.

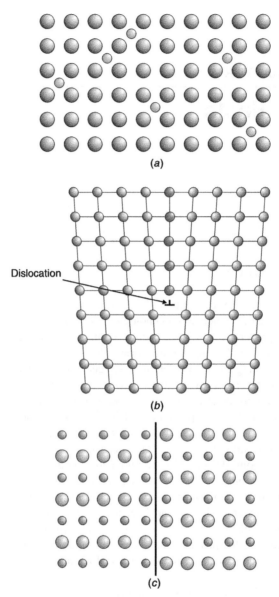

Figure 1.1   Defects in crystalline solids: (*a*) point defects (interstitials); (*b*) a linear defect (edge dislocation); (*c*) a planar defect (antiphase boundary); (*d*) a volume defect (precipitate); (*e*) unit cell (filled) of a structure containing point defects (vacancies); and (*f*) unit cell (filled) of a defect-free structure containing "ordered vacancies."

In addition to the defects listed above, which may be termed structural defects, there are also electronic defects. The first of these are electrons that are in excess of those required for chemical bonding and that, in certain circumstances, constitute charged defects that can carry current. In addition, current in some materials is carried by particles

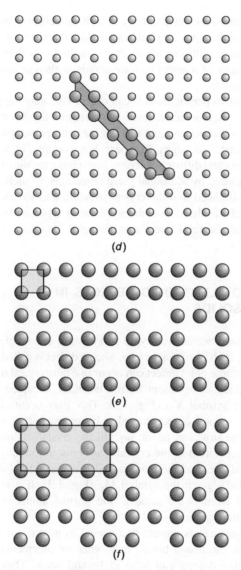

Figure 1.1 (*Continued*).

that behave rather like positive electrons, and these form the second type of electronic defect.[1] They are called electron holes, positive holes, or more often just holes.

Besides the multiplicity of defects that can be envisaged, there is a wide range of host solid phases within which such defects can reside. The differences between an alloy, a metallic sulfide, a crystalline fluoride, or a silicate glass are significant from

[1]These are not the positive equivalent of an electron, a positron, because such a particle would be eliminated instantaneously by combination with an ordinary electron, but are virtual particles equivalent to the absence of an electron. They can be considered to be analogs of a vacancy, which is the absence of an atom.

both a chemical or physical viewpoint. By default, defects have come to be associated with crystalline solids because a crystal has a regular repetition of atoms throughout its volume.[2] A disturbance of this regularity then constitutes the defect (Fig. 1.1*a*–1.1*d*). In this sense, the nature of a defect in, for example, a glass or amorphous polymer, is more difficult to picture.

When circumstances permit, defects can order. This may happen at low temperatures, for example, when defect interactions overcome the disordering effect due to temperature. From a diffraction point of view the ordered sample no longer contains defects, as the characteristic signs disappear. For example, in X-ray analysis broadened diffraction spots and diffuse scattering, both characteristic of disordered crystals, disappear, to be replaced by additional sharp reflections. In such cases the defects are incorporated or assimilated into the crystal structure and so effectively vanish. The original unit cell of the defect-containing phase is replaced by a new unit cell in which the original "defects" are now integral components of the structure (Fig. 1.1*e* and 1.1*f*).

1.2 POINT AND ELECTRONIC DEFECTS IN CRYSTALLINE SOLIDS

The simplest localized defect in a crystal is a mistake at a single atom site in a pure monatomic crystal, such as silicon or iron. Such a defect is called a point defect. Two different types of simple point defect can occur in a pure crystal of an element, M. An atom can be imagined to be absent from a normally occupied position, to leave a vacancy, given the symbol V_M (Fig. 1.2). This may occur, for instance, during crystal growth. A second defect can also be envisaged; namely an extra atom incorporated into the structure, again, say, during crystallization. This extra atom is forced to take up a position in the crystal that is not a normally occupied site: it is called an interstitial site, and the atom is called an interstitial atom (or more often simply an interstitial), given the symbol M_i, (Fig. 1.2). If it is necessary to stress that the interstitial atom is the same as the normal atoms in the structure, it is called a self-interstitial atom.

The various point defects present in a crystal as grown, for example, vacancies, interstitials, excess electrons, holes, and other arrangements, are called *native defects*. These native defects can arise in several ways. They can be introduced during crystal growth, as mentioned above, especially if this is rapid and crystallization is far removed from equilibrium. Defects can also form after the crystal is developed if the crystal is subjected to irradiation by high-energy particles or other forms of radiation. Defects that result from such processes are called *induced defects*. When a pure crystal is heated at a moderate temperature for a long period, a process called annealing, the number of native defects will gradually change. However, no matter how long the sample is annealed, a population of point defects will always remain, even in the purest crystal. These point defects are in thermodynamic equilibrium

---

[2]An introduction to crystal structures and nomenclature is given in the Supplementary Material Section S1.

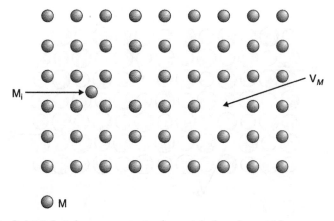

Figure 1.2   Point defects in a pure monatomic crystal of an element M, a vacancy, V_M, and a self-interstitial, M_i.

(Chapter 2) and cannot be eliminated from the solid. They are called *intrinsic* point defects. This residual population is also temperature dependent, and, as treated later (Chapter 2), heating at progressively higher temperature increases the number of defects present.

As well as these intrinsic structural defect populations, electronic defects (excess electrons and holes) will always be found. These are also intrinsic defects and are present even in the purest material. When the equilibria among defects are considered, it is necessary to include both structural and electronic defects.

Turning to pure *compounds*, such as CaO, $MgAl_2O_4$, or FeS, the same intrinsic defects as described above can occur, but in these cases there is more than one set of atoms that can be affected. For example, in a crystal of formula MX, vacancies might occur on metal atom positions, written V_M, or on nonmetal atom positions, given the symbol V_X, or both. Similarly, it is possible to imagine that interstitial metal atoms, written M_i, or nonmetal atoms, written X_i, might occur (Fig. 1.3). The different sets of atom types are frequently called a sublattice, so that one might speak of vacancies on the metal sublattice or on the nonmetal sublattice.

No material is completely pure, and some foreign atoms will invariably be present. If these are undesirable or accidental, they are termed impurities, but if they have been added deliberately, to change the properties of the material on purpose, they are called dopant atoms. Impurities can form point defects when present in low concentrations, the simplest of which are analogs of vacancies and interstitials. For example, an impurity atom A in a crystal of a metal M can occupy atom sites normally occupied by the parent atoms, to form substitutional point defects, written A_M, or can occupy interstitial sites, to form interstitial point defects, written A_i (Fig. 1.4). The doping of aluminum into silicon creates substitutional point defects as the aluminum atoms occupy sites normally filled by silicon atoms. In compounds, the impurities can affect one or all sublattices. For instance, natural sodium chloride often contains

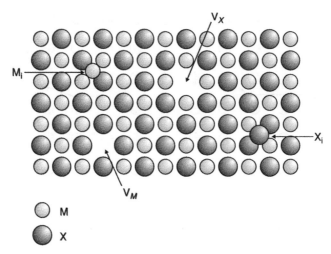

Figure 1.3 Point defects in a crystal of a pure compound, MX, V_M, a metal vacancy; V_X a nonmetal vacancy; M_i, a metal (self-)interstitial; and X_i a nonmetal (self-)interstitial.

potassium impurities as substitutional defects on sites normally occupied by sodium ions, written K_{Na}, that is, the impurities are associated with the metal sublattice. Impurities are called extrinsic defects. In principle, extrinsic defects can be removed by careful processing, but in practice this is very difficult to achieve completely.

Impurities can carry a charge relative to the host structure, as, for example, with a Ca^{2+} ion substituted on a Na^+ site in NaCl or F^- substituted for O^{2-} in CaO. In essence, this means that the impurity carries a different chemical valence, that is,

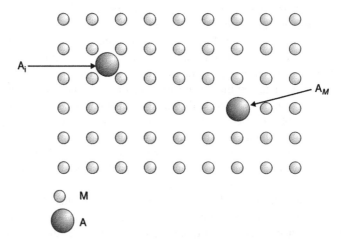

Figure 1.4 Impurity or dopant (A) point defects in a crystal of material M, substitutional, A_M; interstitial, A_i.

it comes from a different group of the periodic table than the host atoms. Deliberate introduction of such impurities is called aliovalent or altervalent doping. The introduction of charged impurities will upset the charge neutrality of the solid, and this must be balanced in some way so as to restore the electronic neutrality of the crystal. One way in which this compensation can be achieved is by the incorporation of other vacancies or interstitial atoms that carry a balancing charge. The substitution of a Ca^{2+} ion for Na^+ in NaCl can be balanced, for example, by the introduction at the same time, of a sodium ion vacancy, V_{Na}, at the same time. This mechanism is called ionic compensation, structural compensation, or less often self-compensation. However, compensation can also take place electronically by the introduction of appropriate numbers of electrons or holes. For example, the substitution of F^- for O^{2-} in CaO could be balanced by the introduction of an equivalent number of electrons, one per added F^-. Impurities that are compensated by excess electrons in this way are called donors or donor impurities, while those that are compensated by addition of holes are called acceptors or acceptor impurities. One consequence of these alternatives is that the electrons and holes present in the solid must be included in the overall accounting system used for assessing the defects present in a solid.

The importance of point defects in a crystal cannot be overstated. They can change the physical properties of a solid significantly. To introduce the range of changes possible, Sections 1.3–1.6 outline some of the physical properties that are influenced in this way.

1.3 ELECTRONIC PROPERTIES: DOPED SILICON AND GERMANIUM AS EXAMPLES

Silicon, Si, lies at the heart of most current electronic devices. Both silicon and the similar semiconductor element germanium, Ge, crystallize with the diamond structure (Fig. 1.5). In this structure each atom is surrounded by four others arranged at the corners of a tetrahedron. Each atom has four outer electrons available for chemical bonding, and these are completely taken up by creating a network of tetrahedrally oriented sp^3-hybrid bonds. Although pure silicon and germanium are intrinsic semiconductors, these properties are inadequate for the creation of sophisticated electronic materials for which selected impurity doping is necessary. The resulting materials are called extrinsic semiconductors because of the extrinsic nature of the defects that give rise to the important conductivity changes.

Impurity atoms will upset the orderly arrangement of bonding electrons and this changes the electronic properties. Doping with a very small amount of an impurity from the next *higher* neighboring group of the periodic table, phosphorus (P), arsenic (As), or antimony (Sb), results in the formation of substitutional point defects in which the impurities occupy normal sites, for example, a phosphorus atom on a site normally occupied by a silicon atom, P_{Si} (Fig. 1.6a). Each of these atoms has five valence electrons available for bonding, and after using four to form the four sp^3-hybrid bonds, one electron per impurity atom is left over. These electrons are easily liberated from the impurity atoms by thermal energy and are

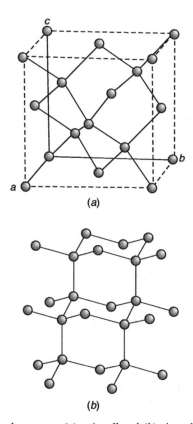

Figure 1.5   Diamond structure: (*a*) unit cell and (*b*) viewed with [111] vertical.

then (more or less) free to move through the crystal under the influence of an external electric field. Each dopant atom contributes one extra electron to the crystal. They are defects, and are given the symbol e′ (the superscript indicates a single negative charge relative to the surroundings) to differentiate them from ordinary electrons. The doped material conducts mainly using these electrons, and, as they are negatively charged, the solid is called is called an *n*-type semiconductor. The atoms P, As, or Sb in silicon or germanium are called donors as they donate extra electrons to the crystal. In terms of band theory they are said to occupy states in the conduction band (Supplementary Material S2), the donors themselves being represented by localized energy levels just below the conduction band in the crystal (Fig. 1.6*b*).

An analogous situation arises on doping silicon or germanium with elements from the next *lower* neighboring periodic table group, aluminum (Al), gallium (Ga), and indium (In). The impurity atoms again form substitutional defects, such as Al_{Si} (Fig. 1.6*c*). In this case the impurities have only three outer bonding electrons available, which are not sufficient to complete four bonds to the surrounding atoms. One bond is an electron short. It simplifies understanding if the missing electron is regarded as a hole, represented by the symbol h•. The superscript indicates that the hole carries a

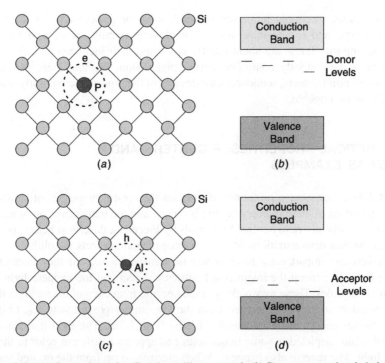

Figure 1.6 (*a*) Donor impurity (P$_{Si}$) in a silicon crystal. (*b*) Donor energy levels below the conduction band. (*c*) Acceptor impurity (Al$_{Si}$) in a silicon crystal. (*d*) Acceptor energy levels above the valence band.

positive charge relative to the surroundings. Each impurity atom introduces one positive hole into the array of bonds within the crystal. Thermal energy is sufficient to allow the holes to leave the impurity atom, and these can also move quite freely throughout the crystal. The doped material conducts mainly using these holes, and, as they are positively charged, the solid is called a *p*-type semiconductor. The impurities are termed acceptors because they can be thought of as accepting electrons from the otherwise full valence band. On an energy band diagram the acceptors are represented by normally vacant energy levels just above the top of the valence band (Fig. 1.6*d*), which become occupied by electrons from that band on thermal excitation. The holes thereby created in the valence band provide the means for conductivity.

The idea of a hole is widely used throughout electronics, but its exact definition varies with circumstances. It is frequently convenient to think of a hole as a real particle able to move throughout the crystal, a sort of positive electron. It is equally possible to think of a hole as an electron missing from a localized site such as an atom or a covalent bond. Thus a metal *cation* M^{2+} could be regarded as a metal *atom* plus two localized holes (M + 2 h$^{\bullet}$). In band theory, a hole is generally regarded as an electron missing from the top of the valence band. All of these designations are more or less equivalent, and the most convenient of them will be adopted in any particular case.

The unique electronic properties of semiconductor devices arise at the regions where p-type and n-type materials are in close proximity, as in $p-n$ junctions. Typical impurity levels are about 0.0001 at %, and their inclusion and distribution need to be very strictly controlled during preparation. Without these deliberately introduced point defects, semiconductor devices of the type now commonly available would not be possible.

1.4   OPTICAL PROPERTIES: F CENTERS AND RUBY AS EXAMPLES

Point defects can have a profound effect upon the optical properties of solids. The most important of these in everyday life is color,[3] and the transformation of transparent ionic solids into richly colored materials by F centers, described below, provided one of the first demonstrations of the existence of point defects in solids.

Defects can impart color to an otherwise transparent solid if they interact with white light. In general the interaction between a defect and the incident light is via electrons or holes. These may pick up some specific frequency of the incident illumination and in so doing are excited from the low-energy ground state, E_0, to one or more higher energy excited states E_1, E_2, E_3 (Fig. 1.7a). The light that leaves the crystal is thus depleted in some frequencies and appears a different color to the incident light. The reverse also happens. When electrons drop from the excited states to the ground state E_0, they release this energy and the same light frequencies will be emitted (Fig. 1.7b). The relationship between the energy gained or lost, ΔE_n, and the frequency, ν, or the wavelength, λ, of the light absorbed or emitted, is

$$E_n - E_0 = \Delta E_n = h\nu_n = \frac{hc}{\lambda_n}$$

where E_n is the energy of the higher energy level ($n = 1, 2, 3, \ldots$), h is Planck's constant, and c is the speed of light.

The first experiments that connected color with defects were carried out in the 1920s and 1930s by Pohl, who studied synthetic alkali halide crystals. A number of ways were discovered by which the colorless starting materials could be made to display intense colors. These included irradiation by X rays, electrolysis (with color moving into the crystal from the cathode), or heating the crystals at high temperatures in the vapor of an alkali metal. Pohl was a strict empiricist who did not openly speculate upon the mechanics of color formation, which he simply attributed to the presence of Farbzentren (lit. color centers), later abbreviated to F centers.

Leading theoreticians were, however, attracted to the phenomenon and soon suggested models for F centers. In 1930 Frenkel suggested that an F center was an electron trapped in a distorted region of crystal structure, an idea that was incorrect in this instance but led directly to development of the concepts of excitons and

---

[3]Color is the name given to the perception of radiation in the electromagnetic spectrum with a wavelength of between 400 and 700 nm, for an average eye. If all wavelengths in this range are present in a distribution similar to that of radiation from the sun, the light is called white light.

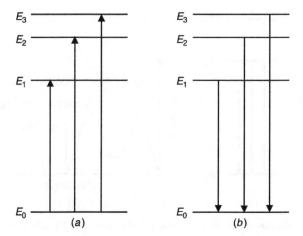

Figure 1.7   (a) Light is absorbed when a defect is excited from the ground state, E_0, to a higher energy state, E_1, E_2, or E_3. (b) Light is emitted when a defect drops in energy from a higher energy state E_1, E_2, or E_3 to the ground state, E_0.

polarons (Section 7.2). In 1934 Schottky suggested that an F center was an anion vacancy together with a trapped electron, and this model was put onto a firm quantum mechanical footing by Gurney and Mott in 1937 and later by Seitz, who extended the work to other types of color centers.

The origin of the color is as follows. The electron trapped at an anion vacancy in an alkali halide crystal is an analog of a hydrogen atom. The electron can occupy one of a number of orbitals, and transitions between some of these levels absorb light and hence endow the solid with a characteristic color. F centers and related defects are discussed further in Chapter 9.

Color can also be induced into colorless crystals by the incorporation of impurity atoms. The mineral corundum, $\alpha\text{-Al}_2\text{O}_3$, is a colorless solid. Rubies are crystals of Al_2O_3 containing atomically dispersed traces of Cr_2O_3 impurity. The formula of the crystal can be written $(\text{Cr}_x\text{Al}_{1-x})_2\text{O}_3$. In the solid the Al^{3+} and Cr^{3+} cations randomly occupy sites between the oxygen ions, so that the Cr^{3+} cations are impurity substitutional, Cr_{Al}, defects. When x takes very small values close to 0.005, the crystal is colored a rich "ruby" red.

The Cr_{Al} defects are responsible for the remarkable change. The color production is mediated by the three $3d$ electrons located on the Cr^{3+} defects. When a Cr^{3+} ion is in a vacuum, these all have the same energy. When the ion is inserted into a corundum crystal, the d electrons interact with the surrounding oxygen ions. This causes the d-electron orbitals to split into two groups, one at a slightly greater energy than the other, a feature called crystal field or ligand field splitting. An important consequence of this is that three *new* energy levels,[4] ^2E, $^4T_2$, and $^4T_1$, are introduced above the ground state, $^4A_2$, which are not present in pure Al_2O_3.

---

[4]Each separate energy level is given a label called a (spectroscopic) term symbol. For the moment it is sufficient to regard these merely as labels. See Further Reading at the end of this chapter for more information.

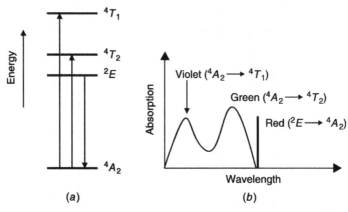

Figure 1.8 Light absorption and emission in ruby: (*a*) energy levels of Cr^{3+} ions and (*b*) absorption spectrum of ruby.

When white light falls onto a crystal of ruby the Cr^{3+} ions selectively absorb some of the radiation and are excited from the ground state $^4A_2$ to energy levels $^4T_2$ and $^4T_1$ (Fig. 1.8*a*). The resulting *absorption spectrum* (Fig. 1.8*b*), consists of two overlapping bell-shaped curves due to these transitions. The absorption curves show that the wavelengths corresponding to violet and green–yellow are strongly absorbed. This means that the color transmitted by the ruby will be red with something of a blue–purple undertone.

At the same time as the excitation is occurring, many of the higher energy Cr^{3+} ions return to the ground state by emitting exactly the same amount of energy as was absorbed, so as to drop back to the ground state from either $^4T_2$ or $^4T_1$. Some ions, however, lose energy to the crystal lattice, warming it slightly, dropping back only to the ^2E energy level. (For quantum mechanical reasons the probability that an ion will pass directly from the ground state to the ^2E state by absorbing energy is low, and so ^2E only becomes filled by this roundabout process.) The ions that return to the ground state from the energy level ^2E emit red light (Fig. 1.8*b*). The color of the best rubies is enhanced by this extra red component. At compositions close to $Cr_{0.005}Al_{0.995}O_3$ this emission can be made to dominate light emission, and the result is laser action (Section 9.9).

The color of rubies (as well as the action of ruby lasers) is thus totally dependent upon substitutional defects.

1.5 BULK PROPERTIES

The mechanical consequences of defect populations are less frequently considered than optical or electronic aspects, but they are of importance in many ways, especially when thin films or nanoparticles are considered.

1.5.1 Unit Cell Dimensions

Defects change the dimensions of the unit cell of a crystal. This can be illustrated by reference to ruby, described in the previous section. Rubies represent just a small part of the system spanning the composition range from pure Al_2O_3 to pure Cr_2O_3. The composition of these mixed crystals, $(Cr_xAl_{1-x})_2O_3$, can cover the range $0 < x < 1$, The unit cell parameters of the solid will vary as the composition changes. Vegard's law, first stated in (1921), is an empirical correlation to the effect that the lattice parameters of members of a solid solution formed between two isostructural phases, such as NaCl and KCl, will fall on a straight line joining the lattice parameters of these two parent structures (Fig. 1.9a):

$$x = \frac{a_{ss} - a_1}{a_2 - a_1}$$

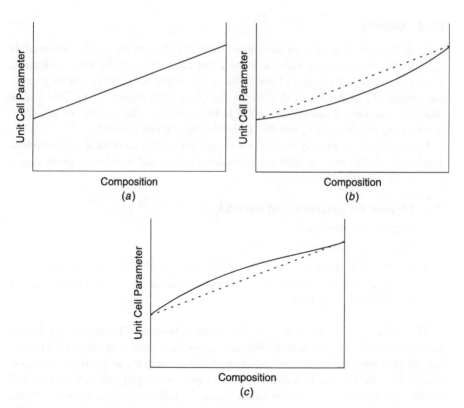

Figure 1.9 Vegard's law relating unit cell parameters to composition for solid solutions and alloys: (a) ideal Vegard's law behavior; (b) negative deviation from Vegard's law; and (c) positive deviation from Vegard's law.

that is,

$$a_{ss} = a_1 + x(a_2 - a_1)$$

where a_1 and a_2 are the lattice parameters of the parent phases, a_{ss} is the lattice parameter of the solid solution, and x is the mole fraction of the parent phase with lattice parameter a_2. (The relationship holds for all unit cell parameters, a, b, and c and any interaxial angles.) This "law" is simply an expression of the idea that the cell parameters are a direct consequence of the sizes of the component atoms in the solid solution. Vegard's law, in its ideal form, is almost never obeyed exactly. A plot of cell parameters that lies below the ideal line (Fig. 1.9b) is said to show a *negative* deviation from Vegard's law, and a plot that lies above the ideal line (Fig. 1.9c) is said to show a *positive* deviation from Vegard's law. In these cases, atomic interactions, which modify the size effects, are responsible for the deviations. In all cases, a plot of composition versus cell parameters can be used to determine the formulas of intermediate structures in a solid solution.

1.5.2 Density

X-ray diffraction allows the dimensions of the unit cell to be accurately measured. If the structure type of the material is known, the ideal cell contents are also known. Thus, the unit cell of a crystal of composition M_2O_3 that adopts the corundum structure contains 12 M atoms and 18 O atoms (Supplementary Material, S1). This readily allows the theoretical density of a solid to be calculated. The weights of all of the atoms in the cell are added, and this is divided by the cell volume.

It was realized at an early stage that a comparison of the theoretical and measured density of a solid can be used to determine the notional species of point defect present. The general procedure is:

1. Measure the composition of the solid.
2. Measure the density.
3. Measure the unit cell parameters.
4. Calculate the theoretical density for alternative point defect populations.
5. Compare the theoretical and experimental densities to see which point defect model best fits the data.

The method can be illustrated by reference to a classical 1933 study of the defects present in wüstite, iron monoxide. Wüstite adopts the sodium chloride (NaCl) structure, and the unit cell should contain 4 Fe and 4 O atoms in the unit cell, with an ideal composition $FeO_{1.0}$, but in reality the composition is oxygen rich and the unit cell dimensions also vary with composition (Table 1.1). Because there is more oxygen present than iron, the real composition can be obtained by assuming either that there are extra oxygen atoms in the unit cell (interstitial defects) to give a composition FeO_{1+x}, or that there are iron vacancies present, to give a formula $Fe_{1-x}O$. It is

TABLE 1.1   Experimental Data for Wüstite, FeO_x

O/Fe Ratio	Fe/O Ratio	Lattice Parameter/nm	Density/kg m^{-3}		
			Observed	Calculated for Interstitial Oxygen	Calculated for Iron Vacancies
1.058	0.945	0.4301	5728	6076	5740
1.075	0.930	0.4292	5658	6136	5706
1.087	0.920	0.4285	5624	6181	5687
1.099	0.910	0.4282	5613	6210	5652

Source: Adapted from E. R. Jette and F. Foote, *J. Chem. Phys.*, **1**, 29 (1933).

possible to determine which of these suppositions is correct by comparing the real and theoretical density of the material.

For example, consider the sample specified in the top line of Table 1.1, with an oxygen : iron ratio of 1.058, a measured density of 5728 kg m^{-3} and a cubic lattice parameter, a, of 0.4301 nm.

1. Assume that the iron atoms in the crystal are in a perfect array, identical to the metal atoms in the sodium chloride structure, and that the 0.058 excess of oxygen is due to interstitial oxygen atoms being present, over and above those on the normal anion positions. The unit cell of the structure now contains 4 Fe and (4×1.058) O. The density is calculated to be 6076 kg m^{-3}.

2. Assume that the oxygen array is perfect and identical to the nonmetal atom array in the sodium chloride structure and that the unit cell contains some vacancies on the iron positions. In this case, one unit cell will contain 4 atoms of oxygen and $(4/1.058)$ atoms of iron, that is, $4 Fe_{0.945}O$. The density is calculated to 5741 kg m^{-3}.

The difference in the two values is surprisingly large and is well within the accuracy of density determinations. The experimental value is in accord with a model that assumes vacancies on the iron positions, as are all results in Table 1.1, indicating that a formula $Fe_{1-x}O$, in which there are vacancies at some of the Fe positions, better reflects the structure.

Although this analysis is correct, it is a macroscopic method that does not give any true crystallographic information. The oxygen vacancies may be arranged in any number of ways. In fact, it is found that the vacancies form clusters that can be regarded as fragments of the next higher oxide, Fe_3O_4, with the spinel structure embedded in a sodium chloride structure matrix (Section 4.4.2).

Because of Vegard's law, it is clear that the density of a solid solution will also be expected to be a linear function of the densities of the parent phases that make up the limits of a solid solution.

1.5.3 Volume

The ambient pressure surrounding a solid can directly influence the point defects present (Chapter 7). This is most often encountered with respect to oxygen pressure, but water vapor and the partial pressure of volatile metals are also of importance in high-temperature applications. Changes in defect populations have a direct bearing upon unit cell dimensions and hence upon the overall dimensions of the solid. These dimensional changes, like changes in density, can be used to infer the type of point defect present, albeit with a low degree of discrimination. For example, uranium dioxide, UO_2, can gain oxygen to form a hyperstoichiometric phase UO_{2+x}. The partial pressure of oxygen that is in equilibrium with the dioxide, $UO_{2.0}$ is about 10^{-9} Pa, depending upon the temperature. When the oxygen partial pressure surrounding the oxide is increased above this value, the solid gains oxygen and expands. This expansion is not uniform but peaks and then decreases (Fig. 1.10). A possible explanation of this change is that interstitial oxygen ions cause initial expansion. When these reach higher concentrations, clustering might cause the volume to fall. X ray or other structural studies are needed to explore such models further (Section 4.4.3).

Changes in density, unit cell dimensions, and macroscopic volume have serious effects. In an environment where point defects (or aggregates of point defects) are generated, such as in the components of nuclear reactors, or in vessels used for the storage of nuclear waste, where point defects are produced as a result of irradiation, dimensional changes can cause components to seize or rupture.

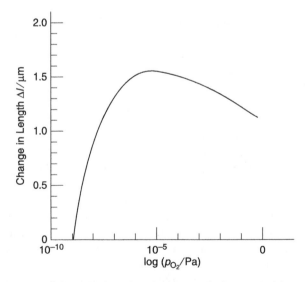

Figure 1.10   Variation of the length of a single crystal of oxygen-rich uranium dioxide, UO_{2+x}, with oxygen partial pressure at 1200°C. [Data redrawn from L. Desgranges, M. Gramond, C. Petot, G. Petot-Ervas, P. Ruello, and B. Saadi, *J. Eur. Ceram. Soc.*, **25**, 2683–2686 (2005).]

Such changes in the defect population can be critical in device manufacture and operation. For example, a thin film of an oxide such as SiO laid down in a vacuum may have a large population of anion vacancy point defects present. Similarly, a film deposited by sputtering in an inert atmosphere may incorporate both vacancies and inert gas interstitial atoms into the structure. When these films are subsequently exposed to different conditions, for example, moist air at high temperatures, changes in the point defect population will result in dimensional changes that can cause the film to buckle or tear.

Ionic conductors, used in electrochemical cells and batteries (Chapter 6), have high point defect populations. Slabs of solid ceramic electrolytes in fuel cells, for instance, often operate under conditions in which one side of the electrolyte is held in oxidizing conditions and the other side in reducing conditions. A significant change in the point defect population over the ceramic can be anticipated in these conditions, which may cause the electrolyte to bow or fracture.

These effects can all be enhanced if the point defects interact to form defect clusters or similar structures, as in $Fe_{1-x}O$ above or UO_{2+x} (Section 4.4). Such clusters can suppress phase changes at low temperatures. Under circumstances in which the clusters dissociate, such as those found in solid oxide fuel cells, the volume change can be considerable, leading to failure of the component.

1.5.4 Young's Modulus (the Elastic Modulus)

The response of a solid to an applied force (a stress, σ) in terms of a change in dimensions (a strain, ε) is given by *Young's modulus, E*:

$$\sigma = E\varepsilon$$

(Hooke's law). [Young's modulus is often called the elastic modulus, but as a number of different (symmetry dependent) elastic moduli are defined for a solid, the generic term, i.e., *the* elastic modulus, is not sufficiently precise.] The value of Young's modulus of a solid containing point defects is noticeably different from a solid that is defect free. This is important is some applications. For example, solid oxide fuel cells (Chapter 6) use oxides such as calcia-stabilized or yttria-stabilized zirconia as the electrolyte. These cells operate at between 700 and 1000°C, and substantial populations of oxygen ion vacancies occur in the electrolyte at these temperatures. Mechanical failure of the cells remains a problem due to the thermal cycling that occurs during operation. It has been found that the Young's modulus of the electrolyte decreases significantly as the number of oxygen ion vacancies increases, exacerbating the mechanical problems encountered.

These examples indicate that it is necessary to keep the possible effect of point defects on bulk and mechanical properties in mind. Although less definitive than electronic and optical properties, they may make the difference in the success or failure of device operation.

1.6 THERMOELECTRIC PROPERTIES: THE SEEBECK COEFFICIENT AS AN EXAMPLE

When the two ends of a material containing mobile charge carriers, holes or electrons, are held at different temperatures, a voltage is produced, a phenomenon called the Seebeck effect (Fig. 1.11). The Seebeck coefficient of a material, α, is defined as the ratio of the electric potential produced when no current flows to the temperature difference present across a material:[5]

$$\alpha = \pm \frac{\phi_H - \phi_C}{T_H - T_C} = \pm \frac{\Delta\phi}{\Delta T}$$

where ϕ_H and ϕ_C are the potentials and T_H and T_C are the temperatures at the hot end and the cold end of the sample, respectively. The main virtue of the Seebeck coefficient in the context of this book is that its sign and magnitude can provide a measure of the concentration of charge carriers, the nature of the charge carriers, and, with some simple assumptions, the number of defects present that give rise to the charge carriers.

In the case of materials that have mobile electrons, that is, n-type semiconductors, the colder end of the rod will be *negative* with respect to the hotter end and the sign of the Seebeck coefficient is negative. In the case where the mobile charge carriers are positive holes, that is, p-type semiconductors, the colder end of the rod will be *positive* with respect to the hotter end, making the Seebeck coefficient positive. For example, the nonstoichiometric forms of NiO, CoO, and FeO all show positive values for α, indicating that conductivity is by way of holes, whereas nonstoichiometric ZnO has a negative value of α, indicating conductivity by way of electrons. In the case of materials with both types of charge carrier present, the one that is present in greatest numbers dominates the measurement.

The magnitude of the Seebeck coefficient is related to the concentration of mobile charge carriers present and is greatest at low defect concentrations, when other methods of analyzing defect populations give least precision. The relationship between the number of defects and the Seebeck coefficient is obtained by estimating the configurational entropy of the defect-containing material. A number of forms for this estimate are found, each depending upon slightly different approximations (Supplementary Material S3). The most direct is

$$\alpha = \pm \left(\frac{k}{e}\right)\left[\ln\left(\frac{n_0}{n_d}\right)\right]$$

---

[5]The Seebeck coefficient is frequently called the thermoelectric power or thermopower, and labeled Q or S. Neither of these alternatives is a good choice. The units of the Seebeck coefficient are not those of power. The symbol Q is most often used to signify heat transfer in materials. The designation S can easily be confused with the entropy of the mobile charge carriers, which is important because the Seebeck coefficient is equivalent to the entropy per mobile charge carrier (see Supplementary Material S3).

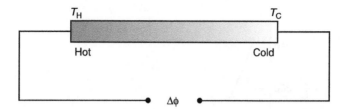

Figure 1.11 Seebeck effect. A sample with one end maintained at a high-temperature T_H and the other at a low-temperature T_C will develop a potential difference $\Delta\phi$.

where n_0 is the number of sites in the sublattice containing defects and n_d is the number of defects giving rise to mobile electrons or holes and $k/e = 86.17\ \mu\text{V K}^{-1}$. The positive version applies to p-type materials and the negative expression to n-type materials. Note that n_0/n_d increases as the number of defects falls, and so the value of α is expected to be greatest for lowest defect populations.

This equation is formally equivalent to the Heikes equation:

$$\alpha = -\left(\frac{k}{e}\right)\ln\left(\frac{1-c}{c}\right)\quad\text{for electrons}$$

$$= +\left(\frac{k}{e}\right)\ln\left(\frac{1-c}{c}\right)\quad\text{for holes}^{6}$$

where c is the fraction of defects (or mobile charge carriers) present (Supplementary Material S3). This form is useful because the value of c is directly related to the composition of the sample. For example, the material $\text{LaNi}_x\text{Co}_{1-x}\text{O}_3$, which has the perovskite structure, can be analyzed in this way. Each Ni^{2+} ion that replaces a Co^{3+} ion in the parent compound $\text{La}^{3+}\text{Co}^{3+}\text{O}_3$ forces one of the other Co^{3+} ions to transform to Co^{4+} to maintain charge neutrality. Each Co^{4+} ion can be considered to be a Co^{3+} ion plus a trapped hole. Electronic conductivity can then be considered to occur by the migration of holes from one Co^{4+} ion to a neighboring Co^{3+} ion:

$$\text{Co}^{4+} + \text{Co}^{3+} \rightleftharpoons \text{Co}^{3+} + \text{Co}^{4+}$$

that is,

$$(\text{Co}^{3+} + \text{h}^{\bullet}) + \text{Co}^{3+} \rightleftharpoons \text{Co}^{3+} + (\text{Co}^{3+} + \text{h}^{\bullet})$$

---

[6]Note that the form of the equation for holes is often written

$$\alpha = -\left(\frac{k}{e}\right)\ln\left(\frac{c}{1-c}\right)$$

which is identical to that given as $\ln x = -\ln(1/x)$.

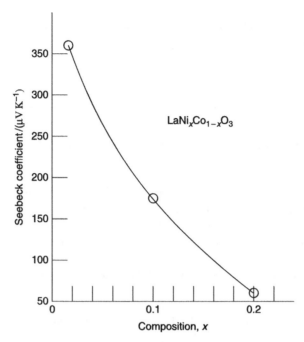

Figure 1.12 Seebeck coefficient of the oxide $LaNi_xCo_{1-x}O_3$ as a function of the composition, x. [Data adapted from R. Robert, L. Becker, M. Trottmann, A. Reller, and A. Weidenkraft, *J. Solid State Chem.*, **179**, 3893–3899, (2006).]

The number of mobile holes is equal to the number of impurity Ni^{2+} ions, and so the fraction c in the Heikes equation is equal to x in $LaNi_xCo_{1-x}O_3$. In accord with the theory, the Seebeck coefficient, α, is positive and greatest at low values of x and decreases as x increase (Fig. 1.12). Substituting a value of $c = 0.02$ into the equation yields a value of $\alpha = +335 \ \mu V \ K^{-1}$, in good agreement with the experimental value of $360 \ \mu V \ K^{-1}$ (Robert *et al.*, 2006). Note that the above example also shows that an experimentally determined value of the Seebeck coefficient can be used to estimate the concentration of impurity defects in a doped oxide.

1.7 POINT DEFECT NOTATION

Point defect populations profoundly affect both the physical and chemical properties of materials. In order to describe these consequences a simple and self-consistent set of symbols is required. The most widely employed system is *the Kröger–Vink notation*. Using this formalism, it is possible to incorporate defect formation into chemical equations and hence use the powerful methods of chemical thermodynamics to treat defect equilibria.

In the Kröger–Vink notation, empty atom positions, that is, vacancies, are indicated by the symbol V. Acknowledging that V is the chemical symbol for the element vanadium, it is necessary to add that, where confusion may occur, the symbol for a vacancy is written Va. The atom that is absent from a normally occupied site is specified by the normal chemical symbol for the element, written as a subscript. Thus in NiO, for example, the symbol V_O would represent an oxygen atom vacancy and V_{Ni} a nickel atom vacancy.

The position of a defect that has been substituted for another atom in the structure is represented by a subscript that is the chemical symbol of the atom normally found at the site occupied by the defect impurity atom. The impurity is given its normal chemical symbol, and the site occupied is written as a subscript, using the chemical symbol for the atom that normally occupies the site. Thus, an Mg atom on a Ni site in NiO would be written as Mg_{Ni}. The same nomenclature is used if an atom in a crystal occupies the wrong site. For example, antisite defects in GaN would be written as Ga_N and N_{Ga}.

Interstitial positions, positions in a crystal not normally occupied by an atom, are denoted by the subscript i. For example, F_i would represent an interstitial fluorine atom in, say, a crystal of fluorite, CaF_2.

It is possible for one or more lattice defects to associate with one another, that is, to cluster together. These are indicated by enclosing the components of such a cluster in parentheses. As an example, $(V_M V_X)$ would represent a defect in which a vacancy on a metal site and a vacancy on a nonmetal site are associated as a vacancy pair.

1.8 CHARGES ON DEFECTS

One of the most difficult problems when working with defects, especially in ionic crystals, is to decide on the charge on the ions and atoms of importance. The Kröger–Vink notation bypasses the problem of deciding on the *real charges* on defects (z_d), by considering only *effective charges* (q_e). The effective charge on a defect is the charge that the defect has *with respect to the charge that would be present at the same site in a perfect crystal*. Thus, the effective charge is a relative charge. For atomic or ionic species, this is equal to the difference between the real charges on the defect species, z_d, minus the real charge at the site occupied in a perfect crystal, z_s:

$$q_e = z_d - z_s \qquad (1.1)$$

To distinguish effective charges from real charges, the superscript $'$ is used for each unit of effective negative charge and the superscript $\bullet$ is used for each unit of effective positive charge. The real charges on a defect are still given the superscript symbols $-$ and $+$.

1.8.1 Electrons and Electron Holes

The charged defects that most readily come to mind are electrons. In a crystal containing defects, some fraction of the electrons may be free to move through the matrix. These are denoted by the symbol e'. The superscript $'$ represents the effective negative charge on the electron (q_e), and it is written in this way to emphasize that it is considered relative to the surroundings rather than as an isolated real point charge. The concentration of electrons that are free to carry current through a crystal is frequently given the symbol n in semiconductor physics.

The counterparts to electrons in semiconducting solids are holes, represented by the symbol $h^\bullet$. Each hole will bear an effective positive charge, q_e, of $+1$, which is represented by the superscript $\bullet$ to emphasize that it is considered relative to the surrounding structure. The concentration of holes that are free to carry current through a crystal is often given the symbol p in semiconductor physics.

1.8.2   Atomic and Ionic Defects

Point defects can carry a charge. In ionic crystals, this may be considered to be the normal state of affairs. The effective charge on these defects is the charge that the defect has with respect to the charge present or expected at the same point in the perfect crystal structure. To illustrate this concept, consider the situation in an ionic material such as NaCl, which it is convenient to consider as made up of the charged ions Na^+ and Cl^-. What is the effective charge on a sodium vacancy, V_{Na}, in the NaCl structure? The result is easily obtained using Eq. (1.1). The real charge on the vacancy, z_d, is 0. The real charge at the site in a perfect crystal, z_s, is due to the presence of Na^+, that is, $+1$, hence:

$$q_e = z_d - z_s = (0 - 1) = -1$$

Relative to the normal situation at the site, the vacancy appears to bear an effective negative charge equivalent to -1. Hence, a vacancy at a sodium ion (Na^+) site in NaCl would be written as V'_{Na}.

In general, the absence of a positive ion will leave a vacancy with a negative effective charge relative to the normally occupied site. Multiple effective negative charges can exist and are written using superscript n'. A Ca^{2+} ion vacancy in a crystal of CaO will bear an effective negative charge of $2'$, and the vacancy has the symbol V''_{Ca} as can be seen from Eq. (1.1):

$$q_e = z_d - z_s = (0 - 2) = -2$$

The same reasoning indicates that the absence of a negative ion will leave a positive effective charge relative to a normal site occupied by a negative ion. A vacancy at a chloride ion (Cl^-) site is positively charged relative to the normal situation prevailing at an anion site in the crystal. Using Eq. (1.1):

$$q_e = z_d - z_s = (0 - (-1)) = +1$$

Hence, the vacancy has an effective charge of $+1$, which would be written $V^\bullet_{Cl}$.

In general, the absence of a negative ion will endow a site with a positive effective charge. Multiple effective positive charges can exist and are written using superscript $n\bullet$. An oxide ion (O^{2-}) vacancy in a crystal of CaO will bear an effective positive charge of

$$q_e = z_d - z_s = (0 - (-2)) = +2$$

and the vacancy has the symbol $V_O^{2\bullet}$.

An effective charge relative to the host lattice is possible with any defect. These are added as superscripts to the appropriate symbol: V_M, V_X, M_i, M_X and associated defects such as $(V_M V_X)$.

Interstitial sites, which are normally unoccupied in a crystal, will have no preexisting charge. When an atom or an ion occupies an interstitial site, its real charge is the same as the effective charge. Thus, for a Zn^{2+} ion at an interstitial site, from Eq. (1.1):

$$q_e = z_d - z_s = (2 - 0) = 2$$

The defect is given the symbol $Zn_i^{2\bullet}$.

Substitution of an ion with one valence by another with a different valence, aliovalent substitution, will create a charged defect. For example, a divalent ion such as Ca^{2+} substituted for a monovalent Na^+ on a sodium site in NaCl gives a local electronic charge augmented by one extra positive charge:

$$q_e = z_d - z_s = (2 - 1) = 1$$

The defect has an effective charge of 1 and it is represented by the symbol $Ca_{Na}^{\bullet}$.

Not all defects carry effective charges. Frequently, this need not be noted. For instance, suppose that a sodium ion in NaCl, represented by Na_{Na}, is substituted by a potassium ion, represented by K_{Na}. Clearly, the defect will have no effective charge:

$$q_e = z_d - z_s = (1 - 1) = 0$$

This defect is therefore neutral in terms of effective charge. The same could be said of a neutral lithium atom introduced into an interstitial site in titanium disulfide, TiS_2, which would be written Li_i. However, it is sometimes important to emphasize that the defect is neutral in terms of effective charge. This is made clear by the use of a superscript x. Thus a K^+ ion substituted for a Na^+ ion could be written K_{Na}^x when the effective charge situation needs to be specified. Similarly, an interstitial Li atom could be represented as Li_i^x to emphasize the lack of an effective charge on the defect when it is essential to do so.

Some defects are termed amphoteric[7] defects. These are defects that can take on either a positive or negative effective charge, depending upon circumstances. For

---

[7]Amphoteric is a chemical term used to describe (mainly) oxides or hydroxides that are able to react both with acids and alkalis. In the present use, it is not related to acid–base properties.

TABLE 1.2 Kröger–Vink Notation for Defects in Crystals[a]

Defect Type	Notation	Defect Type	Notation
Metal vacancy at metal (M) site	V_M	Nonmetal vacancy at nonmetal (Y) site	V_Y
Impurity metal (A) at metal (M) site	A_M	Impurity non-metal (Z) at nonmetal site	Z_Y
Interstitial metal (M)	M_i	Interstitial nonmetal (Y)	Y_i
Neutral metal (M) vacancy	V_M^x	Neutral nonmetal (Y) vacancy	V_Y^x
Metal (M) vacancy with negative effective charge	V_M'	Nonmetal (Y) vacancy with positive effective charge	$V_Y^\bullet$
Interstitial metal (M) with positive effective charge	$M_i^\bullet$	Interstitial nonmetal (X) with negative effective charge	X_i'
Interstitial metal (M) with n positive effective charges	$M_i^{n\bullet}$	Interstitial nonmetal (Y) with n negative effective charges	$Y_i^{n\prime}$
Free electron[b]	e'	Free hole[b]	$h^\bullet$
Associated defects (vacancy pair)	$(V_M V_Y)$	Associated defects with positive effective charge	$(V_M V_Y)^\bullet$

[a]The definitive definitions of this nomenclature and further examples are to be found in the *IUPAC Red Book on the Nomenclature of Inorganic Chemistry*, Chapter I.6.
[b]The concentrations of these defects are frequently designated by n and p respectively.

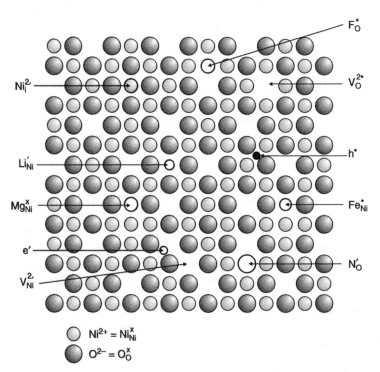

Figure 1.13 Point defects in nickel oxide, NiO (schematic): Ni^{2+} vacancy; Ni^{2+} interstitial; Li^+ on a Ni^{2+} site; Mg^{2+} on a Ni^{2+} site; Fe^{3+} on a Ni^{2+} site; O^{2-} vacancy; N^{3-} on an O^{2-} site; F^- on an O^{2-} site; free electron; free hole.

example, the incorporation of lithium ions, Li^+, into zinc selenide, ZnSe, can lead to lithium interstitials, $Li_i^\bullet$, in which the real charge and the effective charge are identical, as stated above. Alternatively, the ions can form substitution defects in which the Li^+ ions occupy Zn^{2+} sites, in which case the designation is Li'_{Zn}. As the populations of these two alternatives changes, for instance, with lithium concentration, so does the electronic character of the defect population.

The effective charges on an ionic defect can be considered to be linked to the defect by an imaginary bond. If the bond is weak, the effective charge can be liberated, say by thermal energy, so that it becomes free to move in an applied electric field and so contribute to the electronic conductivity of the material. Whether the effective charge on a defect is considered to be strongly associated with the defect or free depends upon the results obtained when the physical properties of the solid are measured.

The main features of the Kröger–Vink notation are summarized in Table 1.2 and are illustrated with respect to point defects in a crystal containing Ni^{2+} and O^{2-} ions in Figure 1.13.

1.9 BALANCED POPULATIONS OF POINT DEFECTS: SCHOTTKY AND FRENKEL DEFECTS

1.9.1 Schottky Defects

Compounds are made up of atoms of more than one chemical element. The point defects that can occur in pure compounds parallel those that occur in monatomic materials, but there is an added complication in this case concerning the composition of the material. In this chapter discussion is confined to the situation in which the composition of the crystal is (virtually) fixed. Such solids are called stoichiometric compounds. (The situations that arise when the composition is allowed to vary are considered in Chapter 4 and throughout much of the rest of this book. This latter type of solid is called a nonstoichiometric compound.) The composition problem can be illustrated with respect to a simple compound such as sodium chloride.

Sodium chloride, also known as rock salt or halite, is composed of equal numbers of sodium (Na) and chlorine (Cl) atoms, has a chemical formula NaCl and a simple structure (Fig. 1.14). It is a good first approximation to regard this material as being composed of ions. The introduction of vacancies on the cation sublattice will upset both the composition and the charge balance. If x such vacancies occur, the formula of the crystal will now be $Na_{1-x}Cl$, and the overall material will have an excess negative charge of $x-$ because the number of chloride ions is greater than the number of sodium ions by this amount. The compound should be written $[Na_{1-x}Cl]^{x-}$. The same will be true for the anion sublattice. If x vacancies are placed on the anion sublattice, the material will take on an overall positive charge because the number of sodium ions now outnumbers the chlorine ions, and the formula becomes $[NaCl_{1-x}]^{x+}$. Ordinary crystals of sodium chloride do not show an overall negative or positive charge or have a formula different to NaCl. Thus, if vacancy defects occur in these crystals, the numbers on the anion and cation

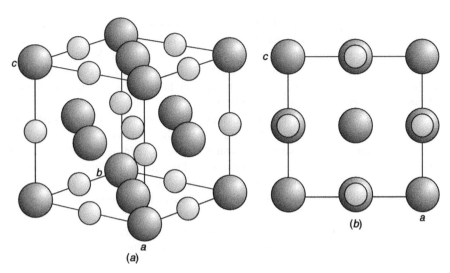

Figure 1.14   Crystal structure of sodium chloride, NaCl: (*a*) a perspective view of one unit cell and (*b*) projection down [010], the *b* axis.

sublattices must be balanced to maintain the correct formula and preserve electrical neutrality. This means that we must introduce equal numbers of vacancies onto both sublattices.

Such a situation was envisaged by Schottky and Wagner, whose ideas were presented in 1931. The defects arising from balanced populations of cation and anion vacancies in any crystal, not just NaCl, are now known as Schottky defects. For example, if the crystal has a formula MX, then the number of cation vacancies will be equal to the number of anion vacancies, in order to maintain the composition and electrical neutrality (or electroneutrality). In such a crystal, one Schottky defect consists of one cation vacancy together with one anion vacancy, although these vacancies are not necessarily imagined to be near to each other in the crystal. It is necessary to remember that the number of Schottky defects in a crystal of formula MX is equal to one-half of *total* the number of vacancies. Schottky defects are frequently represented diagrammatically by a drawing of the sort shown in Figure 1.15*a*. In a real crystal the situation will be more complex, as atoms in the vicinity of the defects will move slightly (relax) due to the changed situation.

In crystals of more complex formula, such as titanium dioxide, TiO_2, a Schottky defect will consist of two anion vacancies and one cation vacancy. This is because it is necessary to counterbalance the loss of one Ti^{4+} ion from the crystal by the absence of two O^{2-} ions in order to maintain composition and electroneutrality. This ratio of two anion vacancies per one cation vacancy will hold in all ionic compounds of formula MX_2. In crystals like Al_2O_3, two Al^{3+} vacancies must be balanced by three O^{2-} vacancies. Thus, in crystals with a formula M_2X_3, a Schottky defect will consist of two vacancies on the cation sublattice and three vacancies on the anion sublattice. These vacancies are not considered to be clustered together but are distributed

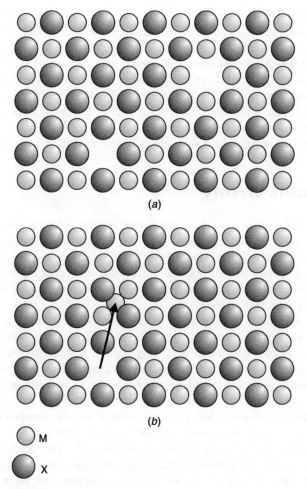

Figure 1.15 Balanced populations of point defects in an ionic crystal of formula MX (schematic): (*a*) Schottky defects and (*b*) Frenkel defects.

at random throughout the crystal structure, in the relative numbers needed to keep the crystals electrically neutral.

The formation energy of Schottky defects is described further in Chapter 2.

1.9.2 Frenkel Defects

As already illustrated, it is also possible to imagine a defect related to ions in interstices, that is, interstitials. Such defects were first suggested as being of importance by Frenkel and are known as *Frenkel defects*. In this case, an atom or ion from one sublattice moves to a normally empty site (an interstitial site) in the crystal, leaving a vacancy behind (Fig. 1.15*b*). A Frenkel defect may involve either the cation or the anion sublattice. Thus, in any crystal of formula MX, a Frenkel defect consists of one interstitial ion plus one vacant site in the sublattice where that ion would normally be found. Because ions are being moved internally, no problem of electronic charge balance occurs. This means that the number of interstitials and vacancies forming a Frenkel defect population is not connected to the formula of the compound. For example, if we have Frenkel defects on the anion sublattice in CaF_2, we can think of just one F^- ion being displaced; it is not necessary to displace two F^- ions to form the Frenkel defect.

The formation energy of Frenkel defects is described further in Chapter 2.

1.10 ANTISITE DEFECTS

An antisite defect is an atom on a site normally occupied by a different chemical species that exists in the compound. Antisite defects are a feature of a number of important materials, especially weakly ionic or covalently bonded ones. In a compound of formula AB the antisite defects that can occur are an A atom on a site normally occupied by a B atom (Fig. 1.16a), or a B atom on a site normally occupied by an A atom (Fig. 1.16b).

Antisite defects are not important in binary ionic compounds, as the resultant increase in electrostatic energy is generally prohibitive. For example, if an Na^+ ion were to replace a Cl^- ion in NaCl (Fig. 1.16c), it would be surrounded by an octahedron of Na^+ cations, leading to strong electrostatic repulsion. Similarly, a Cl^- anion on an Na^+ site would be surrounded by an octahedron of Cl^- anions (Fig. 1.16d) again, leading to strong electrostatic repulsion. This is not true when cations of different elements are present together in an ionic compound, as antistite defects can then be accommodated without the electrostatic penalty. An example is the oxide spinel, $MgAl_2O_4$. Notionally all of the Mg^{2+} ions occupy tetrahedrally coordinated sites in the structure, written (Mg), and Al^{3+} ions occupy octahedrally coordinated sites, written $[Al_2]$, so that the structure can be represented as $(Mg)[Al_2]O_4$. In most such crystals it is found that a small number of Mg^{2+} and Al^{3+} cations exchange sites to form antisite defects $(Mg_{1-x}Al_x)[Al_{2-x}Mg_x]O_4$:

$$Mg_{Mg} + Al_{Al} \longrightarrow Mg'_{Al} + Al^{\bullet}_{Mg}$$

where the effective charges are derived assuming the compound is ionic. This exchange of cations between octahedral and tetrahedral sites is generally designated in terms of normal and inverse spinel structures (Supplementary Material S1). Similarly, in $Er_2Ti_2O_7$, which has the pyrochlore structure, antisite defects occur between the two metal atoms. In this structure, in the absence of defects, the larger atom (Er) is coordinated by eight oxygen atoms in an approximately cubic arrangement, while the smaller atom (Ti) is in distorted octahedral coordination. The antisite defects can be represented as forming by the reaction:

$$Er_{Er} + Ti_{Ti} \longrightarrow Er'_{Ti} + Ti^{\bullet}_{Er}$$

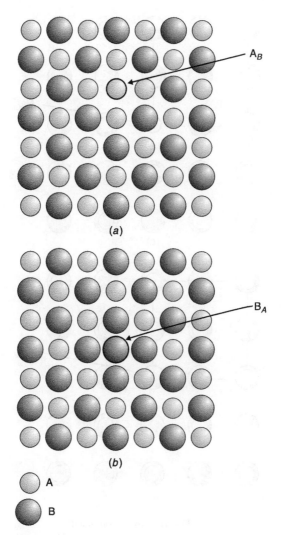

Figure 1.16 Antisite point defects in an ionic crystal of formula MX (schematic): (*a*) A on B sites, A_B; (*b*) B on A sites, B_A; (*c*) Na^+ on a Cl^- site in sodium chloride, $Na_{Cl}^{2\bullet}$; and (*d*) Cl^- on an Na^+ site in sodium chloride, $Cl_{Na}^{2\prime}$.

where the effective charges indicate an ionic model. About 2.9 at % of these defects are found in normal preparations (see also Section 4.4).

In metallic and many semiconducting crystals, the valence electrons are delocalized throughout the solid, so that antisite defects are not accompanied by prohibitive energy costs and are rather common. For example, an important defect in the semiconducting material GaAs, which has the zinc blend structure (Supplementary Material S1), is the antisite defect formed when an As atom occupies a Ga site.

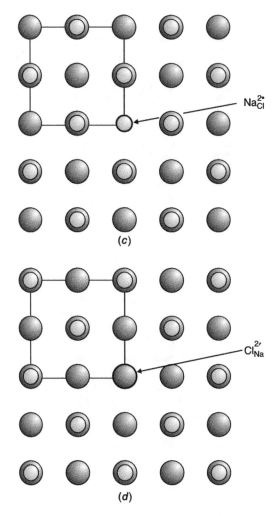

Figure 1.16   (*Continued*).

In cases where the antisite defects are balanced, such as a Ga atom on an As site balanced by an As atom on a Ga site, the composition of the compound is unaltered. In cases where this is not so, the composition of the material will drift away from the stoichiometric formula unless a population of compensating defects is also present. For example, the alloy FeAl contains antisite defects consisting of iron atoms on aluminum sites without a balancing population of aluminum atoms on iron sites. The composition will be iron rich unless compensating defects such as Al interstitials or Fe vacancies are also present in numbers sufficient to restore the stoichiometry. Experiments show that iron vacancies (V_{Fe}) are the compensating defects when the composition is maintained at FeAl.

The creation of antisite defects can occur during crystal growth, when atoms are misplaced on the surface of the growing crystal. Alternatively, they can be created by internal mechanisms once the crystal is formed, provided that sufficient energy is applied to allow for atom movement.

1.11 DEFECT FORMATION AND REACTION EQUATIONS

Defects are often deliberately introduced into a solid in order to modify physical or chemical properties. However, defects do not occur in the balance of reactants expressed in traditional chemical equations, and so these important components are lost to the chemical accounting system that the equations represent. Fortunately, traditional chemical equations can be easily modified so as to include defect formation. The incorporation of defects into normal chemical equations allows a strict account of these important entities to be kept and at the same time facilitates the application of chemical thermodynamics to the system. In this sense it is possible to build up a *defect chemistry* in which the defects play a role analogous to that of the chemical atoms themselves. The Kröger–Vink notation allows this to be done provided the normal rules that apply to balanced chemical equations are preserved.

1.11.1 Addition and Subtraction of Atoms

When writing defect formation equations, the strategy involved is always to add or subtract elements to or from a crystal via *electrically neutral atoms*. When ionic crystals are involved, this requires that electrons are considered separately. Thus, if one considers NiO to be ionic, formation of a V_{Ni} would imply the removal of a neutral Ni atom, that is, removal of a Ni^{2+} ion together with two electrons. Similarly, formation of a V_O would imply removal of a neutral oxygen atom, that is, removal of an O^{2-} ion, followed by the addition of two electrons to the crystal. An alternative way to express this is to say the removal of an O^{2-} ion together with $2h^{\bullet}$. Similarly, only neutral atoms are added to interstitial positions. If ions are considered to be present, the requisite number of electrons must be added or subtracted as well. Thus, the formation of an interstitial Zn^{2+} defect would involve the addition of a neutral Zn atom and the removal of two electrons.

1.11.2 Equation Formalism

The formation of defects can be considered as the *reaction* of a nominally perfect crystal with dopant. The rules for writing equations including defects are similar to those of elementary chemistry, but as the matrix is a crystal structure, quantities must be specified with respect to crystallographic sites rather than molecules or moles.

1. *The number of metal atom sites must always be in the correct proportion to the number of nonmetal atom sites in the crystal.* This is in essence a crystal structure

constraint. In a crystal structure there are a certain number of atoms that occupy specified positions within the space group of the solid. Provided that the crystal structure is (even only loosely) specified, the number of atom sites must remain fixed. Thus, in CaS, which adopts the sodium chloride structure, formula MX, there must always be equal numbers of metal and nonmetal atom positions in the equation. For an oxide such as TiO_2, with the rutile structure, there must always be twice as many anion sites as cation sites, or, in general, for a compound M_aX_b, there must be an a metal atom sites for every b nonmetal atom sites. As long as this proportion is maintained, the *total* number of sites can vary, as this simply corresponds to more or less substance present. If the crystal contains vacancies, these must be counted as part of the total number of sites, as each vacancy can be considered to occupy a site just as legally as an atom. Interstitial atoms do not occupy normal sites and so do not count when this rule is being applied.

2. *The total number of atoms on one side of the equation must balance the total number of atoms on the other side.* This rule is simply an expression of the well-known chemical fact that atoms are neither created nor destroyed during a chemical reaction. Remember that subscripts and superscripts are labels describing charges and sites and are not counted in evaluating the atom balance.

3. *The crystal must always be electrically neutral.* This means not only that the total charge on one side of the equation must be equal to the total charge on the other side, but also that the sum of the charges on each side of the equation must equal zero. In this assessment, both effective and real charges must be counted if both sorts are present.

Recall that only neutral atoms are involved in reactions. After reaction, neutral atoms can dissociate into charged species if this is thought to represent the real situation in the crystal, provided that electroneutrality, as described, is maintained.

Reactions involving the creation, destruction, and elimination of defects can appear mysterious. In such cases it is useful to break the reaction down into hypothetical steps that can be represented by partial equations, rather akin to the half-reactions used to simplify redox reactions in chemistry. The complete defect formation equation is found by adding the partial equations together. The rules described above can be interpreted more flexibly in these partial equations but must be rigorously obeyed in the final equation. Finally, it is necessary to mention that a defect formation equation can often be written in terms of just structural (i.e., ionic) defects such as interstitials and vacancies or in terms of just electronic defects, electrons, and holes. Which of these alternatives is preferred will depend upon the physical properties of the solid. An insulator such as MgO is likely to utilize structural defects to compensate for the changes taking place, whereas a semiconducting transition-metal oxide with several easily accessible valence states is likely to prefer electronic compensation.

To illustrate exactly how these rules work, a number of examples follow. In the first, the formation of antisite defects, a simple example that does not involve changes in atom numbers or charges on defects, is described. Secondly, two reactions involving oxides, nickel oxide and cadmium oxide, both of which are nonstoichiometric, but for opposite reasons, indicate how to deal with a solid–gas interaction in terms of electronic compensation. Thirdly, the formation of calcia-stabilized zirconia, $ZrO_2 \cdot xCaO$, where x is of the order of $10-15$ mol % outlines the method for a reaction between two solids utilizing structural compensation. Finally, defect formation in more complex oxides containing two cation species is outlined. None of these reactions can be represented by conventional chemical equations.

1.11.3 Formation of Antisite Defects

The creation of a complementary pair of antisite defects consisting of an A atom on a B atom site, A_B, and a B atom on an A atom site, B_A, can be written in terms of a chemical equation:

$$A_A + B_B \longrightarrow A_B + B_A$$

Antisite defects can be created via the intermediate formation of a Frenkel defect by the following rules:

1. Frenkel defect formation on the A sublattice:

$$A_A \longrightarrow A_i + V_A$$

2. Exchange of the interstitial A_i atom with a B atom on its correct site:

$$A_i + B_B \longrightarrow A_B + B_i$$

3. The interstitial B_i atom can eliminate the vacancy to form a B atom antisite defect:

$$B_i + V_A \longrightarrow B_A$$

4. If these three equations are added, the result is

$$A_A + B_B \longrightarrow A_B + B_A$$

The same equations can also be written in terms of interstitial B atom formation and exchange of the interstitial B_i with an A_A atom, and so forth.

1.11.4 Nickel Oxide

Nickel oxide, NiO, which adopts the sodium chloride structure (Fig. 1.14), can readily be made slightly oxygen rich, and, because the solid then contains more oxygen than nickel, the crystal must also contain a population of point defects. This situation can formally be considered as a reaction of oxygen gas with stoichiometric NiO, and the simplest assumption is to suppose that the extra oxygen extends the crystal by adding extra oxygen sites. Atoms are added as neutral atoms, and initially it is reasonable to suppose that NiO is built of neutral atoms. Because of rule 1 above, each oxygen added must be balanced by the creation of a corresponding nickel site, so that the reaction equation is

$$\tfrac{x}{2}O_2(NiO) \longrightarrow x\,O_O + x\,V_{Ni} \qquad (1.2)$$

The reactant, oxygen gas, is written as in a normal equation. Strictly speaking the entry (NiO) to specify the host crystal is superfluous, but its inclusion avoids confusion (see later examples). The host crystal is written in parentheses to emphasize that it is not really part of the equation.[8] It might be considered to be more realistic if NiO is supposed to be ionic, containing Ni^{2+} and O^{2-} ions. In an ionic crystal each Ni vacancy must carry two effective negative charges, written V_{Ni}'', because a normal Ni^{2+} ion has been removed. The neutral oxygen atom introduced onto the oxygen ion site will be missing two electrons compared to O^{2-}, so that it carries two effective positive charges, written $O_O^{2\bullet}$. This transformation can be designated by extending Eq. (1.2) in the following way:

$$x\,O_O + x\,V_{Ni} \longrightarrow x\,O_O^{2\bullet} + x\,V_{Ni}'' \qquad (1.3)$$

The charges balance, in accordance with rule 3 above. Finally, it may be regarded as reasonable to assume that the added oxygen ends up as O^{2-}. In this case each added atom of oxygen must gain two electrons. These will be taken from another source in the crystal, thereby generating two positive holes, $h^\bullet$:

$$x\,O_O^{2\bullet} \longrightarrow x\,O_O + 2x\,h^\bullet \qquad (1.4)$$

The complete equation is derived by adding Eqs. (1.2), (1.3), and (1.4):

$$\tfrac{x}{2}O_2(NiO) \longrightarrow x\,O_O + x\,V_{Ni}'' + 2x\,h^\bullet \qquad (1.5)$$

The creation of each vacancy is accompanied by the creation of a hole. If the ionic assumption is correct, the solid would therefore be expected to behave as a *p*-type semiconductor. This is, in fact, the case for NiO. However, if for some analogous case this is not confirmed experimentally, the equation is not valid.

Chemically, it might be preferable to specify exactly the source of the electrons donated to form O^{2-} ions in NiO. The physical properties of the solid suggest that the two electrons actually come from separate Ni^{2+} ions, converting each of them to the ion Ni^{3+}. The defect is then a Ni^{3+} ion located on a Ni^{2+} site. This defect has an effective positive charge of one unit compared to the Ni^{2+} ion, so it would be written $Ni_{Ni}^\bullet$. Taking this into account, the reaction Eq. (1.4) needs to be

[8]Frequently, in defect formation equations the host crystal is written above the reaction arrows. Throughout this book the convention of placing the host crystal in parentheses on the left side of the equation will be followed.

written as

$$x\, O_O^{2\bullet} + 2x\, Ni_{Ni} \longrightarrow x\, O_O + 2x\, Ni_{Ni}^{\bullet} \tag{1.6}$$

Adding Eqs. (1.2), (1.3), and (1.6) gives

$$\tfrac{x}{2} O_2(NiO) + 2x\, Ni_{Ni} \longrightarrow x\, O_O + x\, V_{Ni}^{2\prime} + 2x\, Ni_{Ni}^{\bullet} \tag{1.7}$$

This reveals that two alternative defect structures can be imagined, one with free holes and one with Ni^{3+} defects. A further possibility is that the hole may be lightly bound to an Ni^{2+} ion to give a defect complex that could be written $(Ni_{Ni} + h^{\bullet})$. All of these descriptions are valid. The one adopted would be the one most consistent with the measured properties of the solid.

1.11.5 Cadmium Oxide

Cadmium oxide, CdO, like nickel oxide, also adopts the sodium chloride structure (Fig. 1.14). However, unlike nickel oxide, this compound can be made to contain more metal than oxygen. The defects that cause this metal excess are usually considered to be interstitial Cd atoms or ions. In this case the reaction is one in which the solid formally loses oxygen. Because of the rules of equation writing, this must involve the removal of neutral oxygen atoms. Each oxygen lost results in the loss of a nonmetal site. In order to keep the site ratio correct, a metal site must also be lost, forcing the metal into interstitial sites:

$$x\, CdO \longrightarrow \tfrac{x}{2} O_2 + x\, Cd_i$$

If CdO is considered to be ionic, and built of Cd^{2+} and O^{2-} ions, the cadmium interstitial will show two effective positive charges, written $Cd_i^{2\bullet}$. At the same time, removal of an oxygen atom will require that the two electrons that are part of the O^{2-} ion are left in the crystal:

$$x\, CdO \longrightarrow \tfrac{x}{2} O_2 + x Cd_i^{2\bullet} + 2x\, e'$$

The presence of free electrons means that metal-rich CdO should be an n-type semiconductor. It is possible to imagine that the divalent cadmium interstitials can take up an electron to form monovalent interstitials:

$$x\, CdO \longrightarrow \tfrac{x}{2} O_2 + x\, Cd_i^{\bullet} + x\, e'$$

As in the case of nickel oxide, the equation that best represents the real situation can only be determined experimentally. In this case one possible method would be to measure the way in which the electronic conductivity varies with the partial pressure of the surrounding oxygen atmosphere (Chapter 7).

1.11.6 Calcia-stabilized Zirconia

The same principles apply when two solids react, as can be illustrated by the reaction of a crystal of zirconia, ZrO_2, with a small amount of calcia, CaO, to produce a crystal of calcia-stabilized zirconia. There are two principal ways that this reaction can be imagined to occur: either the Ca atoms occupy Zr sites or they occupy interstitial sites.

Suppose that on reaction Ca atoms are located on normal metal sites in ZrO_2. In order to comply with rule 1 in Section 1.11.2, it is necessary to create two anion sites per Ca atom introduced. Initially, these are vacant. However, as an oxygen atom from the CaO must also be accommodated in the crystal, it is reasonable to place it in one of these sites. The other site remains vacant. If we take the ZrO_2 to be ionic and composed of Zr^{4+} and O^{2-} ions, the added (neutral) atoms will acquire effective charges. The neutral Ca atom at the Zr^{4+} displays an effective charge of 4 with respect to the charge experienced when a normal Zr^{4+} ion is present, so it is written as $Ca_{Zr}^{4\prime}$. Similarly, for ionic $(Zr^{4+}O_2^{2-})$ a neutral oxygen atom at an occupied oxygen site acquires an effective charge of $2\bullet$, and is hence written $O_O^{2\bullet}$. The same is true of the oxygen vacancy. Compared to a site occupied by an O^{2-} ion, the site has an effective positive charge of $2\bullet$, so the vacancy is written $V_O^{2\bullet}$. Formally, one can therefore write

$$CaO(ZrO_2) \longrightarrow Ca_{Zr}^{4\prime} + V_O^{2\bullet} + O_O^{2\bullet} \qquad (1.8)$$

The host crystal, ZrO_2, is written in parentheses (ZrO_2). Note that the equation conserves mass balance, electrical charge balance, and site numbers in accordance with the rules given in Section 1.11.2. Now CaO is normally regarded as an ionic compound, as implied by the foregoing allocation of charges, so that *ions* should occupy the sites, not neutral atoms. To achieve this, two electrons are transferred from the neutral Ca atom to create Ca^{2+} and lodged upon the neutral O atom to create O^{2-}. The effective charge on the Ca ion will now be just two negative units instead of four, and there will be no effective charge on the oxygen ion as all normal oxygen sites contain O^{2-}. To stress this electroneutrality, it is given the symbol O_O^x. There is no change in the status of the oxygen vacancy. The process is summarized by

$$Ca_{Zr}^{4\prime} + V_O^{2\bullet} + O_O^{2\bullet} \longrightarrow Ca_{Zr}^{2\prime} + V_O^{2\bullet} + O_O^x \qquad (1.9)$$

The final outcome for the reaction between the two *ionic* crystals is obtained by adding Eqs. (1.8) and (1.9) to give

$$CaO(ZrO_2) \longrightarrow Ca_{Zr}^{2\prime} + V_O^{2\bullet} + O_O^x$$

For chemical reasons it might be argued that the Ca^{2+} ions do not occupy Zr^{4+} sites but prefer interstitial positions, while the oxygen atoms occupy newly formed sites. The Ca interstitial atoms do not affect site numbers, but the oxygen atoms must maintain the site ratio of the ZrO_2 matrix, so that one Zr vacancy must be created for each pair of oxygen atoms added. Because neutral atoms are added, the vacancy and oxygen atoms will carry effective charges as above. The reaction is

$$2CaO(ZrO_2) \longrightarrow 2Ca_i + 2O_O^{2\bullet} + V_{Zr}^{4\prime} \tag{1.10}$$

If the added species are ions (Ca^{2+} and O^{2-}) the Ca^{2+} ions will have an effective charge of $2\bullet$, written $Ca_i^{2\bullet}$. Each oxide ion has no effective charge compared to a normal oxide anion, written O_O^x. There is no change as far as the Zr vacancy is concerned. The transformation can formally be expressed as

$$2Ca_i + 2O_O^{2\bullet} + V_{Zr}^{4\prime} \longrightarrow 2Ca_i^{2\bullet} + 2O_O^x + V_{Zr}^{4\prime} \tag{1.11}$$

The reaction of ionic CaO with ionic ZrO_2 can be obtained by addition of Eqs. (1.10) and (1.11):

$$2CaO(ZrO_2) \longrightarrow 2Ca_i^{2\bullet} + 2O_O^x + V_{Zr}^{4\prime}$$

Experimental evidence is needed to decide which (if any) of these formal equations represents the true situation in the material (see Section 4.4.5).

1.11.7 Ternary Oxides

The principles described above apply equally well to oxides with more complex formulas. In these materials, however, there are generally a number of different cations or anions present. Generally, only one of the ionic species will be affected by the defect forming reaction while (ideally) others will remain unaltered. The reactant, on the other hand, can be introduced into any of the suitable ion sites. This leads to a certain amount of complexity in writing the defect equations that apply. The simplest way to bypass this difficulty is to "decompose" the complex oxide into its major components and treat these separately. Two examples, using the perovskite structure, can illustrate this.

The perovskite structure, ABO_3 (where A represents a large cation and B a medium-size cation) is adopted by many solids and solid solutions between them can readily be prepared. Vacancy-containing systems with the perovskite structure are of interest as electrolytes in solid-state batteries and fuel cells. Typical representatives of this type of material can be made by introducing a higher valence cation into the A sites or a lower valance cation into the B sites.

The first example is the substitution of La^{3+} for the alkaline earth Ca^{2+} cation in $CaTiO_3$. In this reaction, the Ti sites are unaffected, and it can be written simply in terms of the reaction between the impurity La_2O_3 and a CaO matrix. The site maintenance rule means that three oxygen anions added (from La_2O_3) must be balanced by the creation of three cation sites. Assuming ionic materials:

$$La_2O_3(CaO) \longrightarrow 2La_{Ca}^{\bullet} + V_{Ca}^{2\prime} + 3O_O$$

One vacancy is generated for every two La^{3+} substituents. If one wishes to include the TiO_2, which is a "sleeping partner," it is only necessary to maintain the correct stoichiometry. The creation of three new A cation sites $(2La_{Ca}^{\bullet} + V_{Ca}^{2\prime})$ requires the inclusion of three Ti sites and the corresponding number of anion sites:

$$La_2O_3(CaTiO_3) \longrightarrow 2La_{Ca}^{\bullet} + V_{Ca}^{2\prime} + 3Ti_{Ti} + 9O_O$$

Exactly the same approach can be employed for substitution on the B sites, and this can provide our second example. Assuming ionic compounds, consider the substitution of Cr^{3+} for Ti^{4+} in $CaTiO_3$. This time the CaO component is the sleeping partner and can be ignored so the reaction can be considered to be between the Cr_2O_3 impurity and the TiO_2 matrix. The site maintenance rule requires that inclusion of two Cr^{3+} ions on Ti^{4+} sites requires the creation of four O^{2-} sites:

$$Cr_2O_3(TiO_2) \longrightarrow 2Cr_{Ti}^{\prime} + 3O_O + V_O^{2\bullet}$$

One anion vacancy is generated for every two Cr^{3+} substituents. As before, if it is helpful, include the CaO, simply maintain the stoichiometry. The creation of two new B cation sites $(2Cr_{Ti}^{\prime})$ requires the inclusion of two Ca sites and two anion sites:

$$Cr_2O_3(CaTiO_3) \longrightarrow 2Ca_{Ca} + 2Cr_{Ti}^{\prime} + 5O_O + V_O^{2\bullet}$$

1.12   COMBINATIONS OF POINT DEFECTS IN PURE MATERIALS

There is no obvious reason why only one defect type should occur in a crystal, and several different species would be expected to be present. However, the formation energy of each defect type is different, and it is often a reasonable approximation to assume that only one or a small number of defect types will dominate the chemical and physical properties of the solid.

For example, the formation of an intrinsic interstitial defect requires the simultaneous creation of a vacancy. These may not remain close together in the crystal, and it is legitimate to consider that the two defects occur in equal numbers. Thus, in silicon it is possible to write the formation equation for silicon self-interstitials, Si_i, as

$$Si_{Si} \longrightarrow Si_i + V_{Si}$$

where Si_{Si} represents the perfect crystal before defect formation, Si_i represents the self-interstitial, and V_{Si} represents a vacancy on a normally occupied silicon site. Similarly, in the alloy FeAl, antisite defects consisting of Fe atoms on Al sites,

Fe_{Al}, coexist with vacancies on Fe sites, V_{Fe} (Section 1.10). The iron sites, therefore, contain populations of two defects, vacancies and aluminum atoms.

Calculations of defect formation energies (Section 2.10) will allow estimates of the numbers of each kind of defect present to be made using formulas of the type:

$$n_d \propto \exp\left(\frac{-E_d}{RT}\right)$$

where n_d is the number of defects present, E_d the molar formation energy of the defect, R the gas constant, and T the temperature (K).

1.13 STRUCTURAL CONSEQUENCES OF POINT DEFECT POPULATIONS

The various types of point defect found in pure or almost pure stoichiometric solids are summarized in Figure 1.17. It is not easy to imagine the three-dimensional consequences of the presence of any of these defects from two-dimensional diagrams, but it is important to remember that the real structure of the crystal surrounding a defect can be important. If it is at all possible, try to consult or build crystal models. This will reveal that it is easier to create vacancies at some atom sites than others, and that it is easier to introduce interstitials into the more open parts of the structure.

Moreover, point defects will undoubtedly cause significant crystallographic distortions in the region of the affected site, which endow the defects with a

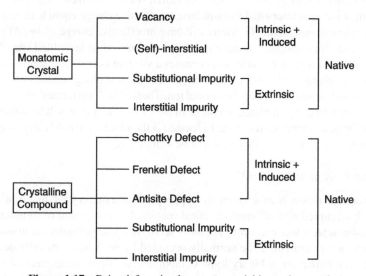

Figure 1.17 Point defects in almost pure stoichiometric crystals.

spatial extension greater than that implied in the term point defect. The resulting strain field plays an important part in the aggregation of point defects into clusters and other extended defects. The fraction of point defects that aggregate will depend upon the relative energy gain when a cluster forms, as well as the mobility of the defects in the crystal. The electronic and other interactions between defects can be calculated using approaches analogous to the Debye–Hückel treatment used for electrolytes or the Lorentz theory of the polarizability of insulating solids. In fact, simulations and quantum mechanical calculations (Chapter 2) suggest that defect association is the rule rather than the exception.

1.14 ANSWERS TO INTRODUCTORY QUESTIONS

What is a point defect?

A point defect is a localized defect that consists of a mistake at a single atom site in a solid. The simplest point defects that can occur in pure crystals are missing atoms, called vacancies, or atoms displaced from the correct site into positions not normally occupied in the crystal, called self-interstitials. Additionally atoms of an impurity can occupy a normal atom site to form substitutional defects or can occupy a normally vacant position in the crystal structure to form an interstitial. Other point defects can be characterized in pure compounds that contain more than one atom. The best known of these are Frenkel defects, Schottky defects, and antisite defects.

What is the effective charge on a defect?

The effective charge on a defect is the charge with respect to that normally occurring at the same site in a perfect crystal. Interstitial sites are normally unoccupied, so an atom or ion at an interstitial site will have an effective charge equal to its real charge. For example, interstitial Ti^{4+} ions will have an effective charge of $4\bullet$. Any neutral atom, Zn, for example, will have no effective charge at an interstitial site. When an impurity atom or an impurity ion occupies a site that normally contains an ion, the effective charge will be the difference between the real charge on the impurity atom or ion and the real charge on the normal ion. Thus, a Ca^{2+} ion located on a Na^+ site in a crystal will carry an effective charge of $2 - 1 = 1$, that is, it will be written $Ca_{Na}^{\bullet}$. Similarly, a neutral atom such as Li located at the same site would carry an effective charge of $0 - 1 = -1$, that is, it will be written $Li_{Na}^{'}$.

What is an antisite defect?

An antisite defect is an atom on an inappropriate site in a crystal, that is, a site normally occupied by a different chemical species. In a compound of formula AB the antisite defects that can occur are an A atom on a site normally occupied by a B atom, or a B atom on a site normally occupied by an A atom. Antisite defects are not very important in binary ionic compounds, as the misplacement of an ion is energetically costly, and so unfavorable. In ternary ionic compounds, however, such as spinels, AB_2O_4, the transfer of A ions to B sites and vice versa, is not expensive on energy and can occur. In metallic and many semiconducting crystals, the valence electrons are delocalized throughout the solid, so that also in these cases antisite defects are not accompanied by prohibitive energy costs. For example, most important defect in the semiconducting material GaAs is the antisite defect formed when an As atom occupies a Ga site.

PROBLEMS AND EXERCISES

Quick Quiz

1. Point defects that form in a crystal exposed to radioactivity are called:
   (a) Intrinsic defects
   (b) Induced defects
   (c) Native defects

2. Intrinsic defects in crystals:
   (a) Are always present
   (b) Only form if two or more different atom species occur
   (c) Only form in the presence of impurities

3. Impurity atoms are:
   (a) Intrinsic defects
   (b) Extrinsic defects
   (c) Native defects

4. An impurity P atom on a Si site in a crystal of silicon is a:
   (a) Interstitial defect
   (b) Antisite defect
   (c) Substitutional defect

5. The number of Schottky defects in a crystal of formula MX is equal to:
   (a) Half the total number of vacancies present
   (b) The total number of vacancies present
   (c) Double the total number of vacancies present

6. A single anion Frenkel defect in an ionic crystal of formula MX_2 needs to be balanced by:
   (a) A cation Frenkel defect
   (b) Two cation Frenkel defects
   (c) Nothing

7. A metal atom N on an M site in a crystal MX would be written:
   (a) N_M
   (b) M_N
   (c) N_{MX}

8. An anion vacancy in an ionic oxide has the symbol:

   (a) V_O^x

   (b) $V_O^{2\prime}$

   (c) $V_O^{2\bullet}$

9. A Ca^{2+} ion on a Na^+ site in a crystal is written:

   (a) Ca_{Ca}^x

   (b) $Ca_{Na}^{\bullet}$

   (c) $Ca_{Na}^{\prime}$

10. The equation representing doping of La_2O_3 into CaO is:

    (a) $La_2O_3\ (CaO) \longrightarrow 2La_{Ca}^{\bullet} + V_{Ca}^{2\prime} + 3O_O$

    (b) $La_2O_3\ (CaO) \longrightarrow 2La_{Ca}^{3\bullet} + V_{Ca}^{2\prime} + 3O_O$

    (c) $La_2O_3\ (CaO) \longrightarrow 2La_{Ca}^{3\prime} + V_{Ca}^{2\prime} + 3O_O$

Calculations and Questions

1. (a) A ruby crystal has a composition $(Al_{0.99}Cr_{0.01})_2O_3$. How many Cr^{3+} ions are there in a ruby of dimensions $1\ cm^3$? (The unit cell dimensions are $a = 0.4763\ nm$, $c = 1.3009\ nm$; Z (the unit cell contents) $= 6Al_2O_3$; the volume of the unit cell can be taken as $0.966a^2c$.)

   (b) A solid solution between the spinel structure phases $MgGa_2O_4$ ($a = 0.82780\ nm$) and $MgGaMnO_4$ ($a = 0.83645\ nm$) obeys Vegard's law. What is the composition of the phase with a lattice parameter of $0.83000\ nm$?

2. (a) The unit cell of zirconium sulfide, ZrS, is sodium chloride type, $a = 0.514\ nm$. The measured density is $4800\ kg\ m^{-3}$. Suggest a possible defect structure and formula for the solid. There are four ZrS units per unit cell, relative molar mass: Zr, $91.22\ g\ mol^{-1}$; S, $32.07\ g\ mol^{-1}$.

   (b) A sample of a cubic calcia-stabilized zirconia ceramic has a density of $5720\ kg\ m^{-3}$ and a lattice parameter of $0.5130\ nm$. Assuming $V_O^{2\bullet}$ are the dominant point defects, suggest a composition for the solid. There are four MO_2 units in the unit cell. Molar mass: Zr, $91.22\ g\ mol^{-1}$; Ca, $40.08\ g\ mol^{-1}$; O, $16.00\ g\ mol^{-1}$.

3. The Seebeck coefficient for pure $LaCoO_3$ is $+600\ \mu V\ K^{-1}$. (a) What are the mobile charge carriers? (b) Suppose these occur because the crystal contains a trace of an impurity, Co^{4+}, calculate the defect concentration and the formula of the material (data from Robert et al., 2006).

4. Estimate the Seebeck coefficient for the reduced oxide $TiO_{1.94}$, assuming that the defects are Ti^{3+} ions and the parent phase is TiO_2.

5. Plot a graph of the expected Seebeck coefficient of $La_{1-x}Sr_xMnO_3$ for $x = 0-0.2$.

6. A complex Hg-Sr-Ca-Co containing oxide has a Seebeck coefficient of $115\,\mu V\,K^{-1}$, due to the presence of both Co^{3+} and Co^{4+}. If the defects are the Co^{4+} ions, what is the average valence of the Co ions (data from Pelloquin *et al.*, 2003).

7. Write defect equations for the following reactions:

(a) MgO doped with Li_2O

(b) Slight reduction of PrO_2

(c) TiO_2 doped with Ga_2O_3

(d) $LaCoO_3$ doped with CaO

(e) $MgAl_2O_4$ doped with V_2O_3 to form $Mg(Al_{2-x}V_x)_2O_4$

8. Nickel oxide, NiO, is doped with lithium oxide, Li_2O, to form $Li_xNi_{1-x}O$ with the sodium chloride structure. (a) Derive the form of the Heikes equation for the variation of Seebeck coefficient, α, with the degree of doping, x. The following table gives values of α versus $\log[(1-x)/x]$ for this material. (b) Are the current carriers holes or electrons? (c) Estimate the value of the constant term k/e.

$\alpha/mV\,K^{-1}$	$\log[(1-x)/x]$
0.12	0.2
0.18	0.5
0.28	1.0
0.39	1.5
0.48	2.0
0.59	2.5
0.69	3.0

Data adapted from E. Antolini, *Mater. Chem. Phys.*, **82**, 937–948 (2003).

9. The oxides with general formulas A_2O, AO, A_2O_3, AO_2, and A_2O_5 are doped into MgO so that the cation substitutes for Mg. Write general defect equations for the reactions, assuming cation vacancies rather than electronic compensation occurs.

10. The crystallographic data for Al_2O_3 are $a = 0.47628$ nm, $c = 13.0032$ nm, and for V_2O_3 are $a = 0.4955$ nm, $c = 14.003$ nm. The structure of both oxides is the corundum type, the volume of the hexagonal unit cell can be taken as $0.966a^2c$ and Z (the unit cell contents) $= 6$ formula units. (a) Calculate the density of Al_2O_3. (b) Calculate the density of V_2O_3. (c) Provided Vegard's law is obeyed, estimate the density of $V_{0.6}Al_{1.4}O_3$. (d) The measured density is $3620\,kg\,m^{-3}$. What does this suggest about Vegard's law dependence?

REFERENCES

D. Pelloguin, A. Maignan, S. Hébert, C. Michel, and B. Raveau, *J. Solid State Chem.*, **170**, 374–381 (2003).

R. Robert, L. Becker, M. Trottman, A. Reller, and A. Weidenkraft, *J. Solid State Chem.*, **179**, 3893–3899 (2006).

FURTHER READING

Much of the material in this chapter is introductory. Background information on solids is given in:

R. J. D. Tilley, *Understanding Solids*, Wiley, Chichester, 2004.

R. J. D. Tilley, *Crystals and Crystal Structures*, Wiley, Chichester, 2006.

A. F. Wells, *Structural Inorganic Chemistry*, 5th ed., Oxford University Press, Oxford, United Kingdom, 1984.

A. R. West, *Solid State Chemistry*, Wiley, Chichester, 1984.

Expansion and further explanation with respect to defects and defect chemistry and physics will be found in:

F. Agullo-Lopez, C. R. A. Catlow, and P. D. Townsend *Point Defects in Materials*, Academic, New York, 1988.

D. M. Smyth *The Defect Chemistry of Metal Oxides*, Oxford University Press, Oxford, United Kingdom, 2000.

A. M. Stoneham, *The Theory of Defects in Solids*, Oxford University Press, Oxford, United Kingdom, 1985.

Various authors *Mat. Res. Soc. Bull.*, **XVI**, November (1991) and December (1991).

Many aspects of the study and importance of defects in solids presented from a historical and materials perspective are to be found in:

R. W. Cahn, *The Coming of Materials Science*, Pergamon/Elsevier, Oxford, United Kingdom, 2001.

Details about the Heikes formula is found in:

R. R. Heikes and R. W. Ure, Jr., *Thermoelectricity; Science and Engineering*, Wiley/Interscience, New York, 1961, p. 40.

Intrinsic Point Defects in Stoichiometric Compounds

What is an intrinsic defect?

What point defects are vital for the operation of LiI pacemaker batteries?

What is the basis of "atomistic simulation" calculations of point defect formation energies?

Intrinsic point defects are always present in a crystal as an inescapable property of the solid. For this to be so the intrinsic defect must be stable from a thermodynamic point of view. In this chapter the consequences of this thermodynamic aspect will be considered in more detail.

2.1 EQUILIBRIUM POPULATION OF VACANCIES IN A MONATOMIC CRYSTAL

In order to estimate the likelihood that vacancies are present in a crystal, it is necessary to determine the energy needed to form the defects in an otherwise perfect solid. Thermodynamics, which gives information about systems when they are at equilibrium, is able to provide guidance on this problem. In general, the first law of thermodynamics indicates that there will be an energy cost in introducing the defects, which can loosely be equated with changes in the chemical bonds around the defect. The second law of thermodynamics says that this energy cost may be recouped via the disorder introduced into the crystal. These imprecise concepts can be quantified by considering the Gibbs energy of a crystal, G:

$$G = H - TS$$

where H is the enthalpy, S is the entropy, and T the temperature (K) of the crystal.

If vacancies exist in the crystal under equilibrium, the Gibbs energy must be less than it is in a perfect crystal. However, if the introduction of more and more defects causes the Gibbs energy to continually fall, then ultimately no crystal will remain.

Defects in Solids, by Richard J. D. Tilley
Copyright © 2008 John Wiley & Sons, Inc.

This implies that the form of the Gibbs energy curve must pass through a minimum if defects are to be present at equilibrium (Fig. 2.1a).

The change in the Gibbs energy of the crystal by an amount ΔG_V, due to the introduction of n_V vacancies distributed over N possible atom sites is given by

$$\Delta G_V = \Delta H_V - T \, \Delta S_V \tag{2.1}$$

where ΔH_V is the associated change in enthalpy and ΔS_V the change in the entropy of the crystal. In fact, the ΔH term is approximately equivalent to the bond energy that

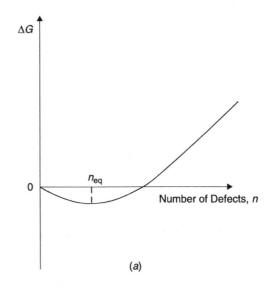

(a)

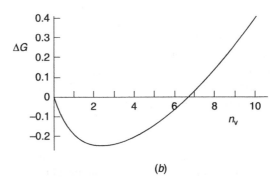

(b)

Figure 2.1 Change in Gibbs energy, ΔG, of a crystal as a function of the number of point defects present: (a) Variation of ΔG with number of point defects (schematic); at equilibrium, n_{eq} defects are present in the crystal; (b) calculated variation of ΔG for $h_V = 0.6$ eV, $kT = 0.1$, $N = 1000$; the equilibrium number of point defects is 2.5 per 1000.

must be expended in forming the defects, and the ΔS term is equivalent to the additional randomness in the crystal due to the defects. These correspond to the first law and second law contributions mentioned. In general, the energy increase due to the enthalpy term, ΔH_V, will be offset by the energy decrease due to the $T \Delta S_V$ term. To find ΔG_V, it is necessary to determine the change in the enthalpy ΔH_V, the change in entropy ΔS_V, or both. Because the enthalpy tends to be associated more with the bonding energy between nearest neighbors, as a first approximation this term can be regarded as constant. The change in entropy consists of terms due to the vibration of the atoms around the defects and terms due to the arrangements of the defects in the crystal. This latter quantity, called the configurational entropy, is relatively easy to assess using the well-established methods of statistical mechanics.[1]

As a first approximation, assume that the total entropy change, ΔS_V, is made up solely of configurational entropy. This can be determined by using the Boltzmann equation for the entropy of a disordered system:

$$S = k \ln W$$

where S is the entropy of a system in which W is the number of ways of distributing n defects over N sites at random, and k is the Boltzmann constant. Probability theory shows that W is given by

$$W = \frac{N!}{(N-n)!\,n!}$$

where the symbol $N!$, called *factorial N*, is mathematical shorthand for the expression:

$$N \times (N-1) \times (N-2), \ldots, 1$$

Similar expressions can be written for $n!$ and $(N - n)!$ Thus, the number of ways that n_V vacancies can be arranged over the available N sites in the crystal is

$$W = \frac{N!}{(N-n_V)!\,n_V!}$$

Therefore the change in configurational entropy caused by introducing these defects is

$$\Delta S_V = k \ln W = k \ln \left[\frac{N!}{(N-n_V)!\,n_V!} \right]$$

---

[1]What follows is a simplified version of the discussion given by Wagner and Schottky (1931).

To proceed, it is necessary to eliminate the factorials. This is usually done by employing the approximation:

$$\ln N! \approx N \ln N - N \tag{2.2}$$

which is referred to as Stirling's approximation.[2]

Substituting from Eq. (2.2):

$$\Delta S_V = k\{N \ln N - (N - n_V)\ln(N - n_V) - n_V \ln n_V\} \tag{2.3}$$

The enthalpy change involved, ΔH_V, is not explicitly calculated. It is assumed that the enthalpy to form one vacancy, Δh_V, is constant over the temperature range of interest, so that the total enthalpy change, ΔH_V, is given by

$$\Delta H_V = n_V \Delta h_V$$

Substituting this, and the value of ΔS_V from Eq. (2.3) into Eq. (2.1), gives

$$\Delta G_V = n_V \Delta h_V - kT\{N \ln N - (N - n_V)\ln(N - n_V) - n_V \ln n_V\}$$

A plot of this equation (Fig. 2.1b) closely resembles Figure 2.1a. The minimum in the curve gives the equilibrium number of vacancies present and confirms that vacancies exist in *all* crystals at temperatures above 0 K. For this reason these defects cannot be removed by thermal treatment but are always present in a crystal. Such defects are thus intrinsic defects. At equilibrium, ΔG_V will be equal to zero and the minimum in the ΔG_V versus n_V curve is given by

$$\left(\frac{d\Delta G_V}{dn_V}\right)_T = 0$$

that is,

$$\left(\frac{d\Delta G_V}{dn_V}\right)_T = \frac{d}{dn_V}\{n_V \Delta h_V - kT[N \ln N - (N - n_V)\ln(N - n_V) - n_V\ln n_V]\}$$

$$= 0$$

$$\Delta h_V - kT\left\{\frac{d}{dn_V}[N \ln N - (N - n_V)\ln(N - n_V) - n_V \ln n_V]\right\} = 0$$

[2]In fact this last approximation is not very accurate and is several percent in error even for values of N as large as 10^{10}. The correct expression for Stirling's approximation is

$$\ln N! \approx N \ln N - N + \tfrac{1}{2}\ln(2\pi N)$$

(Supplementary Material S4).

Remembering that $(N \ln N)$ is constant, so its differential is zero, the differential of $(\ln x)$ is $(1/x)$ and of $(x \ln x)$ is $(1 + \ln x)$, on differentiating:

$$\Delta h_V - kT \left[\ln (N - n_V) + \frac{(N - n_V)}{(N - n_V)} - \ln n_V - \frac{n_V}{n_V} \right] = 0$$

hence:

$$\Delta h_V = kT \ln \left[\frac{(N - n_V)}{n_V} \right]$$

Rearranging:

$$n_V = (N - n_V) e^{-\Delta h_V / kT}$$

If N is considered to be very much greater than n_V:

$$n_V \approx N e^{-\Delta h_V / kT} \tag{2.4}$$

In the knowledge that the Gibbs energy of a crystal containing a small number of intrinsic defects is lower than that of a perfect crystal, the defect population can be treated as a chemical equilibrium. In the case of vacancies we can write

$$\text{null} \rightleftharpoons \text{vacancy}$$
$$0 \rightleftharpoons V_M$$

where null represents a normal occupied site in a crystal and V_M represents a vacancy on the position normally occupied by an atom M. Chemically, the reaction can be imagined as the migration of an atom within a crystal to the surface to extend the volume of the solid or to evaporate, while leaving a vacancy in its place. The law of mass action in its simplest form, with concentrations instead of activities, can be applied if there is no interaction between the defects to give an expression for the equilibrium constant for vacancy generation, K_V:

$$K_V = \frac{n_V}{N} = \exp\left(\frac{-\Delta G_V}{RT}\right)$$

where n_V is the number of vacant sites, N is the total number of normally occupied atom sites, and the term n_V/N, the fraction of vacancies in the crystal thus represents concentration (defects per atom site), ΔG_V is the molar Gibbs energy of formation of the vacancies, R is the gas constant, and T the temperature (K). This equation is often written

$$n_V = N \exp\left(\frac{-\Delta G_V}{RT}\right) \tag{2.5a}$$

where n_V is taken as the number of vacant sites per unit volume, and N is the total number of normally occupied atom sites per unit volume. When the Gibbs energy for the formation of a single defect is used, Eq. (2.5a) is written in the form:

$$n_V = N \exp\left(\frac{-\Delta g_V}{kT}\right) \tag{2.5b}$$

where the Gibbs energy $-\Delta g_V$ is often given the units joules or electron volts,[3] (per defect), and is paired with the Bolzmann constant, k with units $J\ K^{-1}$ or $eV\ K^{-1}$.

The fraction of vacancies in the crystal, n_V/N, is given by

$$\frac{n_V}{N} = \exp\left(\frac{-\Delta G_V}{RT}\right) \tag{2.6}$$

The Gibbs energy, ΔG_V, is often split into two terms using Eq. (2.1) to give

$$\frac{n_V}{N} = \exp\left(\frac{-\Delta H_V}{RT}\right) + \exp\left(\frac{\Delta S_V}{T}\right)$$

and the entropy term neglected, so that:

$$\frac{n_V}{N} \approx \exp\left(\frac{-\Delta H_V}{RT}\right) \tag{2.7}$$

When written in terms of energy per defect, Eqs. (2.6) and (2.7) become

$$\frac{n_V}{N} = \exp\left(\frac{-\Delta g_V}{kT}\right)$$

$$\approx \exp\left(\frac{-\Delta h_V}{kT}\right)$$

2.2 EQUILIBRIUM POPULATION OF SELF-INTERSTITIALS IN A MONATOMIC CRYSTAL

Because the configurational entropy of interstitial defects has the same form as that of vacancies, a population of self-interstitial atoms is also thermodynamically stable. The creation of these defects can then also be treated as a pseudochemical equilibrium, and an equation for the relationship between the number of self-interstitials and the appropriate equilibrium constant for interstitial generation, K_i, is readily

---

[3]In the literature, the calculations are almost always expressed in units of electron volt (eV), where 1 eV is equivalent to $96.485\ kJ\ mol^{-1}$ or $1.60218 \times 10^{-19}\ J$.

obtained. The creation of a self-interstitial on one of N_i possible interstitial sites leaves a vacancy on one of the N normal lattice sites:

$$\text{null} \rightleftharpoons \text{interstitial} + \text{vacancy}$$
$$0 \rightleftharpoons M_i + V_M$$

where M_i represents an interstitial M atom and V_M a vacancy on a normally occupied atom site in the crystal. The law of mass action in its simplest form, with concentrations instead of activities, can be applied if there is no interaction between the defects to give an expression for the equilibrium constant for self-interstitial generation, K_i:

$$K_i = \left(\frac{n_i}{N_i}\right)\left(\frac{n_V}{N}\right) = \exp\left(\frac{-\Delta G_i}{RT}\right)$$

where n_i is the number of interstitials, n_V the number of vacancies, N_i the number of suitable interstitial sites, N the number of normally occupied sites, ΔG_i is the molar Gibbs energy of formation of the self-interstitials, R is the gas constant, and T the absolute temperature. As the number of interstitials and vacancies is the same in a monatomic crystal, $n_i = n_V$:

$$n_i = \sqrt{NN_i}\exp\left(\frac{-\Delta G_i}{2RT}\right) \tag{2.8}$$

where n_i is the number of interstitials per unit volume, n_V the number of vacancies per unit volume, N_i the number of suitable interstitial sites per unit volume, and N the number of normally occupied sites per unit volume. The fraction of interstitial defects is given by

$$\frac{n_i}{\sqrt{NN_i}} = \exp\left(\frac{-\Delta G_i}{2RT}\right)$$

In the case where the number of normal and interstitial sites is the same, that is, $N = N_i$:

$$n_i = N\exp\left(\frac{-\Delta G_i}{2RT}\right)$$

which is identical to the equation for vacancies, as is the fraction of interstitials in the crystal, n_i/N:

$$\frac{n_i}{N} = \exp\left(\frac{-\Delta G_i}{2RT}\right)$$

The Gibbs energy, ΔG_i, is often split into two terms and the entropy term neglected, as in the previous section, to give

$$n_i \approx \sqrt{NN_i}\exp\left(\frac{-\Delta H_i}{2RT}\right) \tag{2.9}$$

Similar relationships hold if the Gibbs energy of formation per defect, Δg_i, and the enthalpy of formation per defect, Δh_i, is used and R is replaced by k, the Boltzmann constant:

$$n_i = \sqrt{NN_i}\,\exp\left(\frac{-\Delta g_i}{2kT}\right)$$

$$\approx \sqrt{NN_i}\,\exp\left(\frac{-\Delta h_i}{2kT}\right)$$

2.3   EQUILIBRIUM POPULATION OF SCHOTTKY DEFECTS IN A CRYSTAL

A Schottky defect in a crystal consists of a cation and anion vacancy combination that ensures overall electroneutrality in the crystal (Section 1.9). The estimation of the configurational entropy change in creating a population of Schottky defects in a crystal can be obtained in the same way as that of a population of vacancies in a monatomic crystal. The method follows that given in Section 2.1 for the equilibrium concentration of vacancies in a monatomic crystal and is set out in detail in Supplementary Material S4.

Because Schottky defects are present as equilibrium species, the defect population can be treated as a chemical equilibrium. For a crystal of composition MX:

$$\text{null} \rightleftharpoons \text{cation vacancy} + \text{anion vacancy}$$
$$0 \rightleftharpoons V_M + V_X$$

where V_M represents a vacancy on a cation site and V_X a vacancy on an anion site. If there is no interaction between the defects, the law of mass action in its simplest form, with concentrations instead of activities, yields an expression for the equilibrium constant for the formation of Schottky defects, K_S:

$$K_S = \left(\frac{n_{cv}}{N}\right)\left(\frac{n_{av}}{N}\right) = \exp\left(\frac{-\Delta G_S}{RT}\right)$$

where n_{cv} is the number of cation vacancies, n_{av} the number of anion vacancies, N the number of cation sites, equal to the number of anion sites in the crystal, ΔG_S is the molar Gibbs energy of formation of the Schottky defects, R is the gas constant, and T the absolute temperature. Because n_{cv} is equal to n_{av}, it is possible to write the number of Schottky defects, n_S, as

$$n_{cv} = n_{av} = n_S = Ne^{-\Delta G_S/2RT} \tag{2.10}$$

where n_S is the number of Schottky defects per unit volume distributed over the N cation or anion sites per unit volume. The fraction of defects present is given by

$$\frac{n_{cv}}{N} = \frac{n_{av}}{N} = \frac{n_S}{N} = e^{-\Delta G_S/2RT} \tag{2.11}$$

The Gibbs energy, ΔG_S, is often replaced by the enthalpy of Schottky defect formation, ΔH_S, as in previous sections, to give

$$n_S \approx N e^{-\Delta H_S/2RT} \tag{2.12}$$

where ΔH_S is in joules per mole and is the energy required to form 1 mol of Schottky defects, and R is the gas constant. Similar relationships hold if the Gibbs energy of formation per defect, Δg_S, or the enthalpy of formation per defect, Δh_S, is used, thus:

$$n_S = N e^{-\Delta g_S/2kT}$$

$$\approx N e^{-\Delta h_S/2kT}$$

where the units of Δg_S and Δh_S are joules or electron volts per defect, and k is the Boltzmann constant.

Remember that these formulas only apply to materials with a composition MX.

Some values for the enthalpy of formation of Schottky defects in alkali halides of formula MX that adopt the sodium chloride structure are given in Table 2.1. The experimental determination of these values (obtained mostly from diffusion or ionic conductivity data (Chapters 5 and 6) is not easy, and there is a large scatter of values in the literature. The most reliable data are for the easily purified alkali halides. Currently, values for defect formation energies are more often obtained from calculations (Section 2.10).

Despite the uncertainty in the experimental values given in Table 2.1, they reveal that more energy is required to form defects as the systems are traversed in the direction LiX to KX and that the energy falls on passing from MF toward MI (Fig. 2.2).

TABLE 2.1 Formation Enthalpy of Schottky Defects in Some Alkali Halide Compounds of Formula MX[a]

Compound	$\Delta H_S/\text{kJ mol}^{-1}$	$\Delta h_S/\text{eV}$
LiF	225.2	2.33
LiCl	204.1	2.12
LiBr	173.4	1.80
LiI	102.4	1.06
NaF	233.1	2.42
NaCl	225.8	2.34
NaBr	203.0	2.10
NaI	140.9	1.46
KF	262.0	2.72
KCl	244.5	2.53
KBr	224.6	2.33
KI	153.0	1.59

[a]All compounds have the sodium chloride structure.

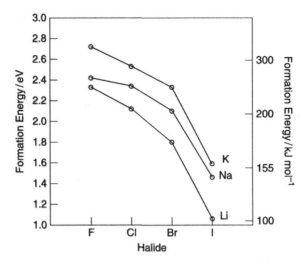

Figure 2.2   Variation of Schottky formation energy of the sodium chloride structure halides MX; M = Li, Na, K; X = F, Cl, Br, I.

2.4   LITHIUM IODIDE BATTERY

Batteries, almost indispensable in the modern world, are discussed in more detail in Chapter 6. In essence, a battery consists of an electrode (the anode), where electrons are moved out of the cell, and an electrode (the cathode), where electrons move into the cell. The electrons are generated in the region of the anode and then move around an external circuit, carrying out a useful function, before entering the cathode. The anode and cathode are separated inside the battery by an electrolyte, and the circuit is completed inside the battery by ions moving across it. A key component in battery construction is an electrolyte that can support ionic conduction but *not* electronic conduction.

Lithium iodide is the electrolyte in a number of specialist batteries, especially in implanted cardiac pacemakers. In this battery the anode is made of lithium metal. A conducting polymer of iodine and poly-2-vinyl pyridine (P2VP) is employed as cathode because iodine itself is not a good enough electronic conductor (Fig. 2.3*a*). The cell is fabricated by placing the Li anode in contact with the poly-vinyl pyridine–iodine polymer. The lithium, being a reactive metal, immediately combines with the iodine in the polymer to form a thin layer of lithium iodide, LiI, which acts as the electrolyte:

$$\text{Li(s)} + \tfrac{1}{2}\text{I}_2(\text{s}) \longrightarrow \text{LiI(s)}$$

In order for the battery to function, the lithium iodide must be able to transfer ions. LiI adopts the sodium chloride structure, and there are no open channels for ions to use. In fact, the cell operation is sustained by the Schottky defect population in the

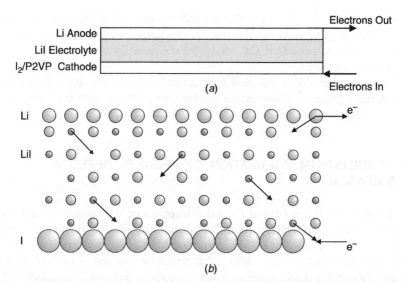

Figure 2.3 Basis of a lithium iodide cell (schematic): (*a*) electrons are liberated at the lithium metal anode and re-enter the cell via the I_2/P2PV cathode; (*b*) lithium ion transport across the electrolyte via Li vacancies, to form LiI at the anode. The number of vacancies (Schottky defects) has been grossly exaggerated.

solid. On closing the external circuit, the Li atoms in the anode surface become Li^+ ions at the anode–electrolyte interface.

$$\text{Anode reaction} \quad Li(s) \longrightarrow Li^+ + e^-$$

These diffuse through the lithium iodide via cation vacancies that form as part of the intrinsic Schottky defects in the crystals, to reach the iodine in the cathode (Fig. 2.3*b*). The electrons lost by the lithium metal on ionizing traverse the external circuit and arrive at the interface between the cathode and the electrolyte. Here they react with the iodine and the incoming Li^+ ions to form more lithium iodide.

$$\text{Cathode reaction} \quad 2Li^+ + I_2 + 2e^- \longrightarrow 2LiI$$

During use the thickness of the LiI electrolyte gradually increases because of this reaction.

The Schottky defect population in the electrolyte is rather too low for practical purposes. To overcome this problem the LiI is sometimes doped with CaI_2. The extra I^- ions extend the LiI structure, and the Ca^{2+} ions form substitutional impurity defects on sites normally reserved for Li^+ ions. The consequence of this is that each Ca^{2+} ion in LiI will form one cation vacancy over and above those present due to Schottky defects in order to maintain charge neutrality. This can be written

$$CaI_2(LiI) \longrightarrow Ca_{Li}^{\bullet} + 2I_I + V_{Li}'$$

where the term $CaI_2(LiI)$ means that a small amount of CaI_2 is doped into an LiI matrix. Because this increases the number of cation vacancies substantially—one vacancy is created per added Ca^{2+} ion—the conductivity of the electrolyte can be substantially increased.

The actual current drain that these batteries can support is low and is limited by the point defect population. However, the cell has a long life and high reliability, making it ideal for medical use in heart pacemakers.

2.5 EQUILIBRIUM POPULATION OF FRENKEL DEFECTS IN A CRYSTAL

The estimation of the number of Frenkel defects in a crystal can proceed along lines parallel to those for Schottky defects by estimating the configurational entropy (Supplementary Material S4). This approach confirms that Frenkel defects are thermodynamically stable intrinsic defects that cannot be removed by thermal treatment. Because of this, the defect population can be treated as a chemical equilibrium. For a crystal of composition MX, the appropriate chemical equilibrium for Frenkel defects on the cation sublattice is

$$\text{null} \rightleftharpoons \text{cation vacancy} + \text{cation interstitial}$$
$$0 \rightleftharpoons V_M + M_i$$

where V_M represents a vacancy on a cation site and M_i a cation interstitial. If there is no interaction between the defects, the law of mass action in its simplest form, with concentrations instead of activities, yields an expression for the equilibrium constant for the formation of Frenkel defects on cation sites, K_{cF}, given by

$$K_{cF} = \left(\frac{n_{cv}}{N}\right)\left(\frac{n_{ci}}{N_i}\right) = \exp\left(\frac{-\Delta G_{cF}}{RT}\right)$$

where n_{cv} is the number of cation vacancies, n_{ci} the number of cation interstitials, N the number of occupied cation sites that support Frenkel defects, N_i is the number of available interstitial sites, ΔG_{cF} is the molar Gibbs energy of formation of the cation Frenkel defects, R is the gas constant, and T the absolute temperature. Because n_{cv} is equal to n_{ci} and both are equal to the number of Frenkel defects, n_F, it is possible to write

$$n_{cv} = n_{ci} = n_{cF} = \sqrt{NN_i}\exp\left(\frac{-\Delta G_{cF}}{2RT}\right) \qquad (2.13)$$

where n_{cF} is the number of cation Frenkel defects per unit volume distributed over the N cation and N_i interstitial sites per unit volume. The fraction of Frenkel defects present is given by

$$\frac{n_{cv}}{\sqrt{NN_i}} = \frac{n_{ci}}{\sqrt{NN_i}} = \frac{n_{cF}}{\sqrt{NN_i}} = \exp\left(\frac{-\Delta G_{cF}}{2RT}\right) \tag{2.14a}$$

In the case where the number of normal and interstitial sites is the same, that is, $N = N_i$:

$$n_{cF} = N \exp\left(\frac{-\Delta G_i}{2RT}\right)$$

which is identical to the equation for Schottky defects, as is the fraction of Frenkel defects, n_{cF}/N:

$$\frac{n_{cF}}{N} = \exp\left(\frac{-\Delta G_i}{2RT}\right)$$

The Gibbs energy, ΔG_{cF}, is often replaced by the enthalpy of Frenkel defect formation, ΔH_{cF}, as described above, to give

$$n_{cF} \approx \sqrt{NN_i} \exp\left(\frac{-\Delta H_{cF}}{2RT}\right) \tag{2.15a}$$

where ΔH_{cF} is in joules per mole and is the energy required to form 1 mol of Frenkel defects, and R is the gas constant. Similar relationships hold if the Gibbs energy of formation per defect, Δg_{cF}, or the enthalpy of formation per defect, Δh_{cF}, is used, thus:

$$n_{cF} = \sqrt{NN_i} \exp\left(\frac{-\Delta g_{cF}}{2kT}\right)$$

$$\approx \sqrt{NN_i} \exp\left(\frac{-\Delta h_{cF}}{2kT}\right)$$

where the units of Δg_{cF} and Δh_{cF} are Joules or electron volts per defect, and k is the Boltzmann constant.

A similar equation can be written for Frenkel defects on the anion positions:

$$\text{null} \rightleftharpoons \text{anion vacancy} + \text{anion interstitial}$$
$$0 \rightleftharpoons V_X + X_i$$

where V_X represents a vacancy on an anion site and X_i an anion interstitial. Following the same procedure as before:

$$n_{av} = n_{ai} = n_{aF} = \sqrt{NN_i} \exp\left(\frac{-\Delta G_{aF}}{2RT}\right) \tag{2.14b}$$

TABLE 2.2 Formation Enthalpy of Frenkel Defects in Some Compounds of Formula MX and MX$_2$

Compound	ΔH_F/kJ mol^{-1}	Δh_F/eV	Compound	ΔH_F/kJ mol^{-1}	Δh_F/eV
AgCl^a	139.7	1.45	CaF$_2{}^b$	261.4	2.71
AgBr^a	109.0	1.13	SrF$_2{}^b$	167.4	1.74
β-AgI^a	57.8	0.60	BaF$_2{}^b$	184.3	1.91

^{a}Frenkel defects on the cation sublattice of a sodium chloride structure compound.
^{b}Frenkel defects on the anion sublattice of a fluorite structure compound.

$$n_{aF} \approx \sqrt{NN_i} \exp\left(\frac{-\Delta H_{aF}}{2RT}\right) \tag{2.15b}$$

$$= \sqrt{NN_i} \exp\left(\frac{-\Delta g_{aF}}{2kT}\right)$$

$$\approx \sqrt{NN_i} \exp\left(\frac{-\Delta h_{aF}}{2kT}\right)$$

where ΔG_{aF} is the molar Gibbs energy of formation of the anion defects, ΔH_{aF} is the molar enthalpy of formation of the anion defects, Δg_{aF} is the Gibbs energy required to form one defect, Δh_{aF} is the enthalpy of formation of one defect, R is the gas constant, and k is the Boltzmann constant. In the case where the number of interstitial positions used, N_i, is equal to the number of normally occupied cation sites, N, the equations are identical to those for Schottky defects.

Some experimental values for the formation enthalpy of Frenkel defects are given in Table 2.2. As with Schottky defects, it is not easy to determine these values experimentally and there is a large scatter in the values found in the literature. (Calculated values of the defect formation energies for AgCl and AgBr, which differ a little from those in Table 2.2, can be found in Fig. 2.5.)

2.6 PHOTOGRAPHIC FILM

Photographic film was the most widely used storage method for images throughout the twentieth century and still has an important part to play in image capture and storage, although digital data recording is now widely used. Both black-and-white and color photography using film are possible because of the presence of point defects in the crystals used. The light-sensitive materials employed are silver halides, notably AgBr, which are dispersed in gelatine to form the photographic emulsion. To ensure that the crystals are free of macroscopic defects such as dislocations, which degrade the perfection of the photographic images produced, the silver halide crystals are carefully grown within the gelatine matrix itself. The crystals so formed are usually thin triangular or hexagonal plates, varying between 0.01 and 10 μm in size, and in photographic parlance are known as grains.

When the emulsion is exposed to light, a latent image is said to form. After illumination each grain will contain a latent image, that is, it will have interacted with the light photons, or it will have remained unchanged. The film is then put into a developer. A grain that contains a latent image is totally reduced to metallic silver. A crystallite with no latent image remains unchanged. The reactions taking place can be written down schematically as:

$$\text{AgBr (crystal)} + \text{light photons} \rightarrow [\text{AgBr crystal} + \text{latent image}]$$

$$[\text{AgBr crystal} + \text{latent image}] + \text{developer} \rightarrow \text{Ag crystal}$$

It has been found that only a few photons, maybe as little as six, are needed to form the latent image. Photographic film is a very sensitive light detector. The final step in the photographic process, fixing, removes the unreacted silver bromide crystals from the emulsion, thus stabilizing the image (Fig. 2.4).

Despite the fact that not all details of the photographic process are completely understood, the overall mechanism for the production of the latent image is well known. Silver chloride, AgBr, crystallizes with the sodium chloride structure. While Schottky defects are the major structural point defect type present in most crystals with this structure, it is found that the silver halides, including AgBr, favor Frenkel defects (Fig. 2.5).

The formation of latent images is a multistage process, involving the Frenkel defect population. The major steps in the formation of the latent image follow a path similar to that originally suggested by Gurney and Mott (1938):

1. Interaction of a light photon with a halogen ion in the AgBr crystal. The energy from the photon ($h\nu$) liberates an electron from this ion:

$$h\nu + \text{Br}^- \longrightarrow \text{e}' + \text{Br}^\bullet$$

   Note that this is often written as the generation of an electron and a hole pair by excitation of an electron from the valence band to the conduction band:

$$h\nu \longrightarrow \text{e}' + \text{h}^\bullet$$

   As the valence band is largely made up of contributions from bromine orbitals, the two descriptions are formally identical.

2. The liberated electron is free to move in the structure and migrates to an interstitial silver ion, $\text{Ag}_i^\bullet$, which is part of a Frenkel defect, to form a neutral silver atom Ag_i^x:

$$\text{Ag}_i^\bullet + \text{e}' \longrightarrow \text{Ag}_i^x$$

3. In many instances, the above reaction will then take place in the reverse direction, and the silver atom will revert to the normal stable state as a Frenkel

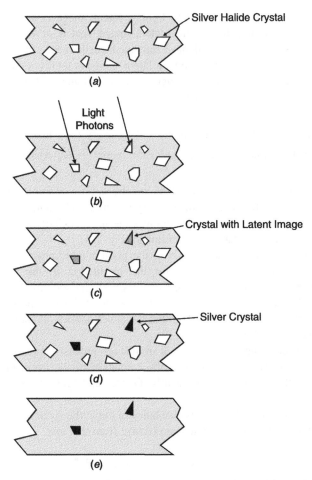

Figure 2.4 Production of a (negative) image in a photographic emulsion: (*a*) film emulsion containing crystallites of silver halides in gelatine; (*b*) interaction of some crystallites with light; (*c*) crystallites containing latent images (gray); (*d*) development transforms crystallites containing latent images into silver crystallites; and (*e*) fixing removes all unreacted silver halides leaving silver crystallites, which make up the image.

defect. However, the metal atom seems to be stabilized if another photon activates a nearby region of the crystal before the reverse reaction can take place. This stabilization may take place in either of two ways. It is possible that the interstitial silver atom can trap the electron liberated by the second photon to form the unusual Ag_i' ion, thus:

$$Ag_i^x + e' \longrightarrow Ag_i'$$

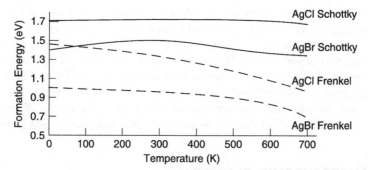

Figure 2.5 Calculated variation of the formation energy of Schottky and Frenkel defects in the halides AgCl and AgBr as a function of temperature. [Redrawn from data in C. R. A. Catlow, *Mat. Res. Soc. Bull.*, **XIV**, 23 (1989).]

The silver ion produced in this reaction is then neutralized by association with another interstitial silver atom, thus:

$$Ag_i^{\bullet} + Ag_i' \longrightarrow 2Ag_i^x$$

to produce a cluster of two neutral silver atoms.

The second possibility is that the second electron could interact with an interstitial ion to yield a second silver atom that would then diffuse to the first silver atom to form an identical cluster of two, namely:

$$Ag_i^{\bullet} + e' \longrightarrow Ag_i^x$$
$$Ag_i^x + Ag_i^x \longrightarrow 2Ag_i^x$$

4. Further aggregation of Ag atoms occurs by a similar mechanism.

Studies of small metal clusters have shown that a latent image consists of a minimum of four silver atoms. These small clusters of silver atoms completely control the chemistry of photography. During development of the film, only those crystallites containing a latent image react, and these are completely reduced to metallic silver. It becomes apparent that the presence of just one small cluster of four silver atoms determines whether the crystallite can react with developer or not.

Clearly, successful latent image formation depends upon a reasonable concentration of Frenkel defects in the halide crystals and the ability of the defects to diffuse through the crystal matrix. The enthalpy of formation of a Frenkel defect in AgBr is about 1 eV or 2×10^{-19} J (Table 2.2). The number of Frenkel defects in a crystal of AgBr at room temperature can be estimated using the equations given in Section 2.5. The ratio n_{cF}/NN^* turns out to be about 10^{-11}. This is not a very high population, and it seems reasonable to wonder whether such a concentration is sufficient to allow latent image formation to take place at all. In fact, it is much too small. However, research has shown that the surface of a silver halide grain has a net negative charge that is balanced by an enhanced number of interstitial silver ions located within a few tens of nanometers of the surface. As the halide grains are so small, the ratio of surface to bulk is large, so that the total population of interstitial silver ions in the grains is much higher than the figure given above, which applies to large crystals with relatively small surfaces. The surface concentration also implies that diffusion distances are relatively short. These factors are of importance in the overall efficiency of the process and make practical photography possible.

2.7 PHOTOCHROMIC GLASSES

Like photographic emulsions, photochromic glass is another material that is sensitive to light. Although many types of photochromic glass have been fabricated, the best known are those that darken on exposure to high-intensity visible or ultraviolet light and regain their transparency when the light intensity falls. Such glasses are widely used in sunglasses, sunroofs, and for architectural purposes.

The mechanism of the darkening transformation is similar to that involved in the photographic process. Photochromic glasses are complex materials that usually contain silver halides as the light-sensitive medium. The glass for this use would typically be an aluminoborosilicate (Pyrex type) material containing about 0.2 wt % of silver bromide or chloride. In addition, a small amount of a copper chloride is also added. When the glass is first fabricated, it is cooled rapidly. Under these conditions the silver and copper halides remain dissolved in the matrix, and the glass produced is transparent and does not show any photochromic behavior at all (Fig. 2.6a and 2.6b). This material is transformed into the photochromic state by

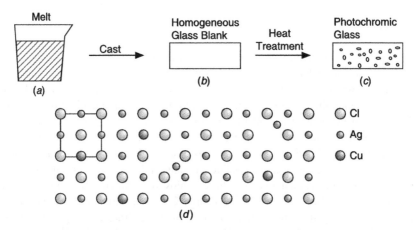

Figure 2.6 Photochromic glass: (a) glass melt containing dissolved CuCl and AgCl; (b) melt is cast into a homogeneous glass blank; (c) heat treatment precipitates crystallites (much exaggerated in size here) in the blank; and (d) sodium chloride structure of AgCl containing copper impurities and Frenkel defects.

heating under carefully controlled conditions of temperature and time, which might be, for example, 550°C for 30 min. followed by 650°C for 30 min. The heat treatment is chosen so that the halides crystallize in the glass matrix (Fig. 2.6c). Care must be taken to ensure that the crystals do not become too large and that they do not aggregate. A desirable size is about 10 nm diameter, and the individual crystallites should be about 100 nm apart. These crystals form volume defects throughout the homogeneous glass matrix.

It is important that the copper is in the monovalent state and incorporated into the silver halide crystals as an impurity. Because the Cu^+ has the same valence as the Ag^+, some Cu^+ will replace Ag^+ in the AgX crystal, to form a dilute solid solution $Cu_xAg_{1-x}X$ (Fig. 2.6d). The defects in this material are substitutional Cu_{Ag} point defects and cation Frenkel defects. These crystallites are precipitated in the complete absence of light, after which a finished glass blank will look clear because the silver halide grains are so small that they do not scatter light.

The influence of light causes changes similar to that occurring in a photographic emulsion. The photons liberate electrons and these are trapped by interstitial silver ions, which exist as Frenkel defects, to form neutral silver atoms. Unlike the photographic process, the electrons are liberated by the Cu^+ ions, which are converted to Cu^{2+} ions ($Cu^{\bullet}_{Ag}$) in the process:

$$hv + Cu_{Ag} \longrightarrow Cu^{\bullet}_{Ag} + e'$$
$$e' + Ag^{\bullet}_i \longrightarrow Ag^x_i$$
$$Ag^x_i + Ag^x_i \longrightarrow 2Ag_i$$

This process continues until a small speck of silver is created. It is these clusters of silver that absorb the light falling on the glass. The absorption characteristics of the silver specks depend quite critically upon their size and shape. Photochromic glass production is carefully controlled so as to produce a wide variety of shapes and sizes of the silver specks, ensuring that the glass darkens uniformly.

In a photographic emulsion the halide atoms produced when electrons are released under the influence of light can diffuse away from the silver or react with the emulsion, and the process becomes irreversible. In the photochromic material the copper ions remain trapped near to the silver particles. This means that the silver particles can release electrons to the Cu^{2+} ions when the light is turned off, reforming Cu^+ ions and making the whole process reversible. This *bleaching process* is the reverse of the darkening process. In fact the darkening and bleaching reactions are taking place simultaneously under normal circumstances, in dynamic equilibrium. When the amount of incident light is high, a large number of silver specks are present in the glass, leading to hence a high degree of darkening. At low light intensity the number of silver particles present decreases and the glass becomes clear again.

Photochromic behavior depends critically upon the interaction of two point defect types with light: Frenkel defects in the silver halide together with substitutional Cu^+ impurity point defects in the silver halide matrix. It is these two defects together that constitute the photochromic phase.

Commercially useful materials require that the *rate* of the combined reaction is rapid. If the darkening takes place too slowly, or if the subsequent fading of the color is too slow, the materials will not be useful. The presence of the copper halide is essential in ensuring that the kinetics of the reaction are appropriate and that the process is reversible.

2.8 EQUILIBRIUM POPULATION OF ANTISITE DEFECTS IN A CRYSTAL

Antisite defect equilibria can be treated in the same way as the other point defect equilibria. The creation of a complementary pair of antisite defects consisting of an A atom on a B atom site, A_B, and a B atom on an A atom site, B_A, can then be written:

$$\text{nul} \rightleftharpoons A_B + B_A$$

The relationship between defect number and free energy of formation can then be derived as in the examples above.

The creation of antisite defects can occur during crystal growth, when wrong atoms are misplaced on the growing surface. Alternatively, they can be created by internal mechanisms once the crystal is formed, provided that sufficient energy is applied to allow for atom movement. For example, antisite defects can be created via the intermediate formation of a Frenkel defect in the following way:

1. Frenkel defect formation on the A sublattice:

$$\text{nul} \rightleftharpoons A_i + V_A$$

2. Exchange of the interstitial A_i atom with a B atom on its correct site, B_B:

$$A_i + B_B \rightleftharpoons A_B + B_i$$

3. The interstitial B_i atom can eliminate the vacancy to form a B atom antisite defect:

$$B_i + V_A \rightleftharpoons B_A$$

4. If these three equations are added, the result is

$$\text{nul} \rightleftharpoons A_B + B_A$$

The same equations can also be written in terms of interstitial B atom formation:

1. Frenkel defect formation on the B sublattice:

$$\text{nul} \rightleftharpoons B_i + V_B$$

2. Exchange of the interstitial B_i atom with an A atom on its correct site, A_A:

$$B_i + A_A \rightleftharpoons B_A + A_i$$

3. The interstitial A_i atom can eliminate the vacancy to form an A atom antisite defect:

$$A_i + V_B \rightleftharpoons A_B$$

4. If these three equations are added, the result is again

$$\text{nul} \rightleftharpoons A_B + B_A$$

These indirect mechanisms may be energetically preferable to a direct exchange of atoms.

2.9 INTRINSIC DEFECTS: TRENDS AND FURTHER CONSIDERATIONS

At all temperatures above $0°K$ Schottky, Frenkel, and antisite point defects are present in thermodynamic equilibrium, and it will not be possible to remove them by annealing or other thermal treatments. Unfortunately, it is not possible to predict, from knowledge of crystal structure alone, which defect type will be present in any crystal. However, it is possible to say that rather close-packed compounds, such as those with the NaCl structure, tend to contain Schottky defects. The important exceptions are the silver halides. More open structures, on the other hand, will be more receptive to the presence of Frenkel defects. Semiconductor crystals are more amenable to antisite defects.

Despite the utility of the formulas given in Eqs. (2.4)–(2.15), they suffer from a number of limitations which it is useful to collect together here.

1. First of all, the formulas apply to materials of composition MX. In order to discuss crystals of different composition, such as M_2X_3, MX_2, and so on, different, though similar, formulas will result.

2. Only one sort of defect is supposed to be found in a crystal. Broadly speaking, this assumption is based upon the fact that the number of defects, n_d, is related to the Gibbs energy of formation, ΔG, by an equation of the form:

$$n_d = N_s \exp\left(\frac{-\Delta G}{RT}\right)$$

where N_s is the number of sites affected by the defect in question, ΔG the defect formation energy, R the gas constant, and T the temperature (K). One would anticipate that only the defect with the lower value of ΔG would be present in important quantities. However, this assumption is often poor, and minority defects may have an important role to play when properties are considered, especially for semiconductor crystals. In fact, calculations show (Section 2.10) that significant populations of many defects are invariably present in a crystal.

3. The treatment assumes that the point defects do not interact with each other. This is not a very good assumption because point defect interactions are important, and it is possible to take such interactions into account in more general formulas. For example, high-purity silicon carbide, SiC, appears to have important populations of carbon and silicon vacancies, $V_C^{\bullet}$ and V_{Si}^{x}, which are equivalent to Schottky defects, together with a large population of divacancy pairs.

4. The important quantities ΔH and ΔS are assumed to be temperature independent. This is often quite a good approximation, but the vibrational component of the entropy, which has been neglected altogether, will become increasingly important at high temperatures. The effects of these factors can cause the major defect type present to change as the temperature increases. Near to the transition temperature a complex equilibrium between both defect types will be present.

These limitations are largely eliminated in sophisticated defect calculations described in the following section. This approach can also include more sophisticated site exclusion rules, which allow defects to either cluster or keep apart from each other. Nevertheless, the formulas quoted are a very good starting point for an exploration of the role of defects in solids and do apply well when defect concentrations are small and at temperatures that are not too high.

2.10 COMPUTATION OF DEFECT ENERGIES

2.10.1 Defect Calculations

It is difficult to measure the formation and migration energy of point defects experimentally. Moreover, questions such as which mechanism is favored for the formation of, for example, antisite defects, are not easily resolved via experiment. The increasing application of computers during the second half of the twentieth century has been vitally important in the resolution of such questions. This approach was first developed as part of the Los Alamos work concerning nuclear fission. It was appreciated that experiments to ascertain neutron diffusion were too difficult to carry out, and the mathematical equations describing the process could not be solved analytically to give meaningful results. The answer was the development of the so-called Monte Carlo method of computation of neutron diffusion by Metropolis and Ulam in which the movement of the particles was essentially decided by random number generation. This method, in which an initial configuration is decided by a series of random choices, is still a widely used computational tool. The second development of relevance to defect studies was the inception of the study of the defects produced when an energetic particle collides with a solid. The initial impact displaces an atom that in turn displaces other atoms in a cascade. These computer studies, published in 1960, eventually evolved into molecular dynamics studies.

From these early beginnings, computer studies have developed into sophisticated tools for the understanding of defects in solids. There are two principal methods used in routine investigations: atomistic simulation and quantum mechanics. In simulation, the properties of a solid are calculated using theories such as classical electrostatics, which are applied to arrays of *atoms*. On the other hand, the calculation of the properties of a solid via quantum mechanics essentially involves solving the Schrödinger equation for the *electrons* in the material.

In simulation methods, the key requisite is a function to express the interaction between the atoms. These interatomic potentials are either empirical or else calculated by quantum mechanics. Having formulated the necessary interatomic potentials, the process to be explored, say vacancy formation, is set out in terms that are considered reasonable from a chemical and physical view. Finally, the method of attack on the problem is chosen. The simplest, in principle, is to minimize the energy of a static structure containing the defects of interest. A more complex approach uses equations of motion for all of the atoms in the solid that are combined with the interatomic forces to calculate how a solid evolves with time and temperature. This molecular dynamics technique is widely used to study diffusion in ionic conductors. A third way in which the simulation can be implicated is by starting from a random configuration—the Monte Carlo approach, and then to compute the effects of small displacements of the atoms. Those new configurations that are energetically unfavorable are eliminated until a final configuration remains that is considered to be optimal with respect to the conditions being investigated.

Quantum mechanical methods follow a similar path, except that the starting point is the solution of the Schrödinger equation for the system under investigation. The most successful and widely used method is that of Density Functional Theory. Once again, a key point is the development of a realistic model that can serve as the input to the computer investigation. Energy minimization, molecular dynamics, and Monte Carlo methods can all be employed in this process.

There are two other methods in which computers can be used to give information about defects in solids, often setting out from atomistic simulations or quantum mechanical foundations. Statistical methods, which can be applied to the generation of random walks, of relevance to diffusion of defects in solids or over surfaces, are well suited to a small computer. Similarly, the generation of patterns, such as the aggregation of atoms by diffusion, or superlattice arrays of defects, or defects formed by radiation damage, can be depicted visually, which leads to a better understanding of atomic processes.

Calculations are now carried out routinely using a wide variety of programs, many of which are freely available. In particular, the charge on a defect can be included so that the formation energies, interactions, and relative importance of two defects such as a charged interstitial as against a neutral interstitial are now accessible. Similarly, computation is not restricted to intrinsic defects, and the energy of formation of defects in doped or impure solids is equally available. Indeed, these two latter aspects are among the most valuable assets of defect computation.

The purpose of the following sections is not to describe the calculations but to give an idea of the basics of the methods, particularly that of atomistic simulation, because quantum mechanical calculations remain largely mathematical in nature and often cannot be described in visual terms. The starting point is a simple estimation of defect interaction energy.

2.10.2 Point Defect Interactions

Defects in a crystal can carry effective charges and, because of this, it is to be expected that the defects would interact with each other quite strongly. It is easy to make a simple estimate of the approximate magnitude of these energy terms.

Classical electrostatic theory gives the energy of interaction of two charges, E_e (sometimes expressed as the work needed to separate them or the Coulomb energy) as

$$E_e = \frac{(-e)(+e)}{4\pi\varepsilon_0 r} = \frac{-e^2}{4\pi\varepsilon_0 r}$$

where the charges have a magnitude of $\pm e$ and so attract each other, r is the separation of the charges, and ε_0 is the permittivity of vacuum. In this formulation, the energy of the pair is zero at infinite separation and decreases as the separation between them falls, so that a negative energy indicates a stable pair. In a vacuum, of course, the charges would come together and be annihilated, unless a counterforce becomes important at low charge separation.

If we apply this formula to defects in a crystal, and again assume that the defects are oppositely charged, so that they attract each other, the energy term will be roughly equivalent to the enthalpy of formation of a defect pair, ΔH_p. The closest separation of the defects will normally be equivalent to the spacing between two adjacent lattice sites.

In order to allow for the crystal structure itself, which will modify the interaction energy, it is possible, as a first approximation, to assume that the force of attraction is "diluted" in the crystal by an amount equal to its relative permittivity. The modified formula is then

$$E_e \approx \Delta H_p \approx \frac{(Z_1 e)(-Z_2 e)}{4\pi\varepsilon_r \varepsilon_0 r} \tag{2.16}$$

where ΔH_p is the enthalpy of interaction, Z_1 and $-Z_2$ are the effective charges on the defects, ε_r is the static relative permittivity of the crystal, and the other symbols have the same meaning as above.

Simple as this theory is, it is good enough to give an indication of whether association of defects is likely to occur. Consider, as an example, a Schottky defect, consisting of a cation vacancy and an anion vacancy, in a crystal of a monovalent metal MX with the sodium chloride structure. These vacancies will have effective charges of $+e$ and $-e$. Their interaction will be greatest when they are closest to each other, that is, when they occupy neighboring sites in the crystal. The separation of these sites is about 3×10^{-10} m. An approximate value for the relative permittivity of a sodium chloride structure crystal is 10. Substituting these values gives a value for the interaction energy, which is attractive:

$$\Delta H_p \approx -7.69 \times 10^{-20} \, \text{J}$$

This value is for one pair of vacancies. To obtain the molar quantity, we multiply ΔH_p above by N_A, the Avogadro constant, to yield

$$\Delta H_p \approx -46.3 \, \text{kJ mol}^{-1}$$

This is similar in magnitude to the energy of formation of Schottky defects, and so it could be anticipated that a reasonable proportion of the vacancies would be associated into pairs.

It is possible to assess the fraction of defects in a crystal that is associated using the approximate interaction energy calculated above using the Boltzmann law. This gives the information about the distribution of a population of defects between two energy states. The fraction f of the population in the upper energy state, when the energy difference between the states is ΔE, is

$$f = \exp\left(\frac{-\Delta E}{kT}\right)$$

where k is the Boltzmann constant and T the temperature (K). Using the example above, the energy separation of the associated and unassociated states is 7.69×10^{-20} J, the value of f at a temperature of 1000 K will be 0.0038. Now in this example the upper energy state corresponds to the separated vacancies. That is, the vast majority of the defects will be associated into pairs at 1000 K.

Although this estimate of the interaction energy between defects is simplistic, it demonstrates that a fair number of defects may cluster together rather than remain as isolated "point defects," provided, of course, that they can diffuse through the crystal. It is difficult, experimentally, to determine the absolute numbers of point defects present in a crystal, and doubly so to determine the percentage that might be associated rather than separate. It is in both of these areas that theoretical calculations are able to bear fruit.

As the above illustration shows, calculations using classical theories, especially electrostatics, are relatively easy to picture. However, matter at an atomic scale is best described by quantum mechanics. Quantum theory calculations are more difficult and are preferred when electronic properties are of prime importance. Both techniques are described in the following sections.

2.10.3   Atomistic Simulation

Atomistic simulation is the rigorous application of the simplistic method sketched in the previous section. The properties of a solid are calculated using theories such as classical electrostatics, which are applied to arrays of atoms, (rather than electrons, which form the focus of quantum mechanical approaches). Moreover, the atoms are frequently represented as point charges. At the outset, the positions of the atoms in the solid are defined. The more atoms specified, the closer the solid will approach a macroscopic crystal, but the greater will be the computer time needed to obtain the answers. In practice, a balance is struck between atom numbers and computer resources. The energy of interaction between the atoms is then expressed algebraically, generally as a function of the position of the atoms. Broadly speaking, these pieces of information allow the total energy of the collection to be determined. The atoms are then displaced and a new energy is calculated. The preferred state of the system is that with the lowest energy. A comparison of the energy of a perfect crystal with that of a crystal containing a defect allows the formation energy of the defect to be estimated. The calculations are referred to two nominal thermodynamic states, constant volume, in which the unit cell of the crystal is fixed, and constant pressure, in which the unit cell dimensions are allowed to vary during the course of the calculation.

These calculations require less computer time than quantum mechanical calculations, and, more importantly, they can be applied to much larger collections of atoms. This means that many problems can be tackled using atomistic simulations that remain impossible to approach (at the time of this writing) using quantum theory. Although the method is simple in principle, the development of the modern programs that are used for atomistic simulation studies required great skill from the authors and contain ingenious algorithms to make the computations feasible. However, the ultimate accuracy of atomistic simulations depends upon how well the interatomic potentials chosen represent the real atomic interactions.

The most critical aspect of atomistic simulations is thus the representation of the interactions between atoms by an algebraic function. If covalency is important, a part of the expression should contain details of how the interaction changes with angle, to mimic directional covalent bonds. In cases where a simulation is used to predict the location of a cluster of atoms within or at the surface of a solid, interactions between the atoms in the cluster, interactions between the atoms in the solid, and interactions between the atoms in the cluster and those in the solid must all be included.

Results have shown that the properties of solids can usually be modeled effectively if the interactions are expressed in terms of those between just pairs of atoms. The resulting potential expressions are termed pair potentials. The number and form of the pair potentials varies with the system chosen, and metals require a different set of potentials than semiconductors or molecules bound by van der Waals forces. To illustrate this consider the method employed with nominally ionic compounds, typically used to calculate the properties of perfect crystals and defect formation energies in these materials.

In an ionic material, the ions interact via long-range electrostatic (Coulombic) forces, as set out in the previous section. Instead of the simple expression used above, it is now necessary to sum these interactions over the whole of the solid. The electrostatic energy of the ionic crystal, E_e, is given by

$$E_e = \frac{1}{2}\left(\sum_i^n \sum_{j \neq i}^n \frac{Z_i e Z_j e}{(4\pi\varepsilon_0)(|r_i - r_j|)}\right) \tag{2.17}$$

where $Z_i e$ and $Z_j e$ are the charges on the ions i and j, the separation of the ions is given by the absolute difference between the position vectors $\mathbf{r}_i$ and $\mathbf{r}_j$, and the constant ε_0 is the permittivity of free space. The summation is over all of the n ions in the crystal, excluding those between the same atom, that is, i is never equal to j. Note that the total interaction is made up of the sum of the interactions between pairs of ions i and j. The factor of $\frac{1}{2}$ is needed as the summation counts each pair interaction twice, and the relative permittivity of the solid is no longer needed as a diluent of the forces.

The form of this expression shows that the energy decreases continuously as the separation of the ions decreases (Fig. 2.7). It is found, of course, that the (negative) interaction between the positive and negative ions predominates, and it is this that holds the solid together and provides the net cohesive energy. If the repulsion between ions of the same charge predominates, the solid is not stable. The use of normal chemical valence charges on the ions, $1+$ for the alkali metals, $2+$ for the alkaline earths, $2-$ for the chalcogenides, and $1-$ for the halogens, give perfectly satisfactory results. This electrostatic term counts for about 80% of the energy of the solid.

Ions of opposite charge will continually approach, and to offset this tendency, a repulsive energy term, which is only important at short range, must be added.

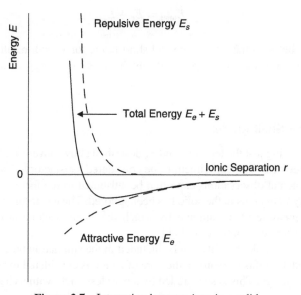

Figure 2.7 Interaction between ions in a solid.

Another short-range force that occurs in a solid is the weak van der Waals attraction between electron orbitals. There are a number of expressions for the short-range potential that takes both these factors into account. One commonly used expression is the Buckingham potential:

$$E_s = \frac{1}{2} \sum_i^n \sum_{i \neq j}^n \left(A \exp\left[\frac{-|r_i - r_j|}{\rho}\right] - \frac{C}{|r_i - r_j|^6} \right) \tag{2.18}$$

where A, ρ, and C are adjustable parameters, and the separation of the ions is given by the absolute difference between the position vectors r_i and r_j. The summation is over all of the n ions in the crystal, excluding those between the same atom, that is, i is never equal to j. Note that the total interaction is again made up of the sum of the interactions between pairs of ions i and j. The factor of $\frac{1}{2}$ is needed as the summation counts each pair interaction twice. The short-range forces act so as to oppose the long-range interactions (Fig. 2.7).

The first (exponential) term represents repulsion between electron orbitals on the atoms. The second term can be seen to be opposite in sign to the first and so represents an attraction—the weak van der Waals interaction between the electron orbitals on approaching atoms. The adjustable parameters can sometimes be calculated using quantum mechanics, but in other systems they are derived empirically by comparing the measured physical properties of a crystal, relative permittivity, elastic constants, and so on, with those calculated with varying parameters until the best fit is obtained. Some parameters obtained in this way, relevant to the calculation of the stability of phases in the system $SrO-SrTiO_3$, are given in Table 2.3.

The total energy of the solid is given by

$$E_{\text{total}} = E_e + E_s$$

The balance between the long-range and short-range forces will result in a relatively shallow minimum, the lowest point of which gives the energy of the stable solid (Fig. 2.7).

2.10.4 The Shell Model

Calculations using just the terms E_e and E_s described above have been found to be too far from agreement with experiment to be considered satisfactory. The disagreement with experimental observations can often be attributed to the fact that the electronic polarizability of the ions in the solid has been ignored. The electronic polarizability of an ion is a measure of the amount by which the electron cloud around an atomic nucleus is distorted in an electric field.

The electronic polarizability can be included in the simulation procedure by use of the shell model. In this formalism, the core of an ion is considered to be massive, and to take a charge $z|e|$. This is surrounded by a massless shell, with a charge $z|e|$, where

TABLE 2.3 Parameters for Use in Buckingham Equation for SrTiO₃

Ion Pair	A/eV^a	$\rho/\text{Å}$	$C/eV\,\text{Å}^{-6}$
$Sr^{2+}-O^{2-}$	682.172	0.39450	—
$Ti^{4+}-O^{2-}$	2179.122	0.30384	8.986
$O^{2-}-O^{2-}$	9547.960	0.29120	32.00

[a]Values from M. A. McCoy, R. W. Grimes, and W. E. Lee, *Philos. Mag.*, **75**, 833–846 (1997).

$|e|$ is the modulus of the electron charge (the magnitude of the charge without the sign) (Fig. 2.8a). The sum of the charges on the shell and core ($Z + z$) is equal to the formal ionic charge of the ion. Note that the model is still not concerned with the behavior of electrons, only with that of atoms. The atoms are now represented by two components, a pointlike core, with a certain charge, surrounded by a massless shell with a certain charge, which models the electron density around the cores. If the ion is regarded as nonpolarizable, all of the charge resides on the core and the shell is centered upon the core. Thus, the small Si^{4+} ion may be regarded as nonpolarizable and allocated a core charge of $+4e$ and the shell a charge of 0, so that $Z = +4$ and $z = 0$. Generally, anions and large cations are regarded as polarizable, and an oxygen ion, for example, might be allocated a core charge of $+0.9e$ and the shell a charge of $-2.9e$, so that $Z = +0.9$ and $z = -2.9$.

The core and the shell are bound together by a "spring." When an ion becomes polarized, the shell is displaced with respect to the core so that the center of the shell is no longer coincident with the core (Fig. 2.8b). The spring is assumed to

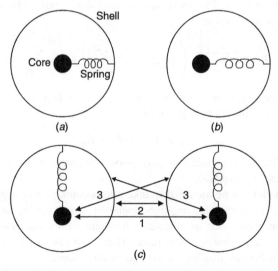

Figure 2.8 Shell model of ionic polarizability: (a) unpolarized ion (no displacement of shell); (b) polarized (displaced shell); (c) interactions: 1, core–core; 2, shell–shell; 3, core–shell.

obey Hooke's law, with a spring or force constant k. (Such a spring is also called a harmonic spring because a mass suspended by such a spring will oscillate with simple harmonic motion if displaced and then released.) The (free ion) polarizability of an ion is then $(z^2)/4\pi k\varepsilon_0$, where ε_0 is the permittivity of free space. The potential energy due to the displacement of the shell with respect to the core, E_{sp}, is given by the classical equation for the potential energy of an extended spring, which becomes

$$E_{sp} = \frac{1}{2}k\delta r_i^2$$

where δr_i is the amount by which the spring on ion i is stretched, that is, it is the distance between the center of the shell and the center of the core on ion i. Note that in general both the core and shell will be displaced (but by different amounts) under the influence of the other ions in the surroundings.

The single pair interactions are now each replaced by four pair interactions: core ion i–core ion j, shell ion i–shell ion j, shell ion i–core ion j, and core ion i–shell ion j (Fig. 2.8c). Equation (2.17) now becomes

$$E_e = \frac{1}{2}\left(\sum_{i}^{n}\sum_{i\neq j}^{n} \frac{Z_i|e|Z_j|e|}{4\pi\varepsilon_0|r_{ci} - r_{cj}|} + \sum_{i}^{n}\sum_{i\neq j}^{n} \frac{z_i|e|z_j|e|}{4\pi\varepsilon_0|r_{si} - r_{sj}|} \right.$$
$$\left. + \sum_{i}^{n}\sum_{i\neq j}^{n} \frac{z_i|e|Z_j|e|}{4\pi\varepsilon_0|r_{si} - r_{cj}|} + \sum_{i}^{n}\sum_{i\neq j}^{n} \frac{Z_i|e|z_j|e|}{4\pi\varepsilon_0|r_{ci} - r_{sj}|} \right) \qquad (2.19)$$

where $z_i|e|$ is the shell charge on ion i, $Z_i|e|$ is the core charge on ion i, r_{si} is the position vector of the shell on ion i and r_{ci} is the position vector of the core on ion i, and similarly for ion j. The short-range potential acts now between the shells, so that Eq. (2.18) is modified to

$$E_s = \frac{1}{2}\sum_{i}^{n}\sum_{i\neq j}^{n}\left(A\exp\left[-\frac{|r_{si} - r_{sj}|}{\rho}\right] - \frac{C}{|r_{si} - r_{sj}|^6} \right) \qquad (2.20)$$

where the symbols represent the same quantities as in Eqs. (2.18) and (2.19). The total energy of the solid is now given by

$$E_{total} = E_e + E_s + E_{sp}$$

With this modification, simulations give good agreement with experimental quantities where these are available.

The shell model can be modified to take into account further aspects of the solid being modeled. In particular, angular-dependent terms can be added. These are important in materials such as silicates where the tetrahedral geometry about the silicon atoms is an important constraint. These aspects can be explored via the sources listed in Further Reading at the end of this chapter.

To calculate the properties of a perfect crystal, the pair potentials required are set up, and the ions and shells are allocated to positions in the unit cell. The forces on the ions are calculated and the ions and shells are moved under the influence of these forces to new positions. The energy of the new situation is then computed, either under the assumption of constant volume or constant pressure. This procedure is iterated until the minimum energy configuration is found. Although this is a simple idea, considerable computational skills are required to derive efficient energy minimization routines, and these are at the heart of the available computer codes for structure simulation.

2.10.5 Defect Formation Energy

The calculation of defect formation energy follows from that of the perfect structure. A defect, say an interstitial ion or a vacancy, is introduced into the structure. The ions around the defect are subjected to a force from the defect and so will attempt to relax to a position of lower energy. The relaxations are calculated and iterated until a minimum energy configuration is obtained. To keep calculations to a reasonable size, the procedure usually adopted is to divide the crystal into two regions: I and II, an approach first used by Mott and Littleton in 1938. To increase the validity of the computations, region II is further subdivided into IIa and IIb (Fig. 2.9). The ions in region I are treated atomistically, while those in region II are treated using classical continuum methods.

In region I the forces on the ions are generally regarded as being large, and all the ions are individually relaxed using the explicit potentials described above, usually making use of the shell model approach. This procedure is iterated until the configuration of minimum energy is achieved, in practice, when the forces on the

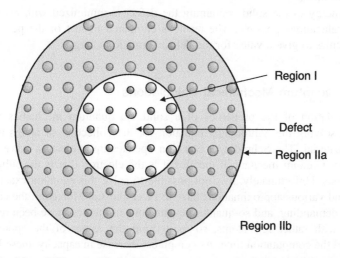

Figure 2.9 Mott–Littleton method of dividing a crystal matrix for the purpose of calculating defect energies. Region I contains the defect and surrounding atoms, which are treated explicitly. Region IIb is treated as a continuum reaching to infinity. Region IIa is a transition region interposed between regions I and IIb.

ions are zero. The computation is expensive and the region is chosen to be as small as possible, while still giving reliable results. Usually, a volume with a radius of $1-2$ nm or so around the defect is considered to be sufficient, containing a few hundred ions.

The forces from the defect are regarded as being weaker in region II, which allows the region to be treated as a continuous dielectric medium without too much loss of precision. The consequence of the presence of the defect is that this region becomes polarized in the same way as a classical slab of insulator becomes polarized in the presence of an electric charge. The energy is evaluated using classical continuum theories. The displacement of an ion, at a position defined by vector $\mathbf{r}$, is assumed to be due to the presence of a charge q on the defect. The displacement of each ion is expressed in terms of the bulk polarization of the structure, $\mathbf{P}$:

$$\mathbf{P} = \left(1 - \frac{1}{\varepsilon}\right)\left(\frac{q}{4\pi\mathbf{r}}\right)$$

where ε is the permittivity of the material, given by ($\varepsilon_r \, \varepsilon_0$), in which ε_r is the relative permittivity of the solid and ε_0 the permittivity of free space. For ease of computation, the energy of this region is supposed to be a harmonic function of the displacement of the ions present, which arise due to the total charge q on the defect in region I. To smooth out the rather abrupt change from region I to region II, region IIa is taken as the inner part of region II. In this volume, the displacements are calculated via the Mott–Littleton (1938) approach, but the interactions with the ions in region I are summed explicitly. This region is usually of the order of 3 nm in radius and contains several thousand ions. The remainder of region II, called region IIb, extends to infinity.

The energy of the solid containing the defect is minimized with respect to the various relaxations possible. The result is compared to that of the perfect defect-free structure to give a value for the defect formation energy.

2.10.6 Quantum Mechanical Calculations

The calculation of the properties of a solid via quantum mechanics essentially involves solving the Schrödinger equation for the collection of atoms that makes up the material. The Schrödinger equation operates upon electron wave functions, and so in quantum mechanical theories it is the electron that is the subject of the calculations. Unfortunately, it is not possible to solve this equation exactly for real solids, and various approximations have to be employed. Moreover, the calculations are very demanding, and so quantum evaluations in the past have been restricted to systems with rather few atoms, so as to limit the extent of the approximations made and the computation time. As computers increase in capacity, these limitations are becoming superseded.

The use of quantum mechanical calculations of solid properties was initially the province of solid-state physics, and the calculation of electron energy levels in metals and semiconductors is well established. Chemical quantum mechanical calculations have tended, for computational reasons, to center upon molecules. However, as theory has advanced, the sizes of the "molecules" have increased, and clusters of several hundred atoms can now be handled. For chemical work, the Schrödinger equation is often recast in the Hartree–Fock formalism. The Hartree–Fock equations can be solved using no input except the fundamental constants, techniques referred to as *ab initio* methods. The computationally expensive *ab initio* procedure is often replaced by using a semi-empirical Hartree–Fock formalism. The equations are solved using methods that introduce approximations based upon observed properties. Both these techniques are used to compute interatomic potentials for use in simulations. Similarly, the techniques are sometimes applied to an embedded cluster. This is a combination of quantum mechanics and classical theory. The properties of a cluster of atoms are computed using Hartree–Fock equations. The cluster is imagined to be embedded in a surrounding solid, which is treated in more approximate terms, often in terms of pair potentials.

The inherent problems associated with the computation of the properties of solids have been reduced by a computational technique called Density Functional Theory. This approach to the calculation of the properties of solids again stems from solid-state physics. In Hartree–Fock equations the N electrons need to be specified by $3N$ variables, indicating the position of each electron in space. The density functional theory replaces these with just the electron density at a point, specified by just three variables. In the commonest formalism of the theory, due to Kohn and Sham, called the local density approximation (LDA), noninteracting electrons move in an effective potential that is described in terms of a uniform electron gas. Density functional theory is now widely used for many chemical calculations, including the stabilities and bulk properties of solids, as well as defect formation energies and configurations in materials such as silicon, GaN, and AgI. At present, the excited states of solids are not well treated in this way.

Quantum mechanical calculations are essentially mathematical, and further discussion is not merited here. More information can be gained from the sources cited in Further Reading at the end of this chapter.

2.11 ANSWERS TO INTRODUCTORY QUESTIONS

What is an intrinsic defect?

An intrinsic defect is one that is in thermodynamic equilibrium in the crystal. This means that a population of these defects cannot be removed by any forms of physical or chemical processing. Schottky, Frenkel, and antisite defects are the best characterized intrinsic defects. A totally defect-free crystal, if warmed to a temperature that allows a certain degree of atom movement, will adjust to allow for the generation of intrinsic defects. The type of intrinsic defects that form will depend upon the relative formation energies of all of the possibilities. The defect with the lowest formation energy will be present in the greatest numbers. This can change with temperature.

What point defects are vital for the operation of LiI pacemaker batteries?

Lithium iodide pacemaker batteries use lithum iodide as the electrolyte, separating the lithium anode and the iodine anode. The function of the electrolyte is to transport ions but not electrons. Lithium iodide achieves this by the transport of Li^+ ions from the anode to the cathode. This transport is made possible by the presence of Li vacancies that are generated by the intrinsic Schottky defect population present in the solid. Lithium ions jump from vacancy to vacancy during battery operation.

What is the basis of "atomistic simulation" calculations of point defect formation energies?

Atomistic simulation is a computational technique to assess the properties of a solid. The solid is considered to be constructed of atoms (or ions) and electrons are not involved. At the simplest level, solids may be modeled by considering that the component species are ions represented by point charges. The interaction between the ions can then be represented by the formulas of classical electrostatics. The difference between the calculations when the solid is perfect compared with that when the solid contains one or more point defects then provides information about the defect formation energy. Rapidly increasing sophistication in terms of both hardware and software, such as using the shell model to include polarization, result in atomistic simulation calculations that give excellent agreement with experimental results when these are available. Such calculations are frequently relied upon for the accurate assessment of the formation energies of defects in a wide variety of solids when experimental data is not available.

PROBLEMS AND EXERCISES

Quick Quiz

1. The number of different types of point defects found in a pure monatomic crystal is:
   - (a) Three
   - (b) Two
   - (c) One

2. The number of vacancies in a metal crystal is proportional to:
   - (a) $\exp(-\Delta H/RT)$
   - (b) $\exp(\Delta H/RT)$
   - (c) $\exp(-\Delta H/2RT)$

3. The formation energies of point defects in a pure metal are 1.0 eV (vacancies) and 1.1 eV (interstitials). The number of vacancies is:
   - (a) Less than the number of interstitials
   - (b) More than the number of interstitials
   - (c) Equal to the number of interstitials

4. The number of Schottky defects in a crystal of formula MX is equal to:

(a) The total number of vacancies present

(b) Double the total number of vacancies present

(c) Half the total number of vacancies present

5. The ratio of cation to anion vacancies due to Schottky defects in a crystal of formula M_2X_3 is equal to:

(a) $1:1$

(b) $3:2$

(c) $2:3$

6. The formation of a Frenkel defect in CaF_2 is represented by:

(a) $0 \rightarrow V_F^{\bullet} + F_i'$

(b) $0 \rightarrow 2V_F^{\bullet} + F_i'$

(c) $0 \rightarrow 2V_F^{\bullet} + Ca_i^{2\bullet}$

7. The enthalpy of formation of Frenkel defects (kJ mol^{-1}) is AgCl, 140, AgBr, 109, AgI, 58. The compound with the greatest number of Ag$^+$ interstitials is:

(a) AgI

(b) AgBr

(c) AgCl

8. The darkening of photochromic sunglasses is due to the formation of:

(a) Copper crystallites

(b) Silver crystallites

(c) (Ag, Cu) crystallites

9. Calculations described as simulations generally deal with:

(a) Atoms

(b) Electrons

(c) Atoms and electrons

10. The shell model is used to take into account:

(a) Atomic displacement

(b) Atomic vibration

(c) Atomic polarization

Calculations and Questions

1. Plot ΔG vs. n_V using the equation:

$$\Delta G_V = n_V \Delta h_V - kT\{N \ln N - (N - n_V)\ln(N - n_V) - n_V \ln n_V\}$$

for values of $h_V = 1, 0.5, 0.1$ eV, $kT = 0.1$ ($k = 8.61739 \times 10^{-5}$ eV K^{-1}, $T = 1160$ K), $N = 10^4$.

2. The energy of formation of a defect in a typical metal varies from approximately 1×10^{-19} J to 6×10^{-19} J. (a) Calculate the variation in the fraction of defects present, n_d/N, in a crystal as a function of the defect formation energy. (b) Calculate the variation in the fraction of defects present as a function of temperature if the defect formation energy is 3.5×10^{-19} J.

3. The enthalpy of formation of vacancies ΔH_V, in pure silver is 105.2 kJ mol^{-1}. (a) Estimate the fraction of vacancies, n_V/N, at 950°C (approximately 12°C below the melting point of silver). (b) Calculate the number of vacancies per cubic meter of silver at 950°C temperature using the crystallographic data: cubic unit cell with edge length $a = 0.4086$ nm, containing four atoms of silver. (c) Repeat the calculation using the density of silver, 10,500 kg m^{-3}, and molar mass, 107.87 g mol^{-1}.

4. Following the method set out in Supplementary Material S4, derive a formula for the number of Schottky defects in a crystal of formula MX_2.

5. The energy of formation of defects in PbF_2 are: anion Frenkel defect, 0.69 eV; cation Frenkel defect, 4.53 eV; Schottky defect, 1.96 eV. (a) What point defects do these consist of? (b) What are (approximately) the relative numbers of these defects in a crystal at 300 K? (Data from H. Jiang et al., 2000).

6. Pure potassium bromide, KBr, which adopts the sodium chloride structure, has the fraction of empty cation sites due to Schottky defects, n_{cv}/N_c, equal to 9.159×10^{-21} at 20°C. (a) Estimate the enthalpy of formation of a Schottky defect, Δh_S. (b) Calculate the number of anion vacancies per cubic meter of KBr at 730°C (just below the melting point of KBr). The unit cell of KBr is cubic with edge length $a = 0.6600$ nm and contains four formula units of KBr.

7. The favored defect type in strontium fluoride, which adopts the fluorite structure, are Frenkel defects on the anion sublattice. The enthalpy of formation of an anion Frenkel defect is estimated to be 167.88 kJ mol^{-1}. Calculate the number of F^- interstitials and vacancies due to anion Frenkel defects per cubic meter in SrF_2 at 1000°C. The unit cell is cubic, with a cell edge of 0.57996 nm and contains four formula units of SrF_2. It is reasonable to assume that the number of suitable interstitial sites is half that of the number of anion sites.

8. Write out the explicit form of Eq. (2.17) for three atoms at r_1, r_2, r_3, with charges q_1, q_2, q_3.

9. The formation energy of Schottky defects in NiO has been estimated at 198 kJ mol^{-1}. The lattice parameter of the sodium chloride structure unit cell is 0.417 nm. (a) Calculate the number of Schottky defects per cubic meter in NiO at 1000°C. (b) How many vacancies are there at this temperature? (c) Estimate the density of NiO and hence the number of Schottky defects per gram of NiO.

10. **(a)** Calculate the relative number of interstitial Ag^+ ions that exist in a crystal of AgCl at 300 K, assuming that all are derived from Frenkel defects, formation enthalpy 2.69×10^{-19} J. **(b)** Knowing that AgCl adopts the sodium chloride structure, with lattice parameter $a = 0.5550$ nm, calculate the absolute number of interstitials present per cubic meter at 300 K.

11. The following table gives the values of the fraction of Schottky defects, n_S/N, in a crystal of NaBr, with the sodium chloride structure, as a function of temperature. Estimate the formation enthalpy of the defects.

Temperature/°C	Fraction of Defects
300	7.66×10^{-10}
400	1.73×10^{-8}
500	1.75×10^{-7}
600	1.04×10^{-6}
700	4.28×10^{-6}

REFERENCES

R. W. Gurney and N. F. Mott, *Proc. Roy. Soc. Lond.*, **164A**, 151 (1938).

H. Jiang, A. Costales, M. A. Blanes, M. Gu, R. Pandy, and J. D. Gale, *Phys. Rev. B*, **62**, 803–809 (2000).

N. F. Mott and M. J. Littleton, *Trans. Farad. Soc.*, **34**, 485–499 (1938).

C. Wagner and W. Schottky, *Zeit. Phys. Chem.*, **11B**, 163–210 (1931).

FURTHER READING

Information on point defect equilibria are given in:

D. M. Smyth, *The Defect Chemistry of Metal Oxides*, Oxford University Press, Oxford, United Kingdom, 2000.

A. M. Stoneham, *The Theory of Defects in Solids*, Oxford University Press, Oxford, United Kingdom, 1985.

F. Agullo-Lopez, C. R. Catlow, and P. D. Townsend *Point Defects in Materials*, Academic, New York, 1988.

The photographic process is described by:

F. C. Brown, The Photographic Process, in *Treatise on Solid State Chemistry*, Vol. 4, *Reactivity of Solids*, N. B. Hannay, Ed., Plenum, New York, 1976, Chapter 7.

Photochromic glass from the inventor's point of view is described in:

S. D. Stookey, *Explorations in Glass*, American Ceramic Society, Westerville, Ohio, 2000.

The Monte Carlo method is described by:

C. P. Robert and G. Casella, *Monte Carlo Statistical Methods*, Springer, New York, 2004.

N. Metropolis and S. Ulam, *Amer. Statistic Assoc.* **44**, 335 (1949).

The history of the Monte Carlo method and molecular dynamics simulations is given in:

R. W. Cahn, *The Coming of Materials Science*, Pergamon/Elsevier, Oxford, United Kingdom, 2001, 465–471.

The development of computer modeling can be seen by comparing:

C. R. A. Catlow, Computational Techniques and Simulation of Crystal Structures, in *Solid State Chemistry Techniques*, A. K. Cheetham and P. Day, Eds., Oxford University Press, Oxford, United Kingdom, 1987.

C. R. A. Catlow, Computer Modelling of Solids, in *Encyclopedia of Inorganic Chemistry*, 2nd ed., R. B. King, Ed., Wiley, Hoboken, NJ, 2006.

C. R. A. Catlow, J. D. Gale, and R. W. Grimes, *J. Solid State Chem.*, **106**, 13–26 (1993).

Information about computational methods can be found in:

Atomistic simulations: Various authors *Mat. Res. Soc. Bull.*, **21**, February (1996).

Density functional theory: Various authors *Mat. Res. Soc. Bull.*, **31**, September (2006).

Applications of density functional theory to surface studies is given by:

N. D. Lang, *Surf. Sci.*, **299/300**, 284–297 (1994).

A good starting place for information about Monte Carlo and density functional methods is:

Wikipedia: *http://en.wikipedia.org/wiki/Monte_Carlo_method.*

http://en.wikipedia.org/wiki/Density_functional_theory.

A widely used progam for atomistic simulation (GULP) is at:

http://gulp.curtin.edu.au.

Four classic papers are:

Y. I. Frenkel, *Z. Physik*, **35**, 652–669 (1926).

R. W. Gurney and N. F. Mott, *Proc. Roy. Soc. Lond.*, **164A**, 151 (1938).

N. F. Mott, M. J. Littleton *Trans. Farad. Soc.*, **34**, 485–499 (1938).

C. Wagner and W. Schottky *Z. Phys., Chem.* **11B**, 163–211 (1931).

CHAPTER 3

Extended Defects

What are dislocations?
What is a magnetic domain?
What defects are important in a PCT thermistor?

Point defects can, for the sake of cataloging, be considered to be zero dimensional. Extended defects with higher dimensionality can also be described. One-dimensional defects extend along a line, two-dimensional defects extend along a plane, and three-dimensional defects occupy a volume. In this chapter these extended defects are introduced.

3.1 DISLOCATIONS

Dislocations are linear defects (Fig. 3.1). Like point defects, dislocations were postulated theoretically before they were observed experimentally. Dislocations help to account for the fact that metals are generally far weaker than theoretical estimates of strength suggest. The deformation of materials is frequently brought about by dislocation movement, which requires much less energy than would be required if the material were dislocation free. Similarly, if dislocation movement is impeded or impossible, a material becomes hard and brittle. This brittleness is normal in ceramics, a property that can, in large part, be attributed to the blocking of dislocation movement rather than a lack of dislocations. Nevertheless, the deformation of ceramics at high temperatures, when dislocation movement is not greatly impeded, is closely similar to that of metals at lower temperatures.

A dislocation line characterizes a line of disruption in the crystal. This implies that the dislocation must either end on the surface of a crystal, on another dislocation, or else form a closed loop. Dislocation loops have been found to occur frequently in crystals.

The presence of dislocations is able to account for many features of crystal growth that cannot be explained if the growing crystal is assumed to be perfect. In these cases, the dislocation provides a low-energy site for the deposition of new material.

Defects in Solids, by Richard J. D. Tilley
Copyright © 2008 John Wiley & Sons, Inc.

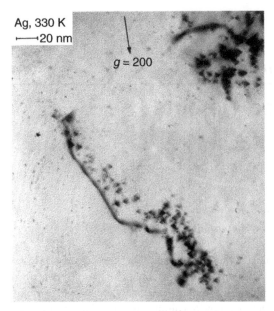

Figure 3.1 Electron micrograph showing a dislocation in silver, imaged as a dark line. The small triangular features that decorate the dislocation are stacking faults formed by the aggregation of point defects. [From W. Sigle, M. L. Jenkins, and J. L. Hutchison, *Phil. Mag. Lett.*, **57** 267 (1988). Reproduced by permission of Taylor and Francis, http://www.informa world.com.]

The vast subject of dislocations, particularly with respect to mechanical properties, will not be considered in this book, and only a few aspects of dislocations, especially interactions with point defects, will be explored.

The disruption to the crystal introduced by a dislocation is characterized by the Burgers vector, **b** (see Supplementary Material S1 for information on directions in crystals). During dislocation motion individual atoms move in a direction parallel to **b**, and the dislocation itself moves in a direction perpendicular to the dislocation line. As the energy of a dislocation is proportional to $\mathbf{b}^2$, dislocations with small Burgers vectors form more readily.

A dislocation represents a disruption of the perfect crystal structure. In the case of Frenkel and Schottky defects the energy to form the defect, the enthalpy, is offset by the entropy of the disordered state. Dislocations also introduce an entropy increase into the system, but the dislocation formation energy is much higher than that of a point defect, and the entropy term is never able to completely balance the enthalpy term, meaning that dislocations are never in thermodynamic equilibrium with their surroundings. They are thus extrinsic defects and, in principle, careful preparation or annealing can result in dislocation-free materials. In practice, this is extremely difficult to achieve, although materials that crystallize in whisker habit, or silicon single crystals prepared under extremely careful conditions, contain few dislocations.

A dislocation is surrounded by displaced atoms and the shifts give rise to a surrounding strain field. The strain field contributes to enhanced chemical reactivity, and crystals react at increased rates near to dislocations. For this reason, dislocations that intersect the surface of a crystal can be revealed by placing the crystal in a solvent. The regions where the dislocations emerge dissolve preferentially to produce small pits called etch figures on the surface. Similarly, precipitates or other defects often aggregate near to dislocations, called dislocation decoration (Fig. 3.1) (see also Section 3.7.3). The disruption that occurs near to a dislocation line also facilitates atomic movement, and diffusion of atoms along a dislocation is normally much higher than diffusion through the bulk solid (Section 5.11).

The strain field around a dislocation means that it will interact strongly with other dislocations as well as other structural defects. The strain field will depend upon the type of dislocation, its width, and its Burgers vector. In fact, the width of a dislocation, w, is defined in terms of the Burgers vector. The width is equated to the distance from the dislocation line over which the atom displacements are greater than $b/4$, where b is the length of the dislocation's Burgers vector, **b**. Metals are usually found to have rather wide dislocations, with w equal to three or four atomic spacings, whereas ceramic materials tend to contain narrow dislocations in which the atomic displacements are only greater than $b/4$ over one or two atomic spacings surrounding the dislocation core.

The Burgers vector of a dislocation can lie at any angle to the dislocation line. Although there are many different types of dislocations, they can all be thought of as combinations of two fundamental types, edge dislocations, which have Burgers vectors perpendicular to the dislocation line, and screw dislocations, with Burgers vectors parallel to the dislocation line.

3.2 EDGE DISLOCATIONS

Edge dislocations, which are primarily linked to mechanical properties, are the simplest to portray on a planar figure. They consist of an extra half plane of atoms inserted into the crystal (Fig. 3.2). The dislocation itself is described by the line, running perpendicular to the plane of the figure, localized at the tip of the inserted atom plane; marked by ⊥ on diagrams. The Burgers vector of the dislocation can be determined by drawing a closed circuit in the crystal, called a Burgers circuit, from atom to atom, in a region of crystal surrounding the dislocation core, proceeding in a clockwise or *right-hand* direction (Fig. 3.3a). An identical circuit is then mapped in a perfect crystal (Fig. 3.3b). The circuit will not close in this case. The vector required to close the loop, running from the last atom, the *finish* atom of the circuit, to the *start* atom, is the Burgers vector, **b**, of the dislocation. This is called the FS/RH (perfect crystal) convention. (Note that this procedure can be reversed. A clockwise circuit is first drawn in a perfect crystal. This is then repeated exactly, in the same direction and the same number of steps, around the dislocation. As before, the circuit will not close. The Burgers vector is now the vector joining the

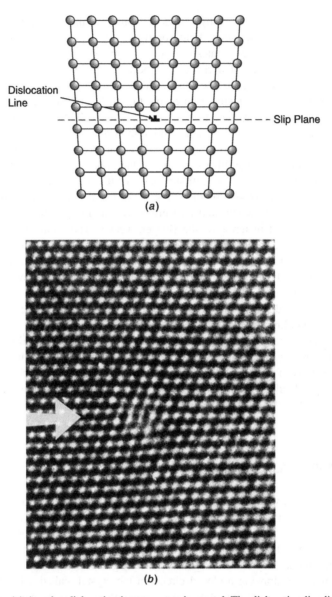

(a)

(b)

Figure 3.2 (a) An edge dislocation in a monatomic crystal. The dislocation line lies upon the slip plane and the arrow indicates the extra half-plane of atoms that produces the fault. (b) An edge dislocation in cadmium telluride, CdTe. The extra half-plane of atoms can be detected when the image is viewed in the direction of the arrow. (Courtesy of Dr J. L. Hutchison, University of Oxford).

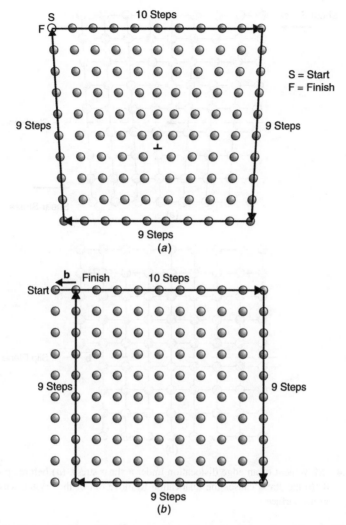

Figure 3.3 Determination of the Burgers vector of an edge dislocation: (*a*) a circuit around an edge dislocation and (*b*) the corresponding circuit in a perfect crystal. The vector linking the finishing atom to the starting atom in (*b*) is the Burgers vector of the dislocation.

start atom to the finish atom in the sequence.) The Burgers vector of an edge dislocation is *perpendicular* to the dislocation line.

The plane containing the dislocation and separating the two parts of the crystal, that with and that without the extra half-plane, is called the slip plane. Slip is the permanent deformation of a crystal due to an applied stress. During slip, parts of the crystal move relative to each other. Dislocation movement, called glide, is generally restricted to planes containing both the Burgers vector and the dislocation. Edge

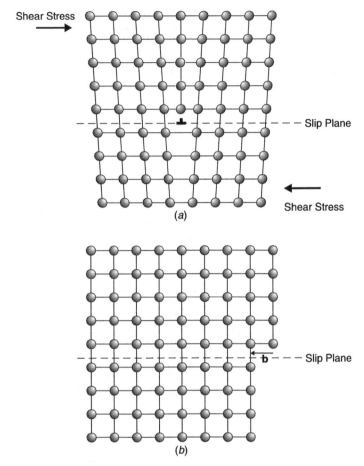

Figure 3.4   Movement of an edge dislocation under a shear stress: (*a*) before application of the stress and (*b*) the stress causes the dislocation to move out of the crystal, leaving a step of height *b* on the surface.

dislocations have the Burgers vector perpendicular to the dislocation line, hence glide occurs on a single plane.

If a shear force is applied to the crystal an edge dislocation glides to reduce the shear. Slip, the permanent deformation, will result if the dislocation is eliminated from the crystal, leaving a step with a height equal to the magnitude of the Burgers vector, **b**, in the crystal profile (Fig. 3.4). It is apparent from these figures that the motion of an edge dislocation is parallel to the Burgers vector. The movement of many dislocations along the same slip plane will result in a slip line on the surface of the crystal that is visible with an optical microscope.

The movement of edge dislocations results in slip at much lower stress levels than that needed in perfect crystals. This is because, in essence, only one line of bonds is broken each time the dislocation is displaced by one atomic spacing, and the stress

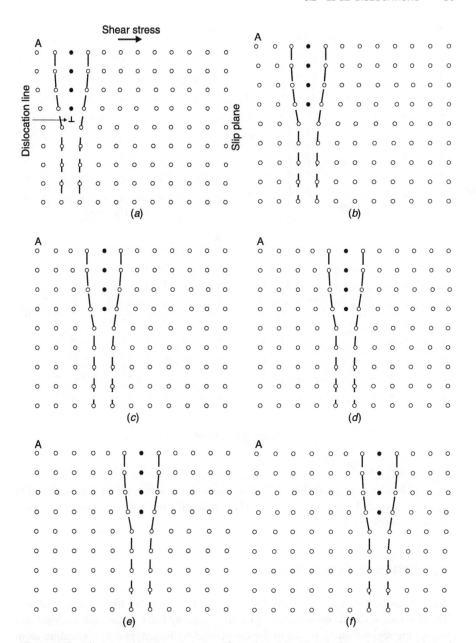

Figure 3.5 Slip caused by the movement of an edge dislocation under a shear stress. At each step (*a*) to (*j*), only a small number of interatomic bonds need to be broken.

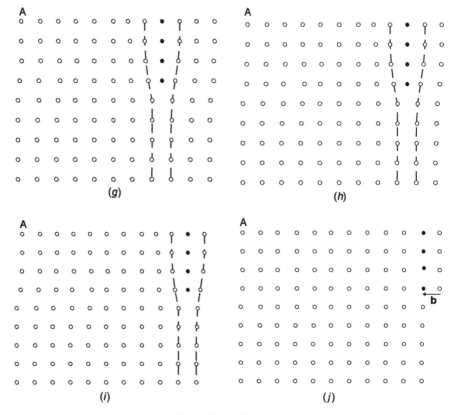

Figure 3.5   (*Continued*).

required is relatively small (Fig. 3.5). If the same deformation were to be produced in a perfect crystal, large numbers of bonds would have to be broken simultaneously, requiring much greater stress.

3.3 SCREW DISLOCATIONS

A screw dislocation creates a fault in a crystal that looks rather like a spiral staircase. The dislocation can be conceptually formed by cutting halfway through a crystal and sliding the regions on each side of the cut parallel to the cut, to create spiraling atom planes (Fig. 3.6). The dislocation line is the central axis of the spiral.

The Burgers vector of a screw dislocation can be determined in exactly the same way as an edge dislocation, following the FS/RH (perfect crystal) convention. A closed Burgers circuit is completed in a clockwise direction around the dislocation (Fig. 3.7*a*). An identical circuit in both direction and number of steps is completed in a perfect crystal (Fig. 3.7*b*). This will not close. The vector needed to close the circuit in the perfect crystal, running from the finish atom to the start atom, is the

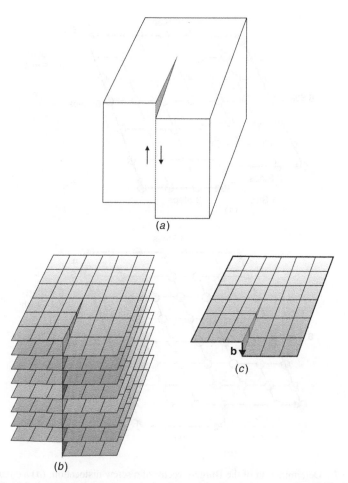

Figure 3.6 (*a*) A screw dislocation can be formed by cutting a crystal and displacing the halves. (*b*) The atom planes spiral around the dislocation line. (*c*) The Burgers vector, **b**, of a screw dislocation is parallel to the dislocation.

Burgers vector of the dislocation (Fig. 3.7*c*). It is found that the Burgers vector of a screw dislocation is parallel to the dislocation line.

The movement of a screw dislocation is perpendicular to the Burgers vector and dislocation glide is along a slip plane that contains both the dislocation line and the Burgers vector. In an edge dislocation this limits glide to a single plane, but in the case of a screw dislocation, in which the Burgers vector is parallel to the dislocation, there is no unique plane defined. A screw dislocation can then move from one slip plane to another, a motion called cross-slip.

Screw dislocations play an important part in crystal growth. The theoretical background to this fact was first developed in 1949 by Frank and colleagues. It was apparent that crystal growth was rapid as long as ledges and similar sites existed on the face of a crystal because these form low-energy positions for the addition of new atoms or

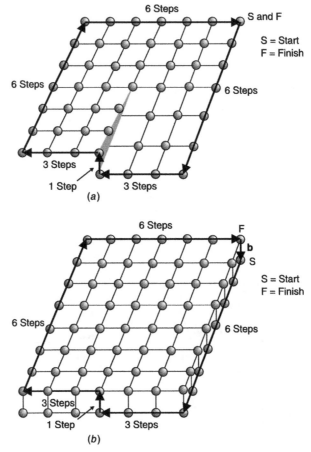

Figure 3.7 Determination of the Burgers vector of a screw dislocation: (*a*) a circuit around a screw dislocation and (*b*) the corresponding circuit in a perfect crystal. The vector linking the finishing atom to the starting atom in (*b*) is the Burgers vector of the dislocation.

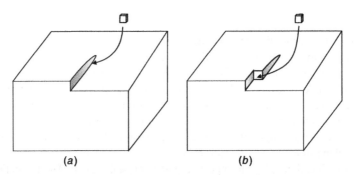

Figure 3.8 Crystal growth at a screw dislocation: (*a*) addition of new material at a step is energetically favored, and (*b*) a step is always present at an emerging screw dislocation.

molecules. However, this rapid growth ensures that the ledges are soon all used up, and all that remains is a smooth surface, at which growth is slow. A screw dislocation penetrating the crystal surface (Fig. 3.8), however, continues to spiral as new material is added to the ledge, so that crystal growth is never slowed. Subsequent to this suggestion many crystals exhibiting growth spirals were observed by optical microscopy.

3.4 MIXED DISLOCATIONS

The Burgers vector of most dislocations is neither perpendicular nor parallel to the dislocation line. Such a dislocation has an intermediate character and is called a mixed dislocation. In this case the atom displacements in the region of the dislocation are a complicated combination of edge and screw components. The mixed edge and screw nature of a dislocation can be illustrated by the structure of a dislocation loop,

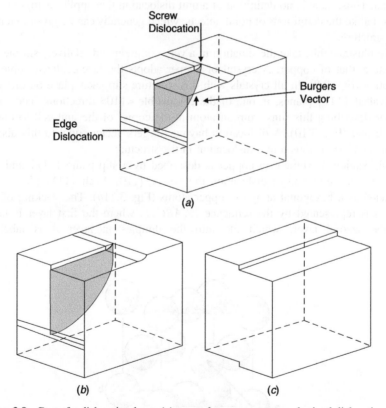

Figure 3.9 Part of a dislocation loop: (*a*) pure edge, pure screw, and mixed dislocation character; (*b*) glide is perpendicular to the Burgers vector, **b**, for the edge component, parallel to **b** for the screw component and at an angle to **b** for the mixed component; and (*c*) continued glide results in removal of the dislocation from the crystal, leaving a step of height **b** on the surface.

which is a dislocation taking on a ring configuration. Such a loop frequently contains both a pure edge and a pure screw part (Fig. 3.9*a*). A dislocation loop can expand or contract by dislocation glide, the edge part moves parallel to the Burgers vector, the screw part perpendicular to the Burgers vector, and the mixed part at an angle to the Burgers vector (Fig. 3.9*b*). Eventually, the dislocation can move out of the crystal, leaving a step of height *b*, the magnitude of the Burgers vector, on the surface (Fig. 3.9*c*).

3.5 UNIT AND PARTIAL DISLOCATIONS

A unit, or perfect, dislocation is defined by a Burgers vector which regenerates the structure perfectly after passage along the slip plane. The dislocations defined above with respect to a simple cubic structure are perfect dislocations. Clearly, then, a unit dislocation is defined in terms of the crystal structure of the host crystal. Thus, there is no definition of a unit dislocation that applies across all structures, unlike the definitions of point defects, which generally can be given in terms of any structure.

To illustrate this, take the situation in a very common and relatively simple metal structure, that of copper. A crystal of copper adopts the face-centered cubic (fcc) structure (Fig. 2.8). In all crystals with this structure slip takes place on one of the equivalent {111} planes, in one of the compatible <110> directions. The shortest vector describing this runs from an atom at the corner of the unit cell to one at a face center (Fig. 3.10). A dislocation having Burgers vector equal to this displacement, $\frac{1}{2}$ <110>, is thus a unit dislocation in the structure.

Dislocation movement in copper is described by a slip plane {111} and a slip direction, the direction of dislocation movement, [110]. Each {111} plane can be depicted as a hexagonal array of copper atoms (Fig. 3.11*a*). The stacking of these planes is represented by the sequence ... *ABC* ... where the first layer is labeled *A*, the second layer, which fits into the dimples in layer *A* is labeled *B*

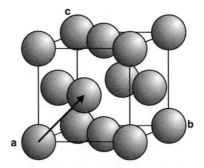

Figure 3.10   Unit cell of face-centered cubic structure of copper. The vector shown represents the Burgers vector of a unit dislocation in this structure.

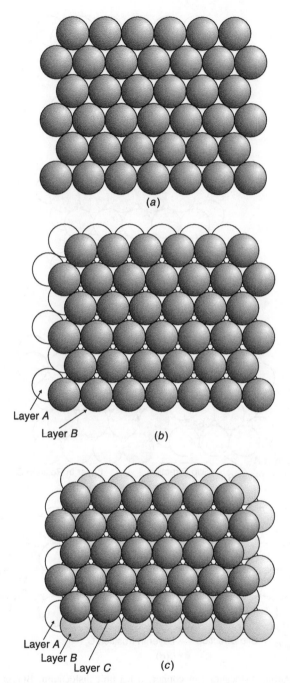

Figure 3.11 Cubic close-packed structure of face-centered cubic crystals such as copper as a packing of atom layers: (*a*) a single close-packed layer of copper atoms; (*b*) two identical layers, layer *B* sits in dimples in layer *A*; (*c*) three identical layers, layer *C* sits in dimples in layer *B* that are not over atoms in layer *A*. The direction normal to these layers is the cubic [111] direction.

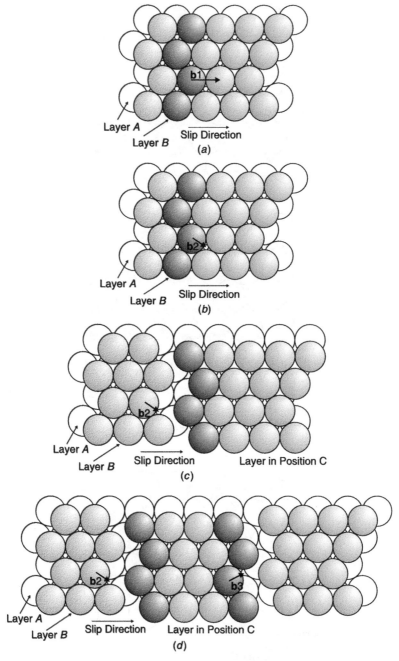

Figure 3.12 Partial dislocations in copper: (*a*) a unit dislocation, Burgers vector **b**1; (*b*) initially slip is easier in the direction represented by the Burgers vector of the partial dislocation **b**2 than **b**1; (*c*) the result of the movement in (*b*) is to generate a stacking fault; and (*d*) the combined effect of displacements by the two partial dislocations **b**2 and **b**3 is identical to that of the unit dislocation, but the partials are separated by a stacking fault.

(Fig. 3.11*b*), and the third layer, which fits into dimples in layer *B* not over atoms in layer *A*, is labeled *C* (Fig. 3.11*c*). Repetition of this sequence then builds up the copper crystal.

The unit edge dislocation corresponds not to a smooth plane of atoms, as illustrated in Figure 3.2, but a corrugated plane. Although the Burgers vector of a unit dislocation, **b**1 (Fig. 3.12*a*), for atoms in a *B* plane is readily defined, it will clearly be difficult to move the atom array in the direction of **b**1 because of the underlying sheet of atoms in the *A* position. Trying this with two layers of balls will make it clear that the easiest movement is to one of the unoccupied dimples, corresponding to the positions of layer *C*. This movement can be accomplished by the passage of a dislocation with Burgers vector **b**2 (Fig. 3.12*b*). This is not a unit vector and so is an imperfect dislocation, usually called a partial dislocation. This Burgers vector is of the form $\frac{1}{6}$ <211>. Now the application of this translation will turn a *B* layer into a *C* layer, and all layers above will be changed so that *A* becomes *B*, *B* becomes *C*, and *C* becomes *A* (Fig. 3.12*c*). The sequence of planes across the slip plane is now . . .*ABCA**CABC. . . where the change takes place at the layer denoted by *. This is a fault called an antiphase boundary (see Section 3.12). The new structure can continue across the slip plane until it reaches the surface. However, it can also propagate only a part of the way across the slip plane. In this case, the original structure can be restored by the movement of another partial dislocation, **b**3 (Fig. 3.12*d*), which is also of the form $\frac{1}{6}$ <211>. The original unit dislocation is said to have dissociated into a pair of partial dislocations called Shockley partial dislocations, each of which can glide independently of the other. The dissociation can be represented by the vector equation:

$$\mathbf{b}1 \longrightarrow \mathbf{b}2 + \mathbf{b}3$$

If both these partial dislocations exist in the crystal, they will be linked by an antiphase boundary (Fig. 3.12*d*).

It is seen, therefore, that after the passage of a perfect dislocation through a crystal, the crystal matrix will be perfect and dislocation free. This will not generally be true for imperfect dislocations, which invariably leave a stacking fault in their wake.

The argument presented above will apply to all structures built of stacked hexagonal planes of atoms. Thus, one would expect to see similar partial dislocations in structures such as graphite, built from hexagonal layers of carbon atoms, boron nitride, BN, built of hexagonal layers of carbon plus boron atoms, and corundum, Al_2O_3, built from hexagonal layers of oxygen atoms. These partials will have different Burgers vectors to those described for copper because the Burgers vector is referred to the unit cell of the structure. However, they will still represent the same transition as depicted for copper, a translation of the layer from, for example, a *B* layer to a *C* layer, in each case.

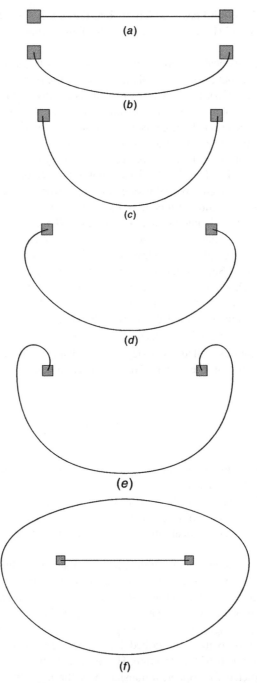

Figure 3.13   Dislocation pinned at each end (*a*) can respond to stress by bowing out (*b, c, d,* and *e*) to form a dislocation loop and reform the pinned dislocation (*f*). The growth of the dislocation represented in (*a*)–(*c*) requires increasing local stress, whereas the steps in (*d*)–(*f*) are spontaneous once the point in (*c*) is passed.

The partial dislocations described in copper and similar materials are not the only ones that can be described, and a number of other types are well known, including Frank partial dislocations, which mediate in a different slip process.

3.6 MULTIPLICATION OF DISLOCATIONS

Dislocation density is measured as the total length of dislocation lines in a unit volume of crystal, meters per meter cubed. However, experimentally it is often simpler to determine the number of dislocations that intersect a surface, so that a common measure of dislocation density is the number of dislocation lines threading a surface, that is, the number per meter squared. In a fairly typical material there will be on the order of 10^8 dislocation lines crossing every square centimeter of solid. However, it is known that if a solid is deformed, the dislocation density rises, perhaps by a factor of 10^3 or 10^4. Clearly, dislocations must be able to multiply under the conditions that lead to deformation.

There are a number of mechanisms by which new dislocations can form, but most require that an existing dislocation becomes trapped, or pinned, in the crystal, so that glide can no longer occur. One of these involves a length of dislocation pinned at each end (Fig. 3.13a). This is called a Frank–Read source. When stress is applied to this defect, the pinned dislocation cannot glide to relieve the stress, but it can bulge out from the pinning centers to achieve the same result (Fig. 3.13b). Further stress increases the degree of bulging, until a critical diameter of the loop forms, which is equal to the separation of the pinning centers (Fig. 3.13c). Up to this point increasing stress is needed to make the dislocation line grow. However, the configuration in Figure 3.13c is that corresponding to maximum stress. Further growth can now proceed spontaneously, as this requires less stress (Fig. 3.13d and 3.13e). Ultimately, the dislocation line can grow so much that both ends can unite to form a dislocation loop, and at the same time a dislocation between the pinning centers is reformed (Fig. 3.13f). As long as stress levels in the pinning region remain high enough, the dislocation loop so formed will continue to grow and a new dislocation loop will be generated. Thus, a Frank–Read source can continually emit dislocation loops during stress, significantly multiplying the dislocation density in a crystal.

3.7 INTERACTION OF DISLOCATIONS AND POINT DEFECTS

3.7.1 Dislocation Loops

The dislocation formation mechanism described in the previous section generates dislocation loops. A dislocation loop can also form by the aggregation of vacancies on a plane in a crystal. Vacancy populations are relatively large at high temperatures, and, if a metal, for example, is held at a temperature near to its melting point, considerable

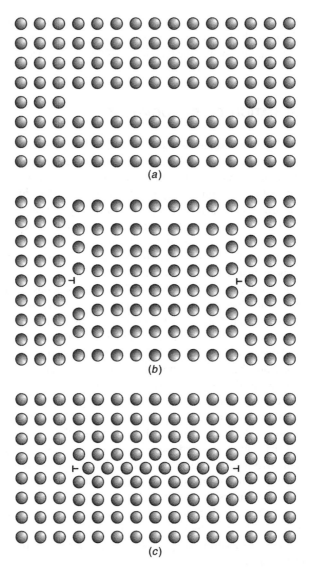

Figure 3.14 Formation of dislocation loops: (*a*) the aggregation of vacancies onto a single plane, (*b*) collapse of the plane to form a dislocation loop, and (*c*) aggregation of interstitials to form a dislocation loop.

numbers of vacancies can form in a crystal. If the metal is now cooled, these vacancies will remain in a nonequilibrium state. Should the metal be held at a temperature that is sufficient for the vacancies to migrate, the strain field around each point defect may be reduced if the defects aggregate. A similar aggregation in ionic crystals may occur because of electrostatic interactions. Thus, it is possible to imagine the vacancies clustering to form di-vacancies, tri-vacancies, and so on. The ultimate shape of the cluster may be spherical, oval, or planar, depending upon the geometry of the structure. A planar aggregation (Fig. 3.14a) may become large enough for the central part to collapse, resulting in a dislocation loop (Fig. 3.14b). The structure is called a prismatic dislocation loop.

An exactly similar situation can be envisaged if the crystal contains a high population of interstitial point defects. Should these aggregate onto a single plane, a dislocation loop will once again form (Fig. 3.14c).

In real crystals, account must be taken of the structural geometry within the crystal. Suppose vacancies aggregate on the close-packed {111} planes in a face-centered cubic crystal such as copper. Collapse to form a prismatic dislocation loop will cause the ...ABC... stacking sequence of planes to be disrupted. For example, if the vacancies aggregate on a B plane (Fig. 3.15) the stacking across the loop will be ...ABCA*CABCABC... and a stacking fault will occur within the loop, indicated by *. A similar argument applies to an interstitial loop (see Section 3.13).

The geometry of these loops means that the Burgers vector of the dislocation does not lie in the glide plane but is normal to the appropriate {111} plane, with a length $\frac{1}{3}$ [111]. Now edge dislocations cannot glide unless the Burgers vector lies in the glide plane, and hence the dislocation is not glissile but immobile, or sessile. Dislocation loops of this type are called Frank sessile loops.

The aggregation of vacancies or interstitials into dislocation loops will depend critically upon the nature of the crystal structure. Thus, ionic crystals such as sodium chloride, NaCl, or moderately ionic crystals such as corundum, Al_2O_3, or rutile, TiO_2, will show different propensities to form dislocation loops, and the most favorable planes will depend upon chemical bonding considerations.

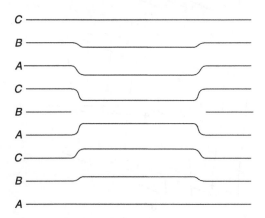

Figure 3.15 Change of stacking across a dislocation loop in a face-centered cubic structure. The structure is that of a Frank sessile dislocation loop.

3.7.2 Dislocation Climb

The edge dislocations bounding the dislocation loops just described cannot glide, but nevertheless the loop can grow by the continued collection of vacancies or interstitials. This method of movement of an edge dislocation, which allows edge

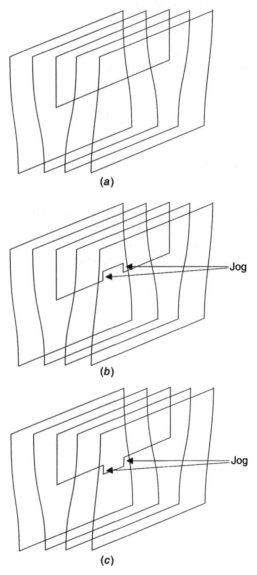

Figure 3.16   Edge dislocation climb: (*a*) an edge dislocation before climb, (*b*) the start of climb by vacancy aggregation on the dislocation, and (*c*) the start of climb by interstitial aggregation on the dislocation.

dislocations to pass obstacles to glide and to expand or shrink, is called climb. The way this comes about is simply by the addition of a vacancy or an interstitial to the termination of the extra half-plane of atoms that forms the dislocation (Fig. 3.16). If an atom migrates to a dislocation, it will leave a vacancy in the structure; while if a vacancy coalesces upon the dislocation, it will leave behind a filled atom site. As vacancies or interstitials will arrive at the end of the dislocation in a sporadic fashion, the termination will not lie in one slip plane but in a number of parallel planes. The short steps in the dislocation that form as a result of climb are called jogs. If sufficient numbers of point defects are amalgamated into the dislocation, it can grow out of the crystal altogether. Dislocations can thus be regarded as a source or a sink of point defects in the structure.

3.7.3 Decoration of Dislocations

The strain field of a dislocation interacts with point defects and can act so as to "attract" them into the dislocation core. If a sufficiently high concentration of point defects occurs in this way, then they can amalgamate to form a precipitate or planar fault, and the dislocation is said to be decorated (Fig. 3.1). This technique was used to reveal dislocations before the widespread use of transmission electron microscopy and served as one of the first proofs that dislocations did indeed exist. Two classic examples are well known. Dash, in 1955, evaporated a thin film of copper onto a thin slice of silicon. Heating the silicon allowed the copper to diffuse into the slice, where it accumulated at dislocations. The slice was photographed in transmission using infrared light, to which silicon is transparent but copper is opaque. The dislocations were revealed as dark lines threading the slice. In the same epoch, Mitchell irradiated crystals of silver chloride, AgCl, with ultraviolet light. This triggers a photographic reaction, leading to the nucleation of tiny crystals of silver (Section 2.6). The preferential sites for nucleation were dislocations threading the crystal. The small crystallites so formed, tracing the courses of the dislocation lines, could be seen using conventional optical microscopy. (Also see Section 3.8.)

3.8 DISLOCATIONS IN NONMETALLIC CRYSTALS

Dislocations occur profusely in nonmetallic materials. As mentioned above, ceramics are brittle at ordinary temperatures, not because of a lack of dislocations but because these cannot easily glide due to strong bonding between the component atoms. Organic crystals, which are usually composed of molecules consisting of strongly bound atoms, linked by weak external bonds, usually glide by movement of molecules rather than atoms, and dislocations can be referred to the molecular array rather than the atom array.

Many ceramic crystals can be described in terms of close-packed planes of oxygen atoms (or ions). For example, corundum, Al_2O_3, can be regarded as consisting of layers of oxide ions, O^{2-}, with small Al^{3+} ions contained in an ordered fraction of

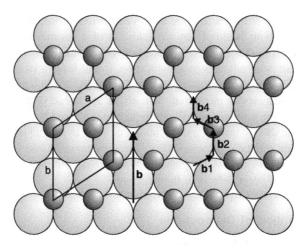

Figure 3.17   Layer of the corundum (Al_2O_3) structure, projected down the c axis. The unit cell is marked. A unit dislocation, Burgers vector **b**, can be decomposed into four partial dislocations **b1**–**b4**.

octahedral interstices lying between these layers (although corundum is not an especially ionic material in reality). Glide takes place on these close-packed planes, in the three equivalent directions <10$\bar{1}$0>. A unit dislocation can be imagined to decompose into four partial dislocations, also called quarter dislocations (Fig. 3.17).

Layered structures are particularly liable to slip between the layers, and such materials are usually easily cleaved in this way. An example is provided by the cadmium iodide, CdI_2, structure. In this (and other isostructural compounds such as tin disulfide, SnS_2, and nickel bromide, $NiBr_2$), the structure is composed of sheets of cation centered anion ocatahedra stacked vertically over one another and linked by weak bonds (Fig. 3.18a). The compounds are hexagonal and the **c** axis runs from a metal atom in one layer to another in the layer above. The iodide ions in CdI_2 are stacked in hexagonal close packing (see Fig. 3.11b) and can be described by the sequence . . .*ABABABAB*. . . . The Cd^{2+} ions occupy octahedral sites between every other layer. These correspond to the positions labeled C in cubic closest packing (see Fig. 3.11c), and to indicate this, the structure of CdI_2 can be written as . . .*AcB AcB AcB*. . . where the I^- positions are represented by A and B and the Cd^{2+} positions by c (Fig. 3.18b).

The lowest energy glide is between neighboring iodide layers that do not contain Cd^{2+} ions. There are three equivalent unit dislocations, with Burgers vector **b**, one of which is drawn in Figure 3.18b. It is seen, however, that glide is more easily accomplished via partial dislocations with Burgers vectors **b1** and **b2** (Fig. 3.18b). The motion of a partial dislocation will introduce a stacking fault into the later sequence, which becomes . . .*AcB AcB* * *CbA CbA*. . . , where the stacking fault is indicated by *. Similar dislocations have been seen in many other layered compounds.

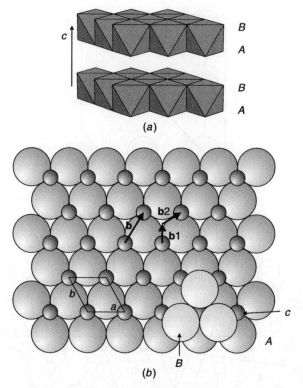

Figure 3.18 Cadmium iodide, CdI_2, structure: (*a*) perspective view of the structure as layers of CdI_6 octahedra; (*b*) one layer of the structure, with the lower I^- anion (*A*) layer and the middle Cd^{2+} (*c*) layer shown complete and just three anions of the upper (*B*) layer indicated. The vector **b** represents the Burgers vector of one of three unit dislocations and **b**1 and **b**2 represent the Burgers vectors of the two equivalent partial dislocations. The unit cell dimensions are *a*, *b*, and *c*.

In both of the previous examples, glide takes place along close-packed layers of atoms. When this does not occur more complex dislocation structures can result. For example, in the sodium chloride structure (Fig. 3.19*a*), the shortest vector connecting two similar atoms joins an atom at a cell corner to one at the center of a unit cell face, written $\frac{1}{2}$<110>. This is the Burgers vector for normal low-energy dislocations. The slip plane in the sodium chloride structure is normally {110} type, rather than {100}, as slip on this latter plane would involve unfavorable interactions between similarly charged ions. An edge dislocation in this structure now needs the insertion of two half-planes in order to make the surrounding structure complete (Fig. 3.19*b*).

In an ionic material such as sodium chloride the dislocation can bear a charge, which will depend upon the half-planes inserted. If the tip has an excess of cations, it will be positively charged; whereas if the tip consists of anions, it will be negatively charged (Fig. 3.19*b*). Similarly, when the dislocation reaches the surface, a charge may be present, which may enhance chemical reactivity at this

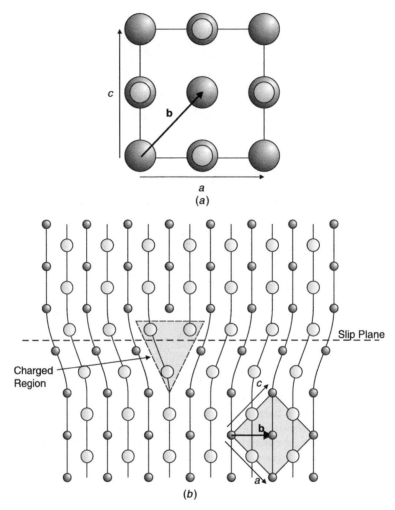

Figure 3.19 (*a*) Projection of the sodium chloride (NaCl) structure down the *b* axis. The Burgers vector **b**, $\frac{1}{2}[101]$, is the shortest vector connecting two identical atoms in this structure. (*b*) An edge dislocation in the NaCl structure, consisting of two extra half-planes of atoms. If these are ionic, the tip will bear a charge, depending upon the ions in the termination, and slip will occur on {110}.

site. Climb of the dislocation via jogs will also involve regions of charge. Dislocations can also attract an "atmosphere" of charges around the tip in the following way. Suppose that climb is by the aggregation of cations to the tip. The dislocation then acts as a cation vacancy source, and the dislocation acquires more positive than negative charge. This will attract an atmosphere of negatively charged point defects to the tip region, again something that may result in enhanced chemical reactivity as well as modified physical interactions.

This is one mechanism by which dislocations can become "decorated" (Section 3.7.3). A classic example is the decoration of dislocations in potassium chloride,

KCl, containing a small addition of silver chloride, AgCl, impurity. The silver ions tend to form positively charged Ag^+ interstitial ions. These will be attracted to a negatively charged dislocation tip, and light will cause these to aggregate into clusters, just as in the photographic process (Section 2.6). The resulting clusters are visible in an optical microscope.

The sodium chloride structure is adopted by a large number of compounds, from the ionic alkali halides NaCl and KCl, to covalent sulfides such as PbS, or metallic oxides such as titanium oxide, TiO. Slip and dislocation structures in these materials will vary according to the type chemical bonding that prevails. Thus, slip on {100} may be preferred when ionic character is suppressed, as it is in the more metallic materials.

3.9   INTERNAL BOUNDARIES

Many crystals contain internal boundaries or interfaces. Although to the eye a crystal may appear as a single crystal, these internal boundaries can often be recognized with diffraction techniques. The earliest practitioners of X-ray structure analysis realized that the width of the diffracted X-ray beams indicated that even excellent single crystals were not perfectly ordered. It was suggested that single crystals consisted of large numbers of "domains," which were taken to be slightly misaligned regions (Fig. 3.20a). Of necessity, the interface between two domains forms a two-dimensional defect. Moreover, layered materials such as clays or micas gave very disordered X-ray diffraction results, indicating that many defects were present in the solid. These were frequently interpreted as planar defects due to mistakes in the stacking of the layers making up the crystal (Fig. 3.20b).

Internal boundaries are important in influencing the properties of single crystals in a number of ways. Impurities and other point defects, such as self-interstitials or vacancies, often congregate near to such interfaces. Moreover, because the regularity of the crystal structure is disrupted at the interface, unusual atom coordination can occur, allowing impurity atoms to be more readily accommodated. This in turn leads to differing, often enhanced, chemical reactivity, dissolution, and other physicochemical properties.

For energetic reasons, internal boundaries are almost always planar in crystals. This is not a rule, though, and in some circumstances curved boundaries can occur. These are frequently found when the boundary is simply a variation in metal atom ordering of the type characterized by antiphase boundaries (see below).

Internal boundaries in a crystal, when disordered, form extended defects. However, if the boundaries become ordered, they simply extend the unit cell of the structure and hence are no longer regarded either as boundaries or defects (Fig. 3.20c). In addition, some boundaries can change the composition of a solid locally and, if present in large numbers, can change the macroscopic composition noticeably. When these are ordered, new series of compounds form. Boundaries that do cause significant composition changes are described in Chapter 4.

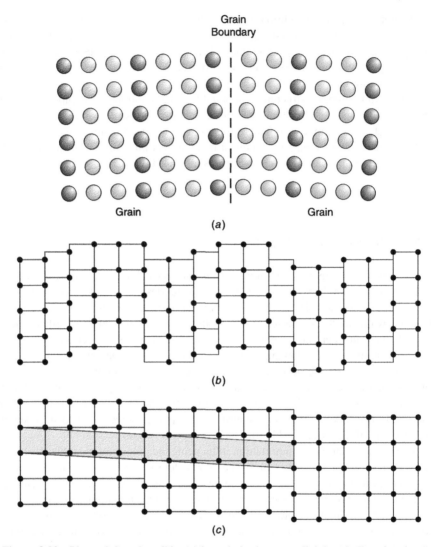

Figure 3.20 Planar defects in solids: (*a*) boundaries between slightly misaligned regions or domains; (*b*) stacking mistakes in solids built of layers, such as the micas or clays; (*c*) "ordered planar faults" assimilated into a crystal to give a new structure and unit cell (shaded).

3.10 LOW-ANGLE GRAIN BOUNDARIES

As remarked in the previous section, single crystals are not perfect but are built of domains of material that are slightly mismatched. The interfaces between these regions are called low-angle or small-angle grain boundaries. These boundaries consist of an array of dislocations arranged so as to remove the misfit between the

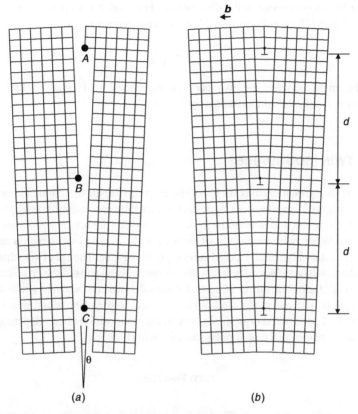

Figure 3.21 Low-angle grain boundary: (*a*) two misaligned parts of the crystal represented as a series of steps and (*b*) a tilt boundary consisting of an array of edge dislocations. The angular orientation difference between the two parts is θ, and the Burgers vector of the dislocations is **b**.

two parts. Two limiting cases are known as low-angle tilt boundaries, which consist of arrays of edge dislocations, and low-angle twist boundaries, which consist of arrays of screw dislocations. Many boundaries show a mixed tilt and twist character.

Low-angle tilt boundaries are the most easily visualized. Two regions of crystal separated by a slight misorientation can be drawn as a set of interlocking steps (Fig. 3.21*a*). The edge dislocations coincide with the steps. The separate parts can be linked to make the edge dislocation array clearer (Fig. 3.21*b*). In the situation in which the misorientation between the two parts of the crystal is θ, the distance between the steps *A* and *C* is given by twice the dislocation separation, 2*d*, where

$$2d = \frac{\mathbf{b}}{\sin(\theta/2)}$$

and **b** is the Burgers vector of the dislocations. For small angles of tilt, $\sin(\theta/2)$ can be replaced by $(\theta/2)$, (θ rad), so that the dislocation separation is

$$d \approx \frac{\mathbf{b}}{\theta}$$

If the angle between the two parts of the crystal is $1°$ (i.e., $\pi/180$ rad), the dislocation separation is approximately 17.2 nm.

3.11   TWIN BOUNDARIES

Twins are intergrown crystals such that the crystallographic directions in one part are related to those in another part by reflection, rotation, or inversion through a center of symmetry across a twin boundary. Twinned crystals are often prized mineralogical specimens. When twins are in contact across a well-defined plane (which is not always so), the boundary is generally called the composition plane. The only twins that are considered here will be reflection twins, where the two related parts of the crystal are mirror images (Fig. 3.22). The mirror plane that relates the two components is called the twin plane. This is frequently, but not always, identical to the plane along which the two mirror-related parts of the crystal join, that is, the composition plane. Repeated parallel composition planes make up a polysynthetic twin (Fig. 3.23).

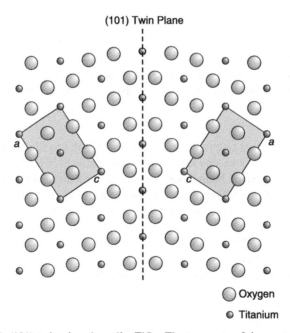

Figure 3.22   A (101) twin plane in rutile, TiO_2. The two parts of the crystal are related by mirror symmetry. The unit cells in the two parts are shaded.

Twin Plane

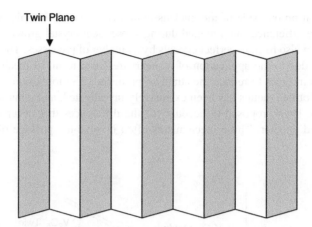

Figure 3.23 Polysynthetic twinning (repeated twinning on a microscale).

Twin geometry may refer to just one set of atoms in the crystal rather than all. This type of twinning is met, for example, in cases where a fraction of octahedral or tetrahedral cation sites in a close-packed array of anions are occupied in an ordered fashion. The close-packed anion array remains unchanged by the twin plane, which applies to the cation array alone (Fig. 3.24). These boundaries are of low energy and are often curved rather than planar. In oxides such as spinel, $MgAl_2O_4$, in which cations are distributed in an ordered fashion over some of the octahedral and tetrahedral sites, boundaries may separate regions that are twinned with respect to the tetrahedral cations only, the octahedral cations only or both.

There are a number of mechanisms by which twins can form. Growth twins are attributed to a mistake occurring when a crystal is nucleated, so that the orientation

Twin Plane

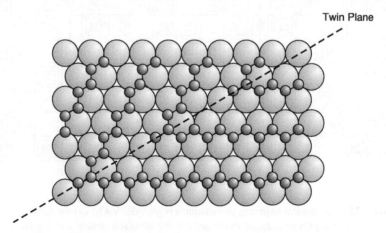

Figure 3.24 Twinning affecting only one set of atoms in a crystal.

of the crystal on one side of the nucleus is in a twinned relationship to that on the other. These differences are retained during subsequent crystal growth.

Twins can also form in perfect crystals by a number of processes. The deformation of a crystal due to the application of a shear stress can result in mechanical twins, which form in order to reduce the strain so produced. The formation of mechanical twins in deformed metals has been extensively investigated, and a number of mechanisms have been proposed to account for the differences that occur between one structure and another. These mechanisms often involve the passage of a sequence

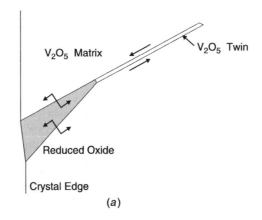

(a)

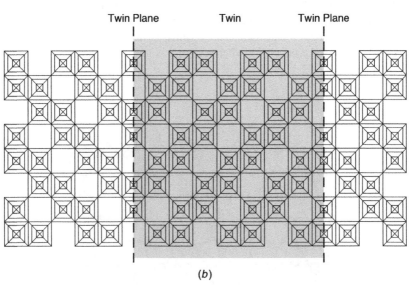

(b)

Figure 3.25 Mechanical twinning in vanadium pentoxide, V_2O_5: (a) the formation of needlelike twins at the tip of reduced oxide and (b) the idealized structure of the twin. Arrows in (a) represent the direction of shear forces.

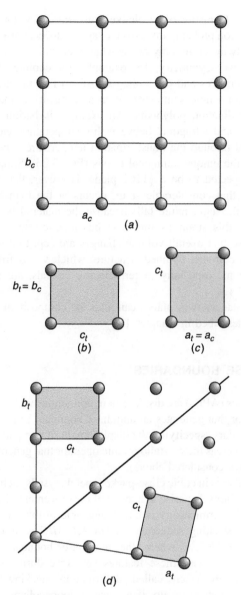

Figure 3.26 Transformation twinning: a cubic crystal (*a*) will become tetragonal on cooling if the length of one axis changes (*b*, *c*); (*d*) the resulting deformation is minimized by twinning.

of dislocations that provide a low-energy means of introducing the required amount of deformation.

Mechanical twins that form to reduce internal stress have also been observed in nonmetallic crystals. Mild reduction of vanadium pentoxide at low temperatures produces regions of reduced oxide that induce considerable shear stress in the matrix that is relieved by the formation of needlelike twins (Fig. 3.25). Once again, the mechanism of transformation probably involves passage of dislocations through the matrix.

Transformation twins commonly occur when a crystal in a high-symmetry form transforms to a lower symmetry, for example, on cooling. Many crystals are cubic at high temperatures and on cooling transform first to tetragonal and then to orthorhombic or monoclinic symmetry, often as a result of small deformations of metal–oxygen coordination polyhedra. An example is barium titanate, which is cubic above 120°C and tetragonal between this temperature and 5°C. On cooling such a cubic crystal through the transformation temperature, any of the three cubic axes may become the unique tetragonal **c** axis (Fig. 3.26). Simplistically, the twin plane would be expected to be a {110} plane. However, the change in unit cell dimensions means that considerable strain occurs at the twin boundary, and this strain increases as the temperature falls because the material becomes "more tetragonal." To relieve this strain as much as possible, crystals tend to twin on a number of planes so that overall volume changes are kept to a minimum, resulting in a crystal with a complex twinned structure, which varies from one material to another. Such regions, especially in ferroelectric crystals, are also referred to as domains (Section 3.13)

In some circumstances twin planes can alter the composition of a crystal. These consequences are described in Chapter 4.

3.12 ANTIPHASE BOUNDARIES

Antiphase boundaries (APBs) are displacement boundaries within a crystal. The crystallographic operator that generates an antiphase boundary in a crystal is a vector **R** *parallel* to the boundary, specifying the displacement of one part with respect to the other (Fig. 3.27), whereas the crystallographic operator that generates a twin is reflection (in the examples considered above).

Antiphase boundaries in cubic close-packed metal crystals such as copper are identical to the stacking faults that arise as a result of the movement of a partial dislocation (Section 3.5) and the terminology is interchangeable in this case. The cubic close-packed metal atom stacking sequence ...*ABCABCABC*... is disrupted to give a sequence ...*ABCACABC*... where the antiphase boundary lies between the layers in bold type. Because these features can arise during plastic deformation of a crystal, they are often called deformation stacking faults, A similar ...*ABCACABC*... sequence occurs if a vacancy loop collapses in a cubic close-packed crystal (Section 3.7). Interstitial atoms in a cubic close-packed metal crystal can also aggregate on a close-packed plane to form an antiphase boundary. In such a case the close-packed sequence will become...*ABCABCACBCABCABC*.... In this case two changes in the stacking sequence occur.

Similar antiphase boundaries form in metals with structures based upon a hexagonal close-packed array of metal atoms, such as magnesium. Condensation of vacancies upon one of the close-packed metal atom planes to form a vacancy loop, followed by subsequent collapse, will result in a hypothetical sequence ...*ABABBABAB*.... This arrangement will be unstable because of the juxtaposition of the two *B* layers. Translation of one part of the crystal by a vector parallel to the fault plane will create an antiphase boundary with the stacking sequence ...*ABABCBCB*.... The condensation of an interstitial population on a close-packed plane will result in an initial sequence ...*ABABAABABAB*..., which will be unstable for a similar reason. Translation of part of the crystal by a vector parallel to the fault can give an antiphase boundary sequence ...*ABABABCBCBC*....

As in the case of twin planes, the antiphase relationship may affect only one part of the structure, for example, the cation substructure, while leaving the anion substructure unchanged. This is particularly common when the anion array can be considered to consist of a close-packed array of ions, which remains unchanged by the antiphase boundary (Fig. 3.28).

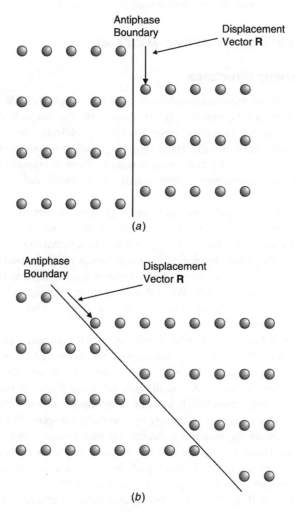

Figure 3.27 Antiphase boundaries: (*a, b*) antiphase boundaries are formed when one part of a crystal is displaced with respect to the other part by a vector parallel to the boundary.

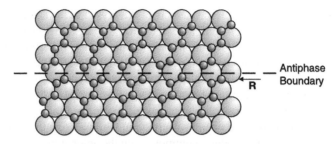

Figure 3.28   Antiphase boundary affecting only one atom type in a crystal.

Because the vector describing an antiphase boundary always lies parallel to the boundary, there is never any composition change involved.

3.13   DOMAINS AND FERROIC MATERIALS

3.13.1   Magnetic Structures

An atom that contains unpaired electrons possesses a magnetic dipole. These dipoles can be thought of, a little imprecisely, as microscopic bar magnets attached to the various atoms present. From the point of view of defects, the most important magnetic structures are those that show long-range order, and magnetic defects can arise in this long-range order that are analogous to those in crystals. In this section planar defects are considered. Other forms of magnetic defect are described in Chapter 9.

There are three forms of long-range order of most importance. Ferromagnetic materials are those in which the magnetic dipoles align parallel to each other over considerable distances in the solid (Fig. 3.29a). In antiferromagnetic compounds the elementary magnetic dipoles align in an antiparallel fashion (Fig. 3.29b). An important group of solids have two different magnetic dipoles present, one of greater magnitude than the other. When these line up in an antiparallel arrangement, they behave rather like ferromagnetic materials (Fig. 3.29c). They are called ferrimagnetic materials.

The interaction between magnetic dipoles is of two principal types: electrostatic and dipole–dipole. When the two interactions are compared, it is found that at short ranges the electrostatic force is the most important, and an arrangement of parallel spins is of lowest energy. As the distance from any dipole increases, the short-range electrostatic interaction falls below that of the dipole–dipole interaction. At this distance, the system can lower its energy by reversing the spins. For this reason the magnetic structure of the material is broken up into magnetic domains known as Weiss domains. These are regions that are magnetically homogeneous, so that the magnetic moments on the individual particles all point in the same direction. However, the magnetic moments point in different directions on passing from one domain to the next (Fig. 3.30). The boundaries between domains form defects in the magnetic arrangement.

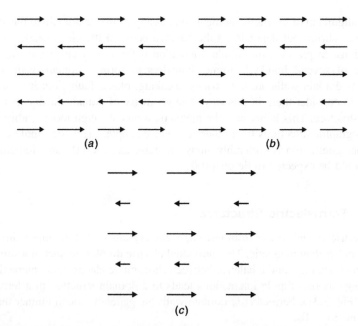

Figure 3.29 Ordered arrays of magnetic dipoles, represented by arrows: (*a*) ferromagnetic ordering, (*b*) antiferromagnetic ordering, and (*c*) ferrimagnetic ordering.

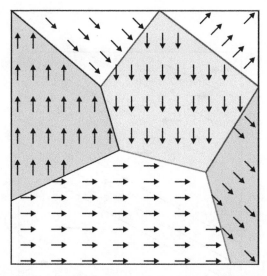

Figure 3.30 Magnetic domains in a ferromagnetic crystal (schematic). The magnetic dipoles, represented by arrows, are aligned parallel in each domain. The domain walls constitute (approximately) planar defects in the structure.

The alignment of magnetic moments on passing from one domain to the adjoining one is not abrupt but depends on the balance between the electrostatic and dipole interactions. In practice, the dipole orientation is found to change gradually over a distance of several hundred atomic diameters to form a domain wall or Bloch wall. The domain walls are not, strictly speaking, planar faults, but it is often convenient to consider them in this way. The geometry of domains also depends upon crystal structure. This is because the magnetic moments align more readily in some crystallographic directions than others do. For example, magnetic dipole alignment in ferromagnetic iron is preferably along the cubic axes, <100> and domain boundaries would be expected to lie on {100}.

3.13.2 Ferroelectric Structures

Ferroelectric crystals differ from ferromagnetic crystals in that the dipole involved is electric rather than magnetic. The individual electric dipoles interact in a similar way to magnetic dipoles, and a balance between short-range electrostatic interactions and long-range dipole–dipole interactions leads to a domain structure in a ferroelectric crystal (Fig. 3.31). Ferroelectric domain walls are generally much thinner than magnetic domain walls.

Antiparallel dipole ordering to produce an antiferroelectric crystal is also commonly encountered. Other ways of ordering electric dipoles are not so well characterized, but parallels with the situation in magnetic materials occur.

Above a temperature called the Curie temperature, T_C, ferroelectric behavior is lost, and the material is said to be in the paraelectric state in which it resembles a normal insulator.

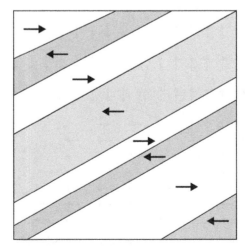

Figure 3.31 Domains in a ferroelectric crystal (schematic). The electric dipoles, represented by arrows, can take one of two directions.

3.13.3 Ferroic Structures

In general, a ferroelectric or ferromagnetic crystal will be composed of an equal number of domains oriented in all the equivalent directions allowed by the crystal symmetry. A defining characteristic of both ferromagnetic and ferroelectric (and related) materials is that the direction of the dipoles can be changed, or switched, by an external field so that the overall magnetization or electric polarization of the sample can be changed by the application of the appropriate magnetic or electric field. In either case, the form that the magnetization or polarization curves take as a function of applied field is that of a hysteresis loop (Fig. 3.32). At the extremes of the loop all dipoles point in the same direction. The practical usefulness of many magnetic and ferroelectric materials is centered upon this switching ability.

Ferromagnetic and ferroelectric materials are only two examples of a wider group that contains domains built up from switchable units. Such solids, which are called ferroic materials, exhibit domain boundaries in the normal state. These include ferroelastic crystals whose domain structure can be switched by the application of mechanical stress. In all such materials, domain walls act as planar defects running throughout the solid.

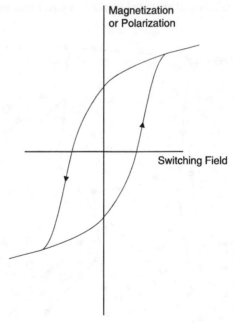

Figure 3.32 Hysteresis behavior of the magnetization of a ferromagnetic crystal or polarization of a ferroelectric crystal with respect to the applied magnetic or electric switching field.

3.14 EXTERNAL SURFACES AND GRAIN BOUNDARIES

Frequently the most important planar defects in a crystal are the external surfaces. These may dominate chemical reactivity, and solids designed as catalysts or, for example, as filters must have large surface areas in order to function. Rates of reaction during corrosion are frequently determined by the amount of surface exposed to the corrosive agent. (A further discussion of surface physics and chemistry, although fascinating, is not possible within the scope of this volume.)

Most metals and ceramics in their normal states are polycrystalline. Polycrystalline solids are composed of many interlocking small crystals, often called grains (Fig. 3.33). The surfaces of the grains that make up the solid are often similar to the external surfaces found on large crystals.

The nature of a surface will depend upon which atoms are exposed. For example, the surface of a crystal with the sodium chloride structure might consist of a mixture of atoms, as on {100} (Fig. 3.34a), or of just one atom type, as on {111} (Fig. 3.34b and 3.34c). However, it must be remembered that no surface is clean and uncontaminated unless it is prepared under very carefully controlled conditions. Absorbed gases, especially water vapor, are invariably present on a surface in air, which leads to changes in chemical and physical properties.

The behavior of polycrystalline materials is often dominated by the boundaries between the crystallites, called grain boundaries. In metals, grain boundaries prevent dislocation motion and reduce the ductility, leading to hard and brittle mechanical properties. Grain boundaries are invariably weaker than the crystal matrix, and

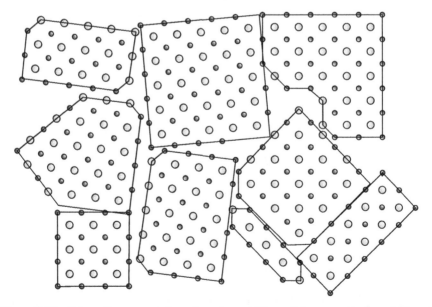

Figure 3.33 Schematic representation of a polycrystalline solid composed of crystallites of one structure.

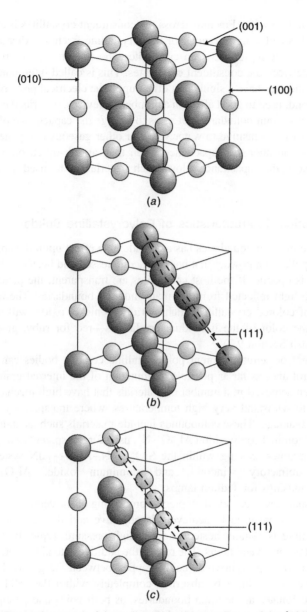

Figure 3.34 Faces of a crystal of the sodium chloride type. Faces of the {100} type contain both atom species (*a*) while planes of {111} type contain one atom type (*b*) or the other (*c*).

the mechanical strength of many brittle solids is dominated by the presence of grain boundaries and does not reflect the intrinsic strength of the crystallites making up the solid. Whereas single crystals frequently fracture by cleavage, separating along crystal planes in which the bonding is relatively weak, a polycrystalline material can fracture in two ways. Fracture across the constituent crystallites is akin to crystal cleavage and is called transgranular or transcrystalline fracture. Alternatively, if the weakest part is the region between crystallites, the fracture surface runs along the boundaries between the constituent crystallites. This is called intergranular fracture.

Grain boundaries have a significant effect upon the electrical properties of a poly-crystalline solid, used to good effect in a number of devices, described below. In insulating materials, grain boundaries act so as to change the capacitance of the ceramic. This effect is often sensitive to water vapor or other gaseous components in the air because they can alter the capacitance when they are absorbed onto the ceramic. Measurement of the capacitance allows such materials to be used as a humidity or gas sensor.

3.14.1   Optical Characteristics of Polycrystalline Solids

The presence of grain boundaries has a considerable effect upon the appearance of a material. A pellet of a powder that has been compressed and heated to bind the mass into a solid is opaque. If the host crystallites are transparent, the pellet will appear white due to light reflected from the many internal boundaries. The appearance of a compact of colored crystallites, such as ruby or nickel oxide, will be an opaque version of the color of the individual crystallites—red for ruby, green for nickel oxide (see also Section 9.7).

Transparent or semitransparent polycrystalline ceramic bodies can be made by careful control and, as far as possible, elimination of the internal grain boundaries. This has been achieved in a number of materials that have high mechanical strength or are able to withstand very high temperatures, where transparency would be an enormous advantage. These compounds include materials such as aluminum oxynitride, with a nominal composition $Al_{23}O_{27}N_5$, used for transparent armored windows, SiAlONs (ceramics existing within the $SiO_2-SiN-Al_2O_3-AlN$ system), used for transparent refractory windows, and aluminum oxide, Al_2O_3, used for transparent housings for sodium lamps.

The transparency is achieved by precise processing of the ceramic body. Carefully chosen additives are used to achieve this objective and the ceramic powder plus additive mixture is usually heated under a high pressure, typically $1500-2000°C$ under 40 MPa, for several hours. The role of the additive is to aid sintering of the particles, especially to prevent voids being trapped between the grains. However, it is important that the additive be absorbed completely within the solid particles and not form precipitates at the grain boundaries, as both voids and precipitates scatter light and reduce transparency. Maximum transparency is achieved if regularly shaped and similar sized crystals make up the final material.

3.14.2   Electronic Properties of Interfaces

The surface between grains is usually a region of increased electrical potential compared to the bulk, so that a barrier to conductivity occurs across the boundary. These grain boundary potentials are often called Schottky barriers. The height and form of the barrier depends sensitively upon the materials in contact.

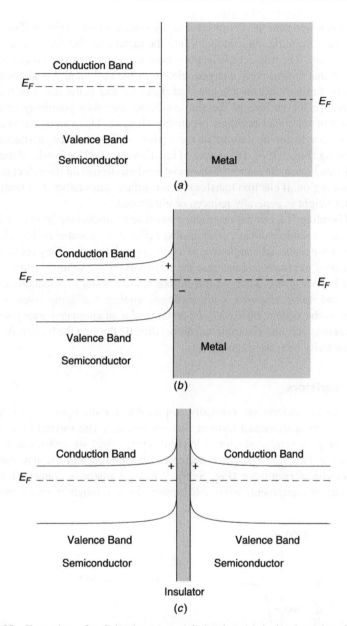

Figure 3.35 Formation of a Schottky (potential) barrier: (a) isolated semiconductor and metal, (b) on contact a Schottky barrier is formed at the interface, and (c) Schottky barriers can also form between semiconducting grains separated by insulating layers.

The best understood barriers are formed between a semiconductor and a metal. When a metal and a semiconductor are brought into contact, electrons will flow from the metal to the semiconductor or *vice versa*, depending upon the relative energy levels across the boundary. Suppose, for example, that the Fermi level of the metal is lower than the Fermi level in the semiconductor (Fig. 3.35*a*). Electrons will now pass from the semiconductor into the metal until the Fermi levels are equal at the boundary. The transfer of electrons from the semiconductor creates a region in the surface that is positively charged relative to the bulk, and this slows and eventually stops further electron transfer. The potential that is the result of this transfer arises because the surface of the semiconductor becomes positively charged and the surface of the metal becomes negatively charged. This causes the energy band in the semiconductor to deform in the region of the interface, a situation called band bending (Fig. 3.35*b*). The degree of band bending and the height of the potential barrier depend upon the source of the transferred electrons and the defect structure in the surface region. If electron transfer is from surface states rather than from the bulk, the barrier height is generally reduced or eliminated.

Band bending is a normal occurrence when semiconducting grains are in contact with a metal, a semiconductor, or an insulator (Fig. 3.35*c*), and grain boundary potential barriers are a normal component of a polycrystalline solid. They act so as to slow the transfer of electrons and increase the resistivity of the solid. Electrons can overcome the barrier in two ways. At low voltages, the transfer is essentially a diffusion process, and some electrons with sufficient energy can jump over the barrier. However, as the voltage increases, at a certain point an alternative transport mechanism comes into play and electrons can tunnel directly through the barrier. At this point the conductivity increases significantly.

3.14.3 Varistors

Grain boundary defects are primarily responsible for the operation of zinc oxide (ZnO) varistors, a shortened form of *variable resistor*. The varistor behaves like an insulator or poor semiconductor at lower electrical field strengths, but at a critical breakdown voltage the resistance decreases enormously and the material behaves like an electrical conductor (Fig. 3.36). When a varistor is connected in parallel with electrical equipment, negligible power flows through it under normal low

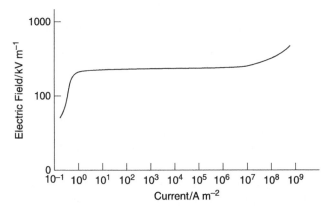

Figure 3.36 Variation of current versus electrical field for a typical zinc oxide varistor.

field conditions and power is provided to the equipment. In the event of a voltage surge, the varistor becomes a good electrical conductor and current is diverted across it, thus protecting the sensitive equipment. This behavior (called nonohmic) is reversible and is controlled by the microstructure of the zinc oxide body. The nonohmic behavior results from the contacts between the zinc oxide grains at the grain boundaries.

Chemically, varistors are principally composed of small grains of zinc oxide that have been doped with 1 or 2% bismuth oxide, Bi_2O_3. The bismuth oxide impurities modify the grain boundary regions. These ions, which can take a valence of $3+$ or $5+$, do not substitute readily into the ZnO matrix. They congregate near to the surface, where relaxation is easier and these large ions can be accommodated with less strain. The bismuth-enriched surface layer has a depth of between 3 and 10 nm and is called the depletion layer, a term borrowed from semiconductor physics. As the name suggests, the surface layer is depleted in mobile charge carriers and is highly insulating, with a resistivity of approximately 10^{10} Ωm. The structure of a varistor can then be described in terms of n-type semiconducting ZnO grains surrounded by a thin insulating shell of Bi_2O_3 structure (Fig. 3.37). In this situation, electrons in the semiconductor near to the insulating layer are trapped in the bismuth-rich surface, leaving a bilayer of ionized donors in the semiconductor. This creates a symmetrical Schottky barrier (see Fig. 3.35c).

When a low voltage is applied to the varistor, the current is due to the thermal transfer of electrons across the barrier by a diffusion process. However, at a certain critical voltage the thin insulating barrier breaks down, and current effectively flows across the insulating layer from the grain interiors, which have a high conductivity. The point at which breakdown occurs is modified by changing the amounts and nature of the chemical dopants used so that protection can be achieved over a wide range of voltages.

The microstructure of commercial varistors is extremely complex, and commercial preparations also contain other dopants, mainly oxides of cobalt, manganese, chromium, and antimony, that are used to fine tune the varistor characteristics. The transition-metal dopants are chemically similar to Zn^{2+} and mainly form substitutional defects within the ZnO grains, such as Co_{Zn}, that modify the n-type behavior of the grain interior. (See also Chapter 8 for further discussion of the electronic

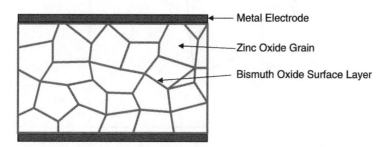

Figure 3.37   Microstructure of a zinc oxide varistor (schematic).

properties of doped oxides.) The antimony oxide tends to form boundary layer phases such as the spinel $Zn_7Sb_2O_{12}$, which play an important role in modifying the breakdown performance of these devices. The actual switching characteriztics of any particular varistor is then a function of both point and grain boundary defects.

3.14.4 Positive Temperature Coefficient Thermistors

Positive temperature coefficient (PCT) thermistors are solids, usually consisting of barium titanate, $BaTiO_3$, in which the electrical resistivity increases dramatically with temperature over a narrow range of temperatures (Fig. 3.38). These devices are used for protection against power, current, and thermal overloads. When turned on, the thermistor has a low resitivity that allows a high current to flow. This in turn heats the thermistor, and if the temperature rise is sufficiently high, the device switches abruptly to the high resisitvity state, which effectively switches off the current flow.

As in the case of varistors, the effect is due to the defect structure of the solid, especially the presence of grain boundaries. There are a considerable number of variables that must be controlled to make a suitable PCT thermistor.

1. The effect only occurs in polycrystalline samples. Single-crystal materials do not show the PTC effect. Because of this, the effect can be attributed to the presence of grain boundaries in the solid. The microstructure of the material,

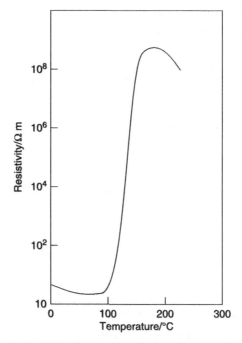

Figure 3.38 PCT effect in doped barium titanate (schematic).

especially the size of the grains in the sample, has a marked effect upon the magnitude of the PTC effect.

2. The $BaTiO_3$ must be doped with higher valence impurities such as La^{3+}, Ce^{3+}, Y^{3+} in place of Ba^{2+} or Nb^{5+} in place of Ti^{4+} and then only in a narrow range of concentrations in the region of 0.2–0.5 mol %. The effect of these impurities is to transform the insulting $BaTiO_3$ into an n-type semiconductor. (These impurities act as donors; see Chapter 8.)

3. The temperature range over which the effect occurs is quite narrow and coincides with the Curie temperature, T_C, of ferroelectric barium titanate (Section 3.13.2). Moreover, the temperature at which the resistivity changes, the switching temperature, is markedly dependent upon the impurities present. The switching temperature is particularly sensitive to replacement of Ba^{2+} by Pb^{2+} and Sr^{2+}, and doping with these ions can then allow switching to occur anywhere in the range of 0–140°C.

4. Barium titanate can be made into an n-type semiconductor by heating in a vacuum. Such materials do not exhibit the PTC effect. [Materials heated in vacuum show negative temperature coefficient (NTC) thermistor behavior; see Chapter 8]. Samples for use in PCT thermistors are heated in oxygen to prevent any chance of reduction.

5. The magnitude of the PTC effect is greatly affected by the way in which the polycrystalline sample is cooled (in oxygen) after preparation.

As this list reveals, the complexity of the phenomenon is considerable, and there is still a degree of uncertainty about the exact roles of the various defect species involved. However, the basic cause of the effect is believed to involve the following steps:

1. Assume for simplicity that the dopant ion is La^{3+}. The La^{3+} ions substitute for Ba^{2+} forming $La_{Ba}^{\bullet}$ defects. The positive charge on these defects must be compensated to maintain charge neutrality. This can be achieved either by ionic or electronic compensation (Section 1.2). As doping with La^{3+} creates n-type material, electronic compensation is believed to occur. The simplest way that this can be achieved is to consider that the counter defects are Ti^{3+} ions. To obtain charge neutrality, one Ti^{3+} ion is created for each La^{3+} ion in the crystal so that the doped material can be written in chemical terms as $Ba_{1-x}^{2+}La_x^{3+}Ti_{1-x}^{4+}Ti_x^{3+}O_3$. The point defects present are $La_{Ba}^{\bullet}$ and Ti_{Ti}'. Electron conductivity than can take place by electron transfer between Ti^{4+} ions:

$$Ti_{Ti} + e' \longrightarrow Ti_{Ti}' \qquad [Ti^{4+} + e' \longrightarrow Ti^{3+}]$$

This leads to the formation of grains of $BaTiO_3$, which are n-type semiconductors.

2. Heating the sample in oxygen will oxidize any Ti^{3+} ions in the grain surfaces to Ti^{4+}, thus causing the material to revert to an insulator. This can be written as

$$x\,Ti_{Ti}' + \tfrac{x}{4}O_2 \longrightarrow x\,Ti_{Ti} + \tfrac{x}{2}O_O$$

or

$$Ba_{1-x}^{2+}La_x^{3+}Ti_{1-x}^{4+}Ti_x^{3+}O_3 + \tfrac{x}{4}O_2 \longrightarrow Ba_{1-x}^{2+}La_x^{3+}Ti^{4+}O_{3+x/2}$$

The extra oxygen can be accommodated in the surface layers but not in the bulk, thus restricting oxidation to the grain boundary region. This accounts for the fact that grain size and cooling rate have an effect on the PCT effect, as both of these factors will influence the amount of insulating layer that is formed on the grains. The insulating layer creates a Schottky barrier, as described above (Fig. 3.35c).

3. Under normal circumstances the Schottky barrier would be expected to reduce the current flow across the sample and hence create a high-resisivity device. However, barium titanate is a ferroelectric below a temperature of approximately 135°C, the Curie temperature, T_C (see Section 3.13.2). It is believed that the action of the ferro-electric domain walls is to reduce the height of the Schottky barrier between certain grains and increase it between others. The overall effect of this is to create low-resistivity pathways through the solid so that the overall measured resisitivity is low. At the Curie temperature the ferroelectric state reverts to a normal insulator. At this point none of the Schottky barriers are reduced, as ferroelectric domains are no longer present, and a high-resistivity state is produced. Thus, the link between the Curie point and the PTC switching temperature is established.

4. The role of dopants such as Sr^{2+} and Pb^{2+} is to alter the Curie temperature of the barium titanate so that devices that switch at a desired temperature can be fabricated.

3.15   VOLUME DEFECTS AND PRECIPITATES

Point defects are only notionally zero dimensional. It is apparent that the atoms around a point defect must relax (move) in response to the defect, and as such the defect occupies a volume of crystal. Atomistic simulations have shown that such volumes of disturbed matrix can be considerable. Moreover, these calculations show that the clustering of point defects is of equal importance. These defect clusters can be small, amounting to a few defects only, or extended over many atoms in non-stoichiometric materials (Section 4.4).

In a similar fashion, the line and planar defects described above are all, strictly speaking, volume defects. For the sake of convenience it is often easiest to ignore this point of view, but it is of importance in real structures, and dislocation tangles, for instance, which certainly affect the mechanical properties of crystals, should be viewed in terms of volume defects.

Apart from these, there are volume defects that cannot conveniently be described in any other terms. The most important of these consist of regions of an impurity phase—precipitates—in the matrix of a material (Fig. 3.39). Precipitates form in a variety of circumstances. Phases that are stable at high temperatures may not be stable at low temperatures, and decreasing the temperature slowly will frequently lead to the formation of precipitates of a new crystal structure within the matrix of the old. Glasses, for example, are inherently unstable, and a glass may slowly recrystallize. In this case precipitates of crystalline material will appear in the non-crystalline matrix.

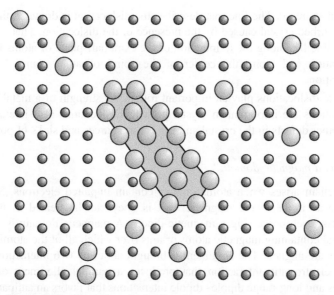

Figure 3.39 Precipitate formed by clustering of atoms in a crystal.

Precipitates have important effects on the mechanical, electronic, and optical properties of solids. Precipitation hardening is an important process used to strengthen metal alloys. In this technique, precipitates are induced to form in the alloy matrix by carefully controlled heat treatment. These precipitates interfere with dislocation movement and have the effect of hardening the alloy significantly.

Precipitates in glass are often encouraged to form to endow the glass with properties that are desired but absent in the pure material. Glass-ceramics are solids that start as glass and are treated to allow considerable degrees of crystallization to occur, leading to enhanced thermal properties. This microstructure is deliberately engineered to form materials widely used in cookware and other applications that need to withstand thermal shock. Opal glass is similarly engineered to contain small precipitates that act so as to scatter light. The important role of precipitates in photochromic glass has already been described (Section 2.7).

Many other examples could be cited from all areas of the solid-state sciences, some of which will be presented in later chapters.

3.16 ANSWERS TO INTRODUCTORY QUESTIONS

What are dislocations?

Dislocations are line defects that occur in crystals. There are many types of dislocation. The easiest to visualize are the edge dislocation, which consists of an extra half-plane of atoms inserted into a crystal and the screw dislocation that resembles a spiral staircase. The line that describes the dislocation is drawn at the center of the disturbed crystal caused by the presence of the dislocation. That is, the dislocation line lies along the termination of the extra half-plane of atoms for an edge dislocation and along the center of the spiral of atom planes for a screw dislocation.

Edge dislocations play an important role in the strength of a metal, and screw dislocations are important in crystal growth. Dislocations also interact strongly with other defects in the crystal and can act as sources and sinks of point defects.

What is a magnetic domain?

Magnetism arises when atoms or ions contain unpaired electrons. Such atoms behave as if they were magnetic, that is, they can be treated as microscopic magnets, each possessing a magnetic dipole. A magnetic domain is a region of a crystal containing magnetic atoms or ions over which all of the atomic magnetic dipoles are aligned. The alignment is brought about by two interactions, a short-range electrostatic interaction that produces a parallel alignment of magnetic dipoles and long-range dipole–dipole interactions that favors an antiparallel alignment. The size of magnetic domains in a crystal is determined by the relative strengths of these interactions.

What defects are important in a PCT thermistor?

A PCT thermistor is an electrical ceramic component that has a low resistivity at the normal operating temperature of the equipment. Under these circumstances current is allowed to flow to the machinery. The current passing through the thermistor heats it. If an unusually high current passes, the temperature rise causes the resistivity of the thermistor to increase manyfold, and the device switches abruptly to a high-resistivity state. This effectively switches off the current flow and so protects the equipment from damage.

The PCT effect is produced by a combination of point defects, grain boundaries, and ferroelectric domains in ceramic blocks of polycrystalline barium titanate. The point defects arise when barium titanate is doped with higher valence cations such as La^{3+}, which replaces Ba^{2+}. These point defects induce compensating defects to maintain charge neutrality and cause the normally insulating barium titanate to become an *n*-type semiconductor. The ceramic is heated in oxygen, and this oxidizes the surface of each grain to form a thin insulating shell. The combination of a thin insulating shell surrounding an *n*-type core produces Schottky barriers between the resulting grains. These are partly removed by interaction with ferroelectric domains to provide a low-resistance pathway through the ceramic. The ceramic switches to a high-resistance state when the ferroelectric behavior is lost. This occurs at the Curie temperature of the doped barium titanate used in the device.

PROBLEMS AND EXERCISES

Quick Quiz

1. Dislocations are:
   (a) Planar defects
   (b) Line defects
   (c) Point defects

2. The Burgers vector and dislocation line are normal to each other in:
   (a) A screw dislocation
   (b) A partial dislocation
   (c) An edge dislocation

3. Dislocation movement occurs by:
   (a) Motion in the slip plane
   (b) Motion perpendicular to the slip plane
   (c) Motion of the slip plane

4. Ceramic materials are brittle compared to metals because:
   (a) They contain no dislocations
   (b) The dislocations cannot move easily
   (c) The dislocations only move on certain planes

5. Transformation twins arise on cooling a crystal because of:
   (a) A decrease in crystallographic symmetry
   (b) An increase in mechanical strain
   (c) A change of chemical composition

6. A stacking sequence of metal atom planes *ABCABCCBACBA* represents:
   (a) An antiphase boundary
   (b) A twin boundary
   (c) A domain boundary

7. A Schottky barrier forms close to:
   (a) Dislocations
   (b) Grain boundaries
   (c) Schottky defects

8. Varistor behavior is principally attributed to:
   (a) Grain boundries
   (b) Dislocations
   (c) Point defects

9. The active material in PTC thermistor is:

    (a) Titanium dioxide

    (b) Zinc oxide

    (c) Barium titanate

10. A precipitate is a:

    (a) Line defect

    (b) Planar defect

    (c) Volume defect

Calculations and Questions

1. The Burgers vector of unit dislocations in copper is equal to the line joining the closest atoms, written $\mathbf{a}/2[110]$. (a) What is the length of this vector in nanometers? This dislocation is often replaced by partial dislocations $\frac{a}{6}[211]$. (b) What is the length of this vector in nanometers? (The copper structure is given in Supplementary Material S1 and drawn in Fig. 3.10. $a = 0.3610$ nm.)

2. The unit Burgers vector in the diamond structure joins two atoms in the $\{100\}$ plane. (a) What is the Burgers vector? This can be split into partial dislocations, which are vectors joining two nearest-neighbor carbon atoms. (b) What is the direction of the vector corresponding to this link? (The diamond structure is given in Supplementary Material S1 and drawn in Figs. 1.5 and 9.27.)

3. Sketch an edge dislocation formed by the insertion of extra material parallel to a cube edge in fluorite, CaF_2, and use the FS/RH convention to determine the Burgers vector. (The fluorite structure is given in Supplementary Material S1 and drawn in Fig. 4.7a.)

4. What are the Burgers vectors of the dislocations shown in Figure 3.17?

5. What are the Burgers vectors of the dislocations shown in Figure 3.18b?

6. A low-angle grain boundary in a single crystal is revealed by a line of dislocations with a separation of 7 nm. The angle between the grains is estimated to be $4°$. What is the Burgers vector of the dislocations?

7. What are the following boundary types?

    (a) $ABCABC * BACBAC$ in fcc metal (Fig. 3.11)

    (b) $ABCABCC * BACBA$ in fcc metal (Fig. 3.11)

    (c) $TiOOTiO * TiOOTiOOTi$ Ti and O planes parallel to (110) in rutile (Fig. 3.22)

    (d) $AcB\ AcB * AcB\ AcB$ in CdI_2 (Fig. 3.18)

    (e) $AcB\ AcB * CbA\ CbA$ in CdI_2 (Fig. 3.18)

    (f) $AcB\ AcB * BcA\ BcA$ in CdI_2 (Fig. 3.18)

8. What is the atomic composition of the (100), (200), (400), (110), and (220) planes in fluorite, CaF_2? (The fluorite structure is given in Supplementary Material S1 and drawn in Fig. 4.7a).

9. (**a**) What is the sequence of planes (in terms of V, VO, VO_2, etc.) across the twin boundaries in Figure 3.25b, V_2O_5? (**b**) Is there a composition change across the boundary?

10. Write the defect equations and the ionic formula of the phases produced by doping of $BaTiO_3$ PCT material with (**a**) Y^{3+} as a Ba^{2+} substituent and (**b**) Nb^{5+} as a Ti^{4+} substituent.

FURTHER READING

Background material on the defects described here is found in:

R. W. Cahn, *The Coming of Materials Science*, Pergamon/Elsevier, Oxford, United Kingdom, 2001.

W. D. Callister, *Materials Science and Engineering, an Introduction*, 5th ed., Wiley, New York, 2000.

M. Catti, Physical Properties of Crystals in *Elements of Crystallography*, 2nd ed., C. Giacovazzo, H. L. Monaco, A. Artioli, D. Viterbo, G. Ferraris, G. Gilli. G. Zanotti, and M. Catti, Eds., Oxford University Press, Oxford, United Kingdom, 2002, Chapter 10.

W. D. Kingery, H. K. Bowen, and D. R. Uhlmann, *Introduction to Ceramics*, 2nd ed., Wiley, New York, 1976.

R. E. Newnham *Structure–Property Relations*, Springer, Berlin, 1975.

R. J. D. Tilley *Understanding Solids*, Wiley, Chichester, 2004.

J. B. Wachtman, Ed. *Ceramic Innovations in the 20th Century*, American Ceramic Society, Westerville, Ohio, 2001.

The history of the development of dislocation theory is given by:

E. Braun, Chapter 5 in *Out of the Crystal Maze*, ed. L. Hoddeson, E. Braun, J. Teichmann, S. Weart, Oxford University Press, New York, 1992.

R. W. Cahn, as above, esp pp. 110–117.

J. W. Dash, *J. Appl. Phys.*, **27**, 1193–1195 (1956).

The early work on dislocation decoration in silver halides is described in:

J. W. Mitchell, *The Beginnings of Solid State Physics*, Royal Society, London, 1980, pp. 140–159.

Information expanding specific areas is in:

D. Cohen and C. B. Carter, *J. Microscopy*, **208**, 84–99 (2002) and references therein.

J. Eckert, G. D. Stucky, and A. K. Cheetham, *Mat. Res. Soc. Bull.* **24** (May), 31–41 (1999).

J. Li, The Mechanics and Physics of Defect Nucleation, *Mat. Res. Soc. Bull.* **32**, 151 (2007).

CHAPTER 4

Structural Aspects of Composition Variation

What are solid solutions?
What are modular structures?
What are incommensurately modulated structures?

4.1 COMPOSITION VARIATION AND NONSTOICHIOMETRY

During the latter part of the nineteenth century and the early years of the twentieth century, there was considerable controversy over the composition of chemical compounds—were compounds strictly stoichiometric, with an immutable composition, or could the composition vary. Indeed, at the turn of the twentieth century, even the existence of atoms was a subject of debate. The principal techniques involved at this epoch were accurate quantitative chemical analysis and metallographic studies of phase equilibria. The advent of X-ray diffraction studies effectively resolved the problem, and the experimental evidence for composition ranges of many solids became incontestable.

4.1.1 Phase Diagrams and the Coexistence of Solids

Probably the most widely employed technique now used in phase studies is powder X-ray diffraction. The X-ray powder pattern of a compound can be used as a fingerprint, and data for many compounds are available. This can be illustrated with reference to the sodium fluoride (NaF)–zinc fluoride (ZnF_2) system. Suppose that pure NaF is mixed with a few percent of pure ZnF_2 and the mixture heated at about 600°C until reaction is complete. An X-ray powder photograph will show the presence of two compounds (or phases), NaF, which will be the major component, and a small amount of a new compound (point A, Fig. 4.1a). A repetition of the experiment, with gradually increasing amounts of ZnF_2 will yield a similar result, but the amount of the new phase will increase relative to the amount of NaF until

Defects in Solids, by Richard J. D. Tilley
Copyright © 2008 John Wiley & Sons, Inc.

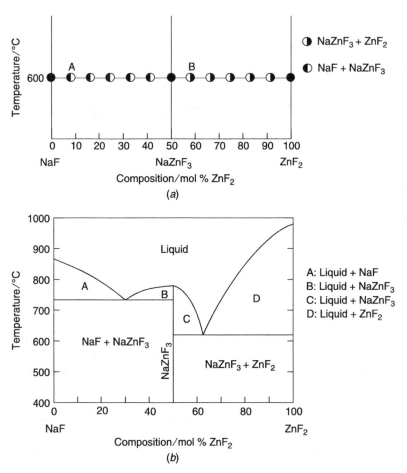

Figure 4.1   The NaF–ZnF$_2$ phase diagram: (*a*) the results of an X-ray investigation into the phases present at 600°C as a function of composition and (*b*) the complete phase diagram. [Redrawn from B. O. Mysen, *Phase Diagrams for Ceramists*, Vol VIII, American Ceramic Society, Westerville, Ohio, 1990, p. 337.]

a mixture of 1NaF plus 1ZnF$_2$ is prepared. At this composition, only one phase is indicated on the X-ray powder diagram. It has the composition NaZnF$_3$.

A slight increase in the amount of ZnF$_2$ in the reaction mixture again yields an X-ray pattern that shows two phases to be present. Now, however, the compounds are NaZnF$_3$ and ZnF$_2$ (point B, Fig. 4.1*a*). This state of affairs continues as more ZnF$_2$ is added to the initial mixture, with the amount of NaZnF$_3$ decreasing and the amount of ZnF$_2$ increasing, until only pure ZnF$_2$ is present, represented at the extreme right in Figure 4.1*a*. Careful preparations reveal the fact that NaF or ZnF$_2$ only appear alone when they are pure, and NaZnF$_3$ only appears alone at the exact composition of 1NaF plus 1ZnF$_2$. In addition, over all the composition range studied, the unit cell dimensions of each of these three phases will be unaltered.

The results of such experiments, carried out over a range of temperatures, are summarized on a phase diagram (Fig. 4.1b), which can be used to predict the outcome of any preparation in the $NaF-ZnF_2$ system. In such diagrams, each phase is drawn as a line, as in the example of $NaZnF_3$ above, because they show no composition range. Such compounds are referred to as stoichiometric compounds and, sometimes, with the appearance of the phase diagram in mind, line phases.

4.1.2 Nonstoichiometric Compounds

Not all systems behave in the same way as the $NaF-ZnF_2$ system, and in many cases there is proof that a compound may have quite a considerable composition range. An example is provided by the $Lu_2O_3-TiO_2$ system. The reaction of TiO_2 with a small amount of Lu_2O_3 at 1400°C gives results entirely analogous to those in the $NaF-ZnF_2$ system, as the reaction mixture consists of TiO_2 plus a small amount of the pyrochlore structure phase $Lu_2Ti_2O_7$. Reaction of TiO_2 with increasing amounts of Lu_2O_3 results in the formation of more pyrochlore at the expense of the TiO_2 component until the ratio of reactants is $2TiO_2 : 1Lu_2O_3$, at which point only $Lu_2Ti_2O_7$ is present after reaction.

Continued increase in the amount of Lu_2O_3 gives a different result to that in the $NaF-ZnF_2$ system because now the pyrochlore phase remains as the only compound present. It persists from $Lu_2Ti_2O_7$ to $Lu_2Ti_{1.17}O_{5.35}$ at 1400°C, and over this composition range the lattice parameter of the unit cell varies smoothly. Thereafter, as more Lu_2O_3 is added, $Lu_2Ti_{1.17}O_{5.35}$ coexists with Lu_2O_3 (Fig. 4.2). The pyrochore phase has a composition range and is nonstoichiometric.

Apart from inorganic "ionic" oxides, sulfides, alloys, many minerals, porous solids, and the like also show composition variation. These materials raise a problem: How does the structure accommodate the alteration in composition? A vast number of different structural ways to account for composition variation are now known.

4.1.3 Phase Diagrams and Composition

It is not always simple to relate the composition ranges shown on conventional phase diagrams to actual compositions or potential defect structures of a crystal. The way in which this is done is outlined in this section, using the MgO phase range in the $MgO-Al_2O_3$ system (Fig. 4.3) for illustration.

At low temperatures magnesium oxide, MgO, which adopts the sodium chloride structure, is virtually a stoichiometric phase, but at high temperatures in the $MgO-Al_2O_3$ system this is not so. At 1800°C the approximate composition range is from pure MgO to 5 mol % Al_2O_3 : 95 mol % MgO. The simplest way to account for this composition range is to assume that point defects are responsible. For this, because both Mg^{2+} and Al^{3+} cations in this system have a fixed valence, electronic compensation is unreasonable. There are then three ways to account for the composition range structurally:

1. Assume that the Mg^{2+}, Al^{3+}, and O^{2-} ions fill the normal sites in the sodium chloride structure. All surplus material is placed into interstitial sites. The limiting

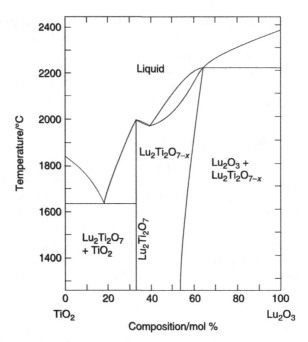

Figure 4.2   The Lu_2O_3–TiO_2 phase diagram. [Adapted from A. V. Shlyakhtina *et al.*, *Solid State Ionics*, **177**, 1149–1155 (2006).]

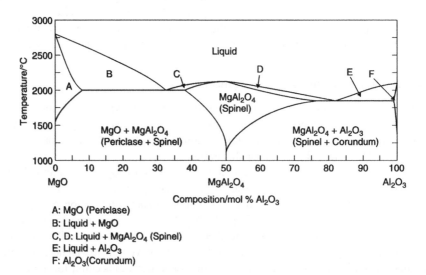

A: MgO (Periclase)
B: Liquid + MgO
C, D: Liquid + $MgAl_2O_4$ (Spinel)
E: Liquid + Al_2O_3
F: Al_2O_3(Corundum)

Figure 4.3   The MgO–Al_2O_3 phase diagram. Both MgO (periclase) and $MgAl_2O_4$ (spinel) have broad composition ranges at high temperatures.

composition is approximately 0.95MgO plus $0.05\text{Al}_2\text{O}_3$ or $\text{Mg}_{0.95}\text{Al}_{0.1}\text{O}_{1.10}$. Filling the cation and anion sites requires $\text{Mg}_{0.95}\text{Al}_{0.05}\text{O}$. The surplus ions present $(\text{Al}_{0.05}\text{O}_{0.10})$ are allocated to interstitial sites.

To generalize this, the composition is given by adding $(1 - x)\,\text{MgO}$ and $x\,(\text{Al}_2\text{O}_3)$ to give $\text{Mg}_{1-x}\text{Al}_{2x}\text{O}_{1+2x}$ ($\text{M}_{1+x}\text{O}_{1+2x}$, where M is the total Mg plus Al).

The defect reaction is

$$x\,\text{Al}_2\text{O}_3\,(\text{MgO}) \longrightarrow x\,\text{Al}_{\text{Mg}}^{\bullet} + x\,\text{Al}_i^{3\bullet} + x\,\text{O}_\text{O} + 2x\,\text{O}_i^{2\prime}$$

where the limiting composition is represented by $x = 0.05$. The point defects present are Al^{3+} substitutional defects, $\text{Al}_{\text{Mg}}^{\bullet}$, Al^{3+} interstitials, $\text{Al}_i^{3\bullet}$, and O^{2-} interstitials, $\text{O}_i^{2\prime}$.

2. Assume that the anions build up a perfect sodium chloride structure matrix. In this case the formula of the nonstoichiometric phase must be expressed as $(\text{Mg}, \text{Al})_z\text{O}_{1.0}$. To make sure that this is in agreement with the measured ratio of MgO and Al_2O_3, use the following procedure:

a. Starting formula: $\text{Mg}_{0.95}\text{Al}_{0.1}\text{O}_{1.10} = (\text{Mg}, \text{Al})_{1.05}\text{O}_{1.10}$

b. Desired formula: $(\text{Mg},\text{Al})_z\text{O}_{1.0} = \text{Mg}_{0.95/1.10}\text{Al}_{0.1/1.10}\text{O}_{1.10/1.10} \approx \text{Mg}_{0.864}$
$\text{Al}_{0.091}\text{O}_{1.0} = (\text{Mg}, \text{Al})_{0.955}\text{O}$

This formula means that there are 0.045 too few cations present compared to those required to build the sodium chloride crystal, which must create an equivalent number of cation vacancies.

The general composition $(1 - x)\,\text{MgO} : x\,(\text{Al}_2\text{O}_3) = \text{Mg}_{1-x}\text{Al}_{2x}\text{O}_{1+2x}$ is now represented by the formula $\text{Mg}_{(1-x)/(1+2x)}\text{Al}_{(2x/(1+2x)}\text{O}_{1.0}$ $([\text{M}_{(1+x)/(1+2x)}\text{O}_{1.0}]$, where M is the total Mg plus Al). The defect reaction is

$$x\,\text{Al}_2\text{O}_3\,(\text{MgO}) \longrightarrow 2x\,\text{Al}_{\text{Mg}}^{\bullet} + 3x\text{O}_\text{O} + x\,\text{V}_{\text{Mg}}^{2\prime}$$

The point defects present are Al^{3+} cations substituted on Mg^{2+} sites, $\text{Al}_{\text{Mg}}^{\bullet}$ and Mg^{2+} vacancies, $\text{V}_{\text{Mg}}^{2\prime}$.

3. Assume that the cations build up a perfect sodium chloride structure matrix. The formula must be expressed as $(\text{Mg}, \text{Al})_{1.0}\text{O}_y$. To make sure that this formula is in agreement with the measured ratio of MgO and Al_2O_3 represented on the phase diagram use the following procedure:

a. Starting formula: $\text{Mg}_{0.95}\text{Al}_{0.1}\text{O}_{1.10} = (\text{Mg}, \text{Al})_{1.05}\text{O}_{1.10}$

b. Desired formula: $(\text{Mg},\text{Al})_{1.0}\text{O}_y = \text{Mg}_{0.95/1.05}\text{Al}_{0.1/1.05}\text{O}_{1.10/1.05} \approx \text{Mg}_{0.905}$
$\text{Al}_{0.095}\text{O}_{1.048} = (\text{Mg}, \text{Al})\text{O}_{1.048}$

TABLE 4.1 Composition Range of MgO in the $MgO-Al_2O_3$ System

Composition	Defect Structure	Formula	Defects
$(1-x)MgO$: $x(Al_2O_3)$	Additional material as interstitials	$M_{1+x}O_{1+2x} = Mg_{1-x}Al_{2x}O_{1+2x}$	$Al_{Mg}^\bullet, Al_i^{3\bullet}, O_i^{2\prime}$
$(1-x)MgO$: $x(Al_2O_3)$	Anion sublattice perfect, cation deficiency as vacancies	$M_{(1+x)/(1+2x)}O_{1.0} = Mg_{(1-x)/(1+2x)} Al_{(2x)/(1+2x)}O_{1.0}$	$Al_{Mg}^\bullet, V_{Mg}^{2\prime}$
$(1-x)MgO$: $x(Al_2O_3)$	Cation sublattice perfect, additional oxygen as interstitials	$M_{1.0}O_{(1+2x)/(1+x)} = Mg_{(1-x)/(1+x)} Al_{(2x)/(1+x)}O_{(1+2x)/(1+x)}$	$Al_{Mg}^\bullet, O_i^{2\prime}$

This formula means that there are 0.48 additional oxygen ions over those required to build the sodium chloride crystal, which must be placed into the structure as interstitials.

The general composition $(1-x)MgO : x(Al_2O_3) = Mg_{1-x}Al_{2x}O_{1+2x} = (Mg, Al)_{1+x}O_{1+2x}$ is now best represented by the formula $Mg_{(1-x)/(1+x)} Al_{(2x)/(1+x)} O_{(1+2x)/(1+x)}(M_{1.0}O_{(1+2x)/(1+x)}$, where M is the total Mg plus Al). The defect reaction is:

$$x\, Al_2O_3 \,(MgO) \longrightarrow 2x\, Al_{Mg}^\bullet + 2x\, O_O + x\, O_i^{2\prime}$$

The point defects present are Al^{3+} cations substituted on Mg^{2+} sites, $Al_{Mg}^\bullet$ and O^{2-} interstitials, $O_i^{2\prime}$.

These alternatives are summarized in Table 4.1. To discover which model turns out to be correct, it is necessary to turn to experiment or calculation, as described in other chapters of this book.

4.2 SUBSTITUTIONAL SOLID SOLUTIONS

Many alloys are substitutional solid solutions, well-studied examples being copper–gold and copper–nickel. In both of these examples, the alloy has the same crystal structure as both parent phases, and the metal atoms simply substitute at random over the available metal atom sites (Fig. 4.4a). The species considered to be the defect is clearly dependent upon which atoms are in the minority.

The likelihood of forming a substitutional solid solution between two metals depends upon a variety of chemical and physical properties. A large number of alloy systems were investigated by Hume-Rothery, in the first part of the last century, with the aim of understanding the principles that controlled alloy formation. His findings with respect to substitutional solid solution formation are summarized in

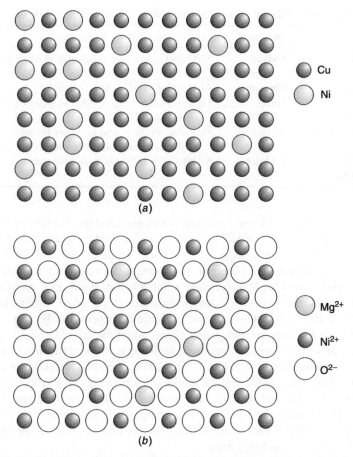

Figure 4.4 (100) planes of substitutional solid solutions: (*a*) substitutional alloys, typified by Ni–Cu and (*b*) inorganic oxides, typified by MgO–NiO.

the empirical Hume-Rothery solubility rules. The likelihood of obtaining a solid solution between two metals is highest when:

1. The crystal structure of each element of the pair is identical.
2. The atomic sizes of the atoms do not differ by more than 15%.
3. The elements do not differ greatly in electronegativity.
4. The elements have the same valence.

Solid solutions occur widely in inorganic and mineral systems (Table 4.2). The Hume-Rothery rules apply well in these cases, but it is important to add:

5. In compounds considered to be ionic the total charge on the ions must be such as to keep the compound electrically neutral.

TABLE 4.2 Some Mineral Systems Showing Broad Solid Solution Ranges

System	Structure	Examples
AO–BO	Sodium chloride	MgO–NiO, complete solid solution often possible
A_2O_3–B_2O_3	Corundum	Al_2O_3–Cr_2O_3, $Al_{1.995}Cr_{0.005}O_3$ ruby
AB_2O_4– $A^*B_2^*O_4$	Spinel	$MgAl_2O_4$–$MgCr_2O_4$, $Mg(Al_{0.99}Cr_{0.01})_2O_4$ ruby spinel, ferrites $(A,B)Fe_2O_4$
A_2SiO_4–B_2SiO_4	Olivine	Mg_2SiO_4–Fe_2SiO_4, end members forsterite and fayalite, solid solution is olivine
$AAlSi_3O_8$– $BAlSi_3O_8$	Feldspar	$KAlSi_3O_8$–$NaAlSi_3O_8$, high temperature

The extent of solid solution formation is well predicted qualitatively by the Hume-Rothery rules. When the "rules" are obeyed well, complete solid solution is likely. When the rules are only poorly obeyed, only small amounts of solid solution (partial solid solution) are observed. For example, isovalent substitution between ions of similar sizes leads to complete solid solution, typified by that formed by the sodium chloride structure oxides magnesium oxide, MgO, and nickel oxide, NiO (Fig. 4.4b). The MgO–CaO system would be expected to be very similar to the MgO–NiO system, as both parent phases adopt the sodium chloride structure. However, only partial solid solutions occur because of the difference in ionic size (Mg^{2+}, 0.086 nm; Ca^{2+}, 0.114 nm). At 2200°C the solid solution range in the crystal $Ca_xMg_{1-x}O$ extends over the composition range $0 < x < 0.05$ near to MgO and $0.88 < x < 1.0$ near to CaO. Note that the extent of these ranges is in good accord with the relative ionic sizes. It is energetically more costly to insert large Ca^{2+} ions into Mg^{2+} sites to form Ca_{Mg} defects than to insert small Mg^{2+} ions into Ca^{2+} sites to form Mg_{Ca} defects.

These days the variables in the Hume-Rothery rules can be made quantitative by making use of computer simulation or quantum mechanical calculations (Section 2.10).

In solids with several atomic constituents, solid solution formation may be restricted to one of the component atoms, as in the pyrochlore structure oxides $Bi_2Pb_2O_7$ and $Tl_2Pb_2O_7$, which form a continuous solid solution with a formula $Bi_{2-x}Tl_xPb_2O_7$ from $x = 0$ to $x = 2$. In some materials only partial replacement is possible. For example, Ca^{2+} can replace Sr^{2+} in $SrCuO_2$ over the range $Ca_xSr_{1-x}CuO_2$ from $x = 0$ to $x = 0.6$, that is, from $SrCuO_2$ to $Ca_{0.6}Sr_{0.4}CuO_2$. In others a double substitution is necessary. For example, mineral samples of pyrochlore have a general formula $(Na, Ca)_2(Nb, Ta)_2O_6(O, OH, F)$. The nominal parent phase has a formula $Na_2(Nb,Ta)_2O_7$. Complete solid solution is possible between the Nb and Ta species, as they are similarly sized M^{5+} cations. The same is not true for the substitution of Na^+ by Ca^{2+} because this will cause charge imbalance. The additional positive charge on the Ca^{2+} ions is compensated for by a variable population of O^{2-}, OH^-, and F^-. Each time an Na^+ is substituted by Ca^{2+} in the parent phase $Na_2(Nb,Ta)_2O_7$, an O^{2-} anion is substituted by either F^- or OH^-.

Interesting effects are found in systems that nominally show point defect populations. The perovskite structure, ABO_3 (where A represents a large cation and B a medium-size cation) is adopted by many oxides, and solid solutions between them can readily be fabricated. Perovskite structure materials containing a high population of vacancies are of interest as electrolytes in solid-state batteries and fuel cells. Typical representatives of this type of material are found in the solid solutions between the perovskite structure phases $CaTiO_3$–$La_{2/3}TiO_3$ and $SrTiO_3$–$La_{2/3}TiO_3$. In both of these, La^{3+} substitutes for the alkaline earth A^{2+} cation, generating one vacancy for every two La^{3+} substituents (Section 1.11.7). For example:

$$La_2O_3 + 3TiO_2 \ (SrTiO_3) \longrightarrow 2La_{Sr}^{\bullet} + V_{Sr}^{2'} + 3Ti_{Ti} + 9O_O$$

This last equation is equivalent to the reaction of $La_2Ti_3O_9$ (i.e., $La_{2/3}TiO_3$) with $SrTiO_3$.

At low concentrations of the lanthanum dopant, the vacancies appear to be distributed at random over the available anion sites. However, as the concentration of La^{3+} and hence of vacancies increases, both tend to order in the crystal.

4.3 POINT DEFECTS AND DEPARTURES FROM STOICHIOMETRY

Many binary inorganic phases with a significant composition range can be listed[1] (Table 4.3). Apart from binary compounds, ternary and other more complex materials may show nonstoichiometry in one or all atom components.

The simplest way to account for composition variation is to include point defect populations into the crystal. This can involve substitution, the incorporation of unbalanced populations of vacancies or by the addition of extra interstitial atoms. This approach has a great advantage in that it allows a crystallographic model to be easily constructed and the formalism of defect reaction equations employed to analyze the situation (Section 1.11). The following sections give examples of this behavior.

4.3.1 Substitution: Gallium Arsenide, GaAs

At 800°C gallium arsenide, GaAs, has a small composition range to the As-rich side of the exact composition GaAs, so that the composition can vary between GaAs and $Ga_{0.9998}As_{1.0002}$. This composition imbalance is believed to be due to unbalanced antisite defects in which the As atoms occupy Ga sites to give a formula $Ga_{1-x}As_{1+x}$. The population of unbalanced antisite defects is a function of temperature, thus generating a temperature-sensitive composition range. The additional As atoms form substitutional point defects, As_{Ga}, distributed throughout the solid.

---

[1] When it is necessary to emphasize the variable composition aspects of a phase, the formula will be preceded by $\sim$.

TABLE 4.3 Binary Nonstoichiometric Phases

Compound	Approximate Formula	Composition Range[a]
	Sodium Chloride Structure Oxides	
TiO_x	$\sim TiO$	$0.65 < x < 1.25$
VO_x	$\sim VO$	$0.79 < x < 1.29$
Mn_xO	$\sim MnO$	$0.848 < x < 1.00$
Fe_xO	$\sim FeO$	$0.833 < x < 0.957$
Co_xO	$\sim CoO$	$0.988 < x < 1.000$
Ni_xO	$\sim NiO$	$0.999 < x < 1.000$
	Fluorite Structure Oxides	
CeO_x	$\sim Ce_2O_3$	$1.50 < x < 1.52$
	$\sim Ce_{32}O_{58}$	$1.805 < x < 1.812$
PrO_x	$\sim PrO_2$	$1.785 < x < 1.830$
UO_x	$\sim UO_2$	$1.65 < x < 2.20$
	$\sim U_4O_9$	$2.20 < x < 2.25$
	Sulfides	
TiS_x	$\sim TiS$	$0.971 < x < 1.064$
	$\sim Ti_8S_9$	$1.112 < x < 1.205$
	$\sim Ti_3S_4$	$1.282 < x < 1.300$
	$\sim Ti_2S_3$	$1.370 < x < 1.587$
	$\sim TiS_2$	$1.818 < x < 1.923$
CrS_x	$\sim Cr_2S_3$	$1.463 < x < 1.500$
	$\sim Cr_3S_4$	$1.286 < x < 1.377$
	$\sim Cr_5S_6$	$1.190 < x < 1.203$
	$\sim Cr_7S_8$	$1.136 < x < 1.142$
Fe_xS	$\sim FeS$	$0.9 < x < 1.0$
Nb_xS	$\sim NbS$	$0.92 < x < 1.00$
Ta_xS_2	$\sim TaS_2$	$1.00 < x < 1.35$

[a]All composition ranges are temperature dependent and the figures given here are only intended as a guide.

There are a number of ways this stoichiometry change can come about. Additional As atoms might be incorporated from the vapor and added to the crystal surface at both As and Ga sites:

$$2x \text{ As (GaAs)} \longrightarrow x \text{ As}_{Ga} + x \text{ As}_{As}$$

and the material formed can be given the formula $(GaAs_x) As_{1+x}$, where the atoms occupying Ga sites are enclosed in parentheses. This model indicates that there are no vacancies in the crystal. Alternatively, some Ga may be lost to the gas phase, leaving vacancies on the gallium sites that are half filled with arsenic atoms:

$$x \text{ Ga (GaAs)} \longrightarrow x \text{ Ga (g)} + x V_{Ga}$$
$$\tfrac{x}{2} V_{Ga} + \tfrac{x}{2} As_{As} \longrightarrow \tfrac{x}{2} As_{Ga}$$

This would produce a crystal with a formula $(Ga_{1-x}As_{x/2})As_{1-x/2}$, where the atoms occupying Ga sites are enclosed in parentheses. This model indicates that there are vacancies on both the Ga and As sublattices in the crystal.

These point defect models need to be regarded as a first approximation. Calculations for *stoichiometric* GaAs suggest that balanced populations of vacancies on both gallium and arsenic sites, V_{Ga} and V_{As}, exist, as well as defect complexes. Calculation for nonstoichiometric materials would undoubtedly throw further light on the most probable defect populations present.

4.3.2 Vacancies: Cobalt Oxide, CoO

The best known inorganic compounds that show small composition variations due to a population of vacancies are probably the transition-metal monoxides, but a large number of other oxides also fall into this class. Cobalt oxide, CoO, which adopts the sodium chloride structure, is a representative example. The oxide is often oxygen rich and is considered to accommodate the composition change by way of a population of vacancies on the normally occupied metal positions, to give true formula $Co_{1-x}O$. In the stoichiometric crystal, the nominal charges on the species are Co^{2+} and O^{2-}. In the crystal containing metal vacancies, these will bear an effective negative charge, $V_{Co}^{2\prime}$. Charge compensation is achieved by the production of holes, which are generally considered to be sited on Co^{2+} ions to generate Co^{3+} defects. The formation of metal ion vacancies can be written in terms of the formal equations:

$$\tfrac{x}{2}O_2(CoO) \; \rightleftharpoons \; xO_O + x\,V_{Co}^{2\prime} + 2x\,h^{\bullet}$$
$$2x\,h^{\bullet} + 2x\,Co_{Co} \; \rightleftharpoons \; 2x\,Co_{Co}^{\bullet}$$

The crystal now contains two-point defect species, Co^{2+} vacancies, and double the number of Co^{3+} ions. The electronic implications of these changes are dealt with in Chapters 7 and 8.

4.3.3 Interstitials: La₂CuO₄ and Sr₂CuO₂F₂

The phase La_2CuO_4 contains trivalent La and divalent Cu and adopts a slightly distorted version of the K_2NiF_4 structure in which the CuO_6 octahedra are lengthened along the **c** axis due to the Jahn–Teller effect.[2] (In the superconductor literature this structure is often called the T or T/O structure.) The La_2CuO_4 structure contains sheets of the perovskite type 1 CuO_6 octahedron in thickness stacked up one on top of the other (Fig. 4.5).

[2]Transition-metal ions Cr^{2+}, Mn^{3+} Co^{2+}, Ni^{3+}, and Cu^{2+} in regular MO_6 octahedral coordination can gain stability by allowing the octahedron to distort so that there are two long and four short bonds, or vice versa. This is called the Jahn–Teller effect.

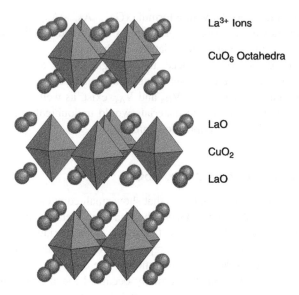

La³⁺ Ions

CuO₆ Octahedra

LaO

CuO₂

LaO

Figure 4.5   Crystal structure of La_2CuO_4 represented as a stacking of CuO_6 distorted octa-hedra and La^{3+} ions. The **c** axis is vertical. The compositions of the planes parallel to **c** are ...LaO, CuO₂, LaO CuO₂ ....

When prepared in air by heating CuO and La_2O_3, the compound is usually stoi-chiometric with a formula exactly La_2CuO_4. However, small changes in the oxygen content can be easily introduced by heating in oxygen at higher pressures to make materials with a formula $La_2CuO_{4+\delta}$. At the temperatures at which these compounds are prepared, the extra oxygen atoms are introduced into the structure as interstitial point defects that occupy sites midway between the LaO planes at random. They are arranged so as to be as far away from the La^{3+} ions as possible, and sit more or less at the centers of La^{3+} tetrahedra.

The incorporation of oxygen can be written:

$$\tfrac{x}{2}O_2 \longrightarrow x\,O_i^x$$

If the interstitial oxygen atoms are transformed to oxide ions, electrons must be pro-vided, generating two holes:

$$x\,O_i^x \longrightarrow x\,O_i^{2\prime} + 2x\,h^\bullet$$

It is convenient to assume that the holes are located on two Cu^{2+} ions; (Cu$_{Cu}$) become Cu^{3+} ions[3] (Cu$_{Cu}^\bullet$):

$$2x\,h^\bullet + 2x\,Cu_{Cu} \longrightarrow 2x\,Cu_{Cu}^\bullet$$

---

[3]The location of the holes is subject to some uncertainty, and location upon O^{2-} ions to form O^- is widely believed to represent the correct situation in many similar materials.

This is the same as saying that two Cu^{3+} ions are created per oxygen interstitial:

$$\tfrac{x}{2}O_2 + 2x\ Cu_{Cu}(La_2CuO_4) \longrightarrow x\ O_i^x + 2x\ Cu_{Cu}^{\bullet}$$

The electronic properties of these oxygen-rich solids change drastically from that of the typically insulating ceramic La_2CuO_4. The material becomes first a metal and then a superconductor (Section 8.5).

The compound $Sr_2CuO_2F_2$ also adopts the K_2NiF_4 structure and in its normal state is a poor conductor. In a similar way to La_2CuO_4, extra F^- ions can be added into (more or less) the same interstitial sites to give an overall composition of $Sr_2CuO_2F_{2+x}$. Once again the change in stoichiometry results in the transformation of the solid into a metal and a superconductor, due to the generation of holes:

$$\tfrac{1}{2}F_2 \longrightarrow F_i' + h^{\bullet}$$

These may be located on the Cu^{2+} ions to form Cu^{3+}.

The interaction between these interstitials is considerable, and when either of these phases is heated at temperatures at which diffusion can occur, they order, thus generating new structures.

4.3.4   Interstitial Impurities: Alloys and Hydrides

There are many metal alloys that contain interstitial atoms embedded in the metal structure. Traditionally, the interstitial alloys most studied are those of the transition metals with carbon and nitrogen, as the addition of these atoms to the crystal structure increases the hardness of the metal considerably. Steel remains the most important traditional interstitial alloy from a world perspective. It consists of carbon atoms distributed at random in interstitial sites within the face-centered cubic structure of iron to form the phase austenite, which exists over the composition range from pure iron to approximately 7 at % carbon.

More recently, hydrogen storage has become important, and interstitial alloys formed by incorporation of hydrogen into metals are of considerable interest. Niobium is typical of these. This metal is able to incorporate interstitial hydrogen up to a limiting composition of approximately $NbH_{0.1}$.

4.3.5   Defect Variation: Zinc Oxide, ZnO

Zinc oxide is normally an *n*-type semiconductor with a narrow stoichiometry range. For many years it was believed that this electronic behavior was due to the presence of $Zn_i^{\bullet}$ (Zn^+) interstitials, but it is now apparent that the defect structure of this simple oxide is more complicated. The main point defects that can be considered to exist are vacancies, V_O and V_{Zn}, interstitials, O_i and Zn_i, and antisite defects, O_{Zn} and Zn_O. Each of these can show various charge states and can occupy several different interstitial sites in the crystal. Combining these, the order of 20 or more defect species can be designated.

Calculations can provide information on the formation energy of these alternatives. Broadly speaking, the number of defects n_d is related to the formation energy by an equation of the sort described in Chapter 2:

$$n_d = N_s \exp\left(\frac{-E}{kT}\right)$$

where N_s is the number of sites affected by the defect in question, E the defect formation energy, k the Boltzmann constant, and T the temperature (K). Using calculated defect formation energies, the general result is that in Zn-rich conditions oxygen vacancies, $V_O^{2\bullet}$, prevail, whereas in the O-rich regime neutral zinc vacancies, V_{Zn}, are preferred. The defect type thus changes as the crystal swings from Zn rich to O rich.

Chemically, this can be expressed in terms of defect formation equations. Taking the partial pressure of the zinc as dominant, zinc-rich material can be formulated by the capture of zinc atoms from the gas phase:

$$Zn\ (ZnO) \longrightarrow Zn_{Zn} + V_O^{2\bullet}$$

where Zn_{Zn} represents a Zn^{2+} ion on a normal cation site and $V_O^{2\bullet}$ an oxygen vacancy on a normal anion site. Similarly, the oxygen-rich material can be generated by loss of zinc atoms to the gas phase:

$$Zn_{Zn} \longrightarrow Zn(g) + V_{Zn}^{2\prime} + 2h^\bullet$$

In this equation, the zinc is lost as atoms, not ions, and two electrons have to be donated to the Zn^{2+} for this to be possible, leaving two holes in the crystal. These holes can sit at the vacancy to create a neutral defect, V_{Zn}:

$$V_{Zn}^{2\prime} + 2h^\bullet \longrightarrow V_{Zn}^x$$

4.3.6 Defect Spinels, M_2O_3

The normal spinel structure has a formula AB_2O_4, where A^{2+} ions occupy tetrahedral sites and B^{3+} ions octahedral sites in a matrix composed of cubic close-packed oxygen ions (Supplementary Material S1). A number of oxides, generally prepared at moderate temperatures, adopt this structure while maintaining a composition M_2O_3. The best known of these are γ-Al_2O_3, γ-Cr_2O_3, γ-Fe_2O_3, and γ-Ga_2O_3. Of these the aluminum oxide is an important surface-active material, and the iron oxide is widely used in magnetic data recording.

Because the phases are prepared at low temperatures, they are poorly crystalline and may contain variable amounts of water, with a consequence that the structures have been inadequately characterized by X-ray diffraction. Nevertheless, these studies have suggested that these compounds have cation vacancies distributed over both octahedral and tetrahedral sites. Thus, the compounds are given a spinel-like formula $V_M M_8 O_{12}$. This may not hold for all materials. Density functional theory calculations suggest that in γ-Al_2O_3, all of the vacancies are located on the octahedral sites so that the structure can be formulated $(Al)[Al_{1.67}V_{0.33}]O_4$, where the tetrahedrally coordinated Al is written in parentheses ( ) and the octahedrally coordinated Al in square brackets [ ]. Similarly, there is evidence that more complex ordering patterns of defects may occur in magnetic γ-Fe_2O_3, perhaps as a result of magnetic interactions.

4.4 DEFECT CLUSTERS

4.4.1 Point Defect Aggregations

Even when the composition range of a nonstoichiometric phase remains small, complex defect structures can occur. Both atomistic simulations and quantum mechanical calculations suggest that point defects tend to cluster. In many systems isolated point defects have been replaced by aggregates of point defects with a well-defined structure. These materials therefore contain a population of volume defects.

At low concentrations, defect clusters can be arranged at random, mimicking point defects but on a larger scale. This seems to be the case in zinc oxide, ZnO, doped with phosphorus, P. The favored defects appear to be phosphorus substituted for Zn, $P_{Zn}^{\bullet}$, and vacancies on zinc sites, $V_{Zn}^{2\prime}$. These defects are not isolated but preferentially form clusters consisting of $(P_{Zn}^{\bullet} + 2V_{Zn}^{2\prime})$.

Defect clusters are similarly prominent in hydrated phases. For example, anatase nanocrystals prepared by sol-gel methods contain high numbers of vacancies on titanium sites, counterbalanced by four protons surrounding the vacancy, making a $(V_{Ti}^{4\prime}\ 4H^{\bullet})$ cluster. In effect the protons are associated with oxygen ions to form OH^- ions, and a vacancy–hydroxyl cluster is an equally valid description. Similar clusters are known in other hydrated systems, the best characterized being Mn^{4+} vacancies plus $4H^{\bullet}$ in γ-MnO_2, known as Reutschi defects.

Antisite defects in the pyrochore structure $Er_2Ti_2O_7$ were mentioned previously (Section 1.10). These defects also occur in the nonstoichiometric compound $Er_{2.09}Ti_{1.94}O_{6.952}$, which is slightly Er_2O_3-rich compared to the stoichiometric parent phase. The formation of the antisite pair is now accompanied by the parallel formation of oxygen vacancies:

$$Er_{Er} + Ti_{Ti} \longrightarrow Er'_{Ti} + Ti_{Er}^{\bullet}$$

$$O_O \longrightarrow V_O^{2\bullet} + O_i^{2\prime}$$

These defects are not found as separate entities but cluster together as (Er'_{Ti}, $Ti^{\bullet}_{Er}$, $V^{2\bullet}_O$, $O^{2\prime}_i$) units.

4.4.2 Iron Oxide, Wüstite

Iron monoxide, $\sim$FeO, often called by its mineral name of wüstite, is always oxygen rich compared to the formula $FeO_{1.0}$, and metallurgical phase diagrams show that it exists between the approximate compositions of 23–25.5 wt % oxygen ($Fe_{0.956}O$–$Fe_{0.865}O$ at 1300°C) (Section 1.5). Although this is one of the earliest examples of cluster geometry, $\sim$FeO is still the subject of continuing investigation. In the stoichiometric crystal, the nominal charges on the species are Fe^{2+} and O^{2-}. In normal crystals, which contain metal vacancies, charge compensation is due to hole formation. These are generally considered to be sited on Fe^{2+} ions to generate Fe^{3+} defects ($Fe^{\bullet}_{Fe}$):

$$\tfrac{x}{2}O_2 + 2Fe_{Fe}(FeO) \rightleftharpoons x\,O_O + x\,V^{2\prime}_{Fe} + 2x\,Fe^{\bullet}_{Fe} \qquad (4.1)$$

Although this is correct in one sense, isolated iron vacancies appear not to occur over much of the composition range. Instead, small groups of atoms and vacancies aggregate into a variety of defect clusters, which are distributed throughout the wüstite matrix (Fig. 4.6). The confirmation of the stability of these clusters compared to isolated point defects was one of the early successes of atomistic simulation techniques.

The most stable cluster consists of an aggregation of four cation vacancies in a tetrahedral geometry surrounding an Fe^{3+} ion, called a 4 : 1 cluster. Cations in the sodium chloride structure normally occupy octahedral sites in which each metal is coordinated to six nonmetal atoms. The central Fe^{3+} ion in the 4 : 1 cluster is displaced into a normally unoccupied tetrahedral site in which the cation is coordinated to four oxygen ions. Because tetrahedral sites in the sodium chloride structure are normally empty, the Fe^{3+} is in an interstitial site. Equation (4.1) can now be written correctly as

$$\tfrac{x}{2}O_2 + 2Fe_{Fe}(FeO) \rightleftharpoons x\,O_O + x\,V^{2\prime}_{Fe} + 2x\,Fe^{\bullet}_i \qquad (4.2)$$

The placement of cations in interstitial positions is stabilized by the presence of the surrounding vacancies and vice versa. This structural fragment is found in the spinel structure adopted by magnetite, Fe_3O_4. The oxygen excess in $\sim$FeO is therefore brought about by the incorporation of fragments of Fe_3O_4, the next higher iron oxide, within the sodium chloride matrix of FeO.

The 4 : 1 structure appears to be foundation for other cluster geometries. These (Fig. 4.6) can be constructed by uniting 4 : 1 clusters via faces to give the 6 : 2 and 8 : 3 clusters, by edges to give a 7 : 2 cluster, or by corners to give an alternative 7 : 2 cluster and the 16 : 4 cluster.

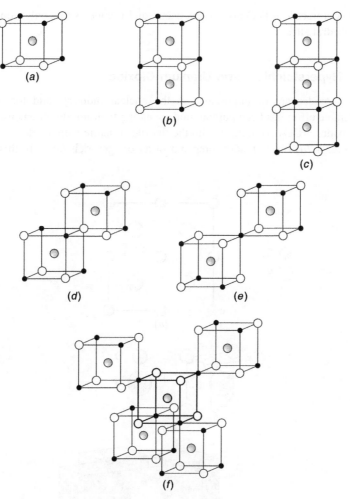

Figure 4.6 Point defect clusters in $Fe_{1-x}O$: (*a*) 4 : 1, (*b*) 6 : 2, (*c*) 8 : 3, (*d*) 7 : 2, (*e*) 7 : 2, and (*f*) 16 : 4. The open circles represent O^{2-} ions on normal sites, the shaded circles Fe^{3+} interstitials in tetrahedral sites, and the small filled circles Fe^{2+} vacancies. In (*f*), four 4 : 1 clusters are arranged tetrahedrally around a central cluster, drawn with heavy outline.

The forces uniting these clusters seem to be relatively weak, as the overall cluster geometry appears to vary both with temperature, composition, and oxygen partial pressure. Broadly speaking, relatively isolated 4 : 1 clusters are found at the highest temperatures and in iron-rich compositions. As the temperature and relative amount of iron falls, more complex clusters are found. To some extent it is as if the 4 : 1 clusters aggregate at lower temperatures and dissociate at higher temperatures.

Despite the numerous studies made on this material, the defect structures found in this system are still not completely clarified. For example, it has been shown that in near stoichiometric ~FeO no clusters occur, and it appears that only iron vacancies on octahedral sites occur.

4.4.3 Hyperstoichiometric Uranium Dioxide

Uranium oxides are of importance in the nuclear industry, and for this reason considerable effort has been put into understanding their nonstoichiometric behavior. The dioxide, $\sim UO_2$ crystallizes with the fluorite structure with an ideal composition MX_2 (Fig. 4.7a) but is readily prepared in an oxygen-rich form. In this state it is

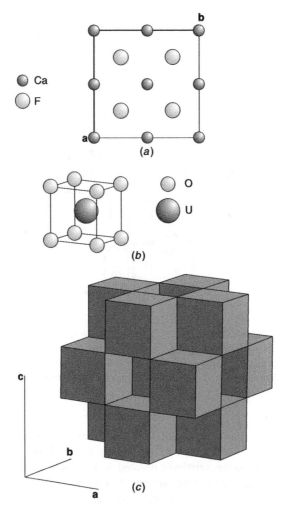

Figure 4.7 Idealized fluorite structure of UO_2: (a) projection of the structure of fluorite, CaF_2, down the cubic **c** axis; (b) a UO_8 cube, which lies at the center of the fluorite unit cell; and (c) stacking of UO_8 cubes to give the idealized structure.

an example of a large class of nonstoichiometric compounds grouped under the heading of anion-excess fluorite structure phases. As with iron monoxide, hyperstoichiometric uranium dioxide, UO_{2+x}, was one of the earliest materials containing clusters to be studied structurally and is still the subject of continuing investigation.

At temperatures above $1100°C$ uranium dioxide can exist between the composition limits of UO_2 and approximately $UO_{2.25}$. The fluorite structure of the parent UO_2 can be imagined to be constructed of edge-shared UO_8 cubes (Fig. 4.7b and 4.7c). At the simplest level, the composition variation can be considered to be due to the presence of interstitial anions. Each cube containing a U ion is adjacent to an empty cube. The incorporation of anions, O^{2-}, within these empty cubes is therefore possible, and it is these interstitial positions that are occupied in oxygen-rich UO_{2+x}.

As in the case of the oxides described above, the incorporation of additional oxygen ions in the solid will upset the charge balance by introducing extra negative charge. This can be corrected by increasing the charge on the parent cations, so that some of the U^{4+} ions must become either U^{5+} or U^{6+}. This can be written as

$$\tfrac{x}{2} \, O_2 \rightleftharpoons x \, O_i^{2\prime} + 2x \, U_U^{\bullet}$$

$$\tfrac{x}{2} \, O_2 \rightleftharpoons x \, O_i^{2\prime} + x \, U_U^{2\bullet}$$

Initial crystallographic studies of oxygen-rich material made it clear that a model for the defect structure consisting of random interstitial point defects was too simple and that point defect clusters were present. One of the earliest cluster geometries to be suggested was that of the so-called *Willis* or *2:2:2 cluster* in UO_{2+x}. The cluster is constructed from four "empty" oxygen cubes, joined in the normal fluorite arrangement (Fig. 4.8a and 4.8b). The four cubes each contain an internal interstitial oxygen ion. These do not sit at the cube centers but are displaced either along a cube body diagonal $<111>$ direction in two of the cubes (Fig. 4.8c) or along a diagonal parallel to $<110>$ in the other two (Fig. 4.8d). The cluster is formed by uniting two peripheral cubes of $<111>$ type with two central cubes of $<110>$ type (Fig. 4.8e). Two of the oxygen ions at the centers of the cubes are derived from the extra oxygen content of the crystal, and two are from normally occupied positions. These two oxygen ions, which should be found at the junctions of three cubes, move into interstitial sites in the middle of the two central cubes (Fig. 4.8e) and in so doing create two oxygen vacancies at these corner junctions. The cluster name comes from the fact that there are two interstitials created by moving normal anions (called $<110>$ interstitials), two normal interstitials (called $<111>$ interstitials), and two vacancies in the grouping. [In some literature, these defects are labeled as O′ ($<110>$) and O″ ($<111>$). The same nomenclature is found with clusters in anion excess fluorides (Section 4.4.3), where these are labeled as F′ ($<110>$) and F″ ($<111>$). The confusion with the Kröger–Vink notation for effective charge is obvious, and this nomenclature will not be used here. Where necessary these sites will be referred to as O1 or F1 ($<110>$) and O2 or F2 ($<111>$)].

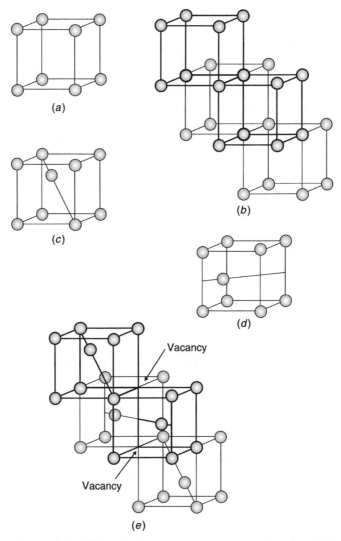

Figure 4.8 Structure of a Willis $2:2:2$ structure: (a) an "empty" O_8 cube; (b) the stacking of four O_8 cubes in the UO_2 structure; (c) an O_8 cube containing a $<111>$ interstitial oxygen ion; (d) an O_8 cube containing a $<110>$ interstitial oxygen ion; and (e) the $2:2:2$ cluster.

As in the case of wüstite, one of the earliest applications of atomistic simulations was to explore the likely stability of this defect cluster. It was found that not only the $2:2:2$ arrangement but also other cluster geometries were preferred over isolated point defects.

It should be remembered that there is still much confusion about the real defect structures occurring over the broad composition range shown by this oxide. For example, there is uncertainty about whether a population of uranium vacancies best explains the data or whether both uranium vacancies and oxygen interstitials are present, at least in some composition regions.

4.4.4 Anion-Excess Fluorite Structures

Anion-excess fluorite structure materials can be typified by the products of the reaction of fluorite, CaF_2, with fluorides of higher valance atoms, such as LaF_3, YF_3, ThF_4, and UF_4. In all cases nonstoichiometric phases form that have considerable composition ranges, and the parent structure can accommodate up to approximately 50 mol% of the added fluoride to form anion-rich fluorite structures. For example, the limiting composition of the nonstoichiometric fluorite structure phase in the CaF_2–YF_3 system is approximately $0.6CaF_2 : 0.4YF_3$ at $1100°C$. The formula can be written as $Ca_xY_{1-x}F_{2+x}$, where x takes values between 0 and 0.4.

Across the phase range of all of these systems the metal atoms La, Y, and the like substitute for calcium on the cation sublattice. Charge balance is ensured by the incorporation of additional F^- ions into the crystals, which, *to a first approximation*, can be regarded as F^- interstitials that occupy the unoccupied (F_8) coordination polyhedra (Fig. 4.7):

$$LaF_3(CaF_2) \longrightarrow La^\bullet_{Ca} + 2F_F + F'_i$$

$$ThF_4(CaF_2) \longrightarrow Th^{2\bullet}_{Ca} + 2F_F + 2F'_i$$

One interstitial ion needs to be accommodated for each M^{3+} incorporated and two for each M^{4+} cation.

There are large numbers of anion excess fluorite-related structures known, a small number of which are listed in Table 4.4. The defect chemistry of these phases is enormously complex, deserving of far more space than can be allocated here. The defect structures can be *roughly* divided into three categories: random interstitials, which in

TABLE 4.4 Some Anion-Excess Fluorite-Related Phases

System	Approximate Composition Range
$Zr(N, O, F)_x$	$2.12 < x < 2.25$
$Nb_2Zr_{x-2}O_{2x+1}$	$7.1 < x < 12$
Ln oxyfluorides (Ln = Y, Sm–Lu)	Typically $MX_{2.12}$–$MX_{2.20}$
$ZrO_2 + ZrF_4$	Typically $MX_{2.12}$–$MX_{2.20}$
$ZrO_2 + UO_2F$	Typically $MX_{2.12}$–$MX_{2.20}$
AF_2–BiF_3 (A = Sr, Ba)	$MX_{2.45}$–$MX_{2.5}$
$(1-x)KF + xBiF_3$ ($K_{1-x}Bi_xF_{1+2x}$)	$0.50 < x < 0.70$
$(1-x)RbF + xBiF_3$ ($Rb_{1-x}Bi_xF_{1+2x}$)	$0.50 < x < 0.60$
U_4O_{9-y}	$UO_{2.235}$–$UO_{2.245}$
$(1-x)PbF_2 + xInF_3$ ($Pb_{1-x}In_xF_{2+x}$)	$0 < x < 0.25$
$LnF_{3-2x}O_x$ (Ln = La–Nd)	$0.58 < x < 1$
$Ca_{1-x}M_xF_{2+2x}$ (M = Th, U)	$0 < x < 0.18$
$Ba_{1-x}Bi_xO_zF_{2+x-2z}$	$0 < x < 0.4; 0 < z < 0.25$

reality almost never exist, defect clusters, which may be ordered or disordered, and modulated structures. These latter are described in Section 4.10, while here some of the cluster geometries are described.

As in the case of UO_{2+x}, crystallographic studies of the anion-excess fluorite phases indicated that there were complex arrangements of interstitials and vacancies, arranged in clusters, present. Some of these clusters are relatively simple to envisage, such as the $2:2:2$ cluster described above, but many are of greater complexity. Indeed, there is still uncertainty of the true defect structure of many of these phases over all composition ranges and temperature regimes, even the most studied, $Ca_{1-x}Y_xF_{2+x}$ in the CaF_2–YF_3 system. The following descriptions must therefore be read in the light that further revisions may be likely.

The foundation of many of the clusters found in the anion-excess fluorites is a square antiprism of anions. A square antiprism can be constructed from a cube by the rotation of one face of anions by $45°$ with respect to the opposite face (Fig. 4.9a and 4.9b). The other principal cluster geometry is that of the cuboctahedron. This can be constructed from a cube by truncating (i.e., removing) the corners until eight equilateral triangles are formed (Fig. 4.9c and 4.9d). In terms of the MF_2 fluorite structure, an MF_8 cube can be transformed into an MF_8 square antiprism by moving four anions from normal positions into the appropriate F1 $<110>$ defect positions, leaving four vacancies at the cube corners (Fig. 4.9e). This operation leaves the cation positions unchanged, but the structure now contains four $<110>$ interstitials and four anion vacancies. If six such operations are performed on a group of six cubes arranged around the faces of a central empty cube, the resulting central polyhedron so formed is a cuboctahedron with a composition of M_6F_{36} (Fig. 4.9f). If a central anion is included in the cluster, the composition becomes M_6F_{37}.

There are many ways in which these square antiprism and cuboctahedral defect clusters can be arranged. A nonstoichiometric composition can be achieved by a random distribution of varying numbers of clusters throughout the crystal matrix. This appears to occur in $Ca_{0.94}Y_{0.06}F_{2.06}$, which contains statistically distributed cuboctahedral clusters.

Alternatively, the clusters can order, in which case new structures that are essentially defect free, are produced. In the PbF_2–ZrF_4 phases $Pb_{1-x}Zr_xF_{2+2x}$, in which x can take the range 0–0.18, pairs of square antiprisms are found. As well as disordered arrangements of the antiprisms, ordering of the clusters is frequent, giving rise to a series of phases $Pb_nZr_2F_{2n+8}$, which includes Pb_3ZrF_{10} ($n = 6$) and Pb_5ZrF_{14} ($n = 10$) (Fig. 4.10). Other members of this series of ordered structures can be constructed by varying the spacing between the rows of square antiprisms.

Ordered cuboctahedra appear in a number of structures, for example, in $Na_7Zr_6F_{31}$, a number of $M_7U_6F_{31}$ and MTh_6F_{31} phases, and in $Ba_4Bi_3F_{17}$, an ordered structure in the anion-excess fluorite range that is found in the BaF_2–BiF_3 system. However, the most important of these ordered structures is that adopted by the oxide $U_4O_{9-\delta}$. This is the next higher oxide to the UO_{2+x} phase described above (Section 4.4.3). The composition is never exactly U_4O_9, a fact accounted for by the structure, which consists of an ordered array of U_6O_{36} cuboctahedra in a fluorite structure matrix.

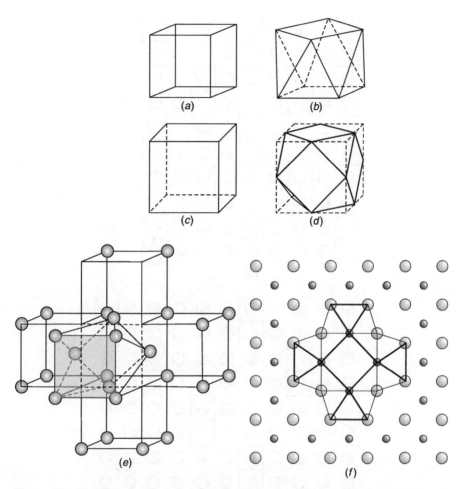

Figure 4.9 Clusters in the fluorite structure: (a, b) transformation of a cube into a square anti-prism; (c, d) transformation of a cube into a cuboctahedron; (e) a single square antiprism formed by the creation of $<110>$ interstitial defects; (f) an M_6F_{36} cluster in a fluorite structure matrix. Cations in the plane of the section are represented by smaller spheres; anions above and below the plane are represented by larger spheres.

Many other clusters have been postulated to occur in these systems, together with a large number of ordered structures in which the clusters are arranged on regular crystallographic positions.

4.4.5 Anion-Deficient Fluorite Structures

Zirconia, ZrO_2, is an important ceramic, possessing high mechanical strength that is maintained to very high temperatures. Unfortunately, pure zirconia is liable to fracture

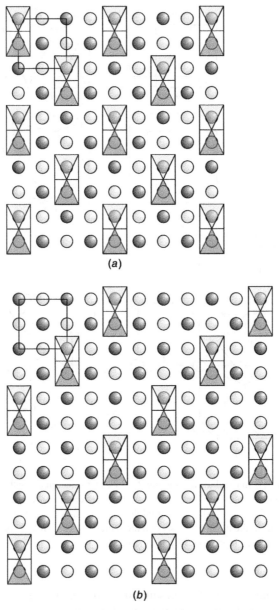

Figure 4.10   Schematic representation of ordered structures in the $Pb_{1-x}Zr_xK_{2+2x}$ system: (*a*) Pb_3ZrF_{10} and (*b*) Pb_5ZrF_{14}. Pairs of square antiprisms, at two levels, are represented by the small square motifs. Spheres represent cations at two levels; anions are omitted for clarity. The fluorite subcell is indicated in the top left of each figure. [Adapted from J. P. Laval, C. Depierrefixe, B. Frit, and G. Roult, *J. Solid State Chem.*, **54**, 260–276 (1984).]

when cycled between high and low temperatures. To overcome this, a cubic form, stabilized zirconia, is preferred. This is made by the addition of other oxides to the zirconia, the two most common being calcia (CaO) and yttria (Y_2O_3), to produce calcia-stabilized zirconia (CSZ) and yttria-stabilized zirconia (YSZ), respectively.

Both of these materials are nonstoichiometric phases with broad composition ranges and are examples of a class of nonstoichiometric compounds all grouped under the heading of anion-deficient fluorite structure phases. The reaction of ZrO_2, with CaO has been described in formal terms in Section 1.11.6 To a first approximation, the structure contains Ca^{2+} cations substituted for Zr^{4+} cations. The charge imbalance is corrected by the introduction of anion vacancies. Each Ca^{2+} added to the ZrO_2 produces an oxygen vacancy at the same time to give a true formula for the crystal of $Ca_x^{2+}Zr_{1-x}^{4+}O_{2-x}$. The same procedure can be used to describe the formation of yttria-stabilized zirconia. Assuming that the materials are ionic:

$$x\,Y_2O_3(ZrO_2) \longrightarrow 2x\,Y'_{Zr} + 3x\,O_O + x\,V_O^{2\bullet}$$

The substitution of $2Zr^{4+}$ cations by $2Y^{3+}$ cations leads to the formation of one oxygen vacancy.

The parent structure of the anion-deficient fluorite structure phases is the cubic fluorite structure (Fig. 4.7). As in the case of the anion-excess fluorite-related phases, diffraction patterns from typical samples reveals that the defect structure is complex, and the true defect structure is still far from resolved for even the most studied materials. For example, in one of the best known of these, yttria-stabilized zirconia, early studies were interpreted as suggesting that the anions around vacancies were displaced along <111> to form local clusters, rather as in the Willis $2:2:2$ cluster described in the previous section, Recently, the structure has been described in terms of anion modulation (Section 4.10). In addition, simulations indicate that oxygen vacancies prefer to be located as second nearest neighbors to Y^{3+} dopant ions, to form triangular clusters (Fig. 4.11). Note that these suggestions are not

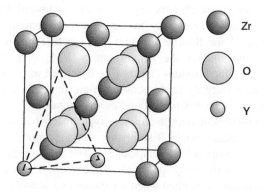

Zr

O

Y

Figure 4.11 Suggested defect cluster in yttria-stabilized zirconia. An oxygen vacancy is paired with two nearest-neighbor Y^{3+} ions. Relaxation of the ions in the cluster is ignored.

mutually exclusive: Each might be correct over some regimes of temperature and pressure.

The Pr–O system was one of the earliest to be investigated, although resolution of the defect structure has been enormously difficult. When prepared in air at about 930°C, the highest oxide, PrO_2, with the fluorite structure, has a composition close to $PrO_{1.833}$. Heating in oxygen allows easy and rapid changes in composition to be obtained. At higher temperatures a broad fluorite-type anion-deficient phase is found close to PrO_2, but at lower temperatures this transforms into a homologous series of stable phases with a formula Pr_nO_{2n-2}, ($n = 7$, 9, 10, 11, 12), each of which has only a very small composition range. Analogs occur in the Tb–O and Ce–O systems. The Pr_7O_{12} structure is well represented, being adopted by Tb_7O_{12}, Y_6UO_{12}, La_6UO_{12}, In_6WO_{12}, $In_4Sn_3O_{12}$, In_5SnSbO_{12}, and oxides $In_{4+x}Sn_{3-2x}Sb_xO_{12}$.

An early model for the structures of these phases was that of ordered oxygen vacancies. A PrO_8 cube of the type found in the fluorite parent phase was modified so that the metal became 6-coordinate by removing the two opposite oxygen ions at the corners of the cube linked by a body diagonal, that is, along the cubic [111] direction. Similarly, removal of just one vacancy leaves a 7-coordinated metal atom. These cubes are ordered so that the vacancies form strings or sheets in the fluorite parent. Around each vacancy considerable structural relaxation takes place, so that defect clusters, with a composition M_7O_{30}, are a superior description of the situation. The structures of these phases are thought to be built by an ordering of these clusters.

An alternative description of these and related phases such as $Zr_3Sc_4O_{12}$, $Zr_{10}Sc_4O_{26}$, and $Zr_{50}Sc_{12}O_{118}$ has been formulated in terms of cluster called a coordination defect (CD). This cluster is derived from the depiction of the fluorite structure not as a set of MX_8 cubes but as two interpenetrating sets of XM_4 tetrahedra, which point along the cubic $<111>$ directions (Fig. 4.12a). The arrangement is clearer if only part of one set of tetrahedra are depicted (Fig. 4.12b). The core of a coordination defect in the oxides mentioned consists of an oxygen vacancy surrounded by four metal atoms, which can be designated as VM_4, where V represents a vacancy and M a lanthanide or related metal atom. That is, the core is simply an empty tetrahedron. Surrounding the core, the nearest six oxygen ions move toward the vacancy and the nearest four metal ions move away to form the coordination defect. It can be described in terms of the fluorite structure as an octahedron of OM_4 tetrahedra, sharing corners, surrounding the central empty VM_4 core (Fig. 4.12c). The complex structures of these oxides can then be described in terms of CD linkages. The CDs retain their integrity in all phases but are linked in a variety of ways to produce the complex superstructures found experimentally.

The final anion-deficient fluorite structure type material to mention is δ-Bi_2O_3. The formula of this phase makes it surprising that a fluorite structure form exists, but such a structure occurs at high temperatures. The resulting phase is an excellent O^{2-} ion conductor with many potential applications. Unfortunately, the high-temperature form is not maintained when the compound is cooled to room temperature. However, fluorite structure anion-deficient phases of the same type can be prepared by reaction with many other oxides, and these are stable at room temperature. The majority of these materials have a modulated anion substructure (Section

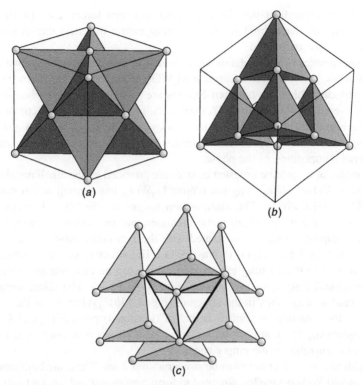

Figure 4.12   Coordination defects in the fluorite structure: (*a*) fluorite structure represented as two interpenetrating sets of XM_4 tetrahedra pointing along the cube $<111>$ directions; (*b*) fragment of one subset of tetrahedra, all pointing in the same direction; and (*c*) coordination defect. The central part of the defect, heavy outline, is the unoccupied tetrahedron core of the cluster. The cubic unit cell in (*a*) and (*b*) is outlined.

4.10). However, ordered phases also form that can be regarded as containing clusters rather than isolated oxygen vacancies. For example, $Bi_{94}Nb_{32}O_{221}$ contains chains of NbO_6 octahedra, while the phases $Pb_5Bi_{17}P_5O_{43}$ and $Pb_5Bi_{18}P_4O_{42}$ contain ordered arrays of PO_4 tetrahedra.

4.5   INTERPOLATION

The difference between interstitial species and interpolated species is one of scale. Interstitial sites are thought of as normally unoccupied positions in relatively closely packed crystal structures. One tends to speak of interpolation when foreign atoms enter larger normally unoccupied volumes, cages, or tunnels in a structure, which otherwise remains topotactically unchanged.

The likelihood of finding that a nonstoichiometric composition range is due to the presence of interpolated atoms in a crystal will depend upon the openness of the structure, the strength of the chemical bonds between layers, and the size of the impurity. Despite these limitations, nonstoichiometric materials that utilize interpolation are many and varied.

Good examples of interpolation are provided by the compounds known as tungsten "bronzes," with a general formula M_xWO_3, where M can be almost any metal in the periodic table. The tungsten bronzes are so called because when they were first discovered, by Wohler in 1837, their metallic luster and high electrical conductivity led him to believe that he had made alloys of tungsten rather than new oxides. The tungsten bronzes adopt several structures, depending upon the foreign metal and the overall composition of the phase.

Examples of perovskite tungsten bronzes are provided by the alkali metal tungsten bronzes, A_xWO_3. Lithium tungsten bronze Li_xWO_3, has a composition range from pure WO_3 to $Li_{0.5}WO_3$. The composition ranges of the other bronzes tend to shrink as the size of the alkali metal increases, but the sodium bronze, Na_xWO_3, has two composition ranges associated with the perovskite bronze structure, WO_3 to $Na_{0.11}WO_3$ and $Na_{0.41}WO_3$ to $Na_{0.95}WO_3$. The structure of these compounds is simply related to that of tungsten trioxide, which can be conveniently described as a corner-shared linkage of (WO_6) octahedra (Fig. 4.13a). The alkali metal atoms occupy random cage sites in the structure (Fig. 4.13b). [When all of the cage sites are filled, the structure so formed is that of (idealized) perovskite, $CaTiO_3$, hence the designation.] The composition range of these phases is a result of the variable degree of occupation of the empty cages.

In addition, two other tungsten bronze structures exist. These are both made up of corner-linked WO_6 octahedra, arranged to form pentagonal and square tunnels in the tetragonal tungsten bronze structure or hexagonal tunnels in the hexagonal tungsten bronze structure. Variable filling of the tunnels by metal atoms gives rise to wide stoichiometry ranges. In the sodium tetragonal tungsten bronze phases, which exist over the composition range $Na_{0.26}WO_3 - Na_{0.38}WO_3$, both the pentagonal and square tunnels are partly filled (Fig. 4.14a). However, in many tetragonal tungsten

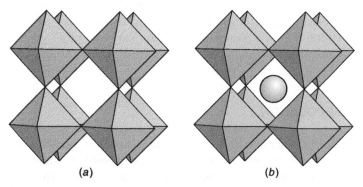

(a)   **(b)**

Figure 4.13   Perovskite tungsten bronze structure: (a) idealized structure of WO_3, composed of corner-linked WO_6 octahedra and (b) interpolation of large A cations into the cages of the WO_3 structure generates the perovskite tungsten bronze structures, A_xWO_3.

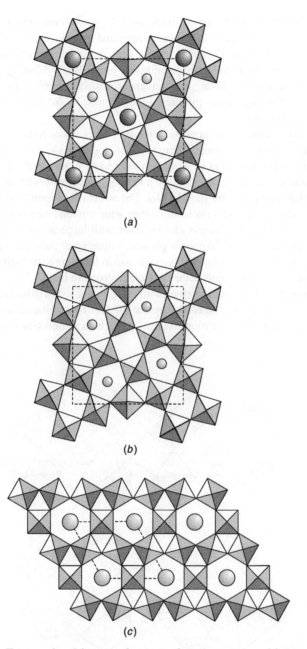

Figure 4.14 Tetragonal and hexagonal tungsten bronze structures: (*a*) tetragonal tungsten bronze structure adopted by Na_xWO_3 in which both pentagonal and square tunnels are partly occupied; (*b*) tetragonal tungsten bronze structure adopted by Pb_xWO_3 in which only the pentagonal tunnels are partly occupied; and (*c*) hexagonal tungsten bronze structure adopted by In_xWO_3 in which hexagonal tunnels are partly occupied. The shaded squares represent WO_6 octahedra, and the shaded circles represent interpolated cations. The unit cells are outlined.

bronzes, typified by Pb_xWO_3 and Sn_xWO_3, only the pentagonal tunnels are occupied (Fig. 4.14b). (The tunnels, especially the pentagonal tunnels, can also be filled with alternate $M-O-M-O-M-O$ chains, to give fully oxidized forms with a fixed composition, such as the high dielectric permittivity phase $Ba_6Ti_2Nb_8O_{30}$.) In the hexagonal tungsten bronzes the hexagonal tunnels are partly occupied (Fig. 4.14c). The composition of the hexagonal tungsten bronze phase K_xWO_3 extends from $x = 0.19$ to 0.33 in this way.

The principal composition variation in these bronzes is due to the interpolation of large metal atoms into the tunnels. There is considerable evidence to suggest that the tunnel filling can be ordered both within an individual tunnel and between tunnels.

The hollandite structure is adopted by a variety of minerals with an overall composition $A_2M_8X_{16}$, typified by $Ba_2Mn_8O_{16}$. The structure is composed of ribbons of edge-sharing MnO_6 octahedra enclosing rather wide tunnels, which are occupied by Ba (Fig. 4.15). In fact, the tunnels can be filled with large metal atoms in variable proportions to create nonstoichiometric phases. A number of these have received considerable attention in recent years for possible use in the storage of high-level nuclear waste. Typical of these materials is the hollandite structure $Ba_xAl_{2x}Ti_{8-2x}O_{16}$, formulated to immobilize large radioactive atoms such as Cs and Rb. Because of the open-tunnel structure, these can easily replace the Ba. To maintain charge neutrality, replacement of the Ba^{2+} by a monovalent ion such as radioactive Cs^+ requires an

Figure 4.15   Hollandite structure. The shaded diamonds represent chains of edge-shared MnO_6 octahedra and the shaded circles Ba^{2+}. The light and heavy shading represent octahedra and atoms at two different heights. The unit cell is outlined.

adjustment in the Al : Ti ratio, so that a fabricated composition might typically be $BaCs_{0.28}Al_{2.28}Ti_{5.72}O_{16}$.

4.6 INTERCALATION

Intercalation is mainly associated with the insertion of atoms or molecules between the layers of a structure in such a way that the layers maintain their integrity. The first material to be studied extensively, and which is still the subject of enquiry, is graphite.

Intrinsically, intercalation compounds are characterized by strong chemical bonding within the layers (intralayer bonding) and weak chemical bonding between the layers (interlayer bonding). There are many examples, ranging from clays and hydroxides to sulfides and oxides. Such compounds are of increasing importance in many applications, particularly in battery technology.

The prototypes for this latter application are derived from layer structure disulfides, such as TiS_2 and NbS_2. Take TiS_2 as an example. The structure of TiS_2 is of the CdI_2 type (Fig. 4.16a) and can be thought of as a hexagonal packing of sulfur layers in which the octahedral positions between every alternate pair of layers is occupied by titanium to form layers of composition TiS_2. The layers can also be visualized as a stacking of sheets of edge-sharing TiS_6 octahedra. These are held together by weak van der Waals bonds.

A series of nonstoichiometric phases can be generated by gradually filling the vacant octahedral sites in TiS_2 with Ti itself. Ultimately, all of the available octahedral sites are filled and the material has a composition TiS, with the nickel arsenide NiAs, structure (Fig. 4.16b). The formula of this nonstoichiometric phase range can be written $Ti_{1+x}S_2$, with x taking all values from 0 to 1.0. Similar phases can be generated with the use of other transition metals, such as Cr.

One of the most widely explored systems is derived from the interpolation of Li between the TiS_2 layers in varying amounts to form nonstoichiometric phases with a general formula Li_xTiS_2. Because the bonding between the layers is weak, this process is easily reversible. The open nature of the structure allows the Li atoms to move readily in and out of the crystals, and these compounds can act as convenient alkali metal reservoirs in batteries and other devices. A battery using lithium intercalated into TiS_2 as the cathode was initially developed some 30 years ago.

For a number of reasons the TiS_2 battery was not widely exploited commercially, but similar batteries based upon intercalation into transition-metal oxides Li_xTO_2, where T is a 3d transition metal such as Ni, Co, or Mn (or a solid solution of these metals, $Li_xT_{1-y}T'_yO_2$) are widely available. The first of these, the "Sony cell," introduced in 1991, employs Li_xCoO_2 as the cathode and the intercalation of Li into graphite as the anode, to form Li_xC_6.

The structure of the parent phase CoO_2 is similar to that of TiS_2 in that it is composed of layers of composition CoO_2 in which the Co^{4+} ions are in octahedral coordination. The stacking of the oxygen ion layers is in a $\ldots ABAB \ldots$ hexagonal

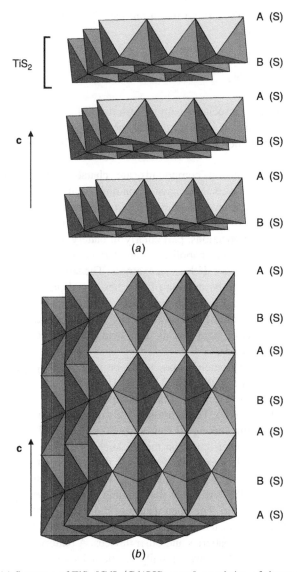

Figure 4.16 (*a*) Structure of TiS_2 [CdI_2/$Cd(OH)_2$-type], consisting of sheets of TiS_2 octahedra. The S atoms, which are in hexagonal packing, lie at the vertices of the octahedra. (*b*) The NiAs structure of TiS, resulting from filling of the empty octahedral interlayer sites in TiS_2.

arrangement (Fig. 4.17*a*). Intercalation of Li causes the CoO_2 layers to be displaced, so that the oxygen stacking becomes cubic . . .*ABCABC*. . . packing (Fig. 4.17*b*). The composition range runs from $x = 0$ to $x = 1$, and over this range several other stacking sequences and possible cation-ordering arrangements have been reported. In

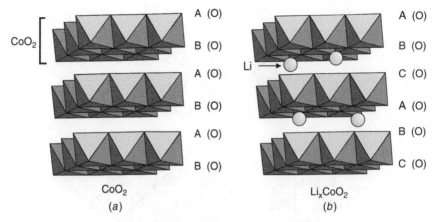

Figure 4.17 Structure of the intercalation phase Li_xCoO_2: (*a*) packing of the oxygen layers in CoO_2 before intercalation is hexagonal closest packing and (*b*) after intercalation the packing changes to cubic closest packing.

addition, at low Li concentrations the Li may only intercalate into a fraction of the available layers. Partly because of these structural limitations, commercial batteries only cycle over a restricted range of Li compositions. A similar rearrangement of the graphite layers also takes place during intercalation of Li. Such changes, brought about by intercalation, are not uncommon. (Also see Section 8.7.)

4.7 LINEAR DEFECTS

Although dislocations can alter the composition of a crystal, the change is generally minute, even when dislocation density is high, and this effect will be ignored here. However, linear defects that contribute to composition variation are found in several groups of phases. Variable tunnel filling is one of these. In the tungsten bronzes the filling of tunnels is irregular, and the distribution of these partly filled tunnels over the total tunnel population can be described in terms of linear defects arranged in a random or partly ordered fashion. A closely related group of compounds uses chains of filled (or partly filled) pentagonal columns (PC's) to accommodate nonstoichiometry. These are a group of mostly fully oxidized transition-metal oxides related to the tetragonal tungsten bronze structure (Fig. 4.14). In them, the pentagonal tunnels formed by a ring of five corner-linked MO_6 octahedra, which contain metal atoms in the true bronzes, are filled with chains of alternating metal and oxygen, ...O–M–O–M–O... converting them to columns of MO_7 pentagonal bipyramids (Fig. 4.18*a*). A group of four pentagonal tunnels can fit coherently into a WO_3 type of matrix, (Fig. 4.18*b*). In many nonstoichiometric phases the tunnels are frequently filled in pairs; the empty pair serving to reduce strain in the

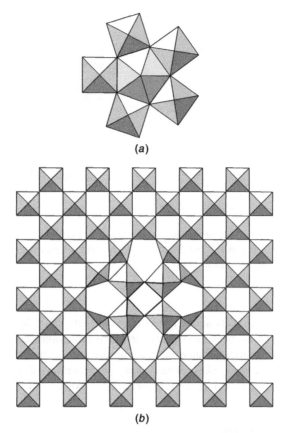

(a)

(b)

Figure 4.18 Pentagonal columns (PCs): (*a*) ideal pentagonal column, composed of five MO_6 octahedra forming the outline of an MO_7 pentagon; (*b*) idealized representation of four PCs coherently intergrown in a WO_3-like structure; (*c*) filling of two opposite PCs; (*d*) electron micrograph showing disordered PCs in a WO_3-like matrix following reaction of Nb_2O_5 and WO_3. The filled PCs are imaged as black blobs. The square grating background shows the underlying WO_3-like substructure, with a spacing of 0.4 nm.

surrounding structure (Fig. 4.18*c*). These groups frequently link up in chains to form sheets of defects running parallel to <110> planes in the WO_3-like surrounding structure.

As in previous discussions, two alternatives exist. If the PC elements are perfectly ordered, then a defect-free structure is formed—one such possibility being the tetragonal tungsten bronze structure. If the PC elements in the host structure are disordered, a nonstoichiometric compound is generated. Composition variation can be accommodated either by varying the number of PC's present or by varying the degree of filling. Such defects occur, for example, when WO_3 is reacted with Nb_2O_5 for short periods of time at temperatures below about 1500 K (Fig. 4.18*d*).

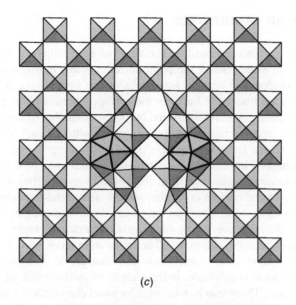

(c)

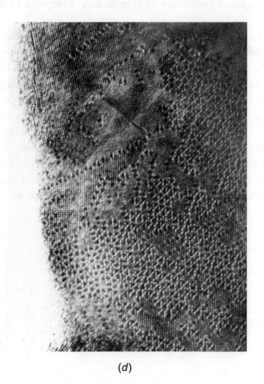

(d)

Figure 4.18 (*Continued*).

4.8 MODULAR STRUCTURES

In the previous sections composition variation has been attributed, more or less, to point defects and extensions of the point defect concept. In this section structures that can be considered to be built from slabs of one or more parent structures are described. They are frequently found in mineral specimens, and the piecemeal way in which early examples were discovered has led to a number of more or less synonymic terms for their description, including "intergrowth phases," "composite structures," "polysynthetic twinned phases," "polysomatic phases," and "tropochemical cell-twinned phases." In general, they are all considered to be modular structures.

Modular structures can be built from slabs of the same or different compositions, the slabs widths can be random or regular, and the slabs themselves can be ordered in a variety of ways (Fig. 4.19). Composition variation can then occur by variation of the amount of each slab type present, by variation in the degree of order present, or, when slabs have the same composition, by changes in atom ratios introduced at the slab boundaries. In addition, the planar boundaries that divide up a modular structure create new coordination polyhedra in the vicinity of the fault that are not present in the parent structure. These may provide sites for novel chemical reactions or introduce changes in the physical properties of significance compared to those of the parent structure.

A number of solids encompass composition variation by way of intergrowth of structures with different formulas, say A and B, which interleave to give a compound of formula A_aB_b. Examples abound, especially in minerals. A classic series is that formed by an ordered intergrowth of slabs of the mica and pyroxene structures. The micas have formulas typified by phlogopite, $KMg_3(OH)_2Si_3AlO_{10}$, where the K^+ ions lie between the silicate layers, the Mg^{2+} ions are in octahedral sites, and the Si^{4+} and Al^{3+} ions occupy tetrahedral sites within the silicate layers. The pyroxenes can be represented by enstatite, $MgSiO_3$, where the Mg^{2+} ions occupy octahedral sites and the Si^{4+} ions occupy tetrahedral sites within the silicate layers. Both these materials have layer structures that fit together parallel to the layer planes. If an idealized mica slab is represented by M, and an idealized pyroxene

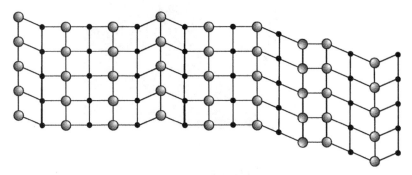

Figure 4.19 Modular crystal containing a random stacking of slabs, equivalent to a random distribution of planar defects.

slab by P, the polysomatic series can span the range from $\ldots MMM\ldots$, representing pure mica, to $\ldots PPP\ldots$, representing pure pyroxene. Many intermediates are known, for example, the sequence $\ldots MMPMMP\ldots$ is found in the mineral jimthompsonite, $(Mg,Fe)_{10}(OH)_4Si_{12}O_{32}$, and the sequence $\ldots MPMMPMPMMP\ldots$ corresponds to the mineral chesterite, $(Mg,Fe)_{17}(OH)_6Si_{20}O_{54}$.

Apart from structures that are built of slabs, modular structures that can be constructed of columns in a jigsawlike assembly are well known. In the complex chemistry of the cuprate superconductors and related inorganic oxides, series of structures that are described as tubular, stairlike, and so on have been characterized. Alloy structures that are built of columns of intersecting structures are also well known. Structures built of linked columns, tunnels, and intersecting slabs are also found in minerals. Only one of these more complex structure types will be described, the niobium oxide block structures, chosen as they played a significant role in the history of nonstoichiometry.

4.8.1 Crystallographic Shear Phases

A crystallographic shear (CS) plane is a fault in which a plane of atoms has been (notionally) removed from the crystal. In oxides, this is frequently a plane of oxygen atoms, eliminated as a result of reduction. In the resulting structures, the slab types are all identical and the same as the parent phase. To illustrate this phenomenon, crystallographic shear in reduced tungsten trioxide will be described.

Although tungsten trioxide has a unit cell that is monoclinic at room temperature, it is convenient, for the present purpose, to idealize it to cubic and view the structure in projection down one of the cubic axes. The structure can then be represented as an array of corner-linked squares (Fig. 4.20a). Very slight reduction to a composition of $WO_{2.9998}$, for example, results in a crystal containing a low concentration of faults that lie on $\{102\}$ planes. Structurally, these faults consist of lamellae made up of blocks of four edge-shared octahedra in a normal WO_3-like matrix of corner-sharing octahedra (Fig. 4.20b). The structure can be regarded as the coherent intergrowth of slabs of a WO_3-like structure and slabs of edge-shared WO_6 octahedra (Fig. 4.20c). (These latter units can be regarded as fragments of a cubic form of WO_2.)

The formation of the CS plane can also be considered as a collapse of the structure. The collapse or CS vector is an octahedron edge. This operation has the effect of eliminating a plane of oxygen atoms from the crystal. One oxygen atom is lost per block of four edge-shared octahedra created. This is most easily appreciated if the sequence of W and O atomic planes in WO_3 is drawn (Fig. 4.21). This figure shows that the sequence of atom planes parallel to (102) is $\ldots WO_2-O-WO_2-O-WO_2-O\ldots$ (Note that in the projection of this figure the metal oxygen planes have a composition WO, but there is an oxygen over each tungsten, making the true composition WO_2, as written.) The crystallographic shear operation can then be envisaged as the complete removal of an oxygen only plane followed by subsequent collapse, giving the sequence $\ldots WO_2-O-WO_2-WO_2-O-WO_2-O\ldots$ in traversing a CS plane (Fig. 4.21). This operation can be easily followed if a

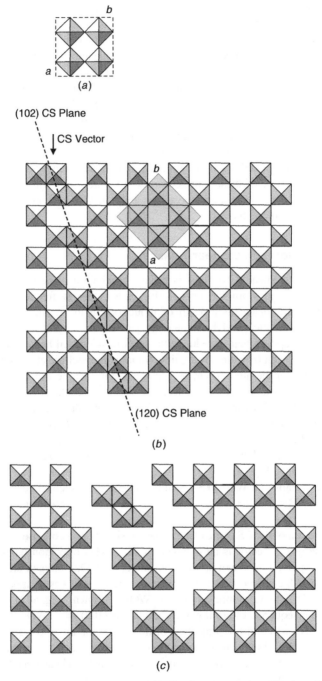

Figure 4.20 (*a*) Idealized cubic structure of WO_3 shown as corner-shared octahedra (shaded squares) projected down [001]; (*b*) an idealized (120) CS plane, with the cubic WO_3 unit cell shaded; and (*c*) the structure in (*b*) separated into three slabs.

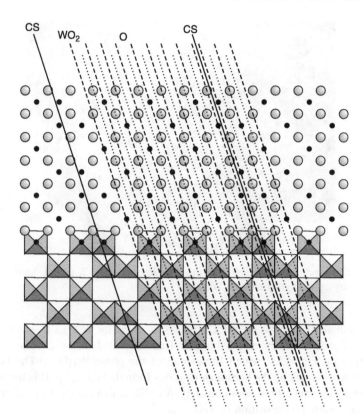

Figure 4.21 Idealized structure of WO_3 containing two (120) CS planes. The top part of the figure shows the oxygen and metal atom array while the lower part shows the structure as polyhedra.

drawing of the perfect structure is made and cut with a scissors along the CS plane, followed by translation of one part by the CS vector.

The mechanism of formation of CS planes is not yet completely unraveled. However, aggregation of oxygen vacancies onto a (102) plane followed by collapse of the structure by the appropriate CS vector will result in a disk-shaped CS plane, bounded by a dislocation loop (Section 3.7.1). The structural distortions present in the region of the bounding loop may act so as to favor further vacancy aggregation and hence expansion of the CS plane by dislocation climb. This mechanism would suggest that CS planes in a slightly reduced crystal should form on all of the possible $\{102\}$ planes. This is found to be so, and slightly reduced crystals contain CS planes lying upon many $\{102\}$ planes (Fig. 4.22).

Continued reduction causes an increase in the density of the CS planes and as the composition approaches $WO_{2.97}$, they tend to become ordered. In the case where the CS planes are perfectly ordered, defects are no longer present. Each CS plane is equivalent to the removal of a $\{102\}$ oxygen-only plane, and the composition of a

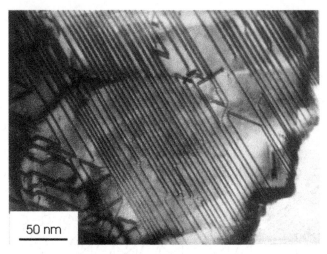

Figure 4.22   Electron micrograph showing disordered $\{102\}$ CS planes, imaged as dark lines, in a crystal of slightly reduced WO_3.

crystal containing ordered $\{120\}$ CS planes is given by W_nO_{3n-1}, where n is the number of octahedra separating the CS planes (counted in the direction between the arrows on Fig. 4.23, which shows part of the phase $W_{11}O_{32}$). The family of oxides is represented by the homologous series formula W_nO_{3n-1}. This homologous series spans the range from approximately $W_{30}O_{89}$, with a composition of $WO_{2.9666}$, to $W_{18}O_{53}$, with a composition of $WO_{2.9444}$.

When the composition falls to about $WO_{2.94}$, $\{102\}$ CS planes are replaced by $\{103\}$ CS planes. The structure of these CS planes consists of blocks of six edge-shared octahedra (Fig. 4.24). (The structure remains an intergrowth of "cubic WO_2" with WO_3, but now the interface between the two regions has changed, and, as a result, the strings of cubic WO_2 have lengthened.) In a $\{103\}$ CS plane two oxygen ions are lost per block of six edge-shared octahedra. A perfect ordering of $\{103\}$ CS planes gives rise to a homologous series of oxides with a general formula of W_nO_{3n-2}. The phases exist over a composition range of approximately $WO_{2.93}$ to $WO_{2.87}$, that is, from $W_{25}O_{73}$ to $W_{16}O_{46}$. As in the previous example, perfect order is rarely found in these oxides.

The change of CS plane appears to be a response to lattice strain and again illustrates the interaction between mechanical properties and defect structure. The cations in the blocks of edge-shared octahedra within the CS planes repel each other and so set up strain in the lattice. The total lattice strain depends upon the number of CS planes present. When the composition of a reduced crystal reaches about $WO_{2.92}$, the $\{102\}$ CS planes are very close, and the lattice strain becomes too great for further oxygen loss to take place. However, if the CS plane changes to $\{103\}$, the same degree of reduction is achieved with about half the number of CS planes and so the total lattice strain falls. This is because each block of six edge-shared octahedra loses two oxygen ions compared to one for a block of four edge-shared octahedra.

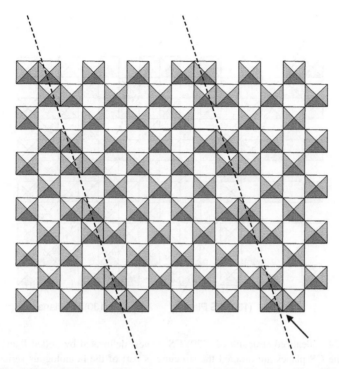

Figure 4.23 Idealized structure of $W_{11}O_{32}$, containing (120) CS planes. The oxide forms part of the homologous series W_nO_{3n-1}, where the value of n is the number of octahedra separating the ordered CS planes counted in the direction arrowed.

The decrease in lattice strain is so great that further reduction, now using {103} CS planes, can take place.

This suggests that under carefully controlled conditions, CS planes between (102) and (103) could form. This is difficult to achieve in the binary W–O system but is found when oxygen loss is achieved by reaction with other oxides, and a large number of CS phenomena are displayed that are not registered in the binary systems. If, for example, WO_3 is reacted with small amounts of Ta_2O_5, each pair of Ta ions that enter the WO_3 lattice require a reduction in the amount of oxygen ions present by one, which is achieved by the incorporation of CS planes. These are not only on {102} or {103}, as one would expect from pure WO_3, but CS planes lying on planes between these extremes also form. At some compositions between these two regions wavy CS planes are found. In the seemingly similar case in which WO_3 is reacted with small amounts of Nb_2O_5, the CS planes are not on {102} or {103} but form initially on {104} planes. These change to {001} CS planes as the Nb concentration increases. At compositions between these two regions wavy CS planes are found. Other impurities also produce CS planes that are not found in the binary tungsten–oxygen system and reveal that complex factors are involved in determining the planes upon which CS actually occurs.

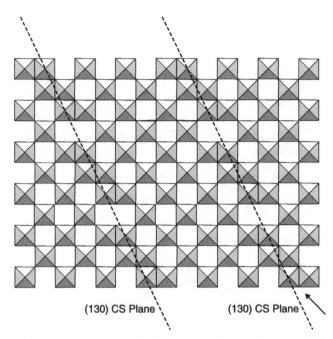

Figure 4.24   Idealized structure of (130) CS planes, delineated by dotted lines, in reduced WO_3. If the CS planes are ordered the structure is part of the homologous series W_nO_{3n-2}, where the value of n is the number of octahedra separating the ordered CS planes counted in the direction arrowed.

Some other transition-metal oxides also accommodate oxygen loss by use of crystallographic shear. These include molybdenum trioxide, MoO_3, which has two different CS systems, dependent upon composition, VO_2, and notably the rutile form of TiO_2, which, like WO_3, changes CS plane type as the degree of oxygen loss increases, from $\{132\}$ at lower degrees of reduction to $\{121\}$ at higher degrees. In all of these cases, the structures may be regarded with equal validity as a parent matrix containing slabs of a lower oxide, Ti_2O_3, for example, in the case of the rutile CS structures.

4.8.2   Twinning and Composition Variation

Twin planes are most frequently internal boundaries across which the crystal matrix is reflected (Section 3.11). Some twin planes do not change the composition of the crystal while at others atoms are lost and a composition change can result. If faults that alter the composition are introduced in considerable numbers, the crystal will take on the aspect of a modular material and show a variable composition.

This feature can be illustrated with respect to frequently observed twinning that occurs on $\{101\}$ planes in the rutile form of titanium dioxide. Large numbers of disordered $\{101\}$ twin planes occur in very slightly reduced rutile, suggesting that these defects form in order to accommodate slight oxygen loss.

The sequence of atom planes parallel to a $\{101\}$ plane in rutile is ...Ti–O–O–Ti–O–O–Ti–O–O–Ti.... When the twin plane is a sheet of Ti atoms, the sequence of atom planes parallel and across the twin plane is unchanged and the composition remains as TiO_2 (Fig. 4.25a). However, twinning across an oxygen plane results in a composition change. In one case (Fig. 4.25b) an oxygen plane is lost and the sequence across the twin plan is ... Ti–O–O–Ti–O–Ti–O–O–Ti .... In formal ionic terms this means that some Ti^{3+} ions need to be generated from Ti^{4+} to balance the oxygen loss. If the twin plane density is high, as in polysynthetic twinning, the composition of the crystal will become Ti_nO_{2n-1}, where n is the number of twin planes in the sample.

In the other case, the sequence becomes ...Ti–O–O–Ti–O–O–O–Ti–O–O–Ti..., and oxygen is gained (Fig. 4.25c). The boundary is equivalent to the insertion of a plane of oxygen atoms. In formal ionic terms this means that some Ti^{4+} ions need to be replaced by M^{5+} to balance the oxygen gain. This can happen if the crystal contains a low concentration of Nb^{5+} ions, for example. If the twin plane density is high, as in polysynthetic twinning, the composition of the crystal will become $(M, Ti)_nO_{2n+1}$, where n is the number of twin planes in the sample and M is an appropriate higher valence ion.

4.8.3 Chemical Twinning

The term chemical twinning (CT) is used to refer to repeated ordered twin planes that change the stoichiometry of the bulk significantly. New coordination polyhedra that do not occur in the parent structure are a feature of chemical twinning, and the occupation of these generates a number of new stuctures. Both aspects are illustrated by the $PbS–Bi_2S_3$ system.

Lead sulfide, PbS, often referred to by its mineral name, galena, has the sodium chloride structure (Fig. 4.26a). It reacts with Bi_2S_3 to form what, at a low level of discrimination, is a solid solution. From a crystal chemistry viewpoint, the problem is how to accommodate Bi_2S_3 into galena. Two chemical problems arise; the Bi^{3+}, a lone-pair ion, will not adopt octahedral coordination readily, and for each pair of Bi^{3+} ions added to the galena, three extra sulfur ions must be fitted in. The solution is given by the structures of the two phases, heyrovskyite, $Pb_{24}Bi_8S_{36}$, and lillianite, $Pb_{12}Bi_8S_{24}$ (Fig. 4.26b and 4.26c). It is seen that the structures of these minerals are made up of strips of PbS in a twinned relationship to each other. The twin planes are $\{113\}$ with respect to the galena structure. Viewed in the directions of the arrows in these figures, the atom sheets are close-packed sulfur arrays. The twinning allows extra sulfur ions to be incorporated into the structure and creates new sites in the twin plane that accommodate the Bi ions.[4]

---

[4]The actual distribution of the Pb and Bi atoms in these structures is still uncertain and not as clear-cut as suggested here.

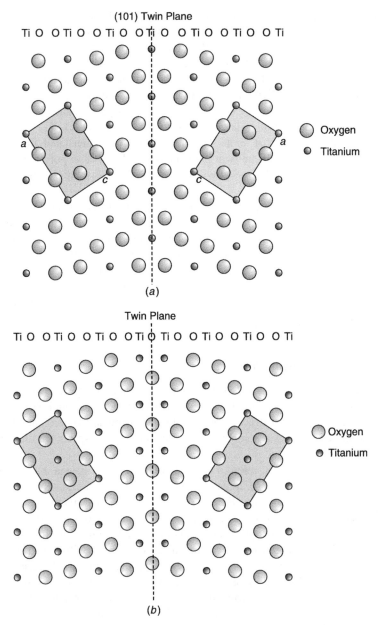

Figure 4.25 $\{101\}$ twinning in TiO_2 (rutile): (*a*) Ti twin plane, no composition change; (*b*) O twin plane, oxygen loss; and (*c*) O twin plane, oxygen gain.

The difference between lillianite and heyrovskyite is in the width of the galenalike slabs. Counting the number of close-packed sulfur planes in each slab gives a stacking sequence of (6, 6) for lillianite and (9, 9) for heyrovskyite. Synthetic and mineral samples contain other members of the series in which the PbS slab regions have other

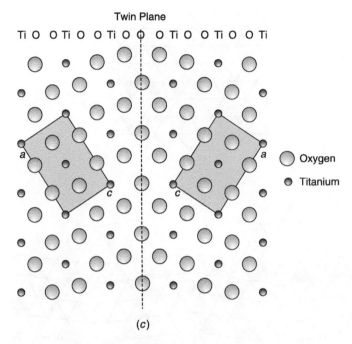

Twin Plane

Ti O   O Ti O   O Ti O ¢   O Ti O   O Ti O   O Ti

a   *a*   ◯ Oxygen

c   *c*   ● Titanium

(*c*)

Figure 4.25   (*Continued*).

thicknesses. In many minerals the slab sequence is made up of a regular repetition of two slab widths, such as (4, 3), or more complex arrangements, such as (4, 4, 3, 4, 4, 3). In addition, disordered phases abound.

4.8.4   Perovskite-Related Structures

The oxides that adopt the perovskite structure, ABO_3, where A is a large cation, typically Sr^{2+}, and B is a medium-sized cation, typically Ti^{4+}, are prone to accommodate composition change by forming modular structures. These are frequently disordered so that in practice the crystals contain large numbers of planar faults. In this section, four such structural series are described, based upon the $Ca_2Nb_2O_7$, Sr_2TiO_4 (Ruddleston–Popper), $KLaNb_2O_7$ (Dion–Jacobson), and $Bi_4Ti_3O_{12}$ (Aurivillius) structures. (Another important group, the high-temperature cuprate superconductors, is described in Section 8.6.)

The parent perovskite-type structure (Fig. 4.13*b*) is composed of corner-linked BO_6 octahedra surrounding large A cations and is conveniently idealized to cubic symmetry (Fig. 4.27*a*). (The real structures have lower symmetry than the idealized structures, mainly due to temperature-sensitive distortions of the BO_6 octahedra.) In the phases related to $Ca_2Nb_2O_7$ the parent structure is broken into slabs parallel to {110} planes. The formula of each slab is $A_nB_nO_{3n+2}$, where n is the number of

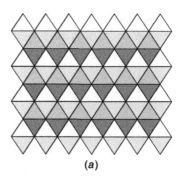

(a)

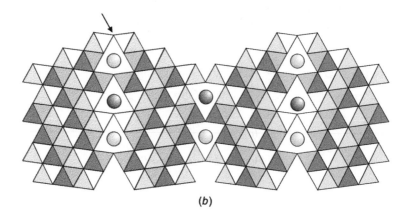

(b)

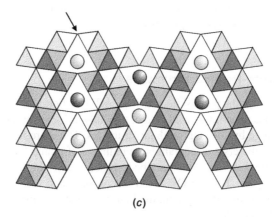

(c)

Figure 4.26 Chemical twinning in the lead bismuth sulfosalts: (a) idealized structure of galena, PbS, projected onto (110); (b) idealized structure of heyrovskyite; and (c) idealized structure of lillianite. Shaded diamonds represent MS_6 octahedra, those at a higher level shown lighter. Bi atoms are represented by shaded spheres, those at a higher level shown lighter. The twin planes are $\{113\}$ with respect to the galena cell, and the arrows indicate planes of close-packed S atoms.

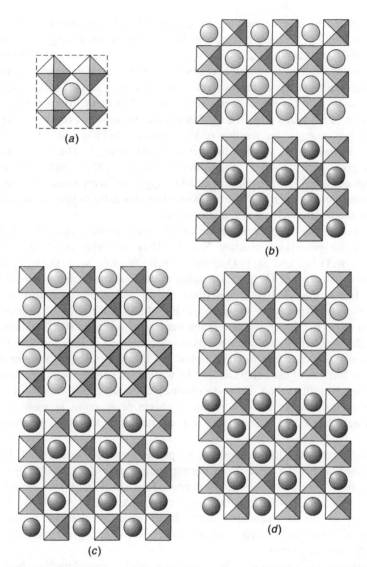

Figure 4.27   Idealized structures of the $A_nB_nO_{3n+2}$ phases: (*a*) ABO_3, (perovskite) unit cell; (*b*) $n = 4$, $Ca_4Nb_4O_{14}$; (*c*) $n = 5$, $Ca_5(Ti, Nb)_5O_{17}$; and (*d*) $n = 4.5$, $Ca_9(Ti, Nb)_9O_{29}$. The shaded squares represent $(Ti, Nb)O_6$ octahedra and the shaded circles represent Ca atoms.

complete BO_6 octahedra in the slice. The first materials of this type to be investigated are found in the system bounded by the end members $Ca_2Nb_2O_7$ and $NaNbO_3$. The structure of $Ca_2Nb_2O_7$ is composed of slabs four octahedra in thickness and for this reason is better written $Ca_4Nb_4O_{14}$, (Fig. 4.27*b*). The oxide $NaNbO_3$ has the perovskite structure. Compositions between $Ca_4Nb_4O_{14}$ and $NaNbO_3$ are composed of wider and wider slabs (Fig. 4.27*c* and 4.27*d*). Close to $Ca_4Nb_4O_{14}$ the slabs are often ordered to give a homologous series of phases described by the formula $(Na, Ca)_nNb_nO_{3n+2}$, where n represents the number of metal–oxygen octahedra in each slab. Thus, $Ca_4Nb_4O_{14}$ is the $n = 4$ member of the series and $NaNbO_3$ is the $n = \infty$ member. Apart from phases corresponding to a single value of n, innumerable repeating sequences involving two values (n_1, n_2), such as (4,4,4,5), (4,4,5), (4,5), (4,5,5), and so on, have been characterized, as well as many disordered sequences.

Similar $\{110\}$ faults arise in a number of other systems as a result of composition change. For example, the reaction of the perovskite $SrTiO_3$ with Nb_2O_5 results in crystals containing randomly distributed $\{110\}$ defects. The incorporation of Nb_2O_5 gives each crystal a composition $Sr(Nbi_xTi_{1-x})O_{3+x}$ and each Nb^{5+} ion that replaces a Ti^{4+} ion requires that extra oxygen is incorporated into the crystal. The planar defects open the structure and allow this extra oxygen to be accommodated without introducing oxygen interstitials.

The Ruddleston–Popper phases are another structurally simple series built from slabs of the perovskite structure $SrTiO_3$. They are represented by the oxides Sr_2TiO_4, $Sr_3Ti_2O_7$, and $Sr_4Ti_3O_{10}$, with general formula $Sr_{n+1}Ti_nO_{3n+1}$. In these materials, slabs of $SrTiO_3$ are cut parallel to the idealized cubic perovskite $\{100\}$ planes (rather than $\{110\}$ as described above) and stacked together, each slab being slightly displaced in the process (Fig. 4.28). Each slice of structure has a formula $(A_{n+1}B_nO_{3n+1})$, where n is the number of complete BO_6 octahedra in the slice. The structure is also described as being composed of intergrowths of varying thicknesses of perovskite $SrTiO_3$, linked by identical slabs of sodium chloride SrO. The formula of each slab can then be written $[AO (A_nB_nO_{3n})]$. For comparison with the Aurivillius and Dion–Jacobson phases described below, it is more convenient to separate the two bounding sheets of A atoms in each perovskite-like sheet, labeled A^*, from the rest of the slab to write the formula as $[A_2^*(A_{n-1}B_nO_{3n+1})]$.

Many examples of Ruddleston–Popper phases have been synthesized. The structure of the first member of the series, corresponding to $n = 1$, is adopted by a number of compounds, including the important phase La_2CuO_4 (Section 4.3.3) and is often referred to as the K_2NiF_4 structure. In practice, synthesis of $A_{n+1}B_nO_{3n+1}$ phases frequently results in disordered materials in which random or partly ordered regions of $\{100\}$ faults occur, and particular efforts have to be made to produce perfectly ordered crystals.

If the pair of A^* atoms at the boundaries of the perovskite-like sheets in the Ruddleston–Popper phases are replaced with just one A^* atom, the series of phases takes the formula $A^*(A_{n-1}B_nO_{3n+1})$, where A and A^* are large ions, typically a $(+1/+3)$ pairing, and B is a medium-sized transition-metal ion, typically Nb^{5+}. These materials are called Dion–Jacobson phases. The majority of examples synthesized to date are of the $n = 2$ phase, typified by $KLaNb_2O_7$, $CsBiNb_2O_7$, and so on (Fig. 4.29). A few examples of the $n = 3$ phase are also known, including $CsCa_2Nb_3O_{10}$.

If the A^* layers in the above series are replaced by a layer of composition Bi_2O_2, a series of phases is formed called Aurivillius phases, with a general formula $(Bi_2O_2)(A_{n-1}B_nO_{3n+1})$, where A is a large cation, B a medium-sized cation, and the index n runs from 1 to ∞. The structure of the Bi_2O_2 layer is similar to that of

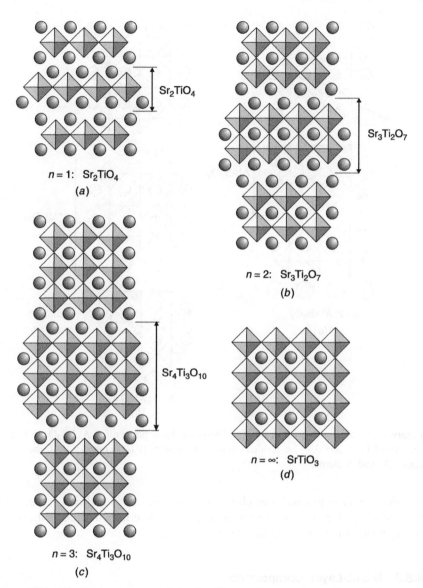

Figure 4.28 Idealized structures of the Ruddleston–Popper phases $Sr_{n+1}Ti_nO_{3n+1}$: (a) Sr_2TiO_4, $n = 1$; (b) $Sr_3Ti_2O_7$, $n = 2$; (c) $Sr_4Ti_3O_{10}$, $n = 3$; and (d) $SrTiO_3$, $n = \infty$. The shaded squares represent TiO_6 octahedra and the shaded circles Sr atoms.

fluorite, so that the series is also represented by an intergrowth of fluorite and perovskite structure elements. The $n = 1$ member of the series is represented by Bi_2WO_6, the $n = 2$ member by $Bi_2SrTa_2O_9$, but the best-known member of this series of phases is the ferroelectric $Bi_4Ti_3O_{12}$, in which $n = 3$ and A is Bi (Fig. 4.30a). Ordered intergrowth of slabs of more than one thickness, especially n and $(n + 1)$,

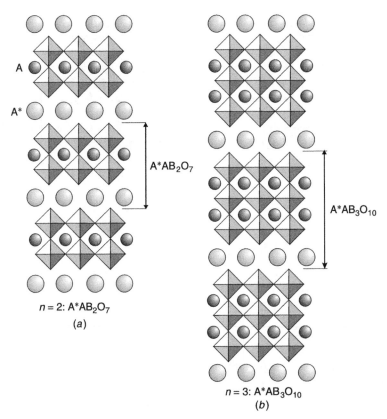

Figure 4.29 Idealized structures of the Dion–Jacobson phases: (*a*) $A^*AB_2O_7$, $n = 2$ and (*b*) $A^*AB_3O_{10}$, $n = 3$. The shaded squares represent BO_6 octahedra and the shaded circles A^* and A atoms.

are often found in this series of phases, for example, $n = 1$ and $n = 2$, typified by $Bi_5TiNbWO_{15}$ (Fig. 4.30*b*). As with the other series described, disorder in the relative arrangement of the slabs is commonplace.

4.8.5 Misfit-Layer Compounds

This group of interesting materials includes metal sulfides related to "$LaCrS_3$," which were originally assigned a simple MTS_3 formula, in which M is typically Sn, Pb, Sb, Bi, or a lanthanide and T is typically Ti, V, or Cr. Other related sulfide phases have been prepared using Nb and Ta, and oxide phases, especially using CoO_2 are also well known. Although commonly called misfit-layer compounds, they are also known as composite crystals, composite structures, or incommensurate intergrowth structures.

The structures of all these phases are built up from TX_2 layers consisting of sheets of octahedra, similar to those in the CdI_2 structure of TiS_2 or sheets of trigonal prisms similar to those found in the β-MoS_2 structure of NbS_2. These are separated by

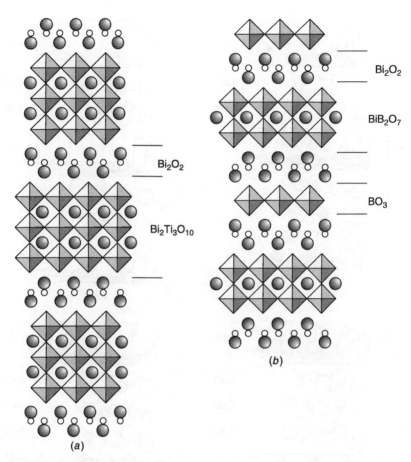

Figure 4.30 Idealized structures of the Aurivillius phases viewed along [110]: (*a*) $Bi_4Ti_3O_{12}$, $n = 3$; and (*b*) $Bi_5TiNbWO_{15}$, ordered $n = 1$ $n = 2$. The shaded squares represent TiO_6 or (Ti, Nb, W)O_6 octahedra and the shaded circles Bi atoms. B represents Ti, Nb, W.

inserted layers that have the sodium chloride structure. The correct general formula of these phases is, therefore, $(MX)_nTX_2$, where n has a nonintegral value, exemplified by the correct formula for LaCrS$_3$, which is $(LaS)_{1.20}CrS_2$.

The most important structural feature of these phases is that the two layer types do not fit with each other in one direction. A typical example is given by $(LaS)_{1.20}CrS_2$. The a-lattice parameters of the sodium chloride structure (LaS) and the CdI$_2$ structure (CrS$_2$) meshes are virtually the same, but the b parameters are quite different, with the ratio b (LaS)$/b$ (CrS$_2$) $\approx \frac{5}{3}$, so that the two only come into register after hundreds of repetitions of these units. The material, therefore, adopts a structure that is *incommensurate* and accounts for the family name of misfit-layer compounds. The structure consists of thin slices of sodium chloride structure LaS two atom layers thick, lying between single octahedral CrS$_2$ layers (Fig. 4.31a). The closely related structure

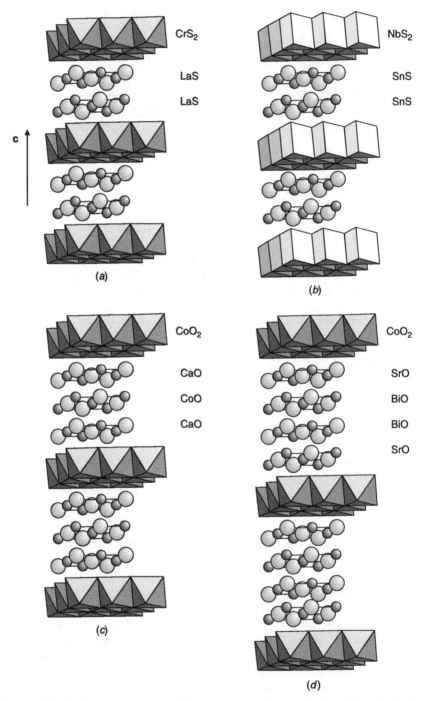

Figure 4.31 Idealized structures of misfit-layer compounds: (a) $(LaS)_{1.20}CrS_2$; (b) $(SnS)_{1.17}$ NbS_2; (c) $(Ca_2CoO_3)_{0.62}(CoO_2)$; and (d) $(Bi_{0.87}SrO_2)_2(CoO_2)_{1.82}$. The shaded polyhedra represent sheets of MX_6 octahedra (a, c, d) or trigonal prisms (b).

of $(SnS)_{1.17}NbS_2$ has the sodium chloride structure SnS layers lying between sheets of NbS_2 trigonal prisms (Figure 4.31b).

Clearly, composition variation can take place in these compounds by changing the number of layers involved in these structures. For example, $(PbS)_{1.14}(NbS_2)_2$ has double layers of NbS_2 trigonal prisms separating the sodium chloride structure layers. The material $(CaOH)_{1.14}(CoO_2)$ is similar to $(LaS)_{1.20}CrS_2$ in structure, while $(Ca_2CoO_3)_{0.62}(CoO_2)$ has three sodium chloride structure sheets separating the CoO_2 octahedral layers (Fig. 4.31c), and $(Bi_{0.87}SrO_2)_2(CoO_2)_{1.82}$ has four sodium chloride structure sheets separating the CoO_2 octahedral layers (Fig. 4.31d). A large number of defects can be envisaged for all these materials, including unfilled and partly filled layers as well as variable layer thickness.

4.8.6 Niobium Pentoxide Block Structures

A remarkable group of oxides that have compositions close to niobium pentoxide, Nb_2O_5, form when the oxide is reduced, either directly or by reaction with lower valence oxides such as TiO_2, or oxidized by reaction with higher valence oxides such as WO_3. The oxides lie between the approximate composition limits of $(M, Nb)O_{2.40}$ and $(M, Nb)O_{2.7}$. All the structures are related to the form of niobium pentoxide stable at high temperatures, designated H-Nb_2O_5 (Fig. 4.32a). It is composed of columns of material with a WO_3-like structure formed by two intersecting sets of CS planes. These break the matrix up into columns, which, in projection, look like rectangular blocks, and hence a common name for these materials is block structures. In H-Nb_2O_5 the blocks have two sizes, (3×4) octahedra and (3×5) octahedra, neatly fitted together to form the structure.

The CS planes are associated with oxygen loss and serve to reduce the composition of the parent corner-shared octahedral structure from NbO_3 to Nb_2O_5, while retaining octahedral coordination for the Nb^{5+} cations. The centers of the blocks have a composition NbO_3 and the block boundaries have a composition NbO_2.

Oxidation and reduction of H-Nb_2O_5 result in innumerable new block structures. Reduction leads to smaller overall blocks as in the binary oxides $Nb_{22}O_{54}$, $Nb_{47}O_{116}$, $Nb_{25}O_{62}$, $Nb_{39}O_{97}$, $Nb_{53}O_{132}$, and $Nb_{12}O_{29}$, and the ternary oxide $Ti_2Nb_{10}O_{29}$, which has blocks (3×4) octahedra in size (Fig. 4.32b). Oxidation leads to increased block sizes, typified by $W_3Nb_{14}O_{44}$, made up of (4×4) blocks (Fig. 4.32c).

While ordered structures such as those illustrated are defect free, defects commonly occur in these phases, especially during formation and reaction. These can take the form of variation in the filling of the tetrahedral sites that exist between block edges or in variable block sizes, these latter often called Wadsley defects (Fig. 4.33). Although it might be thought that reactions between these phases, or the interconversion of one block size to another would be difficult, the reverse is true and the reactions of the block structures take place very rapidly. For example, reaction between Nb_2O_5 and WO_3 produces perfectly ordered phases within 15 minutes at 1400 K and a variety of more disordered nonstoichiometric phases in shorter times.

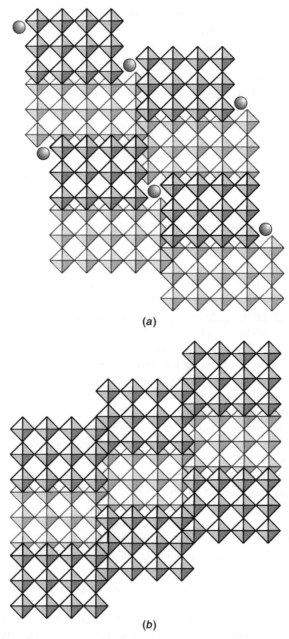

(a)

(b)

Figure 4.32 Niobium pentoxide block structures: (a) H-Nb_2O_5, (b) $Ti_2Nb_{10}O_{29}$ monoclinic, and (c) $W_3Nb_{14}O_{44}$. The shaded squares represent MO_6 octahedra at two different levels and the shaded spheres represent cations in tetrahedral coordination.

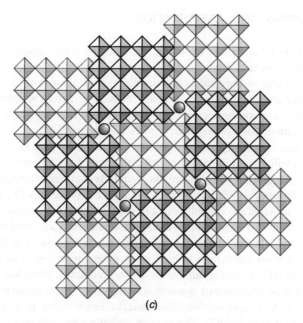

(c)

Figure 4.32 (*Continued*).

Figure 4.33 Electron micrograph showing disorder in a niobium pentoxide block structure. The phase was synthesized from Nb_2O_5 and CoF_3 (which is analogous to WO_3). The region at the lower center of the micrograph consists of blocks of 4×4 octahedra with the $W_3Nb_{14}O_{44}$ structure with a block size of approximately 1.5 nm. (Courtesy of Dr. J. Hutchison, University of Oxford.)

4.9 ORDERING AND ASSIMILATION

If defects in a crystal are perfectly ordered, the situation is best described in terms of a new defect-free structure. From a crystallographic viewpoint a longer unit cell, which is a multiple of that of the "defect-containing" parent phase, will describe the new material. There are many systems in which disordered defect populations occur in some circumstances and ordered phases in others. For example, ordered CS planes give rise to homologous series of new phases such as W_nO_{3n-2}. Each phase will be characterized by the separation between the ordered planar boundaries, n, and each structure will have a fixed composition that will differ in composition from its neighbors by a small but definite amount. Similarly, structures containing perfectly ordered PC elements are well known and include the oxide Mo_5O_{14} and the oxide $Nb_{16}W_{18}O_{64}$, which is an ordered variant of the tetragonal tungsten bronze structure.

Point defects are far more mobile than extended defects and tend to order at low temperatures and disorder at high temperatures. The two phases may then be similar chemically but show important differences in other properties. This aspect can be illustrated with reference to a number of oxygen-deficient perovskite-related structures. One of most investigated systems showing this type of behavior is the phase $SrFeO_{2.5+\delta}$. At high temperatures this material has a cubic perovskite structure, ideally of composition $SrFeO_3$. The oxygen deficit varies with surrounding oxygen partial pressure (Chapter 7) and is due to disordered oxygen vacancies. At temperatures below 900°C these oxygen vacancies order. The best known ordered structure is that of $Sr_2Fe_2O_5$, an example of the brownmillerite type. In this compound half of the Fe cations are in octahedral sites and half in tetrahedral sites. The tetrahedral sites are created by ordering the "oxygen vacancies" in rows (Fig. 4.34a) followed by a slight displacement of the cations nearest to the vacancies so as to move them into the center of the tetrahedra so formed. The structure is now made up of sheets of perovskite structure interleaved with slabs of corner-linked tetrahedra and can be considered to be a modular structure (Fig. 4.34b). The defects have been assimilated into the structure and no longer exist.

The brownmillerite structure type can be described as a sequence ... OTOT ... where O stands for the octahedral sheet and T the tetrahedral sheet. A number of other phases closely related to this structure have been characterized, including $Ca_2LaFe_3O_8$, with a stacking sequence ... OOTOOT ... and $Ca_4Ti_2Fe_2O_{11}$ and $La_4Mn_4O_{11}$, with a stacking sequence ... OOOTOOOT .... These phases are the $n = 3$ and $n = 4$ members of the series $A_nB_nO_{3n-1}$, of which the brownmillerite structure type is the $n = 2$ member.

The oxide $Ba_2In_2O_5$ is another well-studied phase that adopts the brownmillerite structure. This material disorders above 930°C to a perovskite-type structure containing oxygen vacancies. Both the Sr–Fe and Ba–In oxides are of interest for electrochemical applications in fuel cells and similar devices (Section 6.10).

The oxygen composition of doped brownmillerite structure materials is not always exactly 5.0 and depends upon a number of factors, especially the surrounding oxygen partial pressure. Thus, the oxide formed when Fe is replaced by Co has a broad

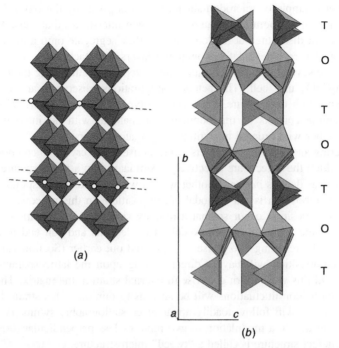

Figure 4.34 Brownmillerite structure: (*a*) ideal perovskite structure (circles indicate oxygen atoms that need to be removed to convert octahedra to tetrahedra in brownmillerite) and (*b*) idealized brownmillerite structure consisting of sheets of octahedra and tetrahedra.

oxygen stoichiometry range $SrCo_{0.8}Fe_{0.2}O_{2.582}-SrCo_{0.8}Fe_{0.2}O_{2.332}$, depending upon the surrounding oxygen partial pressure.

In general, an ordered brownmillerite phase transforms to the disordered state via a two-phase region in which the ordered and disordered forms coexist. However, the metal atoms are unchanged on passing from the brownmillerite structure to the disordered perovskite structure, and the transformation can be viewed as the creation of defect clusters that gradually expand to form a recognizable unit of a new structure coherently embedded in the surrounding matrix. These small units are referred to as microdomains. The defect solid then consists of microdomains randomly distributed throughout a defect-free matrix.

There is, naturally, no real distinction between a large defect cluster and a small microdomain. However, microdomains are generally regarded as larger, 10–20 nm in diameter, and showing a sufficiently different structure to the surrounding matrix so as to be readily discerned. Thus, for example, when CeO_2 is reacted with an oxide such as Y_2O_3 to form the series $(1-x)CeO_2-xY_2O_3$, the mixtures consist of a defect-free fluorite structure matrix and a defective fluorite phase that contains a random array of microdomains consisting of regions in which oxygen vacancies have ordered.

Another example of microdomain formation is given by the oxide $La_2CuO_{4+\delta}$ (Section 4.3.3). At room temperature the oxygen interstitials that cause the oxygen excess are not distributed at random. Instead they segregate onto a limited number of the LaO layers. These defect-containing layers can stack up in several different ways to give several different crystallographic repeat lengths along the **c** axis of the basic La_2CuO_4 unit cell. The actual configurations observed vary with oxygen content and with temperature. On cooling a typical preparation the microstructure of the sample consists of microdomains of material with no interstitial oxygen atoms together with ordered interstitial oxygen atom layers.

Clustering and ordering of defects is a response of the crystal to composition variation in which the defects are effectively assimilated into a new structure and so no longer exist as defects *per se*. Another way in which solids can respond to an inappropriate composition is via spinodal decomposition. In this process, small fluctuations in composition occur so that a roughly sinusoidal composition variation is set up. As time proceeds the sinusoidal curve deepens and can end in a series of lamellae of the resulting two phases. As pointed out earlier (Section 1.5), the variation in composition will have a direct bearing upon the lattice parameter of the material, and this will in turn give rise to internal strain in the matrix. The direction of the composition fluctuations will be such as to minimize this strain. In a crystal these directions will follow readily apparent crystallographic symmetry axes, and, when this results in a modulation in two more or less perpendicular directions, the resultant defect structure is called a "tweed" microstructure.

These and similar transformations are best documented in mineralogical studies (see Further Reading section at the end of this chapter).

4.10 MODULATED STRUCTURES

4.10.1 Structure Modulations

The continuous sinusoidal composition change that occurs during spinodal decomposition can be considered to be a modulation of the solid structure. It is now known that many structures employ modulation in response to compositional or crystallographic variations, and in such cases the material flexibly accommodates changes without recourse to defect populations. (Other modulations, in, for example, magnetic moments or electron spins, although important, will not be discussed here.)

In general, these defect-free modulated structures can, to a first approximation, be divided into two parts. One part is a conventional structure that behaves like a normal crystal, but a second part exists that is modulated[5] in one, two, or three dimensions. The fixed part of the structure might be, for example, the metal atoms, while the anions might be modulated in some fashion. The primary modulation might be in the position of the atoms, called a displacive modulation (Fig. 4.35a). Displacive modulations sometimes occur when a crystal structure is transforming from one

---

[5]Take care to note that a *modulated* structure is *not* the same as a *modular* structure.

stable structure to another as a result of a change in temperature. Alternatively, the modulation might be in the occupancy of a site, called compositional modulation, such as the gradual replacement of O by F in a compound $M(O, F)_2$ (Fig. 4.35b). In such a case the site occupancy factor would vary in a regular way throughout the crystal. The existence of compositionally modulated structures was uncovered little by little so that a number of different names, such as "vernier structures," "chimney-ladder structures," and "infinitely adaptive compounds" can be found in the literature.

Modulations are normally described as waves. The modulation wave can fit exactly with the underlying unmodulated component, or more precisely with the unit cell of the underlying component, in which case the structure is described as a commensurately modulated structure. In cases where the dimensions of the modulation are incommensurate (i.e., do not fit) with the unit cell of the underlying structure, the phase is an incommensurately modulated phase. Modulation changes are normally continuous and reversible.

These descriptions represent somewhat idealized situations. In many real examples the modulated component of the structure can force the notionally fixed part of the structure to become modulated in turn, leading to considerable crystallographic complexity.

4.10.2 Yttrium Oxyfluoride and Related Structures

Some of the earliest examples of modulated structures to be unraveled were the fluorite-related vernier structures. These structures occur in a number of anion-excess fluorite-related phases and use a modulation to accommodate composition variation. They can be illustrated by the orthorhombic phases formed when the oxyfluoride YOF reacts with small amounts of YF_3 to give composition YO_xF_{3-2x}, with x in the range $0.78-0.87$, but similar phases occur in the $Zr(N, O, F)_x$ system with x taking values of $2.12-2.25$ and other systems in which the Zr is replaced by a variety of lanthanides.

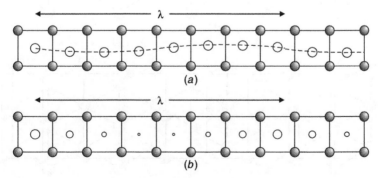

Figure 4.35 Modulated crystal structures: (a) crystal showing a displacive modulation of one set of atoms and (b) a crystal showing a compositional modulation of one set of atoms. (The change in the average chemical nature of the atom is represented by differing circle diameters.)

The YOF phases form in the composition range between $YX_{2.130}$ and $YX_{2.220}$. where X represents the anions (O, F). The complexity of this system was originally described in terms of a large number of different ordered phases forming a homologous series of compounds with a general formula $Y_nO_{n-1}F_{n+2}$. However, it is now recognized that the whole composition range is best described as a modulated structure. The idealized structures of these phases can be thought of as made up of three parts, a fluorite structure cubic cation net and an anion substructure that consist of idealized square anion nets interleaved with hexagonal anion nets (Fig. 4.36a and 4.36b). The two anion nets are in a vernier relationship to one another. This means that a whole number, N, of squares in the square net will fit exactly with the whole number of triangles M in the hexagonal net. The idealized anion nets in the compound $Y_7O_6F_9$, has a relatively small repeat in which seven squares fit with eight triangles (Fig. 4.36c). In the real compounds the three nets are mutually modulated and the hexagonal net can be thought of as expanding or contracting with respect to the square net, so as to allow for composition variation.

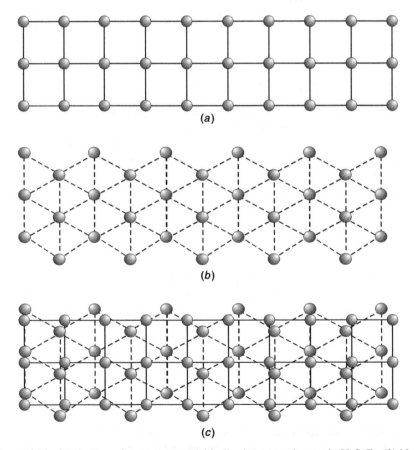

Figure 4.36   Y–O–F vernier structures: (a) idealized square anion net in $Y_7O_6F_9$; (b) idealized hexagonal anion net in $Y_7O_6F_9$; and (c) the two nets superimposed to show that seven square cells fit with eight triangular cells, which represents the idealized anion nets in $Y_7O_6F_9$.

4.10.3 U_3O_8-Related Structures

As well as giving rise to fluorite-related modulated structures described above, ZrO_2 can react with UF_4 to give another series of modulated structures $ZrO_{2-x}F_{2x}$ with x in the range $0.698-0.714$ derived from the α-U_3O_8 structure. However, these structures are best represented by the oxide L-Ta_2O_5 and the series of nonstoichiometric oxides that form when L-Ta_2O_5 is reacted with oxides including Al_2O_3, TiO_2, WO_3, and ZrO_2. The structures of all these phases consist of a sheet of metal and oxygen atoms with a composition of MO_2, in which the M atoms are in a hexagonal array, linked by oxygen atoms lying directly above and below each metal atom to give oxygen only planes of composition $\frac{1}{2}O_2$. The positions of the atoms in the MO_2 planes are described by an incommensurate modulation wave that is a function of both temperature and composition. The modulation is such that the coordination polyhedra around the metal atoms vary from distorted octahedral to almost perfect pentagonal bipyramids.

The chemical source of the modulation lies in the compromise between the composition (i.e., the number of oxygen atoms present) and the coordination preferences of the metal atoms to produce a stable structure. For example, smaller cations such as Al^{3+}, Ti^{4+}, and W^{6+} prefer octahedral coordination, but Ta^{5+} prefers pentagonal bipyramidal coordination. Similarly, reaction with Al_2O_3, TiO_2, and ZrO_2 reduces the oxygen to metal content while reaction with WO_3 increases it. Thus, if L-Ta_2O_5 is reacted with WO_3, the modulation wave changes from that in the pure oxide so as to (i) accommodate more oxygen and (ii) accommodate more metal atoms (mainly W) in octahedral coordination. If ZrO_2 is the dopant, the wave must be such as to take into account less oxygen in the structure and more atoms in pentagonal bypyramidal coordination. However, coordination preferences also vary with temperature, and the modulation changes with temperature to generally increase the relative amount of octahedral coordination polyhedra available. A single composition can thus adopt an infinite number of structures as the temperature varies.

The situation can be illustrated with reference to the structure of $Ta_{74}W_6O_{203}$, $(Ta, W)O_{2.4756}$ (Fig. 4.37a). The departure of the metal substructure in the MO_2 planes from ideal hexagonal is clear (Fig. 4.37b). The combined modulation of the metal and oxygen atoms is such that the positions of the oxygen atoms represent a compromise between providing pentagonal bipyramidal coordination demanded by the size of the large Ta atoms and the octahedral coordination preferred by W and also needed to generate the correct Ta_2O_5 stoichiometry. A percentage of the Ta atoms can reasonably be allocated to regular pentagonal bipyramidal coordination and others to distorted octahedral coordination. Some, however, are polyhedra that are a compromise between these two alternatives.

The structure can be idealized if some license is taken with the classification of which polyhedra are pentagonal bipyramids. In this case, in projection, the structure is seen to be formed of undulating chains of pentagons (Fig. 4.37c). The various structures found in these phases can then be described in terms of undulating chains of varying lengths. Note that there is a disagreement between the experimental and structural composition of these phases. It appears that the oxygen-only layers contain random oxygen vacancies. This aspect of the structures requires further investigation.

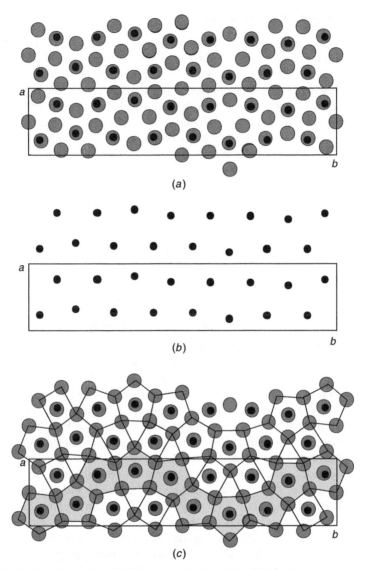

Figure 4.37   Structure of $Ta_{76}W_6O_{203}$: (*a*) projection down $\{001\}$; (*b*) metal only positions; and (*c*) undulating chains of pentagonal bipyramids, shaded. The metal atoms are small circles and the oxygen atoms are large circles. [Data from S. Schmid, J. G. Thompson, A. D. Rae, B. D. Butler, and R. L. Withers, *Acta Crystallogr*, B**51**, 698–708 (1995).]

4.10.4 Sr_xTiS_3 and Related Structures

The strontium titanium sulfides, Sr_xTiS_3, which exist between the x values of 1.05 and 1.22, contain interpenetrating rods of structure, both of which are modulated. In *idealized* Sr_xTiS_3, face-sharing TiS_6 octahedra (Fig. 4.38*a*) form columns with a repeat

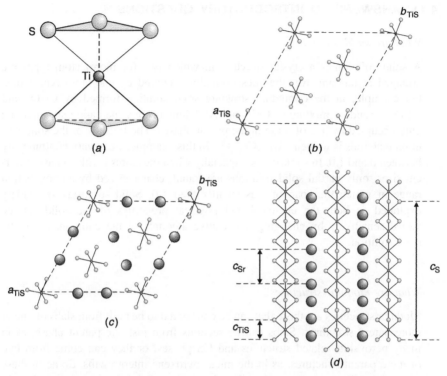

Figure 4.38 Idealized structure of Sr_xTiS_3 modulated phases: (*a*) an idealized TiS_6 octahedron (the shaded faces are shared to form columns); (*b*) idealized hexagonal unit cell formed by columns of face-shared TiS_6 octahedra; (*c*) idealized unit cell formed by columns of face-sharing octahedra and Sr atoms (shaded); and (*d*) idealized structure of $Sr_8(TiS_3)_7$.

composition of $TiS_{6/2}$ or TiS_3, arranged to give a hexagonal unit cell (Fig. 4.38*b*). Chains of Sr atoms lie between these columns to complete the idealized structure (Fig. 4.38*c*). The *a* parameters of the TiS_3 and Sr arrays are equal. The Sr chains are flexible and expand or contract along the **c** axis, as a smooth function of the composition *x* in Sr_xTiS_3. One of the simpler commensurately modulated structures reported is $Sr_8(TiS_3)_7$, the idealized structure of which is drawn in Figure 4.38*d*.

The real structures of these phases are more complex. The coordination of the Ti atoms is always six, but the coordination polyhedron of sulfur atoms around the metal atoms is in turn modulated by the modulations of the Sr chains. The result of this is that some of the TiS_6 polyhedra vary between octahedra and a form some way between an octahedron and a trigonal prism. The vast majority of compositions give incommensurately modulated structures with enormous unit cells. As in the case of the other modulated phases, and the many more not mentioned, composition variation is accommodated without recourse to "defects."

4.11   ANSWERS TO INTRODUCTORY QUESTIONS

What are solid solutions?

A solid solution is a crystal structure in which two (or more) atom types are arranged at random over the sites normally occupied by one atom type alone. For example, in the corundum structure solid solution formed by Cr_2O_3 and Al_2O_3, a random mixture of Cr^{3+} and Al^{3+} ions occupy the cation sites that are only occupied by one of these in the parent phases. The formula of the solid solution materials is written $(Al_{1-x}Cr_x)_2O_3$. In this example, x can vary continuously between 0 and 1.0. In some cases, especially when the atoms involved have different sizes, only partial solid solutions are found, characterized by a composition range in which the span of x is smaller than 1.0. Solid solutions are widely exploited as both the chemical and physical properties of the solid can be varied sensitively by changing the relative amounts of the components of the solid solution.

What are modular structures?

Modular structures are those that can be considered to be built from slabs of one or more parent structures. Slabs can be sections from just one parent phase, as in many perovskite-related structures and CS phases, or they can come from two or more parent structures, as in the mica–pyroxene intergrowths. Some of these crystals possess enormous unit cells, of some hundreds of nanometers in length. In many materials the slab thicknesses may vary widely, in which case the slab boundaries will not fall on a regular lattice and form planar defects.

What are incommensurately modulated structures?

Ideally, incommensurately modulated structures have two fairly distinct parts. One part of the crystal structure is conventional and behaves like a normal crystal. An additional, more or less independent part, exists that is modulated in one, two, or three dimensions. For example, the fixed part of the structure might be the metal atom array, while the modulated part might be the anion array. The modulation might be in the position of the atoms, called a displacive modulation or the occupancy of a site, for example, the gradual replacement of O by F in a compound $M(O, F)_2$, to give a compositional modulation. In some more complex crystals modulation in one part of the structure induces a corresponding modulation in the "fixed" part.

In cases where the wavelength of the modulation fits exactly with the dimensions of the underlying structure, a commensurately modulated crystallographic phase forms. In cases where the dimensions of the modulation are incommensurate (i.e., do not fit) with the underlying (generally small) unit cell of the parent structure, the phase is an incommensurately modulated phase.

PROBLEMS AND EXERCISES

Quick Quiz

1. A nonstoichiometric compound has:
   (a) An indeterminate structure
   (b) A composition range
   (c) A range of components

2. Two spinel structure oxides in the ratio $0.25NiAl_2O_4 : 0.75MgFe_2O_4$ are heated. The formula of the resulting solid solution is:
   (a) $Ni_{0.75}Mg_{0.25}Al_{0.5}Fe_{1.5}O_4$
   (b) $Ni_{0.75}Mg_{0.25}Al_{0.25}Fe_{1.75}O_4$
   (c) $Ni_{0.25}Mg_{0.75}Al_{0.5}Fe_{1.5}O_4$

3. A nonstoichiometric ionic sulfide has a formula $M_{1-x}S$. The material is likely to contain:
   (a) Excess electrons
   (b) Holes
   (c) Neither

4. $0.95NiO + 0.05Fe_2O_3$ react to form a solid solution with the sodium chloride (NaCl) structure. The additional material is all accommodated as interstitials. The formula is:
   (a) $(Ni, Fe)_{1.05}O_{1.10}$
   (b) $(Ni, Fe)_{0.955}O$
   (c) $(Ni, Fe)O_{1.048}$

5. Anion-excess fluorite structure nonstoichiometric phases prepared by heating CaF_2 and LaF_3 contain:
   (a) F^- interstitials
   (b) Ca^{2+} interstitials
   (c) La^{3+} interstitials

6. Compounds with the $A_2B_2O_5$ brownmillerite structure disorder at high temperatures to form a structure of the:
   (a) Fluorite type
   (b) Sodium chloride type
   (c) Perovskite type

7. A family of phases with compositions represented by a general formula such as W_nO_{3n-2} is called a:
   (a) Homogeneous series
   (b) Heterogeneous series
   (c) Homologous series

8. A solid is prepared by heating a $1:1$ mixture of $SrTiO_3$ and $Sr_4Nb_4O_{14}$ to give a member of the $Sr_n(Nb, Ti)_nO_{3n+2}$ series of phases. The value of n of the new material is:

   (a) 2.5
   (b) 5
   (c) 10

9. Ruddleston–Popper phases and Aurivillius phases both contain slabs of structure similar to that of:

   (a) Perovskite
   (b) CdI_2
   (c) Spinel

10. If the positions of a set of atoms in a structure follows a wavelike pattern, the modulation is described as:

   (a) Compositional
   (b) Displacive
   (c) Incommensurate

Problems and Exercises

1. The nonstoichiometric pyrochlore structure phase has a composition range from $Lu_2Ti_2O_7$ to $Lu_2Ti_{1.17}O_{5.35}$ at 1400°C (Fig. 4.2). What point defects might account for this?

2. The lattice parameters of several members of the solid solution series between $CaTiO_3$ and $La_{2/3}TiO_3$ are given in the following table. Assuming that the structure remains unchanged across the whole of the composition range: (a) determine the molar ratio of parent phases $CaTiO_3$ and $La_{2/3}TiO_3$ needed to make the composition $Ca_{0.4}La_{0.4}TiO_3$; (b) estimate the composition of the phase with lattice parameters $a \approx 0.545$ nm, $b \approx 0.545$ nm, $c \approx 0.771$ nm; and (c) estimate the density of $La_{2/3}TiO_3$.

Data for the Solid Solution $CaTiO_3$–$La_{2/3}TiO_3$

Composition	Lattice parameter/nm		
	a	b	c
$CaTiO_3$	—	—	—
$Ca_{0.7}La_{0.2}TiO_3$	0.543	0.545	0.769
$Ca_{0.4}La_{0.4}TiO_3$	0.547	0.546	0.771
$Ca_{0.2}La_{0.53}TiO_3$	0.547	0.547	0.775
$La_{2/3}TiO_3$	—	—	—

Source: Adapted from data given by Z. Zhang et al., J. Solid State Chem., **180**, 1083–1092 (2007).

3. At high temperatures the spinel $MgAl_2O_4$ can take in excess alumina to a composition of approximately $70\,mol\%$ Al_2O_3 (Fig. 4.5). (a) What are the possible formulas that fit the composition of this spinel? Write the defect formation equation for the reaction if the excess Al is (b) distributed over both magnesium and aluminum sites and (c) only over aluminum sites. Assume that there is no electronic compensation in the insulating oxide.

4. A hollandite structure phase containing Ba, Ti, Al, and O is fabricated with the aim of immobilizing radioactive nuclides. (a) What is the formula of the hollandite? (b) It is desired to replace 10% of the Ba with radioactive K. What would the formula of the new phase be? (c) It is desired to replace 35% of the Ba with radioactive Sr. What would the formula of the new phase be? (d) It is desired to replace 17% of the Ba with radioactive La. What would the formula of the new phase be?

5. By counting the planes parallel to the $\{113\}$ Bi planes in the twinned phases lillianite and heyrovskyite (Fig. 4.26) or otherwise, determine the homologous series formula that includes these phases.

6. The perovskite-related phases similar to $Ca_2Nb_2O_7$ are composed of slabs of perovskite-like structure n octahedra thick. What is the chemical formula of the oxides composed of slabs of thickness (a) $n = 5$; (b) $(n_1, n_2) = 4,4,5$; (c) $(n_1, n_2) = 4,5,5$; and (d) $(n_1, n_2) = 5,6$.

7. A sample of a strontium titanium sulfide Sr_xTiS_3 was found to have a unit cell in which the repeat distance of the TiS_3 chains, c_{TiS}, was found to be 0.30 nm and that of the Sr chains, c_{Sr}, was found to be 0.54 nm (Fig. 4.38). It is found that 5.5 Sr units fit with 10 TiS_3 units. What is (a) the composition and (b) the approximate unit cell c parameter? Another preparation has a composition $Sr_{1.12}TiS_3$. (c) What is the new approximate unit cell c parameter?

8. The CaF_2–UF_4 phase diagram shows that the CaF_2 (fluorite) structure can take in UF_4 up to a limiting composition of 58 wt % CaF_2 : 42 wt % UF_4 at 1000°C. What is the formula of the limiting composition if (a) the metal lattice in the fluorite parent structure is perfect, as in CaF_2; (b) the F lattice in the fluorite parent structure is perfect as in CaF_2.

9. The composition of the important solid electrolyte phase β-alumina runs from $NaAl_5O_8$ to $NaAl_{11}O_{17}$ (see Section 6.6). Assuming that the structure and formula $NaAl_{11}O_{17}$ is ideal, what is the formula and what point defects are present in the limiting $NaAl_5O_8$ composition if (a) the Al lattice is considered to be identical to that in $NaAl_{11}O_{17}$, and (b) the O lattice is considered to be identical to that in $NaAl_{11}O_{17}$?

10. Titanium nitride, TiN, with the sodium chloride structure, has a composition running from 0.5 at % Ti : 0.5 at % N to 0.72 at % Ti : 0.28 at % N. What is the formula and what point defects are present in the limiting titanium-rich composition if (a) the N lattice is perfect; and (b) the Ti lattice

is perfect. In fact, it is believed that there are vacancies present on both the Ti and N sublattices simultaneously and no interstitials at all. (c) Derive a formula for the ratio of the theoretical density of perfect TiN to that of a material containing vacancies on both sites, Ti_pN_r. The measured density of TiN is 5100 kg m^{-3} and the theoretical density is 5370 kg m^{-3}. (d) What is the real formula of the phase?

FURTHER READING

The subject matter of this chapter is basically covered in crystallographic literature. Some useful sources are listed below.

Crystal structures are best viewed as three-dimensional computer images that can be "rotated" and viewed from any "direction." Crystal structures can be displayed, and downloadable programs for graphical presentation of crystal structures can be found at the EPSRC's Chemical Database Service at Daresbury. This can be accessed at: http://cds.dl.ac.uk/cds. See also:

D. A. Fletcher, R. F. McMeeking, and D. Parkin, *J. Chem. Inf. Comput. Sci.*, **36**, 746–749 (1996).

Books with general coverage are:

B. G. Hyde and S. Andersson, *Inorganic Crystal Structures*, Wiley-Interscience, New York, 1989.

M. O'Keeffe and B. G. Hyde, *Crystal Structures, I. Patterns and Symmetry*, Mineralogical Society of America, Washington, DC, 1996.

A. F. Wells, *Structural Inorganic Chemistry*, 5th ed., Oxford University Press, Oxford, United Kingdom, 1984.

A great deal of material relevant to this chapter, including mineralogical transformations, is given in:

A. Putnis, *Introduction to Mineral Sciences*, Cambridge University Press, Cambridge, United Kingdom, 1992.

Modulated and modular structures:

A. Baronet, Polytypism and Stacking Disorder, in *Reviews in Mineralogy*, Vol. 27, P. R. Busek, Ed., Mineralogical Society of America, Washington, DC, 1992, Chapter 7.

G. Ferraris, Mineral and Inorganic Crystals, in *Fundamentals of Crystallography*, 2nd ed., C. Giacovazzo, H. L. Monaco, G. Artioli, D. Viterbo, G. Ferraris, G. Gilli, G. Zanotti and M. Catti, Eds., Oxford University Press, Oxford, United Kingdom, 2002, Chapter 7.

G. Ferraris, E. Mackovicky, and S. Merlino, *Crystallography of Modular Materials*, International Union of Crystallography Monographs on Crystallography No. 15, Oxford University Press, Oxford, United Kingdom, 2004.

C. Giacovazzo, Chapter 4, Beyond Ideal Crystals, in *Fundamentals of Crystallography*, 2nd ed., C. Giacovazzo, H. L. Monaco, G. Artioli, D. Viterbo, G. Ferraris, G. Gilli, G. Zanotti, and M. Catti, Eds., Oxford University Press, Oxford, United Kingdom, 2002.

T. Janssen and A. Janner, Incommensurabilty in Crystals, *Adv. Phys.*, **36**, 519–624 (1987).

E. Makovicky and B. G. Hyde, Incommensurate, Two-Layer Structures with Complex Crystal Chemistry: Minerals and Related Synthetics, *Mat. Sci. Forum*, **100 & 101**, 1–100 (1992).

S. van Smaalen, Incommenurate Crystal Structures, *Crystal. Rev.*, **4**, 79–202 (1995).

D. R. Veblen, Electron Microscopy Applied to Nonstoichiometry, Polysomatism and Replacement Reactions in Minerals, in *Reviews in Mineralogy*, Vol. 27, P. R. Busek, Ed., Mineralogical Society of America, Washington, DC, 1992, Chapter 6.

D. R. Veblen, Polysomatism and Polysomatic Series: A Review and Applications, *Am. Mineral.*, **76**, 801–826 (1991).

G. A. Wiegers, Misfit Layer Compounds: Structures and Physical Properties, *Prog. Solid State Chem.*, **24**, 1–139 (1996).

R. L. Withers, S. Schmid, and J. G. Thompson, Compositionally and/or Displacively Flexible Systems and Their Underlying Crystal Chemistry, *Prog. Solid State Chem.*, **26**, 1–96 (1998).

References to ~FeO:

H. Fjellvåg, H. Grønvold, S. Stølen, and Bjørn Hauback, *J. Solid State Chem.*, **124**, 52–57 (1992).

M. J. Radler, J. B. Cohen, and J. B. Faber, *J. Phys. Chem. Solids*, **51**, 217–228 (1990).

References to calcia-stabilized zirconia:

R. Devanathan, W. J. Weber, S. C. Singhal, and J. D. Gale, *Solid State Ionics*, **177**, 1251–1258 (2006).

T. R. Welberry, R. L. Withers, J. G. Thompson, and R. D. Butler, *J. Solid State Chem.*, **100**, 71–89 (1992).

References to anion-excess fluorite structures:

D. J. M. Bevan, I. E. Grey, and B. T. M. Willis (β-U_4O_{9-y}) *J. Solid State Chem.*, **61**, 1–7 (1986).

J. P. Laval, C. Depierefixe, B. Frit, and G. Roult ($Pb_{1-x}Zr_xF_{2+2x}$: $0 < x < 0.18$) *J. Solid State Chem.*, **54**, 260–276 (1984).

C. Rocanière, J. P. Laval, P. Dehaudt, B. Gaudreau, A. Chotard, and E. Suard, (($U_{0.9}Ce_{0.1})_4O_{9-\delta}$), *J. Solid State Chem.*, **177**, 1758–1767 (2004).

K. Wurst, E. Schweda, D. J. M. Bevan, C. Mohyla, K. S. Wallwork, and M. Hofmann, ($Zr_{50}Sc_{12}O_{118}$), *Solid State Sci.*, **5**, 1491–1497 (2003).

A description of L-Ta_2O_5 as a modulated structure:

S. Schmid, J. G. Thompson, A. D. Rae, B. D. Butler, and R. L. Withers, *Acta Crystallogr.*, **B51**, 698 (1995); A. D. Rae, S. Schmid, J. G. Thompson, and R. Withers, *Acta Crystallogr.*, **B51**, 709 (1995).

S. Schmid, R. L. Withers, and J. G. Thompson, *J. Solid State Chem.*, **99**, 226 (1992).

A description of Sr_xTiS_3 as a modulated structure:

O. Gourdon, V. Petricek, and M. Evain, *Acta Crystallogr.*, **B56**, 409–418 (2000).

M. Saeki, M. Ohta, K. Kurashima, and M. Onoda, *Mat. Res. Bull.*, **37**, 1519–1529 (2002).

Defects and Diffusion

What is volume diffusion?
How does temperature effect diffusion in solids?
How do defects influence diffusion in solids?

5.1 DIFFUSION

Diffusion is a description of the way in which atoms, ions, or molecules flow through a surrounding medium. For example, when two different atom types (which can be different elements or merely different isotopes of the same element) are united along a planar boundary, diffusion leads to the gradual mixing of the species and blurring of the initially sharp interface (Fig. 5.1). It is apparent that diffusion will be an important factor when considering the reaction of solids with other solids, liquids, or gases.

Diffusion is quantified by measuring the concentration of the diffusing species at different distances from the release point after a given time has elapsed at a precise temperature. Raw experimental data thus consists of concentration and distance values. The degree of diffusion is represented by a diffusion coefficient, which is extracted from the concentration–distance results by solution of one of two diffusion equations. For one-dimensional diffusion, along x, they are Fick's first law of diffusion:

$$J = -D\frac{dc}{dx} \tag{5.1}$$

and Fick's second law of diffusion, known more commonly as the diffusion equation:

$$\frac{dc}{dt} = D\frac{d^2c}{dx^2} \tag{5.2}$$

Defects in Solids, by Richard J. D. Tilley
Copyright © 2008 John Wiley & Sons, Inc.

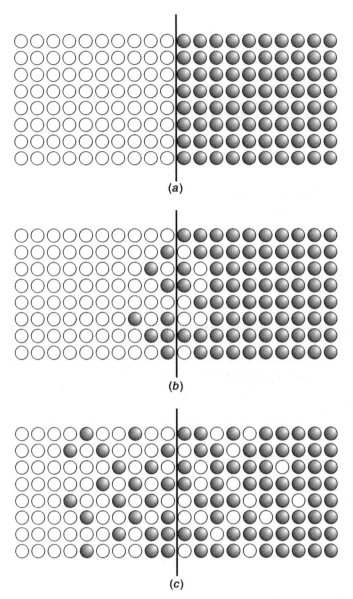

Figure 5.1 Diffusion of atoms across an interface: (*a*) initial configuration and (*b*, *c*) the interface becomes increasingly blurred due to interdiffusion of the two atom species.

In these equations, J is the flux of the diffusing species, with units of amount of substance per unit area per unit time, that is (atoms, grams, or moles), meter squared per second, c is the concentration of the diffusing species, with units of amount of substance per unit volume, that is (atoms, grams, or moles), per cubic meter, at position x, units in meter, after time t, units in seconds, and D is the diffusion coefficient or diffusivity, with units[1] of square meter per second.

There are two overriding considerations to keep in mind when discussing diffusion in solids: the structure of the matrix across which diffusion occurs and the defects present. In a normal crystalline solid, diffusion is mediated by the defects present, and the speed of diffusion will vary significantly if the predominant defect type changes. This is because diffusion involves the movement of a species from a stable position, through some sort of less stable position or bottleneck, to another stable position. Any disorder in the solid due to defects will make this process easier.

Movement through the body of a solid is called volume, lattice, or bulk diffusion. In a gas or liquid, bulk diffusion is usually the same in all directions and the material is described as isotropic. This is also true in amorphous or glassy solids and in cubic crystals. In all other crystals, the rate of bulk diffusion depends upon the direction taken and is anisotropic. Bulk diffusion through a perfect single crystal is dominated by point defects, with both impurity and intrinsic defect populations playing a part.

Diffusion along two- and three-dimensional imperfections in the material is much faster than bulk diffusion, especially at lower temperatures. This process is referred to as short-circuit diffusion. The main imperfections that are of importance for short-circuit diffusion are dislocations and grain boundaries. Dislocations can be thought of as tubes of disordered structure that thread through a crystal, and diffusion along a dislocation core is sometimes called pipe diffusion. Low-angle grain boundaries in crystals are comprised of arrays of dislocations, and diffusion along these dislocations will result in enhanced atom transport compared to the bulk. Grain boundaries are ordinarily regions between individual crystallites in a polycrystalline solid, and diffusion making use of these surfaces is called grain boundary diffusion. Diffusion at external surfaces, surface diffusion, is also important and plays a vital role in many chemical reactions, notably those in heterogeneous catalysis.

In this chapter, to keep the material compact, only the relationship between diffusion and defects in solids will be discussed. Moreover, the diffusion coefficient will be considered as a constant at a fixed temperature, and attention is focused upon the movement of atoms and ions rather than the equally important diffusion of gases or liquids through a solid. Discussion of diffusion *per se*, the extensive literature on classical theories of diffusion, and diffusion when the diffusion coefficient is not a constant will be found in the Further Reading section at the end of this chapter.

5.2 DIFFUSION IN SOLIDS

Diffusion is followed by tracking the movements of tracer species through the solid to obtain the tracer diffusion coefficient, written as D^* when the tracer is identical to one of the components of the crystal, and D_A^* when an impurity or foreign atom A is the tracer. Earlier studies made extensive use of radioactive isotopes because the progress

---

[1]The values of diffusion coefficients in (especially) the older literature are frequently given in square centimeter per second. To convert to square meters per second, multiply by 10^{-4}, that is, $D\,(\mathrm{cm^2\,s^{-1}}) \times 10^{-4} \rightarrow D\,(\mathrm{m^2\,s^{-1}})$.

of the diffusion could be tracked by measuring the penetration of radioactivity within the solid. Other techniques such as secondary ion mass spectroscopy (SIMS) make it possible to track changes in isotope concentration for nonradioactive species. In transparent materials colored ions can be used as tracers, and in biological systems fluorescence provides a useful means of tracking diffusion.

The classical way of following the extent of diffusion through a solid is to apply a thin layer of a tracer species to a carefully polished surface, after which another carefully polished slice is placed on top to form a diffusion couple (Fig. 5.2a). The sample is now heated at an appropriate temperature for some hours during which time the tracer will move into the material (Fig. 5.2b). Several processes occur simultaneously. The tracer will diffuse into the crystal bulk, down grain boundaries and dislocations from the initial layer, and also sideways into the bulk from the grain boundaries and dislocations. After the experiment the sample is carefully sliced parallel to the polished surface, and the overall concentration of the tracer in each slice is measured, allowing a graph of concentration versus penetration depth to be drawn. Such a plot is called a diffusion profile, penetration profile, or concentration profile. The results obtained in this way are schematically illustrated in Figure 5.3.

For a typical polycrystalline material the concentration profile is divided into three segments. Near the original surface the tracer distribution will be characteristic of volume diffusion and show a typical bell shape. The average tracer concentration

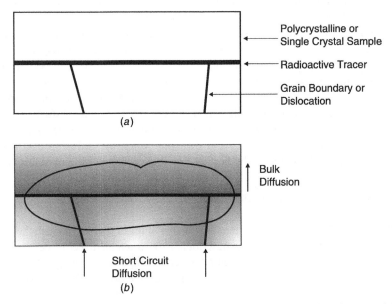

Figure 5.2 Diffusion couple formed by two crystals separated by radioactive material: (a) initially and (b) after heating. Short-circuit diffusion occurs down extended defects, and diffusion into the bulk occurs both from the surface and laterally from extended defects.

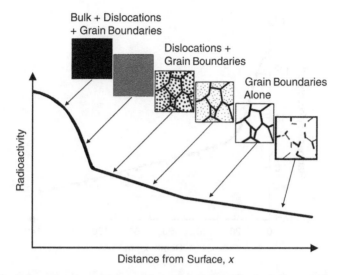

Figure 5.3 Schematic representation of the penetration profile for bulk, grain boundary, and dislocation diffusion in a polycrystalline solid. The initial part of the curve is bell shaped, and the part due to short-circuit diffusion is made up of linear segments. The insets show the distribution of the tracer in the sample.

will be high, and it will not be possible to distinguish the extended defects from the surrounding crystal matrix. However, as the crystal is sliced, the average tracer concentration will fall, and the extended defects start to show higher levels of tracer than the surrounding crystal, dislocations as dots, and grain boundaries as lines. In this region the concentration profile will be linear and is associated with short-circuit diffusion along dislocations and grain boundaries. The segment with the greatest slope is made up of contributions from both dislocations and grain boundaries. The region with lower slope is due to the contribution of the boundaries with the fastest diffusion coefficient. It is not always easy to separate these two regions from one another in practice (Fig. 5.4).

Because of the skills that have been developed during semiconductor device fabrication, a number of elegant ways of exploring diffusion in single crystals of these materials have been invented. Isotope self-diffusion in silicon provides a good example. There are three natural isotopes of silicon, 92.2% of ^{28}Si, 4.7% of ^{29}Si, and 3.1% of ^{30}Si. A single crystal consisting of a sequence of layers of natural silicon interspersed with layers that contain essentially no ^{29}Si or ^{30}Si, a semiconductor "superlattice," can be grown, and the concentration profiles of the individual isotopes measured to obtain diffusion coefficients for these species. By doping the surface with impurities, such as As, simultaneous self-diffusion and impurity diffusion profiles can be obtained. Measurements on similar superlattices of AsSb, using the isotopes ^{69}Ga, ^{71}Ga, ^{121}Sb, and ^{123}Sb shows that the Ga isotopes diffuse rapidly, via bell-shaped concentration profiles, while the Sb is static at 700°C (see Further Reading).

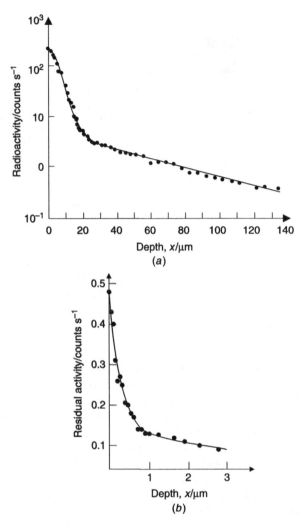

Figure 5.4 Experimental penetration profiles showing short-circuit diffusion. (*a*) Diffusion of radioactive ^{63}Ni into polycrystalline CoO at 953°C. Short-circuit diffusion is associated with grain boundaries. [Redrawn from K. Kowalski, Thesis, University of Nancy (1994)]. (*b*) Diffusion of radioactive ^{26}Al into single crystal α-Al$_2$O$_3$ (corundum) at 1610°C. Short-circuit diffusion is associated with dislocations aligned along low-angle grain boundaries. [Redrawn from M. Le Gall, B. Lesage, and J. Bernardini, *Philos. Mag.* A, **70**, 761–773 (1994).]

To evaluate the bulk diffusion coefficient, using Eq. (5.1) or (5.2), it is necessary to subtract the short-circuit diffusion contribution from the total concentration profile. Ideally, the concentration profile due to bulk diffusion will take on the shape of a bell, gradually flattening as the time of diffusion increases (Fig. 5.5). The solution to the

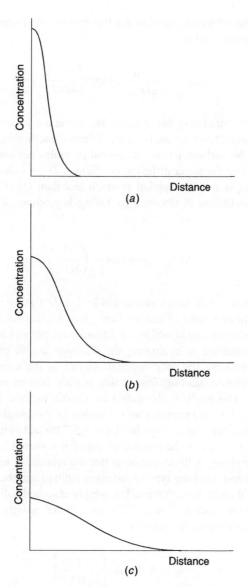

Figure 5.5 Penetration profiles typical of bulk diffusion: (*a*), (*b*), and (*c*) illustrate gradually increasing heating times.

diffusion equation, Eq. (5.2), appropriate to bulk diffusion in a diffusion couple experimental arrangement (Fig. 5.2) is

$$c_x = \left[\frac{c_0}{2(\pi D^* t)^{1/2}} \right] \exp\left(\frac{-x^2}{4D^* t} \right) \qquad (5.3)$$

The solution to the diffusion equation for the experimental situation in which the coated surface is uncovered is

$$c_x = \left[\frac{c_0}{(\pi D^* t)^{1/2}}\right] \exp\left(\frac{-x^2}{4D^* t}\right) \tag{5.4}$$

where c_x is the concentration of the diffusing species at a distance of x from the original surface after time t has elapsed, D^* is the diffusion coefficient, and c_0 is the initial concentration on the surface, usually measured in moles per meter squared. These solutions assume that the tracer diffusion coefficient, D^*, is constant, and the thickness of the thin layer initially applied is much less than $(D^* t)^{1/2}$. A value for the tracer diffusion coefficient is obtained by taking logarithms of both sides of this equation to give

$$\ln c_x = \text{constant} - \left(\frac{x^2}{4D^* t}\right)$$

A plot of $\ln c_x$ versus x^2 will have a gradient of $[-1/(4D^* t)]$ (See also Supplementary Material S5.) A measurement of the gradient gives a value for the tracer diffusion coefficient at the temperature at which the diffusion couple was heated.

An alternative method of evaluating the diffusion coefficient from radioactive tracer data is to measure the total residual activity in the sample after each thin slice has been removed, starting from a thin surface film in an "open-sandwich" type of geometry. This method, also called the Gruzin method, is equivalent to the integration of Eq. (5.4) and provides an estimation of the remaining area under the half of a bell-shaped curve of the type in Figure 5.5. This technique avoids the necessity of careful sectioning, as the amount of material removed in each slice can be determined by weighing. A disadvantage is that the amount of residual radioactivity measured will depend upon the type of radiation emitted and the absorption characteristics of the solid under investigation. For weakly absorbed radiation, such as γ-ray photons from isotopes such as ^{26}Al or ^{65}Zn, the residual activity from a thin surface layer in an open-sandwich experiment is

$$A_r = A_r(0)\,\text{erfc}\left[\frac{x}{2\sqrt{(D^* t)}}\right]$$

where A_r is the residual activity after an amount x of sample has been removed, $A_r(0)$ is the initial activity before sectioning, D^* is the bulk tracer diffusion coefficient, t is the heating time, and erfc[y] is the complementary error function (see Supplementary Material S5).

The aim of many of the studies of diffusion is to relate the measured diffusion coefficient to a *mechanism* of diffusion. By this is meant a model of atomic jumps that accurately reproduces the diffusion coefficient and the measured concentration profile over a wide range of temperatures. This objective has been most pronounced in the study of semiconductors for which diffusion is an important step in the fabrication of device materials. These studies have been facilitated by the fact that semiconductors are generally available as perfect single crystals, and, to some extent, many of the models of diffusion found in the literature have been postulated to account for diffusion in these materials. The following sections in this chapter explore the various models put forward to account for diffusion.

5.3 RANDOM-WALK DIFFUSION IN CRYSTALS

The simplest and most basic model for the diffusion of atoms across the bulk of a solid is to assume that they move by a series of random jumps, due to the fact that all the atoms are being continually jostled by thermal energy. The path followed is called a random (or drunkard's) walk. It is, at first sight, surprising that any diffusion will take place under these circumstances because, intuitively, the distance that an atom will move via random jumps in one direction would be balanced by jumps in the opposite direction, so that the overall displacement would be expected to average out to zero. Nevertheless, this is not so, and a diffusion coefficient for this model can be defined (see Supplementary Material Section S5).

For example, suppose a planar layer of N tracer atoms is the starting point, and suppose that each atom diffuses from the interface by a random walk in a direction perpendicular to the interface, in what is effectively one-dimensional diffusion. The probability of a jump to the right is taken to be equal to the probability of a jump to the left, and each is equal to 0.5. The random-walk model leads to the following result:

$$<x^2> = na^2 = \Gamma t a^2 \tag{5.5}$$

where $<x^2>$ is the average of the square of the distance that each of the N diffusing atoms reaches after carrying out n random steps over the time of the diffusion experiment t. Each jump is of the same distance, a, and Γ is the frequency with which each atom jumps to the next position (Supplementary Material S5).

The surprising result is that net atom displacement will occur due to random movement alone. It is possible to use Eq. (5.5) to define a diffusion coefficient. To agree with Fick's first law, the relationship chosen is

$$<x^2> = 2D_r t \tag{5.6}$$

where D_r is the random-walk diffusion coefficient. This relationship is known as the Einstein (or Einstein–Smoluchowski) diffusion equation. The diffusion constant D_r is often equivalent to the self-diffusion coefficient, D_{self}, which describes the diffusion of atoms in a crystal under no concentration gradient, and in some instances to the tracer diffusion coefficient obtained when concentration gradients are small (see Sections 5.5 and 5.6).

The factor of 2 in Eq. (5.6) arises from the one-dimensional nature of the random walk and, hence, is a result of the geometry of the diffusion process. In the case of random-walk diffusion on a two-dimensional surface:

$$<x^2> = 4D_r t$$

while for random-walk diffusion in a three-dimensional crystal:

$$<x^2> = 6D_r t$$

The random-walk model of diffusion can also be applied to derive the shape of the penetration profile. A plot of the final position reached for each atom (provided the number of diffusing atoms, N, is large) can be approximated by a *continuous* function, the Gaussian or normal distribution curve[2] with a form:

$$N(x) = \frac{2N}{\sqrt{2\pi n}} \exp\left(\frac{-x^2}{4D_r t}\right)$$

where the function $N(x)$ is the number of atoms that reach a position x, N is the number of atoms taking part in the random walk, n is the number of steps in each walk, D_r is the random-walk diffusion coefficient, and t the time of the experiment (Fig. 5.6a). The bell-shaped profile of this curve matches the concentration profile for bulk diffusion but does not explain the linear concentration profile of short-circuit diffusion.

The statistics of the normal distribution can be applied to give more information about random-walk diffusion. The area under the normal distribution curve represents a probability. In the present case, the probability that any particular atom will be found in the region between the starting point of the diffusion and a distance of $\pm\sqrt{<x^2>} = \pm\sqrt{(2D_r t)}$ on either side of it is approximately 68% (Fig. 5.6b). The probability that any particular atom has diffused *further* than this distance is given by the total area under the curve minus the shaded area, which is approximately 32%. The probability that the atoms have diffused further than $\pm 2\sqrt{<x^2>}$, that is, $\pm 2\sqrt{(2D_r t)}$ is equal to about 5%.

It is of considerable practical importance to have some idea of how far an atom or ion will diffuse into a solid during a diffusion experiment. An approximate estimate of the depth to which diffusion is significant is given by the penetration depth, x_p, which is the depth where an appreciable change in the concentration of the tracer can be said to have occurred after a; diffusion time t. A reasonable estimate can be given with respect to the root mean square displacement of the diffusing

---

[2]The normal distribution is generally written as $P(x) = (1/\sqrt{2\pi}\sigma)\exp\left[-(x-\mu)^2/2\sigma^2\right]$ where μ is the mean and σ the standard deviation of the distribution, the whole being normalized so that the area under the curve is equal to 1.0.

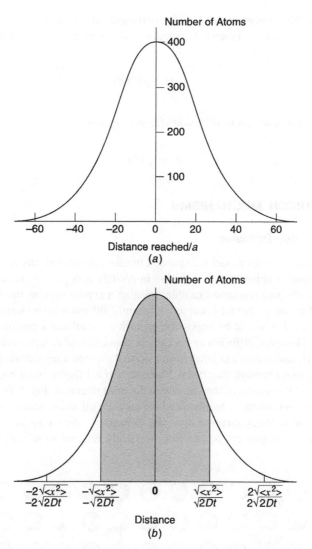

Figure 5.6 (a) Smoothed distribution of the final position reached by 10,000 atoms after each has completed a one-dimensional random walk of 400 steps. The mean of the distribution is zero and the peak is equal to 400 [$2 \times 10,000/\sqrt{(2\pi \times 400)}$]. (b) The probability distribution curve for one-dimensional random-walk diffusion. The shaded area indicates a 68% probability that a diffusing atom will be found between the limits $\pm\sqrt{<x^2>}$.

atoms, $\pm\sqrt{<x^2>}$, so that the penetration depth for one-dimensional diffusion can be expressed as

$$x_p \approx \sqrt{(2Dt)}$$

where D is the appropriate diffusion coefficient and t the time of the diffusion process. This choice is somewhat arbitrary, and sometimes the penetration depth is defined as

$$x_p \approx 2\sqrt{(2Dt)}$$

or as the diffusion length or diffusion distance, L, where

$$L = \sqrt{(Dt)}$$

5.4 DIFFUSION MECHANISMS

5.4.1 Vacancy Diffusion

When the random-walk model is expanded to take into account the real structures of solids, it becomes apparent that diffusion in crystals is dependent upon point defect populations. To give a simple example, imagine a crystal such as that of a metal in which all of the atom sites are occupied. Inherently, diffusion from one normally occupied site to another would be impossible in such a crystal and a random walk cannot occur at all. However, diffusion can occur if a population of defects such as vacancies exists. In this case, atoms can jump from a normal site into a neighboring vacancy and so gradually move through the crystal. Movement of a diffusing atom into a vacant site corresponds to movement of the vacancy in the other direction (Fig. 5.7). In practice, it is often very convenient, in problems where vacancy diffusion occurs, to ignore atom movement and to focus attention upon the diffusion of the vacancies as if they were real particles. This process is therefore frequently referred to as vacancy diffusion

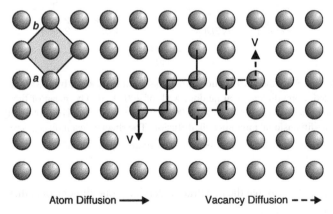

Figure 5.7 Vacancy diffusion in the (001) plane of a cubic crystal. The unit cell is shaded. An atom can jump to a neighboring vacancy but not to an already occupied position. Diffusion parallel to one of the axial directions will be by a zigzag route.

and in general is known as indirect diffusion. Vacancy diffusion appears to be the commonest diffusion mechanism in most pure metals and germanium.

The path that the diffusing atom takes will depend upon the structure of the crystal. For example, the {100} planes of the face-centered cubic structure of elements such as copper are identical to that drawn in Figure 5.7. Direct diffusion of a tracer atom along the cubic axes by vacancy diffusion will require that the moving atom must squeeze between two other atoms. It is more likely that the actual path will be a dog-leg, in <110> directions, shown as a dashed line on Figure 5.7.

5.4.2 Interstitial Diffusion

In the case of interstitials—self-interstitials, impurities, or dopants—two diffusion mechanisms can be envisaged. In the simplest case, an interstitial can jump to a neighboring interstitial position (Fig. 5.8a). This is called interstitial diffusion and is sometimes referred to as direct diffusion to distinguish it from vacancy diffusion (indirect diffusion).

It is important to remember that real diffusion paths are more complex than that suggested by a planar figure. For example, interstitial diffusion is the mechanism by which tool steels are hardened by incorporation of nitrogen or carbon. In the body-centered cubic (bcc) structure adopted by metals such as iron and tungsten, interstitial atoms can occupy either octahedral or tetrahedral sites between the metal atoms (Fig. 5.8b and 5.8c). It is by no means easy to visualize the possible diffusion paths that interstitials can follow in even such a simple structure, but the task can be clarified by plotting the positions of the various interstitial sites within a unit cell while omitting the metal atoms (Fig. 5.8d). There are 12 tetrahedral sites in a body-centered cubic unit cell; 4 on each cube face, and 6 octahedral sites fall at the centers of the cube edges and at the center of each face. (Note, more sites than this are marked because of sharing between neighboring unit cells. For example, 24 tetrahedral sites are marked in total.) Diffusion paths between the tetrahedral sites form the edges of a truncated octahedron. A diagram such as this allows all diffusion paths to be determined. As one example, diffusion parallel to a cubic axis is possible if the diffusing interstitial jumps between alternating octahedral–tetrahedral sites indicated by the dashed line in Figure 5.8d.

A well-documented example of interstitial diffusion is that of silicon self-diffusion in silicon single crystals. Silicon self-diffusion has been extensively studied due to its pivotal importance in semiconductor device technology. Each silicon atom is linked to four other atoms by tetrahedrally arranged strong chemical bonds, and vacancy diffusion would take place along paths delineated by the net joining these atom positions. However, these nets form large channels along <110> (Fig. 5.9), and interstitial diffusion along these channels is a lower energy (preferred) alternative.

5.4.3 Interstitialcy Diffusion

An alternative mechanism by which interstitial atoms can diffuse involves a jump to a normally occupied site together with simultaneous displacement of the occupant into a neighboring interstitial site. This knock-on process is called interstitialcy diffusion.

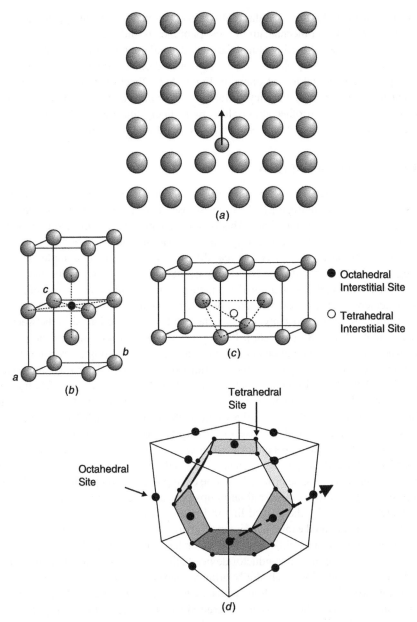

Figure 5.8 Interstitial diffusion: (*a*) interstitial diffusion involving the direct migration of an interstitial atom to an adjacent site in the crystal; (*b, c*) some of the octahedral and tetrahedral interstitial sites in the body-centered cubic structure of metals such as iron and tungsten; and (*d*) the total number of octahedral and tetrahedral sites in a unit cell of the body-centered cubic structure. Diffusion paths parallel to the unit cell edges can occur by a series of alternating octahedral and tetrahedral site jumps, dashed line.

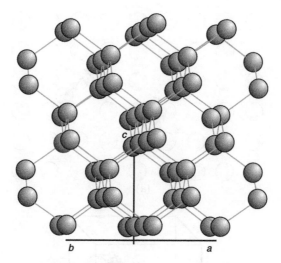

Figure 5.9 Structure of silicon viewed down [110]. The atoms are drawn smaller than the atomic radius suggests for clarity. Channels parallel to <110> facilitate interstitial diffusion in this material.

It can be by way of a "straight" transfer, called colinear interstitialcy diffusion, or by an "elbowed" displacement, called noncolinear interstitialcy diffusion (Fig. 5.10a). In real crystals the path is constrained by local geometry. For example, molecular dynamics simulations strongly suggest that noncolinear interstitialcy diffusion occurs in $RbBiF_4$, which adopts the fluorite structure. In this structure the anions make up an array of cubes, half of which contain cations and half of which remain empty. Interest centers upon the migration of F^- ions. Anion Frenkel defects are the principal point defect type present, and the interstitial F^- components occupy the centers of those cubes that do not contain cations. The diffusion of these F^- interstitials takes place by migration from one empty cube to an adjoining (edge-shared) empty cube by way of a noncolinear interstitialcy mechanism (Fig. 5.10b). The tangled line shows the path of an individual F^- interstitial ion. Starting in the bottom cube, the interstitial is moving about in the interstitial site but never moving far. Eventually, it is able to jump to the adjoining normal site, knocking the ion there into the interstitial site in the upper cube.

5.4.4 Impurity and Cluster Diffusion Mechanisms

Substitutional impurities can move by way of a number of mechanisms. The most usual is the vacancy mechanism described above. Diffusion studies on semiconductors have suggested that a number of additional mechanisms might hold. As well as vacancy diffusion, an impurity can swap places with a neighboring normal atom, exchange diffusion, while in ring diffusion cooperation between several atoms is

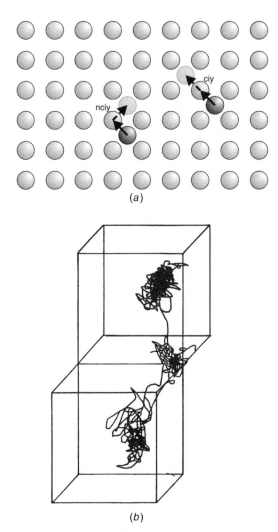

(a)

(b)

Figure 5.10 (a) Colinear (ciy) and noncolinear interstitialcy (nciy) diffusion, schematic; and (b) simulation indicating that a noncolinear diffusion mechanism is responsible for F^- diffusion in RbBiF$_4$. [Redrawn after C. R. A. Catlow, *J. Chem. Soc. Faraday Trans.*, **86**, 1167–1176 (1990).]

needed to make the exchange (Fig. 5.11a). Exchange diffusion has been postulated to take place during the doping of the III–V semiconductor crystals such as GaAs. Finally, an interstitial impurity can move onto a normal lattice site by interstitialcy diffusion, leaving a self-interstitial and a substitutional defect in its place. This is called the kick-out mechanism (Fig. 5.11b and 5.11c) and has been suggested as the mechanism by which gold diffuses in silicon. A number of other mechanisms for diffusion in semiconductors have been postulated, but conclusive proof that they operate is still lacking.

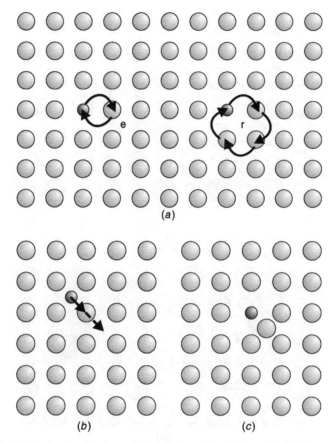

Figure 5.11 Diffusion mechanisms: (*a*) exchange (e) and ring (r) diffusion; (*b*) kick-out diffusion, leading to (*c*) a substitutional defect and a self-interstitial.

When Schottky defects are present in a crystal, vacancies occur on both the cation and anion sublattices, allowing both cation and anion vacancy diffusion to occur (Fig. 5.12*a*). In the case of Frenkel defects interstitial, interstitialcy, and vacancy diffusion can take place in the same crystal with respect to the atoms forming the Frenkel defect population (Fig. 5.12*b*).

Defect clusters can move by a variety of mechanisms. As an example, the idealized diffusion of a cation–anion divacancy within the (100) face of a sodium chloride structure crystal by way of individual cation and anion jumps is shown in Figure 5.13.

5.4.5 Diffusion Paths

To portray cation diffusion paths accurately, it is helpful to remember that the structures can often be regarded as built from layers of close-packed anions with

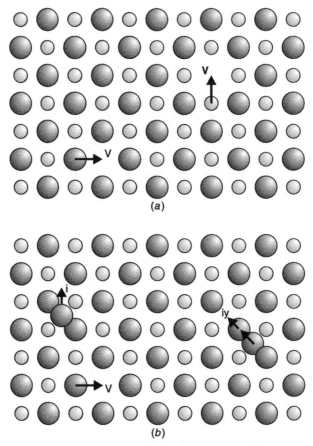

Figure 5.12 Diffusion in crystals of composition MX containing (*a*) Schottky and (*b*) Frenkel defects, schematic; V, vacancy, i, interstitial, iy, interstitialcy.

cations tucked into interstices (spaces) between the layers. These cation sites have a tetrahedral or octahedral geometry (Fig. 5.14). The network of diffusion paths between these sites forms the edges of an anion-centered space-filling polyhedron. The polyhedron appropriate to a cubic close-packed array of anions is an anion-centered rhombic dodecahedron. This can be drawn with respect to the cubic axes (Fig. 5.15*a*), but it is often more convenient to orient the polyhedron with respect to "horizontal" close-packed metal atom layers shown in Figure 5.14 by rotating the polyhedron anticlockwise (Fig. 5.15*b*). (Note that the central anion has been drawn rather small for clarity. In reality, ionic radii suggest that the atoms would fill the polyhedron and be close to contact at the midpoint of each face.) The octahedral sites in these structures are represented by vertices at which four edges meet, while tetrahedral sites are represented by vertices at which three edges meet. The structure of crystals such as MgO and NiO, in which all of the octahedral sites

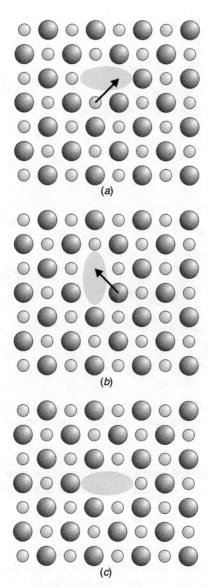

Figure 5.13 Diffusion of a cation–anion divacancy within the (100) plane of a sodium chloride structure crystal: (a–c) shows diffusion by way of individual cation and anion vacancy diffusion.

are filled, to form the sodium chloride structure, is drawn in Figure 5.15c. The corresponding anion-centered polyhedron for the cubic zinc blende (sphalerite) structure of ZnS, in which half of the tetrahedral sites are filled, is shown in Figure 5.15d. The cation diffusion path in both of these structures, and any other

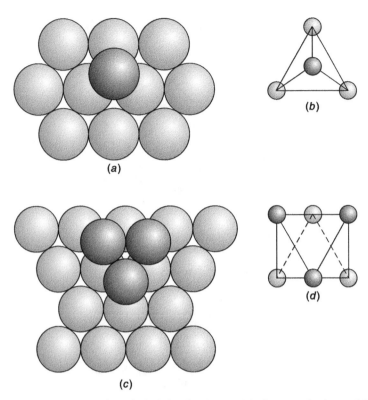

Figure 5.14 Tetrahedral and octahedral sites in closest-packed arrays of spheres: (*a*) a tetrahedral site and (*b*) the same drawn as a polyhedron; (*c*) an octahedral sites and (*d*) the same site drawn as a polyhedron.

structure that can be derived from a cubic close-packed array of anions, is evidently made up of an alternating sequence of octahedral and tetrahedral sites:

$$\cdots oct-tet-oct-tet-oct-tet \cdots$$

Sodium chloride structure crystals have all octahedral sites filled, and so cation diffusion will be dependent upon vacancies on octahedral sites. In the zinc blende (sphalerite) structure, adopted by ZnS, for example, half of the tetrahedral sites are empty, as are all of the octahedral sites, so that self-diffusion can take place without the intervention of a population of defects.

The analogous anion-centered polyhedron for a hexagonal close-packed anion array (Fig. 5.16*a*) is similar to a rhombic dodecahedron but has a mirror plane normal to the vertical axis, which is perpendicular to the close-packed planes of anions and parallel to the hexagonal **c** axis. The octahedral sites in the structure are again found at the vertices where four edges meet, and the tetrahedral sites at the vertices where three edges meet. Common structures formed by an ordered filling these sites are corundum, Al_2O_3 (Fig. 5.16*b*), in which two thirds of the octahedral sites are

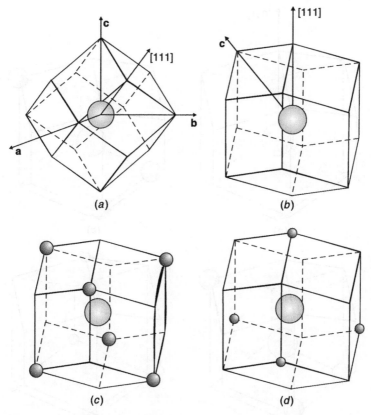

Figure 5.15 Anion-centered polyhedron (rhombic dodecahedron) found in the cubic closest-packed structure: (*a*) oriented with respect to cubic axes, the **c** axis is vertical; (*b*) oriented with [111] vertical; (*c*) cation positions occupied in the sodium chloride, NaCl, structure; and (*d*) cation positions occupied in the zinc blende (sphalerite) cubic ZnS structure.

filled in an ordered manner, and (idealized) rutile, TiO_2 (Fig. 5.16*c*), in which half of the octahedral sites are filled in an ordered manner. Filling of half the tetrahedral sites leads to the hexagonal wurtzite form of ZnS (Fig. 5.16*d*).

In these structures diffusion no longer requires the use of alternating octahedral and tetrahedral sites in all directions. A diffusion path normal to [001] is similar to that in a cubic crystal:

$$\cdots oct-tet-oct-tet-oct-tet \cdots$$

However, octahedral and tetrahedral sites are arranged in chains parallel to the **c** axis, so that diffusion in this direction can use the following paths:

$$\cdots tet-tet-tet-tet-tet \cdots$$

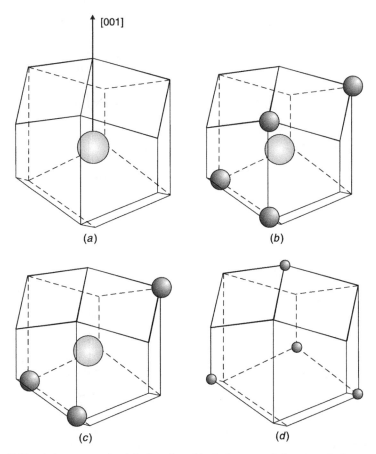

Figure 5.16 Anion-centered polyhedron found in the hexagonal closest-packed structure: (*a*) oriented with the hexagonal **c** axis vertical; (*b*) cation positions occupied in the ideal corundum, Al_2O_3, structure; (*c*) cation positions occupied in the idealized rutile, TiO_2, structure; and (*d*) cation positions occupied in the wurtzite hexagonal ZnS structure; the central cation is omitted for clarity.

or

$$\cdots oct-oct-oct-oct-oct \cdots$$

In all of the structures based upon hexagonal close-packed anions illustrated, continuous diffusion paths through empty sites can be traced, and a population of point defects is not mandatory to facilitate atom transport.

The same atom-centered polyhedra can be used to describe interstitial diffusion in all the many metal structures derived from both face-centered cubic and hexagonal closest packing of atoms. In these cases the polyhedra are centered upon a metal atom and all the tetrahedral and octahedral interstitial sites are empty. The hardening of metals by incorporation of nitrogen or carbon into the surface layers of the material via interstitial diffusion will use these pathways.

5.5 POINT DEFECT CONCENTRATION AND DIFFUSION

The random-walk model of diffusion needs to be modified if it is to accurately represent the mechanism of the diffusion. One important change regards the number of point defects present. It has already been pointed out that vacancy diffusion in, for example, a metal crystal cannot occur without an existing population of vacancies. Because of this the random-walk jump probability must be modified to take vacancy numbers into account. In this case, the probability that a vacancy is available to a diffusing atom can be approximated by the number of vacant sites present in the crystal, $[d]$, expressed as a fraction, that is

$$[d] = \frac{n_V}{N}$$

where n_V is the number of vacancies per unit volume and N the number of normal sites per unit volume, both with respect to the relevant diffusion sublattice. The measured diffusion coefficient, D, which may be a self-diffusion coefficient or a tracer diffusion coefficient, will be related to the random-walk coefficient, D_r, by

$$D = [d]D_r = \tfrac{1}{2}[d]\Gamma a^2$$

via Eq. (5.5) and (5.6), where $[d]$ is the concentration of appropriate defects per unit volume expressed as a fraction of the number of normal sites per unit volume and is unitless. (It is possible to introduce more detailed expressions for $[d]$ if specific crystal geometry and diffusion paths are taken into account.)

More than one point defect species may be present in a crystal at any temperature, and the amount of matter transported by diffusion will depend upon the number of each defect type present. In general, therefore, the overall apparent diffusion coefficient, D, will be the sum of the individual contributions, for example:

$$D = [d_1]D_1 + [d_2]D_2 + [d_3]D_3 + \cdots \tag{5.7}$$

where the terms $[d_1]$ and so forth represent the concentrations of defects present as defect fractions, and the D_n terms refer to the differing random-walk diffusion coefficient contributions relevant to vacancy diffusion, interstitial diffusion, and so on.

Just as one point defect type may dominate the defect population in a crystal, so one diffusion coefficient may be dominant, but the other diffusion coefficients can sometimes make an important contribution to the overall transport of atoms through a solid. It is by no means easy to separate these contributions to a measured value of D, and, as well as theoretical assessments, the way in which the diffusion coefficient varies with temperature can help.

5.6 CORRELATION FACTORS

A second modification to the random-walk model of diffusion is required if motion is not random but correlated in some way with preceding passage through the crystal structure. Suppose the diffusion of a vacancy in a close-packed structure, such as that of a typical metal, is the subject of investigation (Fig. 5.17a). Clearly, the vacancy can jump to any nearest neighboring site. In general, there is no preference, so that the jump is entirely random. The same is true of each succeeding situation. Thus, the vacancy can always move to an adjacent cation site and hence can follow a truly random path, as stated above.

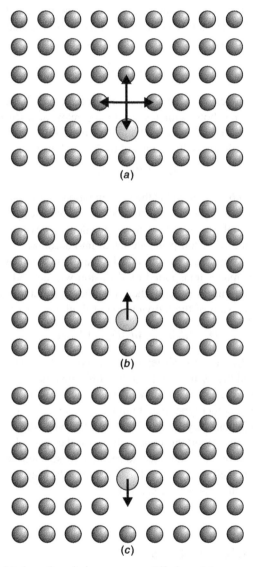

Figure 5.17 Correlated motion during vacancy diffusion: (*a*) vacancy can jump to any surrounding position and its motion follows a random walk; (*b*, *c*) the motion of a tracer atom is correlated, as a jump into a vacancy (*b*) is most likely to be followed by a jump back again (*c*).

However, the diffusion of a tracer atom by the mechanism of vacancy diffusion is different. A tracer can only move if it is next to a vacancy, and in this case, the tracer can only jump to the vacancy (Fig. 5.17b). The possibility of any other jump is excluded. Similarly, when the tracer has made the jump, then it is equally clear that the most likely jump for the tracer is back to the vacancy (Fig. 5.17c). The tracer can only jump to a new position after the vacancy has diffused to an alternative neighboring position.

The vacancy will follow a random-walk diffusion route, while the diffusion of the tracer by a vacancy diffusion mechanism will be constrained. When these processes are considered over many jumps, the mean square displacement of the tracer will be less than that of the vacancy, even though both have taken the same number of jumps. Therefore, it is expected that the observed diffusion coefficient of the tracer will be *less* than that of the vacancy. In these circumstances, the random-walk diffusion equations need to be modified for the tracer. This is done by ascribing a different probability to each of the various jumps that the tracer may make. The result is that the random-walk diffusion expression must be multiplied by a correlation factor, f, which takes the diffusion mechanism into account.

In the case of interstitial diffusion in which we have only a few diffusing interstitial atoms and many available empty interstitial sites, random-walk equations would be accurate, and a correlation factor of 1.0 would be expected. This will be so whether the interstitial is a native atom or a tracer atom. When tracer diffusion by a colinear intersticialcy mechanism is considered, this will not be true and the situation is analogous to that of vacancy diffusion. Consider a tracer atom in an interstitial position (Fig. 5.18a). An initial jump can be in any random direction in the structure. Suppose that the jump shown in Figure 5.18b occurs, leading to the situation in Figure 5.18c. The most likely next jump of the tracer, which must be back to an interstitial site, will be a return jump (Fig. 5.18d). Once again the diffusion of the interstitial is different from that of a completely random walk, and once again a correlation factor, f, is needed to compare the two situations.

The correlation factor, for any mechanism, is given by the ratio of the values of the mean square displacement of the atom (often the tracer) moving in a correlated motion to that of the atom (or vacancy) moving by a random-walk process. If the number of jumps considered is large, the correlation factor f can be written as

$$f = \frac{<x^2> \text{corr}}{<x^2> \text{random}} = \frac{D^*}{D_r}$$

that is,

$$D^* = fD_r$$

where $<x^2>$corr represents the mean square displacement of the correlated (nonrandom) walk by the diffusing atom and $<x^2>$ random the mean square displacement for a truly random diffusion process with the same number of jumps, and D_r and D^* are the random-walk and tracer diffusion coefficients.

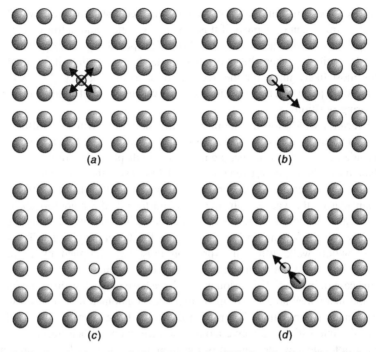

Figure 5.18 Correlated motion during intersticialcy diffusion: (*a, b*) an interstitial can make its first jump at random; and (*c, d*) the second jump is correlated with the initial jump and most likely to reverse the original movement.

Table 5.1 lists some values of the correlation factor for a variety of diffusion mechanisms in some common crystal structure types.

As discussed in the previous section, a number of diffusion mechanisms might operate at the same time. In such cases, Eq. (5.7) needs to take correlation into account, and the measured diffusion coefficient, *D*, can be written:

$$D = f_1[d_1]D_1 + f_2[d_2]D_2 + f_3[d_3]D_3 + \cdots \tag{5.8}$$

where the [*d*] and *D* terms represent the point defect concentrations and random-walk self-diffusion coefficients, as before, and the *f* terms are the appropriate correlation factors.

The diffusion of an impurity atom in a crystal, say K in NaCl, involves other considerations that influence diffusion. In such cases, the probability that the impurity will exchange with the vacancy will depend on factors such as the relative sizes of the impurity compared to the host atoms. In the case of ionic movement, the charge on the diffusing species will also play a part. These factors can also be included in a random-walk analysis by including jump probabilities of the host and impurity atoms and vacancies, all of which are likely to vary from one impurity to another and from one crystal structure to another. All of these alterations can be

TABLE 5.1   Correlation Factors for Self-diffusion

Crystal Structure	Correlation Factor (f)
Vacancy diffusion	
Diamond	0.50
Tungsten (A2, body-centered cubic)	0.73
Copper (A1, face-centered cubic)	0.78
Magnesium (A3, hexagonal close packed), all axes	0.78
Corundum (α-Al_2O_3) (cations); $\parallel$ **a** axis	0.50
Corundum (α-Al_2O_3) (cations); $\parallel$ **c** axis	0.65
Colinear interstitialcy diffusion	
Diamond	0.73
Sodium chloride (NaCl) (cations or anions)	0.33
CsCl (cations or anions)	0.33
Fluorite (CaF_2) cations	0.40
Fluorite (CaF_2) anions	0.74

All data from K. Compaan and Y. Haven, *Trans. Faraday Soc.*, **52**, 786–801 (1956); **54**, 1498–1508 (1958).

assimilated into a multiplying constant appropriate to the diffusion mechanism postulated.

5.7 TEMPERATURE VARIATION OF THE DIFFUSION COEFFICIENT

Diffusion coefficients vary considerably with temperature. This variation is generally expressed in terms of the Arrhenius equation:

$$D = D_0 \exp\left(-\frac{E_a}{RT}\right) \tag{5.9}$$

where R is the gas constant, T is the temperature (K), and D_0 is a constant term referred to as the pre-exponential factor, frequency factor, or (occasionally) as the diffusion coefficient. The term E_a is called the activation energy of diffusion, and D is the measured diffusion coefficient, often the tracer diffusion coefficient. Taking logarithms of both sides of this equation gives

$$\ln D = \ln D_0 - \frac{E_a}{RT}$$

The activation energy can be determined from the gradient of a plot of $\ln D$ versus $1/T$ (Fig. 5.19). Such graphs are known as Arrhenius plots. Diffusion coefficients found in the literature are usually expressed in terms of the Arrhenius equation D_0 and E_a values. Some representative values for self-diffusion coefficients are given in Table 5.2.

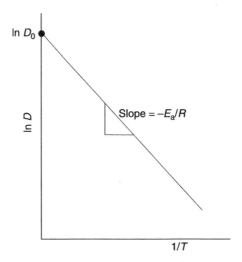

Figure 5.19 Arrhenius plot of diffusion data, $\ln D$ versus $1/T$. The slope of the straight-line graph allows the activation energy of diffusion, E_a, to be determined, and the intercept at $1/T = 0$ gives a value for the pre-exponential factor.

The random-walk model of diffusion does not allow for any temperature variation. To modify the random-walk model to take this into account, it is supposed that each time an atom moves it will have to overcome an energy barrier. This is because the migrating atoms have to leave normally occupied positions that are, by definition, the most stable positions for atoms in the crystal, to pass through less stable normally unoccupied sites. The energy associated with the diffusion path will be characterized by a series of maxima and minima. Diffusion in the sodium chloride structure, for example, is characterized by the sequence ...oct–tet–oct–tet–oct–tet... where the migrating ion jumps from an octahedral site to a tetrahedral site to an octahedral site, and so on. The bottleneck in this sequence is a triangle of three anions, so that the maxima and minima encountered consists of three different peaks (Fig. 5.20a). The overall diffusion rate will be most influenced by the highest of the energy barriers encountered, and for one-dimensional diffusion the energy barrier can be simplified to a series of single peaks and valleys in which the height of the barrier is ΔG_m, the molar Gibbs energy of migration (Fig. 5.20b). The larger the magnitude of ΔG_m the less chance there is that the atom has the necessary energy to make a successful jump.

The probability that an atom will successfully move can be estimated by using Maxwell–Boltzmann statistics. The probability p that an atom will move from one position of minimum energy to an adjacent position is

$$p = \exp\left(-\frac{\Delta g_m}{kT}\right)$$

TABLE 5.2 Representative Values for Self-diffusion Coefficients[a]

Atom	Structure	$D_0/m^2\,s^{-1}$[b]	$E_a/kJ\,mol^{-1}$	E_a/eV
		Metals and Semiconductors		
Cu	Cu (A1)	2.0×10^{-5}	200	2.1
Fe	γ-Fe (A1)	2.0×10^{-5}	270	2.8
Fe	α-Fe (A2)	2.0×10^{-4}	240	2.5
Na	Na (A2)	2.5×10^{-5}	45	0.5
Si	Si (diamond)	0.5×10^{-1}	455	4.7
Au	Si (diamond)	0.5×10^{-1}	460	4.8
Ge	Ge (diamond)	9.5×10^{-4}	290	3.0
		Compounds		
Na^+	NaCl (sodium chloride)	8.5×10^{-8}	190	2.0
Cl^-	NaCl (sodium chloride)	0.5×10^{-4}	245	2.5
K^+	KCl (sodium chloride)	0.5×10^{-4}	255	2.6
Cl^-	KCl (sodium chloride)	1.5×10^{-6}	230	2.4
Mg^{2+}	MgO (sodium chloride)	2.5×10^{-9}	330	3.4
O^{2-}	MgO (sodium chloride)	4.5×10^{-13}	345	3.6
Ni^{2+}	NiO (sodium chloride)	5.0×10^{-10}	255	2.6
O^{2-}	NiO (sodium chloride)	6.0×10^{-12}	240	2.5
Cr^{3+}	CoO (sodium chloride)	8.0×10^{-7}	250	2.6
Ti^{4+}	TiO (sodium chloride)	1.0×10^{-4}	270	2.8
O^{2-}	(Zr,Ca)O_2 (fluorite)	2.5×10^{-16}	510	5.3
O^{2-}	NbO_2 (rutile)	1.5×10^{-6}	200	2.1
Al^{3+}	Al_2O_3 (corundum; $\parallel$ **c** axis)	1.5×10^{-5}	500	5.2
O^{2-}	Al_2O_3 (corundum; $\parallel$ **c** axis)	2.0×10^{-2}	600	6.2
Pb^{2+}	PbS (sodium chloride)	8.5×10^{-13}	145	1.5
S^{2-}	PbS (sodium chloride)	7.0×10^{-13}	135	1.4
Ga	GaAs (sphalerite)	2.0×10^{-10}	400	4.2
As	GaAs (sphalerite)	7.0×10^{-5}	310	3.2
Zn^{2+}	ZnS (sphalerite)	3.0×10^{-12}	145	1.5
S^{2-}	ZnS (sphalerite)	2.0×10^{-4}	305	3.2
Cd^{2+}	CdS (sphalerite)	3.5×10^{-8}	195	2.0
S^{2-}	CdS (sphalerite)	1.5×10^{-10}	200	2.1
		Glasses		
Na	Soda–lime	1.0×10^{-6}	85	0.9
Si	Silica	3.0×10^{-2}	580	6.0
Na	Soda	1.0×10^{-6}	80	0.8

[a]Literature values for self-diffusion coefficients vary widely, indicating the difficulty of making reliable measurements. The values here are meant to be representative only.

[b]The values of diffusion coefficients in the literature are often given in $cm^2\,s^{-1}$. To convert the values given here to $cm^2\,s^{-1}$, multiply by 10^4.

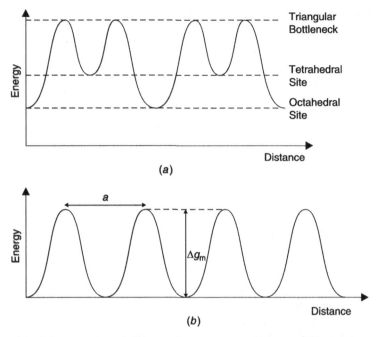

Figure 5.20 Energy barriers during atomic diffusion: (*a*) an atom diffusing in a sodium chloride structure or similar passes through octahedral and tetrahedral sites separated by triangular bottlenecks; and (*b*) a schematic energy barrier with a periodicity equal to *a*, rising to a maximum when the diffusing atom has to pass a "bottleneck" between two stationary atoms.

where Δg_m represents the height of a single energy barrier, k is the Boltzmann constant, and T the absolute temperature. This equation indicates that if Δg_m is very small, the probability that the atom will clear the barrier approaches 1.0; if Δg_m is equal to kT, the probability for a successful jump is about one-third, and if Δg_m increases above kT, the probability that the atom could jump the barrier rapidly decreases.

The atoms in a crystal are vibrating continually with frequency ν, which is usually taken to have a value of about 10^{13} Hz at room temperature. It is reasonable to suppose that the number of attempts at a jump, sometimes called the attempt frequency, will be equal to the frequency with which the atom is vibrating. The number of successful jumps that an atom will make per second, the jump frequency Γ, will be equal to the attempt frequency ν, multiplied by the probability of a successful move, that is,

$$\Gamma = \nu \, \exp\left(-\frac{\Delta g_m}{kT}\right)$$

The relationship between Γ and the diffusion coefficient, for one-dimensional random-walk diffusion, D_r, is [Eqs. (5.5) and (5.6)]:

$$D_r = \tfrac{1}{2}\Gamma a^2$$

Replacing Γ gives

$$D = \tfrac{1}{2}a^2v \, \exp\left(-\frac{\Delta g_m}{kT}\right) \tag{5.10}$$

for one-dimensional diffusion.

In the case of diffusion in a three-dimensional crystal, with equal probability of a jump a along x, y, or z,

$$<x^2> = 6D_t t \quad \text{and} \quad D = \left(\tfrac{1}{6}\right)\Gamma a^2$$

hence

$$D = \left(\tfrac{1}{6}\right)a^2v \, \exp\left(-\frac{\Delta g_m}{kT}\right)$$

The terms $\left(\tfrac{1}{2}\right)$ and $\left(\tfrac{1}{6}\right)$ represent factors that account for the geometry of the diffusion process. In general, the equation for the diffusion coefficient can be written with a geometrical factor, γ, included, so that it becomes

$$D = \gamma a^2 v \, \exp\left(-\frac{\Delta g_m}{kT}\right)$$

A comparison of Eq. (5.10) with the Arrhenius equation, Eq. (5.9), made easier by converting the barrier height Δg_m to ΔG_m (in kJ mol^{-1}) and substituting the gas constant R for the Bolzmann constant k, that is,

$$D = \gamma a^2 v \, \exp\left(-\frac{\Delta G_m}{RT}\right) \tag{5.11}$$

shows that the pre-exponential factor D_0 is equivalent to:

$$D_0 = \gamma a^2 v$$

and the activation energy, E_a, is equivalent to the height of the energy barrier to be surmounted, ΔG_m.

Equation (5.11) ignores the correction factors included in Eqs. (5.7) and (5.8). When these are added in, the measured diffusion coefficient can be written:

$$D = \gamma f[d]a^2 v \, \exp\left(-\frac{\Delta G_m}{RT}\right) \tag{5.12}$$

where f is the correlation factor (if needed) and $[d]$ is the concentration of defects involved in the diffusion.

Clearly, this equation can be made more precise by modifying the terms by reference to the crystal geometry. Additionally, the free energy term $-\Delta G_m$ can be separated into its components, using the following equation:

$$\Delta G_m = \Delta H_m - T \, \Delta S_m$$

where ΔH_m represents the enthalpy of migration and ΔS_m the entropy of migration, so that

$$\exp\left(-\frac{\Delta G_m}{RT}\right) = \exp\left(-\frac{\Delta H_m}{RT}\right) - \exp\left(-\frac{\Delta S_m}{RT}\right)$$

Combining the entropy term with the geometrical factor into a single constant allows Eq. (5.12) to be written as

$$D = \gamma f[d]a^2v \ \exp\left(-\frac{\Delta H_m}{RT}\right) \tag{5.13}$$

where γ is a combined geometrical and entropy factor, f is the correlation factor, $[d]$ is the concentration of defects involved in the diffusion, a is the jump distance from one stable site to the next during diffusion, v is the attempt frequency, $-\Delta H_m$ is the molar enthalpy of migration, R is the gas constant, and T the temperature (K).

Equation (5.13) gives an idea of the structural requirements for a high diffusion coefficient. It is clearly important that the enthalpy of migration is relatively low, which is to say that the diffusing particle should not be linked to neighbors by strong chemical bonds, and atoms do not have to squeeze through bottlenecks. Hence, the structure should ideally have weakly bound atoms and open channels. It is also important that the defect concentration, $[d]$, is high. This factor is discussed in more detail in the following two sections.

The previous discussion has supposed that the height of the potential barrier will be the same at all temperatures. This is probably not so. As the temperature increases, the lattice will expand, and, in general, ΔH_m would be expected to decrease. This will cause an Arrrhenius plot to curve slightly. Moreover, some of the other constant terms in the preceding equations may vary with temperature, also causing the plots to deviate from linearity.

5.8 TEMPERATURE VARIATION AND INTRINSIC DIFFUSION

In most ordinary solids, bulk diffusion is dominated by the impurity content, the number of impurity defects present. Any variation in D_0 from one sample of a material to another is accounted for by the variation of the impurity content. However, the impurity concentration does not affect the activation energy of migration, E_a, so that Arrhenius plots for such crystals will consist of a series of parallel lines (Fig. 5.21a). The value of the preexponential factor D_0 increases as the impurity content increases, in accord with Eq. (5.13).

In very pure crystals, the number of intrinsic defects may be greater than the number of defects due to impurities, especially at high temperatures. Under these circumstances, the value of D_0 will be influenced by the intrinsic defect population and may contribute to the observed value of the activation energy.

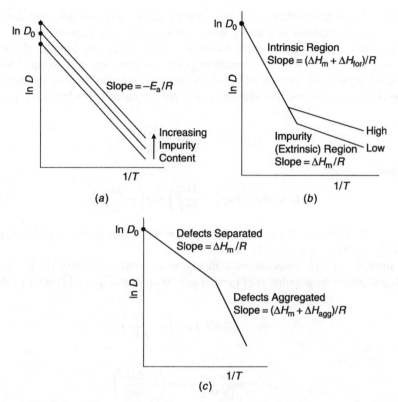

Figure 5.21 Arrhenius plots in crystals: (*a*) variation with impurity content; (*b*) almost pure crystals with low impurity concentrations; and (*c*) low-temperature defect clusters.

Suppose that vacancy diffusion is the principal mechanism involved in atom transport. An expression for the fraction of vacancies in a pure crystal is [Eq. (2.6)]

$$\frac{n_V}{N} = \exp\left[-\frac{\Delta G_V}{RT}\right]$$

Writing [*d*] as n_V/N, the diffusion coefficient can now be written:

$$D = \gamma f a^2 v \, \exp\left(-\frac{\Delta G_V}{RT}\right) \exp\left(-\frac{\Delta G_m}{RT}\right)$$

The activation energy is now made up of two parts, one part being the Gibbs energy to create a vacancy, $-\Delta G_V$, and the other to surmount the energy barrier, $-\Delta G_m$. The entropy terms can be incorporated into the geometrical factor to give

$$D = \gamma f a^2 v \, \exp\left(-\frac{\Delta H_V}{RT}\right) \exp\left(-\frac{\Delta H_m}{RT}\right)$$

where γ is the geometrical factor plus entropy terms, ΔS_V and ΔS_m and ΔH_V and ΔH_m are the enthalpies of vacancy formation and migration, respectively.

The same analysis can be applied to more complex situations. Suppose that cation vacancy diffusion is the predominant migration mechanism, in a sodium chloride structure crystal, of formula MX, which contains Schottky defects as the major type of intrinsic defects. The relevant defect concentration $[d]$ is [Eq. (2.11)]

$$[d] = \frac{n_S}{N} = \exp\left(-\frac{\Delta G_S}{2RT}\right)$$

so that

$$D = \gamma f a^2 v \, \exp\left(-\frac{\Delta H_S}{2RT}\right) \exp\left(-\frac{\Delta H_m}{RT}\right) \tag{5.14}$$

where ΔH_m represents the enthalpy of migration in vacancy diffusion and ΔH_S is the enthalpy of formation Schottky defects.

Similarly, at high temperatures, diffusion in a crystal of formula MX by interstitials will reflect the population of Frenkel defects present [Eqs. (2.13) and (2.14b)] as

$$n_F = \sqrt{(NN^*)} \exp\left(-\frac{\Delta G_F}{RT}\right)$$

so

$$[d] = \frac{n_F}{\sqrt{(NN^*)}} = \exp\left(-\frac{\Delta G_F}{2RT}\right)$$

In these circumstances, it is possible to write Eq. (5.14) in the following form:

$$D = \gamma f a^2 v \, \exp\left(-\frac{\Delta H_F}{2RT}\right) \exp\left(-\frac{\Delta H_m}{RT}\right) \tag{5.15}$$

where ΔH_m represents the enthalpy of the potential barrier to be surmounted by an interstitial atom and ΔH_F is the enthalpy of formation of a Frenkel defect.

Equations (5.14) and (5.15) retain the form of the Arrhenius equation:

$$D = D_0 \, \exp\left(-\frac{E_a}{RT}\right)$$

However, the activation energy, E_a, will consist of two terms, one representing the enthalpy of migration, ΔH_m, and the other enthalpy of defect formation ΔH_{for}.

For Schottky defects:

$$E_a = \Delta H_m + \tfrac{1}{2}\Delta H_S$$

and for Frenkel defects:

$$E_a = \Delta H_m + \tfrac{1}{2}\Delta H_F$$

Thus, Arrhenius plots obtained from very pure materials may then consist of two straight-line parts with differing slopes (Fig. 5.21b). The region corresponding to diffusion at lower temperatures has lower activation energy than the high-temperature region. The point where the two straight lines intersect is called a knee. If a number of different crystals of the same compound are studied, it is found that the position of the knee varies from one crystal to another and depends upon the impurity content. The part of the plot sensitive to impurity content is called the impurity or extrinsic region. The high-temperature part of the plot is unaffected by the impurities present and is called the intrinsic region. A comparison of the two slopes will allow an estimate of both the energy barrier to migration and the relevant defect formation energy to be made. Some values found in this way are listed in Table 5.3.

Note if the predominant defect type changes at a certain temperature, the Arrhenius plot would also be expected to show a knee because both the formation energy and the atom migration mechanism would be different in the high- and low-temperature regimes. This will be so, for example, in the case of defects that aggregate or form clusters at low temperatures and separate at higher temperatures. Following the reasoning above, it can be anticipated that at low temperature, the activation energy will be composed of a term consisting of the energy needed to separate the migrating defect from the aggregate as well as a migration energy:

$$E_a = \Delta H_m + \Delta H_{agg}$$

As defect clusters tend to disassociate at high temperatures, the aggregation enthalpy, ΔH_{agg}, would tend to zero at high temperatures. The high-temperature activation energy would then simply correspond to the migration enethalpy:

$$E_a = \Delta H_m$$

An Arrhenius plot would again be composed of two segments, with the high-temperature part showing a lower slope than the low-temperature part (Fig. 5.21c).

TABLE 5.3 Approximate Enthalpy Values for Formation and Movement of Vacancies in Alkali Halide Crystals

	Schottky Defects		
Material	ΔH_S/kJ mol^{-1}	E_m (cation)/kJ mol^{-1}	E_m (anion) /kJ mol^{-1}
NaCl	192	84	109
NaBr	163	84	113
KCl	230	75	172
KBr	192	64	46

	Frenkel Defects		
Material	ΔH_F/kJ mol^{-1}	E_m (interstitial)/kJ mol^{-1}	E_m (vacancy)/kJ mol^{-1}
AgCl	155	13	36
AgBr	117	11	23

In practice, there is often not a sharp change from one state to another, and the plots are frequently curved over the temperature range in which aggregation becomes important.

5.9 DIFFUSION MECHANISMS AND IMPURITIES

Diffusion in the extrinsic region can readily be modified by doping, although knowledge of the mechanism by which the diffusion takes place is important if this is to be immediately successful. For example, sodium chloride structure materials that conduct by a vacancy mechanism can have the cation conductivity enhanced by doping with divalent cations, as these generate compensating cation vacancies. The inclusion of cadmium chloride into sodium chloride can be written:

$$CdCl_2(NaCl) \longrightarrow Cd_{Na}^{\bullet} + V_{Na}' + 2Cl_{Cl}$$

Each added Cd^{2+} ion generates one cation vacancy. Similarly, each added Ca^{2+} ion into ZrO_2, as in calcia-stabilized zirconia (Section 1.11.6), generates a compensating anion vacancy:

$$CaO(ZrO_2) \longrightarrow Ca_{Zr}^{2'} + V_O^{2\bullet} + O_O$$

In both cases the diffusion coefficient of the ions using the vacancy population will be enhanced considerably.

Diffusion studies coupled with doping data can also yield information about the likely defect types present in a solid. For example, silica, SiO_2, is a common impurity in alumina, Al_2O_3. The Si^{4+} cations are accommodated on Al^{3+} sites as substitutional defects with an effective positive charge, $Si_{Al}^{\bullet}$. As SiO_2 is oxygen rich compared to alumina, the extra oxygen can either be accommodated as oxygen interstitials, $O_i^{2'}$:

$$2SiO_2(Al_2O_3) \longrightarrow 2Si_{Al}^{\bullet} + 3O_O + O_i^{2'}$$

or balanced by Al^{3+} vacancies, $V_{Al}^{3'}$:

$$3SiO_2(Al_2O_3) \longrightarrow 3Si_{Al}^{\bullet} + 6O_O + V_{Al}^{3'}$$

Suppose that it is known that Al diffusion is by a vacancy mechanism. As the SiO_2 impurity content increases, the Al diffusion coefficient should increase if $V_{Al}^{3'}$ defects form. If it does not, the evidence is in favor of the assumption that the incorporation of SiO_2 generates oxygen interstitials. This can be substantiated by measuring the oxygen diffusion coefficient. Oxygen diffusion in Al_2O_3 is believed to be by way of an interstitial mechanism. Clearly, therefore, an increase in oxygen diffusion coefficient with impurity SiO_2 concentration would be expected.

Possible diffusion mechanisms can also be investigated by doping with impurities with greater or smaller ionic charge and comparing the results. For example, ZnO can be doped with Li_2O or Al_2O_3. From the point of view of oxygen diffusion, these should give quite distinctive results. If the impurity atoms are supposed to form substitutional defects, in one case (Li_2O) a deficit of oxygen will lead to oxygen vacancies:

$$Li_2O(ZnO) \longrightarrow 2Li'_{Zn} + O_O + V_O^{2\bullet}$$

while in the other (Al_2O_3) the additional oxygen may lead to oxygen interstitials or aluminum vacancies:

$$Al_2O_3(ZnO) \longrightarrow 2Al^\bullet_{Zn} + 2O_O + O_i^{2\prime}$$
$$Al_2O_3(ZnO) \longrightarrow 2Al^\bullet_{Zn} + 3O_O + V_{Zn}^{2\prime}$$

A comparison of O diffusion in pure ZnO and ZnO doped with varying amounts of Al_2O_3 and Li_2O would indicate whether O diffusion is more likely to be by way of a vacancy or an interstitial mechanism. In fact, oxygen diffusion is similar in pure and Li_2O-doped materials and is much faster in Al_2O_3-doped materials, strongly suggesting that oxygen diffusion is by way of an interstitial mechanism.

Of course, it is possible to write other and more complex defect reactions, which may be equally compatible with the results. In all of these examples, therefore, direct evidence is desirable before the mechanisms can be regarded as proven.

5.10 CHEMICAL AND AMBIPOLAR DIFFUSION

The above sections have focused upon homogeneous systems with a constant composition in which tracer diffusion coefficients give a close approximation to self-diffusion coefficients. However, a diffusion coefficient can be defined for any transport of material across a solid, whether or not such limitations hold. For example, the diffusion processes taking place when a metal A is in contact with a metal B is usually characterized by the interdiffusion coefficient, which provides a measure of the total mixing that has taken place. The mixing that occurs when two chemical compounds, say oxide AO is in contact with oxide BO, is characterized by the chemical diffusion coefficient (see the Further Reading section for more information).

Ambipolar diffusion involves the transport of charged species, and in such cases overall electric charge neutrality must be maintained during diffusion. Moreover, during ambipolar diffusion the difference in the mobilities of the diffusing species sets up a field, the Nernst field, that influences the rates of motion of the particles.

Corrosion reactions of a metal with gaseous species such as oxygen, chlorine, sulfur containing molecules or water vapor to produce a thin layer of product phase are typical of ambipolar diffusion reactions. For example, metal oxidation involves the formation of a metal oxide, the simplest example of which is a monoxide formed by a divalent metal M, such as Mg:

$$M + \tfrac{1}{2}O_2 \longrightarrow MO$$

After initial attack, further reaction can only proceed if ions can diffuse across the product, either from the outside gaseous phase into the inner metal layer or else from the metal out to meet the gas.

The reaction can proceed by diffusion of M^{2+} cations outward from the metal toward the gas atmosphere, allowing the film to grow at the oxide–gas interface. In this case, a large negative charge would be left behind at the metal–metal oxide interface, which would soon slow down the moving cations and bring the reaction to a halt. To maintain electrical neutrality in the system and to allow the reaction to continue, this diffusion must be accompanied by a parallel diffusion of electrons (Fig. 5.22a) or a counterdiffusion of holes (Fig. 5.22b). This disparity of mobilities between the cations and the electrons or holes creates a Nernst field that acts so as to slow down the faster diffusing species and speed up the slower diffusing species

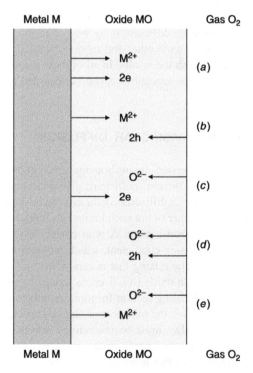

Figure 5.22   Formation of metal oxide films: (*a*) parallel diffusion of cations and electrons; (*b*) counterdiffusion of cations and holes; (*c*) counterdiffusion of anions and electrons; (*d*) parallel diffusion of anions and holes; and (*e*) counterdiffusion of anions and cations.

until charge neutrality is maintained across the oxide layer. It is clear that the process occurring depends on the intrinsic semiconductivity of the oxide film. Electron transport is expected if the oxide is an n-type semiconductor such as ZnO and hole transport if the film consists of a naturally p-type semiconductor, such as NiO.

The diffusion of oxide ions may also dominate an oxidation reaction. In a similar fashion to that described for cations, the anions can be accompanied by a flow of holes or a counterflow of electrons (Fig. 5.22c and 5.22d). The arriving O^{2-} anions extend the oxide film at the metal–metal oxide boundary. Once again, which of these mechanisms is theoretically possible will depend upon whether the oxide is naturally n-type or p-type.

Many metal oxides are insulators. In these cases oxidation can only occur if charge neutrality is maintained by way of a significant counterdiffusion of cations and anions (Fig. 5.22e). Once again, mobilities will be equalized by the Nernst field set up when one species moves faster than the other. When counterdiffusion of ions is involved, the oxide film grows at both the inner and outer surfaces.

5.11 DISLOCATION AND GRAIN BOUNDARY DIFFUSION

The concentration of diffusing material at a depth x from the initial interface in a solid containing dislocations and grain boundaries will be composed of contributions from bulk diffusion in from the initial interface, fast diffusion down the dislocations and grain boundaries, and diffusion into the bulk laterally from the dislocations and grain boundaries. In the part of the crystal corresponding to the linear "tail" of the concentration profile (Fig. 5.3), diffusion from the surface can be ignored and only lateral diffusion from the defect is important. The concentration of the diffusing species in the solid will then depend upon both the bulk diffusion coefficient, for the lateral spread, and the diffusion coefficient along the defect. The form of any equation relating concentration will then involve the relative magnitudes of the diffusion coefficients and the spacing of the defects. It is thus not simple to produce a generalized equation that will fit all circumstances.

The likelihood that Fick's laws will be obeyed in a crystal containing dislocations is dependent upon the spacing between the defects. Provided that this spacing is much greater than the diffusion length $(Dt)^{1/2}$, where D is the bulk diffusion coefficient, Fick's laws are obeyed, with an effective (measured) diffusion coefficient, D_{eff}, given by

$$D_{eff} = D(1 - f_s) + D_d f_s$$

where D_d is the diffusion coefficient along the dislocations and f_s is the fraction of sites that are considered to be involved in dislocation diffusion. (The factor f_s is sometimes written as the fraction of time that a diffusing atom spends in the neighborhood of the dislocation, f_t.) In a normal solid, f_s is of the order of 10^{-8}, and so D_d will only contribute significantly to the measured value of the (effective) diffusion coefficient at low temperatures where volume diffusion is low, even if the value of D_d is 10^3 or 10^4 that of the bulk diffusion coefficient, that is,

$$D_{eff} = D(1 - 10^{-8}) + (D \times 10^4)10^{-8}$$
$$\approx (1 + 10^{-4})D$$

Dislocations can attract a population of impurities, vacancies, or self-interstitials that are bound to the dislocation core by a binding energy Δg_b. These will be liberated and become free to contribute to the overall diffusion at higher temperatures, so that it is possible to write

$$D_{eff} = D(1 - f_s) + D_d f_s \, \exp\left(-\frac{\Delta g_b}{kT}\right)$$

The situation with respect to grain boundary diffusion is similar. The contribution of the grain boundaries to the effective diffusion coefficient can be written as follows:

$$D_{eff} = D(1 - f_s) + D_{gb} f_s$$

where D_{gb} is the grain boundary diffusion coefficient, assuming that the diffusion length $(Dt)^{1/2}$, is much greater than the grain size. The solution to the diffusion equation must now take into account bulk diffusion from the surface, diffusion down the grain boundaries, and bulk diffusion from the grain boundaries into the bulk. In this situation, the one-dimensional solution appropriate to a thin layer of diffusant, Eq. (5.4), becomes

$$D_{gb}\delta = 1.322 \left(\frac{-d \ln c}{dx^{6/5}}\right)^{-5/3} \left(\frac{D}{t}\right)^{1/2}$$

where D_{gb} is the grain boundary diffusion coefficient, δ is the grain boundary width, D the bulk diffusion coefficient, c is the concentration of diffusant in a layer at depth x after time t, when volume diffusion is perpendicular to the grain boundary. The term $(d \ln c/dx^{6/5})$ is the slope of a plot of log c versus $x^{6/5}$, and therefore such a plot allows the value of $D_{gb}\delta$ to be obtained. Note that the grain boundary diffusion coefficient cannot be derived directly, only the product of the diffusion coefficient and the grain boundary width. Moreover, the bulk diffusion coefficient must also be known in order to determine D_{gb}.

Other solutions for the determination of grain boundary diffusion coefficients, using differing assumptions, have also been derived. Further complexity is added when the grain boundary separates two different phases. These specialist topics lie outside the scope of this introduction. (See the Further Reading section for more information.)

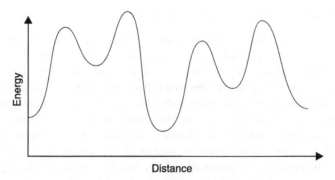

Figure 5.23 Irregular diffusion energy barrier encountered in glassy or amorphous solids.

5.12 DIFFUSION IN AMORPHOUS AND GLASSY SOLIDS

Diffusion coefficients in amorphous solids such as oxide glasses and glasslike amorphous metals can be measured using any of the methods applicable to crystals. In this way it is possible to obtain the diffusion coefficients of, say, alkali and alkaline earth metals in silicate glasses or the diffusion of metal impurities in amorphous alloys. Unlike diffusion in crystals, diffusion coefficients in amorphous solids tend to alter over time, due to relaxation of the amorphous state at the temperature of the diffusion experiment.

Broadly speaking, the diffusion coefficients vary in an Arrhenius-like way as the temperature changes, which is taken as evidence that the diffusing species moves by surmounting an energy barrier, as in the case of the crystals. However, the energy barrier will be less regular than that for a crystal (Fig. 5.23). Moreover, it is difficult to use these diffusion results to assign mechanisms to the atom movement, although atoms in a glass can be described as "interstitial-like" if they occur in a relatively open part of the glassy network or "substitutional-like" if they are bound into the network. Atomistic simulation has been very successful in clarifying diffusion mechanisms in crystals, and it is likely that atomistic studies will elucidate diffusion mechanisms in glasses and amorphous alloys, although problems still remain in defining interatomic potentials and in summing the interactions in a nonrepeating matrix.

5.13 ANSWERS TO INTRODUCTORY QUESTIONS

What is volume diffusion?

Volume diffusion refers to the transport of atoms through the body of a solid. It is also called lattice or bulk diffusion. In amorphous or glassy solids and in cubic crystals, the speed of diffusion in all directions is the same and is said to be isotropic. In all other crystals, the rate of volume diffusion depends upon the direction taken and is anisotropic. Volume diffusion is usually much slower than short-circuit diffusion, which refers to diffusion along two- and three-dimensional imperfections in the material.

How do defects influence diffusion in solids?

The ability of an atom to diffuse through a solid frequently depends upon the number of defects present. For geometrical reasons an atom may be able to jump between sites readily but will be unable to make the move if the target position is occupied by an atom. This means that a population of vacancies is needed for normal atom diffusion to occur at all. (An exception to this is the mechanism of ring diffusion, which appears to operate in some solids.) The diffusion of interstitial atoms clearly requires the presence of defects in the form of interstitials. Impurities, either substitutional or interstitial, are governed by the same constraints. In general, it is reasonable to say that bulk diffusion through a perfect single crystal is dominated by point defects, with both impurity and intrinsic defect populations playing a part. It is also true that the rates of diffusion increase in proportion to the number of defects present.

How does temperature effect diffusion in solids?

Diffusion coefficients invariably increase as the temperature rises. This variation is generally expressed in terms of the Arrhenius equation:

$$D = D_0 \exp\left(-\frac{E_a}{RT}\right)$$

where R is the gas constant, T is the temperature (K), and D_0 is a constant term referred to as the pre-exponential factor, frequency factor, or (occasionally) as the diffusion coefficient. The magnitude of this term is a function of the defect concentrations present. The term E_a is called the activation energy of diffusion and D is the measured diffusion coefficient. A plot of ln D versus $1/T$ will be a straight line if the Arrhenius law is obeyed. Curved plots indicate that complex interactions are taking place between the diffusing species and the surroundings and that diffusion is not a random process.

PROBLEMS AND EXERCISES

Quick Quiz

1. Diffusion through a crystalline structure is called:
   (a) Tracer diffusion
   (b) Volume diffusion
   (c) Self-diffusion

2. Short-circuit diffusion refers to diffusion along:
   (a) Defects
   (b) Tunnels
   (c) Planes

3. A diffusion profile is a graph of:
   (a) Concentration versus time
   (b) Concentration versus distance
   (c) Distance versus time

4. To obtain the diffusion coefficient from a diffusion profile, use:
   (a) Fick's first law
   (b) The Gruzin equation
   (c) The diffusion equation

5. Vacancy diffusion refers to:
   (a) Diffusion of atoms via vacant sites
   (b) Diffusion not involving atoms
   (c) Diffusion of defects

6. Correlation factors compensate for:
   (a) Nonrandom diffusion
   (b) Impurity content
   (c) Defect concentrations

7. The activation energy for diffusion can be determined by a plot of:
   (a) $\ln c$ versus $1/T$
   (b) $\ln D$ versus T
   (c) $\ln D$ versus $1/T$

8. The geometrical factor in diffusion equations compensates for:
   (a) The types of defects present
   (b) The size of the diffusing atoms
   (c) The crystal structure

9. Ti impurities substitute for Mg on doping MgO with a small amount of TiO_2. Will the diffusion coefficient of Mg be expected to:
   (a) Decrease
   (b) Increase
   (c) Stay the same

10. Ambipolar diffusion refers to diffusion involving:
    (a) Polar crystals
    (b) Amorphous solids
    (c) Charged particles

Calculations and Questions

1. Radioactive ^{22}Na was coated onto a glass sample that was made into a diffusion couple (Fig. 5.2) and heated for 4 h at 411°C. The radioactivity perpendicular to the surface is given in the following table. Calculate the tracer diffusion coefficient, D^*, of ^{22}Na in the glass.

Activity/counts s^{-1}	Distance/μm
2.239×10^5	80
1.122×10^5	90
3.311×10^4	100
1.549×10^4	110
3.981×10^3	120
1.479×10^3	130
3.020×10^2	140

Source: Data adapted from E. M. Tanguep Njiokop and H. Mehrer, *Solid State Ionics*, **177**, 2839–2844 (2006).

2. The list below shows the last position reached, in units of the jump step a, during a random walk for 100 atoms, each of which makes 200 jumps. If the jump time is 10^{-3} s and the jump distance, a, is 0.3 nm, estimate the diffusion coefficient (a) in units of a^2 s^{-1} and (b) in units of m^2 s^{-1}:

$$\{0\},\{12\},\{-14\},\{4\},\{4\},\{8\},\{8\},\{-12\},\{-24\},\{16\},\{-8\},$$
$$\{-12\},\{-6\},\{14\},\{14\},\{18\},\{-4\},\{10\},\{6\},\{4\},\{-30\},\{-8\},$$
$$\{16\},\{-14\},\{12\},\{0\},\{-6\},\{0\},\{-6\},\{14\},\{8\},\{-6\},\{6\},$$
$$\{-22\},\{6\},\{-2\},\{-18\},\{-10\},\{18\},\{-6\},\{2\},\{4\},\{-4\},\{-6\},$$
$$\{0\},\{-10\},\{20\},\{-2\},\{12\},\{-10\},\{-4\},\{-12\},\{8\},\{-8\},\{-6\},$$
$$\{-18\},\{8\},\{30\},\{6\},\{0\},\{-12\},\{10\},\{-2\},\{-16\},\{30\},\{-14\},$$
$$\{-12\},\{8\},\{0\},\{-26\},\{4\},\{8\},\{26\},\{14\},\{-18\},\{-4\},\{-4\},$$
$$\{-12\},\{-16\},\{10\},\{14\},\{2\},\{12\},\{0\},\{6\},\{8\},\{20\},\{12\},$$
$$\{10\},\{-16\},\{14\},\{10\},\{22\},\{10\},\{6\},\{12\},\{0\},\{4\},\{0\},\{20\}$$

3. Using the data in Question 2, estimate the activation energy for diffusion if the vibration frequency is 1×10^{13} Hz and the temperature of the diffusion is 1200 K.

4. The fraction of vacancies in a crystal of NaCl, n_V/N due to a population of Schottky defects, is 5×10^{-5} at 1000 K. In a diffusion experiment at this temperature, the activation energy for self-diffusion of Na was found to be 173.2 kJ mol^{-1}. Determine the potential barrier that the diffusing ions have to surmount.

5. The radioactive tracer diffusion coefficient of ^{22}Na in glass is given in the following table. Estimate the activation energy for diffusion.

T/K	$D^*/m^2\ s^{-1}$
400	1.16×10^{-19}
450	2.08×10^{-18}
500	3.63×10^{-17}
550	2.81×10^{-16}
600	2.10×10^{-15}
650	1.03×10^{-14}
700	3.19×10^{-14}

Source: Data adapted from E. M. Tanguep Njiokop and H. Mehrer, *Solid State Ionics*, **177**, 2839–2844 (2006).

6. Using the data in Question 5, at what temperature (°C) will the penetration depth be 5 μm after 2 h of heating?

7. Derive the equation for interstitial diffusion in a pure material equivalent to that for vacancy diffusion:

$$D = \gamma a^2 v\ \exp\left(-\frac{\Delta H_V}{RT}\right) \exp\left(-\frac{\Delta H_m}{RT}\right)$$

8. Radioactive ^{18}O was diffused into a polycrystalline pellet of ZnO at 900°C for 48 h. The diffusion profile had a marked tail showing that extensive grain boundary diffusion had occurred. The variation of the concentration of the radioactive isotope with depth for the tail of the penetration profile is given in the following table. Calculate the grain boundary diffusion coefficient, D_{gb}, of ^{18}O if the bulk diffusion coefficient at 900°C is $5.53 \times 10^{-21}\ m^2\ s^{-1}$ and the grain boundary width is taken as 1 nm.

^{18}O Concentration/%	Distance/μm
1.850	0.21
2.118	0.175
2.387	0.15
2.656	0.125
2.925	0.10

Source: Data adapted from A. C. S. Sabioni et al., *Materials Research*, **6**, 173–178 (2003).

9. Plot the concentration profile over the range 0 to 1×10^{-6} m for Al diffusion into Al_2O_3 (corundum) in a planar sandwich type of diffusion couple using the data: $c_0 = 10$, $t = 100$ h, $D = 8 \times 10^{-20}\ m^2\ s^{-1}$ at 1610°C.

10. Plot the residual activity over the range 0 to 1×10^{-3} m for ^{65}Zn diffusion into ZnO in a planar sandwich type of diffusion couple using the data: $A_r(0) = 10,000$, $t = 29$ h, $D = 2.7 \times 10^{-12}\ m^2\ s^{-1}$ at 1300°C.

11. Plot the concentration profile over the range 0 to 2×10^{-3} m for Fe diffusion into TiO_2 (rutile) in a planar sandwich type of diffusion couple using the data: $c_0 = 600$, $t = 300$ s, $D = 2 \times 10^{-10}$ m^2 s^{-1} at 800°C.

12. Plot the residual activity over the range 0 to 1×10^{-5} m for ^{63}Ni diffusion into CoO in a planar sandwich type of diffusion couple using the data: $A_r(0) = 100$, $t = 30$ min, $D = 1.56 \times 10^{-14}$ m^2 s^{-1} at 953°C.

FURTHER READING

The classical theory of diffusion is covered in:

J. Crank *The Mathematics of Diffusion*, 2nd ed., Oxford University Press, Oxford, United Kingdom, 1975.

J. S. Kirkaldy and D. J. Young *Diffusion in the Condensed State*, Institute of Metals London, 1987.

A. D. Le Claire, Diffusion, in *Treatise on Solid State Chemistry*, Vol. 4, N. B. Hannay, Ed., Plenum, New York, 1976, Chapter 1.

J. Philibert, *Atom Movements: Diffusion and Mass Transport*, translated by S. J. Rothman, Les Editions de Physique, F-91944 Les Ulis, 1991.

Short-circuit diffusion is described in:

A. Atkinson *J. Chem. Soc. Faraday Trans.*, **86**, 1307–1310 (1990).

I. Kaur and W. Gust *Fundamentals of Grain and Interphase Boundary Diffusion*, Ziegler, Stuttgart, 1988.

A discussion of the relation between random walks and diffusion, together with examples of computations, is found in:

S. Wolfram, *A New Kind of Science*, Wolfram Media Inc., Champaign, IL 61820, 2002, Chapter 7.

Diffusion in glass and amorphous materials:

R. S. Averback, *Mat. Res. Soc. Bull.*, **XX**(11), 47–52 (1991).

E. M. Tanguep Njiokep and H. Mehrer, *Solid State Ionics*, **177**, 2839–2844 (2006).

Diffusion in semiconductors, including AsSb, is described by:

H. Bracht, *Mat. Res. Soc. Bull.*, **25**(6), 22–27 (2000).

H. Bracht, E. E. Haller, and R. Clark-Phelps *Phys. Rev. Lett.*, **81**, 393–396 (1998).

H. Bracht, S. P. Nicols, W. Walukiewicz, J. P. Silveira, F. Briones, and E. E. Haller, *Nature*, **408**, 69–72 (2000).

CHAPTER 6

Intrinsic and Extrinsic Defects in Insulators: Ionic Conductivity

Why are ionic conductors vital for battery operation?
How does an oxygen sensor work?
What is a fuel cell?

This chapter mostly discusses defects in insulating crystals. These compounds are characterized by ions with a strong and single predominant valence. They are not easily oxidized or reduced and are typified by the normal group metals [group 1 (Na, K, etc.), group 2, (Mg, Ca, etc.), group 3, (Al, Sc, etc.)]. The compounds of these metals are generally fully stoichiometric. In pure compounds, point defects are limited to populations of vacancies, interstitials, or substituents and, except in exceptional cases, the defect populations are low. Such solids are used in a variety of applications, but in this chapter the focus will be upon ionic conductivity and applications stemming from ionic transport.

6.1 IONIC CONDUCTIVITY

6.1.1 Ionic Conductivity in Solids

Ionic conductivity, in the context of solids, refers to the passage of ions across a solid under the influence of an externally applied electric field. High ionic conductivity is vital to the operation of batteries and related devices. As outlined earlier (Section 2.4), in essence a battery consists of an electrode (the anode), where electrons are moved out of the cell, and an electrode (the cathode), where electrons move into the cell. The electrons are thus generated in the region of the anode and then move around an external circuit, carrying out a useful function, before entering the cathode. The anode and cathode are separated inside the battery by an electrolyte. The circuit is completed inside the battery by *ions* moving across an electrolyte. A key component in battery construction is an electrolyte that can support ionic conduction but *not* electronic conduction.

Defects in Solids, by Richard J. D. Tilley
Copyright © 2008 John Wiley & Sons, Inc.

The output of a battery depends upon the concentrations and Gibbs energy of the components at the anode and cathode. Because of this, battery technology also underpins the operation of many sensors.

For a battery to give a reasonable power output, the ionic conductivity of the electrolyte must be substantial. Historically, this was achieved by the use of liquid electrolytes. However, over the last quarter of a century there has been increasing emphasis on the production of batteries and related devices employing solid electrolytes. These are sturdy and ideal for applications where liquid electrolytes pose problems. The primary technical problem to overcome is that of achieving high ionic conductivity across the solid.

There are three broad categories of materials that have been utilized in this endeavor. In the first, even in fully stoichiometric compounds, the ionic conductivity is high enough to be useful in devices because the cation or anion substructure is mobile and behaves rather like a liquid phase trapped in the solid matrix. A second group have structural features such as open channels that allow easy ion transport. In the third group the ionic conductivity is low and must be increased by the addition of defects, typically impurities. These defects are responsible for the enhancement of ionic transport.

Optimization of all of these categories of materials has produced solids in which the ionic conductivity is as large as that normally found in solutions (Fig. 6.1). Such materials are sometimes called super-ionic conductors, but the terms fast ion conductors or solid electrolytes are to be preferred to avoid confusion with metallic superconductors, which transport electrons and holes, not ions, and by a quite different mechanism.

6.1.2 Fundamental Concepts

The electrical conductivity of a material is made up by the movement of charged particles: cations, anions, electrons, or holes. The conductivity due to charged particles of type i moving through a solid is given by

$$\sigma_i = c_i q_i \mu_i$$

where σ_i is the conductivity ($\Omega^{-1}\,m^{-1} = S\,m^{-1}$),[1] c_i is the number of charged particles per unit volume (particles m^{-3}), q_i is the charge on the particles (C), and μ_i is the mobility of the particles ($m^2\,V^{-1}\,s^{-1}$). When several different species contribute to the overall conductivity, the total conductivity, σ, is simply the sum of the various contributions:

$$\sigma = \sum \sigma_i$$

---

[1]Note: many published studies use centimeters rather than meters when reporting results. To convert conductivity in $\Omega^{-1}\,cm^{-1}$ to $\Omega^{-1}\,m^{-1}$, multiply the value by 100. To convert conductivity in $\Omega^{-1}\,m^{-1}$ to $\Omega^{-1}\,cm^{-1}$, divide the value by 100.

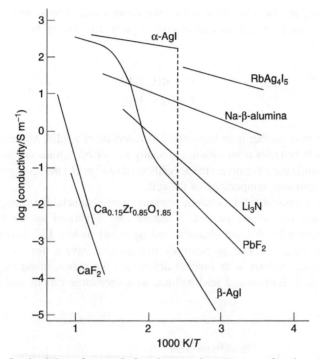

Figure 6.1 Conductivity of several fast ion conductors as a function of reciprocal temperature.

The electrical conductivity of a material that conducts solely by ionic transport is given by

$$\sigma = c_a q_a \mu_a + c_c q_c \mu_c \tag{6.1}$$

where σ is the conductivity, c_a and c_c are the concentrations of mobile anions and cations, q_a and q_c are the charges on the ions, and μ_a and μ_c are the mobilities of the particles. The ionic charge is often written Ze, where Z is the ionic valence and e the electron charge to give

$$\sigma = c_a Z_a e \mu_a + c_c Z_c e \mu_c$$

The fraction of the conductivity that can be apportioned to each ion is called its transport number, defined by

$$\sigma_c = t_c \sigma$$
$$\sigma_a = t_a \sigma$$

where σ_c, σ_a are the conductivities of the cations and anions, and t_c, t_a are the transport numbers for cations and anions, respectively. As can be seen from these relationships:

$$\sigma = \sigma(t_c + t_a)$$
$$t_c + t_a = 1$$

The transport number is an important characteristic of a solid. Materials for use as electrolytes in batteries need cation, especially Li^+ or Na^+, transport numbers to be close to 1, while the electrolytes in fuel cells need O^{2-} or H^+ transport numbers close to 1 at the operating temperature of the cell.

In densely packed solids without obvious open channels, the transport number depends upon the defects present, a feature well illustrated by the mostly ionic halides. Lithium halides are characterized by small mobile Li^+ ions that usually migrate via vacancies due to Schottky defects and have t_c for Li^+ close to 1. Similarly, silver halides with Frenkel defects on the cation sublattice have t_c for Ag^+ close to 1. Barium and lead halides, with very large cations and that contain

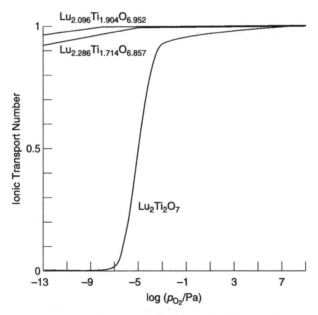

Figure 6.2 Ionic transport number for oxide ion conductivity in the pyrochlore phases $Lu_2Ti_2O_7$, $Lu_{2.096}Ti_{1.904}O_{6.952}$, and $Lu_{2.286}Ti_{1.714}O_{6.857}$. [Data adapted from A. V. Shlyakhtina, J. C. C. Abrantes, A. V. Levchenko, A. V. Knot'ko, O. K. Karyagina, and L. G. Shcherbakova, *Solid State Ionics*, **177**, 1149–1155 (2006).]

Frenkel defects on the anion sublattice show only anion migration and hence have t_a close to 1. The alkali halides NaF, NaCl, NaBr, and KCl in which Schottky defects prevail and in which the cations and anions are of similar sizes have both cation and anion contributions to ionic conductivity and show intermediate values of both anion and cation transport number.

In oxides the situation depends upon the defects present and, because of this, upon the composition, temperature, and the surrounding oxygen partial pressure (Chapter 7). Calcia-stabilized zirconia and similar materials have t_a close to 1 at temperatures greater than about 700°C. Similarly, perovskite structure oxides, including those that adopt the brownmillerite structure (Section 4.9) show oxygen t_a values close to 1 at temperatures above that at which the vacancies disorder. Many oxides that adopt the pyrochlore ($A_2B_2O_7$) structure, show high oxygen ion conductivity with t_a close to 1 at higher oxygen pressures. This latter feature, as well as the complexity that may underlie the magnitude of transport number, is illustrated by the nonstoichiometric phase $Lu_{2+x}Ti_{2-x}O_{7-x/2}$. The oxide ion transport number, t_a, in the stoichiometric phase $Lu_2Ti_2O_7$ is close to 1 at high oxygen partial pressures but falls away at low oxygen partial pressure. The slightly Lu-rich phases $Lu_{2.096}Ti_{1.904}O_{6.952}$ and $Lu_{2.286}Ti_{1.714}O_{6.857}$ are quite different and have anion transport numbers close to 1 independent of the surrounding oxygen partial pressure (Fig. 6.2).

In many transition-metal oxides and sulfides, ionic conductivity is augmented by electronic conductivity, and transport numbers need to include contributions from electrons and holes. These mixed conductors are described in Section 8.8.

6.2 MECHANISMS OF IONIC CONDUCTIVITY

6.2.1 Random-Walk Model

Ionic conductivity can be described by random-walk equations similar to those for diffusion. Taking a one-dimensional model for simplicity, with an electric field applied along the $+x$ direction, an ion can jump either in the field direction or against the field. Unlike the case of simple diffusion, the energy barrier is modified by the applied field so that it is lower in one direction and higher in the other. For ions of nominal charge Ze, the barrier will be reduced by $\frac{1}{2}ZeaE_v$ in one direction and raised by the same amount in the other, where Ze is the ionic charge, a is the jump distance, and E_v is the electric field strength (Fig. 6.3). Note that in this model ions still jump in both directions, it is only the probabilities of a jump to the left or right that change. For positive ions, the likelihood of a jump in the field direction will be greater than that of a jump against the field, while the opposite will be true for negative ions. Analysis of the modified random-walk equations shows that a plot of the final position reached by a large number of ions N starting from a planar interface will still have the form of the normal distribution. However, the peak (or mean) will be shifted away from zero, so that on average positive and negative ions move in opposite directions (Fig. 6.4).

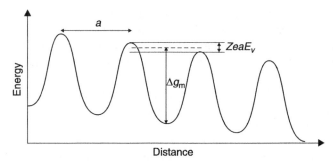

Figure 6.3   Potential barrier to be surmounted by a diffusing ion in the presence of an electric field; schematic. The distance a represents the jump distance between stable sites, and Δg_m is the average height of the potential barrier.

The number of jumps that an ion will make in the favored direction of the field per second can be analyzed as before (Section 5.7) to give:

$$\Gamma_+ = \nu \exp\left[-\frac{(\Delta g_m - \frac{1}{2}ZeaE_v)}{kT}\right]$$

where the potential barrier for migration in the absence of an applied field is Δg_m, the potential barrier in the presence of the field is $(\Delta g_m - \frac{1}{2}ZeaE_v)$, ν is the attempt frequency or vibration frequency of the ion in a stable position, k is the Boltzmann constant, and T the temperature (K). The number of successful jumps in the opposite

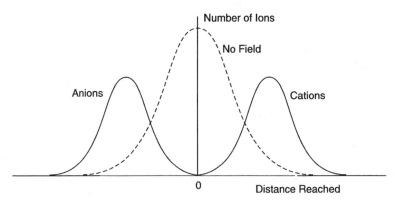

Figure 6.4   Distribution curve for ionic conductivity: random-walk diffusion in the absence of a field is the dashed line. In the presence of the electric field, the distribution splits into two parts, one for cations and one for anions, each shifted with respect to the origin.

direction will be given by

$$\Gamma_- = v \exp\left[-\frac{(\Delta g_m + \frac{1}{2}ZeaE_v)}{kT}\right]$$

where the potential barrier for migration is now $(\Delta g_m + \frac{1}{2}ZeaE_v)$.

The overall jump rate in the direction of the field is $(\Gamma_+ - \Gamma_-)$, and as the net velocity of the ions in the direction of the field, v, is given by the net jump rate multiplied by the distance moved at each jump, we can write

$$v = va\exp\left[-\frac{(\Delta g_m - \frac{1}{2}ZeaE_v)}{kT}\right] - \exp\left[-\frac{(\Delta g_m + \frac{1}{2}ZeaE_v)}{kT}\right]$$

$$= va\exp\left(-\frac{\Delta g_m}{kT}\right)\left[\exp\left(\frac{ZeaE_v}{2kT}\right) - \exp\left(-\frac{ZeaE_v}{2kT}\right)\right]$$

For the field strengths used in conductivity experiments, $ZeaE_v$ is less than kT, so

$$\left[\exp\left(\frac{ZeaE_v}{2kT}\right) - \exp\left(\frac{-ZeaE_v}{2kT}\right)\right] \approx \frac{ZeaE_v}{kT}$$

and v can be written as:

$$v = \left(\frac{vZea^2E_v}{kT}\right)\exp\left(-\frac{\Delta g_m}{kT}\right)$$

The mobility, μ, of the ion is defined as the velocity when the value of E_v is unity, so:

$$\mu = \left(\frac{vZea^2}{kT}\right)\exp\left(-\frac{\Delta g_m}{kT}\right)$$

Often a geometrical factor γ is included to take into account different crystal structures to give

$$\mu = \left(\frac{\gamma vZea^2}{kT}\right)\exp\left(-\frac{\Delta g_m}{kT}\right)$$

Substituting for μ in the equation,

$$\sigma = cZe\mu$$

$$= \left(\frac{\gamma cva^2 Z^2 e^2}{kT}\right) \exp\left(-\frac{\Delta g_m}{kT}\right)$$

where γ is a geometrical factor to take into account different diffusion geometries, c is the concentration of mobile ions present in the crystal, a is the jump distance, Ze the charge on the ions, v is the attempt frequency for a jump, k is the Boltzmann constant, T is the temperature (K), and Δg_m is the height of the barrier to be overcome during migration from one stable position to another.

The barrier height energy term is often split thus:

$$\Delta g_m = \Delta h_m - T \Delta s_m$$

and the entropy term included in the geometrical factor to give

$$\sigma = \left(\frac{\gamma cva^2 Z^2 e^2}{kT}\right) \exp\left(-\frac{\Delta h_m}{kT}\right) \tag{6.2}$$

where Δh_m is the enthalpy of migration. This equation represents Arrhenius-type behavior:

$$\sigma T = \sigma_0 \exp\left(-\frac{\Delta h_m}{kT}\right)$$

where σ_0 is a constant. This equation is often written in more general terms as

$$\sigma T = \sigma_0 \exp\left(-\frac{E_a}{RT}\right) \tag{6.3}$$

where E_a is called the activation energy for conduction and R is the gas constant. The slope of a plot of $\ln(\sigma T)$ versus $(1/T)$ will yield a value for the activation energy (Fig. 6.5). Note that in many published reports, the activation energy is derived from the slope of a plot of $\ln \sigma$ versus $1/T$. Taking logarithms of Eq. (6.3):

$$\ln(\sigma T) = \ln \sigma_0 - \frac{E_a}{RT}$$

$$\ln \sigma + \ln T = \ln \sigma_0 - \frac{E_a}{RT}$$

The term $\ln T$ is not constant but varies across the temperature range. Thus, while a plot of $\ln \sigma$ versus $1/T$ is useful for comparative purposes (Fig. 6.1), the slope of these graphs will not give an activation energy that is a measure of the migration

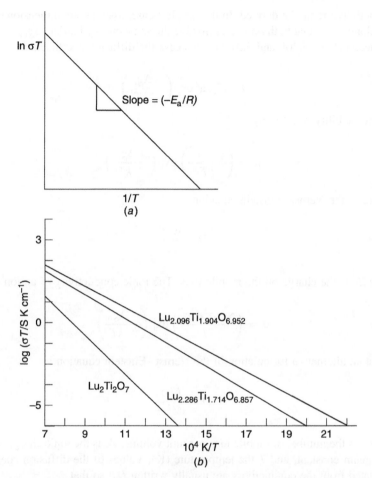

Figure 6.5 Arrhenius plots of $\ln(\sigma T)$ versus $1/T$ for ionic conductivity in a crystal: (*a*) straight-line plot used to obtain the activation energy of migration; (*b*) experimental results for the pyrochlore phases $Lu_2Ti_2O_7$, $Lu_{2.096}Ti_{1.904}O_{6.952}$, and $Lu_{2.286}Ti_{1.714}O_{6.857}$. [Data adapted from A. V. Shlyakhtina, J. C. C. Abrantes, A. V. Levchenko, A. V. Knot'ko, O. K. Karyagina, and L. G. Shcherbakova, *Solid State Ionics*, **177**, 1149–1155 (2006).]

barrier height, or equivalent to that determined from diffusion studies, as described in the following section.

6.2.2 Relationship between Ionic Conductivity and Diffusion Coefficient

If both ionic conductivity and ionic diffusion occur by the same random-walk mechanism, a relationship between the self-diffusion coefficient, D, and the ionic conductivity, σ, can be derived. In the simplest case, assume one-dimensional diffusion along x and that both processes involve the same energy barrier, Δg_m, and jump distance, a (Figs. 5.20b and 6.3). For this case the diffusion coefficient is

$$D = \gamma v a^2 \exp\left(-\frac{\Delta g_m}{kT}\right)$$

and the mobility is given by

$$\mu = \left(\frac{\gamma v Z e a^2}{kT}\right) \exp\left(-\frac{\Delta g_m}{kT}\right)$$

leading to the Nernst–Einstein equation:

$$D = \frac{\mu kT}{Ze}$$

where Ze is the charge on the mobile ions. The ionic conductivity of an ion is

$$\sigma = \left(\frac{\gamma c v a^2 Z^2 e^2}{kT}\right) \exp\left(-\frac{\Delta g_m}{kT}\right)$$

so that an alternative formulation of the Nernst–Einstein equation is

$$D = \frac{kT\sigma}{cZ^2 e^2}$$

where c is the number of mobile ions per unit volume, Ze is the ionic charge, k is the Boltzmann constant, and T the temperature (K). Values of the diffusion coefficient calculated from the conductivity are usually written D_σ, so that

$$D_\sigma = \frac{kT\sigma}{cZ^2 e^2}$$

This equation shows that it is possible to determine the diffusion coefficient from the easier measurement of ionic conductivity. However, D_σ is derived by assuming that the conductivity mechanism utilizes a random-walk mechanism, which may not true.

The Haven ratio, H_R, is defined as

$$H_R = \frac{\text{tracer diffusion coefficient}}{\text{conductivity diffusion coefficient}}$$

$$= \frac{D^*}{D_\sigma}$$

The correlation factor, f, is defined by the ratio of the tracer diffusion coefficient to the random-walk diffusion coefficient (Section 5.6):

$$f = \frac{D^*}{D_r}$$

Taking D_σ to be equal to D_r (which may not always be correct) allows H_R to be equated with the correlation factor:

$$H_R = f = \frac{D^*}{D_\sigma}$$

When ionic conductivity is by way of interstitials, both conductivity and diffusion can occur by random motion, so that the correlation factor and H_R are both equal to 1. In general, the correlation factor for a diffusion mechanism will differ from 1, and in such a case D_σ can be described by the following relationship:

$$D_\sigma = \frac{fkT\sigma}{cZ^2e^2}$$

where f is the correlation factor appropriate to the diffusion mechanism. For vacancy diffusion in a cubic structure,

$$D_\sigma = \frac{f_v kT\sigma}{cZ^2e^2}$$

where f_v is the correlation factor for vacancy self-diffusion, and

$$D_\sigma = \frac{f_{iy} kT\sigma}{cZ^2e^2}$$

for interstitialcy diffusion, where f_{iy} is the appropriate intersticialcy correlation factor.

6.2.3 Ionic Conductivity and Defects

To obtain a solid with a high conductivity, it is clearly important that a large concentration, c, of mobile ions is present in the crystal [Eq. (6.1)]. This entails that a large number of empty sites are available, so that an ion jump is always possible. In addition, a low enthalpy of migration is required, which is to say that there is a low-energy barrier between sites and ions do not have to squeeze through bottlenecks. Hence the structure should ideally have open channels and a high population of vacancy defects.

The close connection between ionic conductivity and diffusion means that the role of defect concentrations will be similar to that discussed in the previous chapter (Sections 5.8 and 5.9). For crystals such as oxides and halides with close-packed structures, where large populations of empty sites do not exist, ionic conductivity will be controlled by the number of point defects present. In very pure crystals intrinsic defect populations may outweigh the impurity defects present at high temperatures. The value of c in the intrinsic region must then reflect the type of intrinsic defect present, which in turn is controlled by the defect formation energy. The equations are then similar in form to those given previously for diffusion. For a sodium chloride structure crystal in which the diffusion mechanism is vacancy diffusion and the intrinsic defects responsible are Schottky defects:

$$\sigma_S = \left(\frac{\gamma v a^2 Z^2 e^2 N}{kT} \right) \exp\left(-\frac{\Delta h_m}{kT} \right) \exp\left(-\frac{\Delta h_S}{2kT} \right)$$

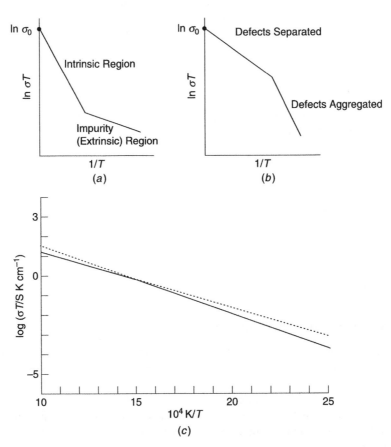

Figure 6.6 Arrhenius plots in crystals: (*a*) almost pure crystals with low impurity concentrations; (*b*) crystals with low-temperature defect clusters; and (*c*) the ionic conductivity of CeO_2 doped with 10 mol % Nd_2O_3, showing defect cluster behavior. [Part (*c*) adapted from data in I. E. L. Stephens and J. A. Kilner, *Solid State Ionics*, **177**, 669–676 (2006).]

while for Frenkel defects:

$$\sigma_F = \left[\frac{\gamma v a^2 Z^2 e^2 (NN^*)^{1/2}}{kT}\right] \exp\left(\frac{-\Delta h_m}{kT}\right) \exp\left(-\frac{\Delta h_F}{2kT}\right)$$

The high temperature value for the activation energy will be composed of two terms, namely:

$$E_{aS} = \Delta h_m + \left(\frac{\Delta h_S}{2}\right)$$

for Schottky defects and

$$E_{aF} = \Delta h_m + \left(\frac{\Delta h_F}{2}\right)$$

for Frenkel defects. At low temperatures the conductivity is due to extrinsic defects, generally impurities. As in the case of diffusion, Arrhenius plots of $\ln \sigma T$ versus $1/T$ will show a convex knee between the low-temperature and high-temperature regions (Fig. 6.6a). Defect aggregation will produce a concave knee rather than convex (Fig. 6.6b and 6.6c). Doping will play similar roles to those described earlier (Sections 5.8 and 5.9). This topic is continued in Section 6.7.

6.3 IMPEDANCE MEASUREMENTS

The conductivity of an ionic conductor can be assessed by direct current (dc) or alternating current (ac) methods. Direct current methods give the resistance R and the capacitance C. The corresponding physical quantity when ac is applied is the impedance, Z, which is the total opposition to the flow of the current. The unit of impedance is the ohm (Ω). The impedance is a function of the frequency of the applied current and is sometimes written $Z(\omega)$ to emphasize this point. Impedance is expressed as a complex quantity:

$$Z = R + iX$$

where R is the real part of the impedance, equated with the resistance of the sample, R, X is imaginary part, called the reactance of the sample, and i is the imaginary operator. The reactance is generally made up of two components. One part is due to inductance, X_L:

$$X_L = 2\pi f L = \omega L$$

where the frequency of the ac is f, the angular frequency is ω, and L is the inductance of the sample. The second part is due to capacitance, X_C:

$$X_C = \frac{1}{2\pi f C} = \frac{1}{\omega C}$$

where C is the capacitance of the sample.

The impedance of a ceramic material such as an oxide is often considered to be made up of a resistive part in parallel with a reactive part (Fig. 6.7a). The impedance of this combination is

$$Z = \frac{RX^2}{R^2 + X^2} + \frac{iR^2 X}{R^2 + X^2}$$

Ideally, the reactance is made up only of a capacitive component in which case the impedance can be written:

$$Z = \frac{R}{1 + R^2 C^2 \omega^2} - \frac{i(R^2 C\omega)}{1 + R^2 C^2 \omega^2}$$

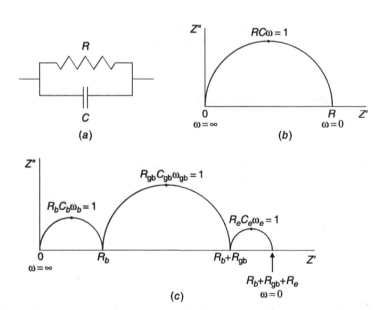

Figure 6.7 Complex impedance of a polycrystalline ceramic sample: (a) representation of the equivalent circuit of a component; (b) the impedance spectrum of the equivalent circuit in (a); (c) the impedance spectrum of a typical ceramic sample. Each semicircular arc represents one component with an equivalent circuit as in (a); that at the highest frequency corresponds to the repose of the bulk, that at middle frequencies to the grain boundary response, and that at lowest frequencies to the electrodes.

If the real part of the impedance is written Z' and the imaginary part as Z'', where

$$Z' = \frac{R}{1 + R^2 C^2 \omega^2}$$

$$-Z'' = \frac{R^2 C \omega}{1 + R^2 C^2 \omega^2}$$

a plot of Z' versus Z'' forms a semicircular arc. The arc passes through the origin for $\omega = \infty$, through the point $(\frac{1}{2}R, \frac{1}{2}R)$ at $RC\omega = 1$, and cuts the Z'' axis at R. The maximum of the arc is given by $RC\omega = 1$ (Fig. 6.7b). The measurement of impedance across a frequency range is called impedance spectroscopy, and the display of the plots in terms of Z' and Z'' are often called complex plane representations, Nyquist plots, or Cole–Cole plots (although these two latter designations are not strictly accurate terminologies).

A typical ceramic sample contains contributions from the bulk, the grain boundaries, and the electrode. Each of these is characterized by a semicircular arc with a maximum at $RC\omega = 1$, where the values of resistance, capacitance, and frequency refer directly to the bulk, grain boundaries, or electrodes (Fig. 6.7c). The separation of resistance due to the bulk from that of the grain boundaries is thus easily achieved using impedance spectroscopy.

6.4 ELECTROCHEMICAL CELLS AND BATTERIES

One of the major applications of ionic conductors is as an electrolyte in an electrochemical cell. An electrochemical cell is a device for the conversion of electrical to chemical energy and vice versa by way of redox reactions. Electrochemical cells consist of two electrodes that are good electron conductors (an anode and a cathode) in contact with an electrolyte that is a good ionic conductor but is not able to conduct electrons. A galvanic cell uses a *spontaneous chemical reaction* to produce an external electric current. Galvanic cells are called batteries in colloquial speech, reflecting the fact that historically several (i.e., a battery of) galvanic cells were generally connected together so as to increase the output power; the phrase "a battery of galvanic cells" became shortened to "a battery."

Batteries are self-contained units in which the chemical reactants and the means of transforming the chemical energy to electrical energy are present (Fig. 6.8a). In a battery, electrons generated as a result of the internal chemical reaction leave via the anode, cross an external circuit, doing useful work, such as powering a computer, and then re-enter the battery via the cathode. The reaction is completed by a transfer of ions across the electrolyte from the anode region to the cathode region or vice versa.

As against this, fuel cells contain only the means for energy conversion. The chemical reactants, or fuel (principally hydrogen or hydrocarbons and oxygen), are supplied from an external source (Fig. 6.8b).

The voltage generated by a battery or fuel cell, the cell potential, is simply related to the Gibbs energy of the cell reaction, ΔG_r, by

$$\Delta G_r = -nE_{cell}F$$

where E_{cell} is the cell potential in volts, defined to be positive, F is the Faraday constant, and n is the number of moles of electrons that migrate from anode to cathode in the cell reaction. When the electrodes are in their standard states, the free energy change is the standard reaction free energy, ΔG_r^0, and the cell voltage is just the standard cell potential, E^0. In this case

$$\Delta G_r^0 = -nE^0F$$

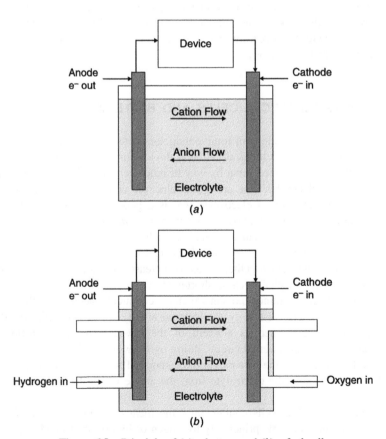

Figure 6.8 Principle of (a) a battery and (b) a fuel cell.

The potential generated by a cell is dependent upon the concentration of the components present. The relationship is given by the following equation:

$$\Delta G_r = -nE_{cell}F = RT\ \ln\left(\frac{Q}{K}\right)$$

where R is the gas constant, T the temperature (K), Q is the reaction quotient, and K is the equilibrium constant for the reaction.[2]

The cell voltage is thus given by

$$-nE_{cell}F = RT\ln\left(\frac{Q}{K}\right)$$

$$E_{cell} = -\left(\frac{RT}{nF}\right)\ln\left(\frac{Q}{K}\right)$$

It is convenient to separate the term due to the equilibrium constant and write the equation for the cell voltage as

$$E_{cell} = \left(\frac{RT}{nF}\right)\ln K - \left(\frac{RT}{nF}\right)\ln Q$$

When all species are in the standard state, $Q = 1$, $\ln Q = 0$, and the cell voltage is E^0, hence:

$$E_{cell} = E^0 = \left(\frac{RT}{nF}\right)\ln K$$

and in general

$$E_{cell} = E^0 - \left(\frac{RT}{nF}\right)\ln Q \qquad (6.4)$$

where E_{cell} is the cell voltage, E^0 is the standard cell voltage, R is the gas constant, T the temperature (K), n is the number of electrons transferred in the cell reaction, F is

---

[2]The reaction quotient, Q_c, of a reaction

$$aA + bB \rightarrow xX + yY$$

is given by

$$Q_c = \frac{[X]^x[Y]^y}{[A]^a[B]^b}$$

where $[A]$ denotes the concentration of component A at any time. Note that the quantity of importance is the activity rather than concentration and for precise work activity must be used. For reactions involving gases, the concentration term can be replaced by the partial pressure of the gaseous reactants to give

$$Q_p = \frac{p_X^x p_Y^y}{p_A^a p_B^b}$$

the Faraday constant, and Q is the reaction quotient of the cell reaction. Equation (6.4) is the Nernst equation.

The relationship expressed by the Nernst equation means that a battery can be used not only as a power supply but also as a tool for the determination of thermodynamic properties and the concentrations of reactants in the electrode regions. Some of these uses are outlined below.

6.5 DISORDERED CATION COMPOUNDS

Disordered cation compounds (Table 6.1) have an unusually high ionic conductivity. For example, the ionic conductivity of a normal ionic halide, such as NaCl or KCl is of the order of $10^{-10}–10^{-14}$ S m^{-1}, while that of the fast ion conductors described in this section, such as α-AgI or α-Ag$_3$SI, is of the order of $10^2–10^1$ S m^{-1}. The first of these materials to be studied, silver sulfide, Ag$_2$S, and lead fluoride, PbF$_2$, were investigated by Faraday in the nineteenth century, and it was Faraday who discovered that at elevated temperatures both showed conductivities similar to that of liquid electrolytes. The cuprous halides CuI and CuBr exhibit similar behavior at temperatures of the order of 400°C. The high ionic conductivity shown by these solids makes them candidates for a number of applications, especially in batteries, but the high temperature onset of the phenomenon is a disadvantage. Because of this there have been considerable efforts to modify the materials in such a way that high ionic conductivity is observed at room temperature.

This strategy has lead to the development of anion-substituted compounds such as Ag$_3$SI and cation-substituted compounds such as RbAg$_4$I$_5$, as well as materials with

TABLE 6.1 Some Disordered Cation Compounds[a]

Silver (Inorganic)	Silver (Organic)	Copper
AgI	[(CH$_3$)$_4$N]$_2$Ag$_{13}$I$_{15}$	α-CuI
Ag$_2$S	(C$_2$H$_5$NH)Ag$_5$I$_6$	β-CuI
Ag$_2$Se	(C$_2$H$_5$NH)$_5$Ag$_{18}$I$_{23}$	α-CuBr
RbAg$_4$I$_5$		β-CuBr
Ag$_3$SI		Cu$_2$Se
Ag$_3$SBr		RbCu$_3$Cl$_4$
Ag$_2$HgI$_4$		CsCu$_9$Br$_{10}$
Ag$_3$HgI$_4$		CsCu$_9$I$_{10}$
Ag$_6$I$_4$WO$_4$		Ag$_3$CuS$_2$
Ag$_7$I$_4$PO$_4$		AgCuS
Ag$_2$CdI$_4$		AgCuSe
Ag$_{1.80}$Hg$_{0.45}$Se$_{0.70}$I$_{1.30}$		Cu$_6$PS$_5$I
Ag$_7$TaS$_6$		Cu$_8$GeS$_6$

[a]These compounds are almost all polymorphic, and high conductivity is generally restricted to the high-temperature form, usually designated α-. Other phases, designated β- or γ-, are lower temperature phases that mostly have normal ionic conductivity. The phases β-CuBr and β-CuI are exceptions to this general rule.

complex compositions in which both anion and cation substitution is involved, as in $Ag_{1.80}Hg_{0.45}Se_{0.70}I_{1.30}$. Other materials incorporating organic groups, such as tetramethyl ammonium silver iodide, $[(CH_3)_4N]Ag_{13}I_{15}$, have also been prepared. In these latter compounds it is anticipated that the large organic molecules will open diffusion channels in the matrix of the solid. There are also economic reasons for replacing silver by copper, and this has lead to the synthesis of phases such as $RbCu_3Cl_4$, as well as mixed copper–silver compounds such as α-Ag_3CuS_2.

The reason for the high conductivity of these and other similar solids was not known until crystallographic studies were completed. The prototype material is silver iodide. When the compound is prepared at room temperature, 2 crystallographic forms, β-AgI and γ-AgI, occur, both with normal ionic conductivity. The phase possessing high ionic conductivity is the high-temperature polymorph, α-AgI, stable above 147°C. The structure of α-AgI consists of a body-centered cubic array of iodide (I^-) ions, with a unit cell edge of 0.5044 nm (Fig. 6.9a). There are 2 iodide (I^-) ions and (on average) 2 silver (Ag^+) in the unit cell. The silver ions occupy tetrahedral positions between the iodide ions in this array (Fig. 6.9b). There are 4 of these positions on each face of the cubic unit cell (Fig. 6.9c), which is equivalent to a total of 12 such positions per unit cell because each site is shared between 2 adjacent unit cells. In the low-temperature polymorphs 2 specific tetrahedral positions are occupied by Ag^+, but in the high-temperature form the Ag^+ ions jump continuously from one tetrahedral site to another. This is described as a "molten sublattice" of Ag^+ ions as they move rather like particles of a fluid through the fixed matrix of I^- ions, which accounts for the remarkable ionic conductivity of the material.

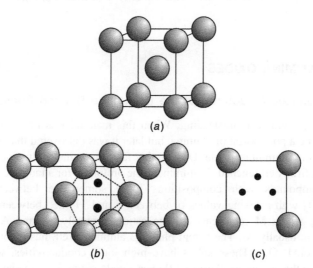

(a)

(b)   (c)

Figure 6.9 Structure of the high-temperature form of AgI (α-AgI): (a) the body-centered cubic arrangement of iodide (I^-) ions; the unit cell is outlined; (b) two (of four) tetrahedral sites on a cube face, indicated by filled circles; and (c) the four tetrahedral sites found on each cube face, indicted by filled circles. Ag^+ ions continuously jump between all of the tetrahedral sites.

In the main, all of the disordered cation compounds in Table 6.1 behave similarly and have low-temperature structures that do not show high conductivity and contain only a low population of point defects. At some higher temperature all or part of the metal atom array becomes mobile with cations able to freely move over some or all of the various allowed cation sites in the unit cell. This is manifested as increasing amplitude of vibration of the cations as the temperature rises. At the transition temperature this vibration is so great that the ions become effectively free. At the same time the anion array, or this and some of the large cations, remains relatively immobile. For example, the compound Ag_3CuS_2 has a distorted body-centered cubic packing of sulfur atoms analogous to the body-centered cubic packing of I^- ions in α-AgI. The Ag^+ ions occupy distorted octahedral or tetrahedral sites and Cu^+ ions are in linear coordination at room temperature. The cation structure disorders at 117°C, giving rise to high conductivity. In this material, although it seems that both Ag^+ and Cu^+ are mobile, the conductivity is dominated by Ag^+ movement.

In the closely related compound AgCuS, the sulfur atoms form a slightly distorted hexagonal close-packed array. The Cu^+ ions are located in positions within this framework to form layers, while the Ag^+ ions lie between the sulfur–copper layers. These Ag^+ ions show a progressively greater anisotropic thermal motion as the temperature rises, until, above 93°C, they are essentially completely mobile, leading to extremely high silver ion conductivity.

In terms of formal point defect terminology, it is possible to think of each silver or copper ion creating an instantaneous interstitial defect and a vacancy, $Ag_i^{\bullet}$ and V'_{Ag}, or $Cu_i^{\bullet}$ and V'_{Cu} as it jumps between two tetrahedral sites. This is equivalent to a high and dynamic concentration of cation Frenkel defects that continuously form and are eliminated. For this to occur, the formation energy of these notional defects must be close to zero.

6.6 β-ALUMINA OXIDES

6.6.1 Idealized Structures of the β-Alumina-Related Phases

The name β-alumina is misleading. When this material was first prepared, it was thought to be a polymorph of alumina, but later it was discovered that the compound was a sodium aluminum oxide. Despite this fact, the name β-alumina was retained and is now firmly entrenched. β-alumina is the prototype for a family of nonstoichiometric compounds with compositions lying somewhere between the limits $A_2O.nM_2O_3$ with n taking values of between 5 and 11 (i.e., between AM_5O_8 and $AM_{11}O_{17}$), where M represents a large cation, typically Na and A represents a trivalent ion, usually Al. The parent phase, β-alumina itself, has a nominal composition of $NaAl_{11}O_{17}$. These solids have high ionic conductivities, with transport numbers for the A cations close to 1.0 over a wide range of temperatures.

The unit cell of β-alumina is hexagonal, with lattice parameters $a = 0.595$ nm, $c = 2.249$ nm. The dominant features of the idealized structure of β-alumina, with composition $NaAl_{11}O_{17}$, are layers called *spinel blocks* stacked perpendicular to the **c** axis (Fig. 6.10a). These blocks are composed of four oxygen layers in a cubic close-packed arrangement with a thickness of approximately 0.66 nm. Within each slab Al^{3+} ions occupy octahedral and tetrahedral positions, so the structure resembles a thin slice of spinel, $MgAl_2O_4$. The formula of a spinel block is Al_9O_{16}, with seven cations in octahedral sites and two in tetrahedral sites, and there are two such sheets in a unit cell. Adjacent sheets are linked by AlO_4 tetrahedra, which share a common apical oxygen atom. These oxygen atoms are often called bridging or column oxygen. The spinel blocks are easily separated, and β-alumina cleaves readily into mica-like foils along these planes.

The Na^+ ions reside in the region between the spinel blocks, called the *conduction planes*, which also contain the apical oxygen atoms of the linking tetrahedra (Fig. 6.10*b* and 6.10*c*). The conduction planes have a thickness of about 0.46 nm. The position occupied by Na^+, called the Beevers–Ross (BR) site, is diagonally opposite to the apical bridging oxygen atom of the tetrahedral groups linking the spinel blocks. Other important positions in the unit cell are the corners, called anti-Beevers–Ross (aBR) positions, and the sites midway between the BR and aBR sites are called mid-oxygen (mO) sites.

The spinel blocks in β-alumina are related by mirror planes that run through the conduction planes; that is, the orientation of one block relative to another is derived by a rotation of 180°. A second form of this compound, called β″-alumina, has similar spinel blocks. However, these are related to each other by a rotation of 120°, so that three spinel block layers are found in the unit cell, not two. The ideal composition of this phase is identical to that of β-alumina, but the unit cell is now rhombohedral. Referred to a hexagonal unit cell, the lattice parameters are $a = 0.614$ nm, $c = 3.385$ nm. The thickness of the spinel blocks and the conduction planes is similar in both structures.[3]

The disposition of the atoms in the conduction plane itself is identical to that in β-alumina (Fig. 6.10*b*). However, the local environment of the conduction planes is different in the two phases because the oxygen ions on both sides of the conduction plane are superimposed in β-alumina and staggered in β″-alumina (Fig. 6.10*d*). This means that Beevers–Ross and anti-Beevers–Ross sites are not present. Instead the sodium sites, which have a tetrahedral geometry, are called Beevers–Ross-type (BR-type). The mid-oxygen (mO) sites, located midway between the BR-type sites, also have a different geometry in the two phases.

In both of these materials the distribution of the ions in the conduction planes changes with temperature. At high temperatures the large cations tend to occupy all suitable sites in a random manner. Thus in β-alumina the BR, aBR, and mO sites, and in β″-alumina the BR-type and mO sites, are occupied statistically.

There are two other phases that are similar to β- and β″-alumina but are built from spinel blocks that are six close-packed oxygen layers thick. The material β‴-alumina is the analog of β-alumina, with the spinel blocks related by 180° mirror-plane (hexagonal) symmetry, while the phase β⁗-alumina, the analog of β″-alumina, has blocks related by 120° rotation and rhombohedral symmetry.

---

[3]Complex structures that have the same chemical formula but differ in the way in which parts of the structure are stacked on top of one another are called polytypes.

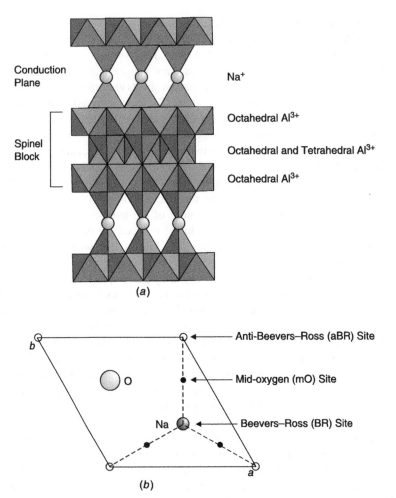

Figure 6.10 Structure of β-alumina: (a) idealized structure projected down [110] drawn as Al^{3+} centered polyhedra; (b) conduction plane, containing one Na^+ and one bridging O^{2-} ion per unit cell, this forming the in-conduction-plane vertex of the tetrahedra in (a).

The open nature of the conduction planes allows the Na^+ ions to be easily replaced in all these phases. In general, the β-alumina phase is less flexible to replacement; the β″-alumina more so, and the sodium can be replaced by almost any monovalent, divalent, or trivalent ion. The idealized formulas of these exchanged solids are $A^+Al_{11}O_{17}$, $A_{0.5}^{2+}Al_{11}O_{17}$, or $A_{0.33}^{3+}Al_{11}O_{17}$.

6.6.2 Defects in β-Alumina

The nonstoichiometric nature of the β-alumina phases lead to considerable difficulties in accurate crystal structure determination. Moreover, the defect structures tend

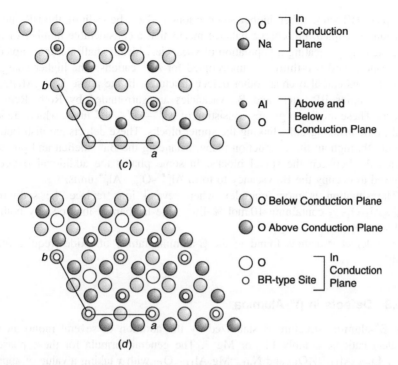

Figure 6.10 (*Continued*) (*c*) conduction plane and the two adjacent oxygen planes. The oxygen atoms are in a hexagonal arrangement and superimposed; (*d*) the conduction plane and two adjacent oxygen planes in β″-alumina. The oxygen atoms are in a hexagonal arrangement and are offset. The Al^{3+} ions lie above and below oxygen in the conduction plane in tetrahedral coordination. All ions are reduced in diameter for clarity.

to differ as the A metal ions in the conduction plane change, so that each phase seems to possess a unique defect structure. However, a number of broad generalizations can be made.

The idealized composition of β-alumina, $NaAl_{11}O_{17}$, is never found and an excess of sodium is necessary to stabilize the phase. The excess sodium is balanced by the incorporation of additional O^{2-} ions into the conduction planes, so that the material has a formula $Na_{1+x}Al_{11}O_{17+x/2}$ with x typically taking values between 0.15 and 0.3. The location of the extra sodium and oxygen varies with temperature and composition. Broadly speaking, the excess sodium forms pairs with an Na^+ ion on a BR site. The pair then occupies two mO sites adjacent to a BR site so that the originally occupied BR site becomes vacant. The additional oxygen atoms, also located in the conduction plane, tend to share sites with normal bridging oxygen atoms. The resulting pairs lie along lines joining the oxygen site to the adjacent corners of the unit cell, in positions rather similar to the mO sites surrounding the sodium ions in BR sites.

The defect structures of β-alumina phases containing cations other than Na^+ are different in detail. For example, in the silver analog of β-alumina, the excess silver occupies aBR sites. When higher valence ions replace the sodium, the BR sites are generally occupied, but charge balance means that a considerable number of these sites are empty, creating a population of vacancies. Ideally, half are unoccupied for A^{2+} cations and two-thirds are unoccupied for A^{3+} cations. The higher charge on these cations can also create other defect structures. In the phase $Ba_{0.75}Al_{11}O_{17.25}$, Ba^{2+} occupies BR sites, and Ba vacancies are surrounded by Roth–Reidinger defects. These are linear groups consisting of $Al_i^{3\bullet} - O_i^{2\prime} - Al_i^{3\bullet}$ units, which are identical to the bridging units linking the spinel blocks. These defects are distributed at random throughout the conduction plane, balancing the Ba^{2+} deficit and providing extra links between the spinel blocks. In some phases the additional oxygen is believed to occupy the Ba vacancy to form $Al_i^{3\bullet} - O_{Ba}^{4\prime} - Al_i^{3\bullet}$ units.

The situation is more complex when several ions replace the sodium. In $Ba_{0.75}Al_{11}O_{17.25}$ containing 10 mol % Eu^{2+}, the lanthanide ions occupy both BR and anti-BR sites.

The defect structures found in the β-alumina family of oxides require further study.

6.6.3 Defects in β″-Alumina

The β″-alumina structure is stabilized by the addition of several monovalent or divalent cations, notably Li^+ or Mg^{2+}. The general formula for these phases is $Na_{1+x}Li_{x/2}Al_{11-x/2}O_{17}$ and $Na_{1+x}Mg_xAl_{11-x}O_{17}$ with x taking a value of approximately 0.67. Additional oxygen is not needed to maintain charge neutrality, which is taken care of by substitution impurity defects $Li_{Al}^{2\prime}$ and Mg_{Al}' within the spinel blocks. The lack of interstitial oxygen in the conduction planes has a considerable effect, and generally the β″-alumina phases are better ionic conductors than their β-alumina counterparts.

There are two identical BR-type sites in the unit cell to accommodate the $1 + x$ Na^+ ions, and there are always vacant BR-type sites in proximity to occupied sites. When cations of higher valence replace sodium, the number of vacant BR-type sites increase in proportion. Although there is no geometrical reason why large cations should occupy other sites, in many compounds, the large cations are located in both BR-type sites and mO sites. As in the case of β-alumina, the defect structure of each compound is uniquely related to the chemical nature of the cations in the conduction layer.

6.6.4 Ionic Conductivity

Both the oxides β- and β″-alumina show extremely high Na^+ ion conductivity. As the structure suggests, the conductivity is anisotropic, and rapid sodium ion transport is limited to the two-dimensional conduction plane. There is almost unimpeded motion in the Na^+ layers, especially in β″-alumina, which lacks interstitial oxygen ion defects in the conduction plane, and the conductivity is of the same order of magnitude as in a strong solution of a sodium salt in water. The conductivity is a thermally activated diffusion-like process as indicted by Arrhenius plots of log σT versus $1/T$ (Fig. 6.11). Note that there is a knee in the Arrhenius plot of the conductivity of β''-alumina, with the high-temperature portion showing lower activation energy than the low-temperature region. This arises from defect interactions in the conduction plane. The mobile sodium ions have a positive charge, and Coulombic repulsion between these ions leads to an ordering of sodium ions and vacancies on aBR sites in the conduction plane. The activation energy for migration then includes a term for the breaking of the forces causing ordering as well as the migration energy. As the temperature rises, thermal vibration increases and the extent of the ordering decreases. The activation energy is now smaller, and at a high enough temperature consists only of the migration energy.

The high conductivity of β-alumina is attributed to the correlated diffusion of pairs of ions in the conduction plane. The sodium excess is accommodated by the displacement of pairs of ions onto mO sites, and these can be considered to be associated defects consisting of pairs of Na^+ ions on mO sites plus a V'_{Na} on a BR site (Fig. 6.12a and 6.12b). A series of atom jumps will then allow the defect to reorient and diffuse through the crystal (Fig. 6.12c and 6.12d). Calculations suggest that this diffusion mechanism has a low activation energy, which would lead to high Na^+ ion conductivity. A similar, but not identical, mechanism can be described for β''-alumina.

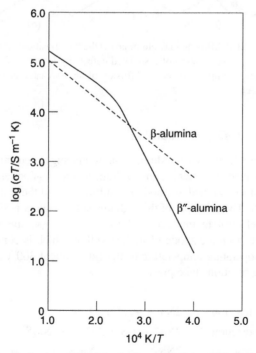

Figure 6.11 Arrhenius plots of the ionic conductivity of β- and β''-alumina.

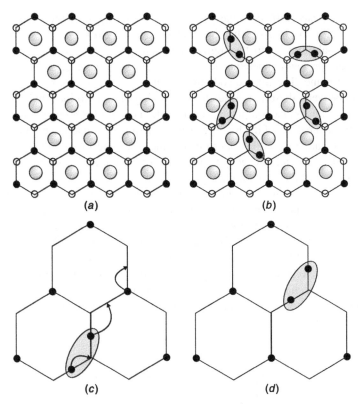

Figure 6.12   Correlated diffusion in β-alumina: (*a*) the Na$^+$ positions in the conduction plane of ideal NaAl$_{11}$O$_{17}$; (*b*) the creation of associated defects by location of pairs of Na$^+$ ions on mO sites; (*c*) the ionic jumps involved in diffusion of an associated defect; and (*d*) the final position of the defect.

6.6.5   Batteries Using β″-Alumina

The high ionic conductivity of sodium β″-alumina suggested that it would form a suitable electrolyte for a battery using sodium as one component. Two such cells have been extensively studied, the sodium–sulfur cell and the sodium–nickel chloride (ZEBRA) cell. The principle of the sodium–sulfur battery is simple (Fig. 6.13*a*). The β″-alumina electrolyte, made in the form of a large test tube, separates an anode of molten sodium from a cathode of molten sulfur, which is contained in a porous carbon felt. The operating temperature of the cell is about 300°C.

The following reactions take place.

$$\text{Anode reaction:} \quad 2\text{Na} \longrightarrow 2\text{Na}^+ + 2\text{e}^-$$

$$\text{Cathode reaction:} \quad 2\text{Na}^+ + x\,\text{S} + 2\text{e}^- \longrightarrow \text{Na}_2\text{S}_x$$

$$\text{Overall cell reaction:} \quad 2\text{Na} + x\,\text{S} \longrightarrow \text{Na}_2\text{S}_x$$

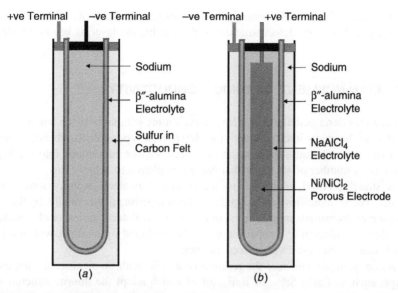

Figure 6.13 Batteries using β″-alumina electrolyte, schematic: (*a*) the sodium–sulfur cell and (*b*) the sodium–nickel chloride (ZEBRA) cell.

The phase Na_2S_x is sodium polysulfide, a material with a sulfur content of between 3 and 5. The anode reaction takes place at the liquid sodium–β″-alumina interface. Here sodium atoms lose an electron and the Na^+ ions formed enter the conduction planes in the electrolyte. The cathode reaction, which occurs at the interface between the β″-alumina and the liquid sulfur forms sodium polysulfides. Despite the desirable properties of the cell, technical and economic considerations have acted so as to curtail large-scale commercial production.

The sodium–nickel chloride cell also employs β″-alumina as an electrolyte. As in the sodium–sulfur cell, the negative electrode is molten sodium metal. The cathode is a porous mixture of Ni metal and the chloride $NiCl_2$. Because the connection between the solid cathode and the solid electrolyte is generally unsatisfactory, a second electrolyte, $NaAlCl_4$, with a melting point of 157°C, is also included in the battery. The cell is operated at a temperature of approximately 300°C. The design concept of the cell is similar to that of the sodium–sulfur battery, but the arrangement of the components is reversed, with the cathode taking the central position and the liquid sodium surrounding the outer face of the β″-alumina electrolyte (Fig. 6.13*b*).

The following reactions take place:

$$\text{Anode reaction:} \quad 2Na \longrightarrow 2Na^+ + 2e^-$$
$$\text{Cathode reaction:} \quad 2Na^+ + NiCl_2 + 2e^- \longrightarrow 2NaCl + Ni$$
$$\text{Overall cell reaction:} \quad 2Na + NiCl_2 \longrightarrow 2NaCl + Ni$$

As with the sodium–sulfur battery, the extraordinary conductivity of the β''-alumina electrolyte, due to the defect structure of the conduction layer, is key to this device.

6.7 ENHANCEMENT OF IONIC CONDUCTIVITY

The features described above that lead to high ionic conductivity are uncommon and most solids have low ionic conductivity. However, it is possible to enhance ionic conductivity if a substantial population of defects can be introduced into the crystal. There are a number of strategies that can be employed to achieve this.

A straightforward method is to incorporate aliovalent impurity ions into the crystal. These impurities can, in principle, be compensated structurally, by the incorporation of interstitials or vacancies, or by electronic defects, holes, or electrons. The possibility of electronic compensation can be excluded by working with insulating solids that contain ions with a fixed valence.

As an example, consider the reaction of a solid with no significant composition range, such as CaF_2, SrF_2, or BaF_2, all of which adopt the fluorite structure-type, with a similar aliovalent compound. Doping with an ion of lower valence will frequently induce charge compensation via anion vacancies. Reaction of CaF_2 with NaF can be written:

$$NaF\ (CaF_2) \longrightarrow Na'_{Ca} + F_F + V_F^{\bullet}$$

The reaction introduces one vacancy into the F^- positions for every Na^+ substituted onto a Ca^{2+} position. The fluoride ion conductivity is substantially increased from that of the parent CaF_2. Reaction with an ion of higher valence can also introduce substantial defect populations and lead to an increase in fluoride ion conductivity. For instance, doping CaF_2 with GdF_3 yields a stable phase with a stoichiometry spanning the range CaF_2 to $(Ca, Gd)F_{2.4}$. This range is due to a significant population of interstitials:

$$GdF_3(CaF_2) \longrightarrow Gd_{Ca}^{\bullet} + 2F_F + F_i'$$

The fluoride ion interstitials again lead to an increase in ionic conductivity. At lower temperatures this increase is modest because the interstitials aggregate into clusters, thus impeding ionic diffusion. At higher temperatures the clusters tend to dissociate, resulting in a substantial increase in conductivity.

The stabilized zirconia family of oxides, especially calcia-stabilized zirconia, are solids in which oxide ion conductivity has been increased to the extent that they are widely used solid electrolytes (Section 1.11.6, Section 4.4.5, and Section 6.8).

In many nonstoichiometric solids high-temperature structures contain disordered defects, but, as in the previous example, these order or form aggregates at lower temperatures. If this aggregation could be suppressed, low-temperature ionic conductivity would be enhanced. An example of this strategy is provided by compounds with the brownmillerite structure, $A_2B_2O_5$ (Section 4.9). The cation arrangement in this structure is identical to that in the perovskite structure, ABO_3. At low temperatures the oxygen atoms are ordered in the brownmillerite pattern. At high temperatures many materials with the brownmillerite structure disorder to form a perovskite-like structure with a formula $ABO_{2.5}$. Compared to normal perovskites, formula ABO_3, the structure contains a large number of disordered oxygen vacancies. These materials show high oxygen ion conductivity while the ordered low-temperature forms have a normal ionic conductivity. The high-temperature structure can often be stabilized to room temperature by the addition of suitable impurity cations.

One such material is the oxide $Ba_2In_2O_5$. This structure disorders above 930°C to a perovskite-type structure containing disordered oxygen vacancies. The phase is of interest for electrochemical applications in fuel cells and similar devices. For these applications high oxygen diffusivity is a requisite. This is found in the high-temperature disordered phases but not in the ordered brownmillerite forms. For example, the oxygen ion conductivity of $Ba_2In_2O_5$ below the disorder order transition is approximately 0.1 S m^{-1}. Above the transition temperature it jumps to 10 S m^{-1} (Fig. 6.14).

The disordered structure can be stabilized to room temperature by inclusion of substitutional impurities on the In sites. Thus the oxide formed when Ga is substituted for In, $Ba_2(In_{1-x}Ga_x)_2O_{5+\delta}$ to form Ga_{In} defects has a disordered cubic perovskite structure even at room temperature for values of x between 0.25 and 0.5, and the similar $Ba_2(In_{1-x}Co_x)_2O_{5+\delta}$ with Co_{In} defects has a disordered cubic perovskite structure at room temperature when x lies between 0.2 and 0.8. The defects present in the In sites hinder oxygen ordering during the timescale over which the samples cool from the

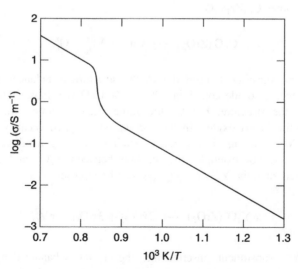

Figure 6.14 Conductivity of $Ba_2In_2O_5$ as a function of temperature. [Data taken from T. Yao, Y. Uchimoto, M. Kinuhata, T. Inagaki, and H. Yoshida, *Solid State Ionics*, **132**, 189–198, (2000).]

preparation temperature. Note that the oxygen composition of these doped materials is not always exactly 5.0 and depends upon a number of factors, especially the surrounding oxygen partial pressure (Chapters 7 and 8).

These two examples illustrate how desirable properties of a solid can be enhanced by selective doping. One of the most important examples follows.

6.8 CALCIA-STABILIZED ZIRCONIA AND RELATED FAST OXYGEN ION CONDUCTORS

6.8.1 Structure and Oxygen Diffusion in Fluorite Structure Oxides

A number of oxides with the fluorite structure are used in solid-state electrochemical systems. They have formulas $AO_2 \cdot xCaO$ or $AO_2 \cdot xM_2O_3$, where A is typically Zr, Hf, and Th, and M is usually La, Sm, Y, Yb, or Sc. Calcia-stabilized zirconia, $ZrO_2.xCaO$, typifies the group. The technological importance of these materials lies in the fact that they are fast ion conductors for oxygen ions at moderate temperatures and are stable to high temperatures. This property is enhanced by the fact that there is negligible cation diffusion or electronic conductivity in these materials, which makes them ideal for use in a diverse variety of batteries and sensors.

The parent phase is a stoichiometric oxide MO_2 with the fluorite structure. Substitution of a lower valence cation for Zr^{4+} is compensated by oxygen vacancies (Section 1.11.6 and Section 4.4.5). Taking calcia-stabilized zirconia as an example, addition of CaO drops the metal to oxygen ratio to below 2.0, and the formula of the oxide becomes $Ca_xZr_{1-x}O_{2-x}$.

$$CaO\,(ZrO_2) \;\longrightarrow\; Ca_{Zr}^{2\prime} + V_O^{2\bullet} + O_O^x$$

Each Ca^{2+} ion substituent inserted into the structure is balanced by one anion vacancy. Hence, an oxide containing 20 mol % CaO will have 20 mol % oxygen vacancies in the structure. Exactly the same situation holds when zirconia is reacted with the M_2O_3 oxides. In these cases one oxygen vacancy will form for every two M^{3+} cations incorporated, leading to a generalized formula of $M_x^{3+}Zr_{1-x}O_{2-x/2}$. For example, the reaction between Y_2O_3 and ZrO_2 to form yttria-stabilized zirconia, $Y_x^{3+}Zr_{1-x}O_{2-x/2}$, can be represented as:

$$x\,Y_2O_3(ZrO_2) \;\longrightarrow\; 2x\,Y_{Zr}' + 3x\,O_O + x\,V_O^{2\bullet}$$

Every two Y^{3+} substituents inserted into the crystal is balanced by one oxygen vacancy.

The result of the oxygen vacancy population is that the diffusion coefficient of oxygen ions is increased by many orders of magnitude compared to a normal oxide.

6.8.2 Free Energy and Stoichiometry of Oxides

Because the potential developed across a stabilized zirconia electrolyte is simply related to the free energy of the reactions taking place in the surrounding cell, the material can be used to measure the free energy of formation of an oxide. (Details of cells and cell types for this task are outside the scope of this book and only principles will be outlined. For information on these techniques see the Further Reading section at the end of this chapter.)

Because the stabilized zirconias transport oxygen ions, cells containing these as electrolytes can readily give information on the formation of oxides. One of the simplest arrangements is that in which the stabilized zirconia separates oxygen gas and a metal–metal oxide mixture. The voltage measured when no current is being drawn is directly related to the free energy of formation of the metal oxide. For example, suppose a stoichiometric metal–metal oxide, M–MO, mixture forms the anode and oxygen the cathode (Fig. 6.15). The cell is written schematically as:

$$\text{Pt, M, MO} \, \| \, \text{stab. zirconia} \, \| \, \text{O}_2(p_{\text{O}_2}), \text{Pt}$$

where the anode is conventionally at the left and the cathode at the right. The reactions taking place are

$$\text{Anode reaction:} \quad \text{M} + \text{O}^{2-} \longrightarrow \text{MO} + 2e^-$$

$$\text{Cathode reaction:} \quad 2e^- + \tfrac{1}{2}\text{O}_2\,(\text{g})\,(p_{\text{O}_2}) \longrightarrow \text{O}^{2-}$$

$$\text{Overall cell reaction:} \quad \text{M} + \tfrac{1}{2}\text{O}_2(\text{g})\,(p_{\text{O}_2}) \longrightarrow \text{MO}$$

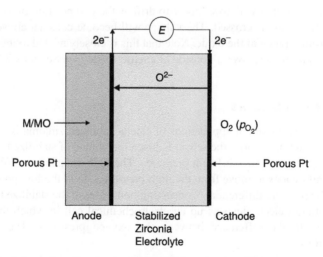

Figure 6.15 Schematic cell using a stabilized zirconia electrolyte to measure the Gibbs energy of formation of an oxide MO. The cell voltage, E, is measured under open-circuit conditions when no current flows.

Oxygen is transferred from the high-pressure side (p_{O_2}) to the low-pressure (M, MO) side of the cell. Two electrons are transferred via the external circuit from the anode to the cathode. The Gibbs energy change of the cell reaction, ΔG_r, which is equal to the Gibbs energy of formation of MO, is given by:

$$\Delta G_r = -2EF$$

where E is the cell voltage and F is the Faraday constant. Note that the cell voltage is measured when the circuit is open and no current is actually flowing. If the reactants are in their standard states, the standard Gibbs energy of reaction, ΔG_r^0, is measured. When the cell is set up, the voltage will be constant provided that there is always an adequate amount of M and MO present.

If the oxide is a nonstoichiometric phase MO_x, it will still be able to gain oxygen. The cell will be

$$\text{Pt, } MO_x \parallel \text{stab. zirconia} \parallel O_2(p_{O_2}), \text{Pt}$$

The reactions taking place are

$$
\begin{aligned}
\text{Anode reaction:} \quad & MO_x + \delta O^{2-} \longrightarrow MO_{x+\delta} + 2\delta e^- \\
\text{Cathode reaction:} \quad & 2\delta e^- + (\delta/2)O_2(g)\,(p_{O_2}) \longrightarrow \delta O^{2-} \\
\text{Overall cell reaction:} \quad & MO_x + (\delta/2)O_2(g)\,(p_{O_2}) \longrightarrow MO_{x+\delta}
\end{aligned}
$$

The voltage measured will now appear to drift as the composition range of the non-stoichiometric oxide is crossed. The voltage will become constant above and below the composition range of the oxide. Note that this is closely related to the variation of oxygen partial pressure over a nonstoichiometric oxide (see Sections 7.3, 7.4).

6.8.3 Oxygen Sensors

One of the most important applications of calcia-stabilized zirconia is as an oxygen sensor. In its simplest form, the sensor is just a membrane of stabilized zirconia separating oxygen gas at two different pressures. The high oxygen ion diffusion coefficient will allow ions to move from the high-pressure side to the low-pressure side to even out the pressure differential. Connecting both sides of the stabilized zirconia via porous platinum electrodes sets up an electrochemical cell in which the voltage is proportional to the difference between the oxygen pressures. The cell can be represented as:

$$\text{Pt, } O_2(p'_{O_2}) \parallel \text{stab. zirconia} \parallel O_2(p''_{O_2}), \text{Pt}$$

The reactions taking place are

Anode reaction: $\quad 2O^{2-} \longrightarrow O_2(g)\,(p'_{O_2}) + 4e^-$

Cathode reaction: $\quad 4e^- + O_2(g)\,(p''_{O_2}) \longrightarrow 2O^{2-}$

Overall cell reaction: $\quad O_2(g)\,(p''_{O_2}) \longrightarrow O_2(g)\,(p'_{O_2})$

Oxygen is transferred from the high-pressure side (p''_{O_2}) to the low-pressure (p'_{O_2}) side of the cell. The cell voltage is related to the oxygen pressures by the Nernst equation, Eq. (6.4):

$$E_{\text{cell}} = E^0 - \left(\frac{RT}{nF}\right) \ln Q$$

In this case, the number of electrons transferred, n, is 4 and the appropriate reaction quotient is

$$Q = \frac{p'_{O_2}}{p''_{O_2}}$$

Noting that E^0 for this cell reaction is zero:

$$E = -\left(\frac{RT}{4F}\right) \ln\left(\frac{p'_{O_2}}{p''_{O_2}}\right)$$

$$= -\left(\frac{RT}{4F}\right) \ln\left[\frac{\text{anode pressure (low)}}{\text{cathode pressure (high)}}\right]$$

Solving this equation for the oxygen partial pressure gives

$$p'_{O_2} = p''_{O_2} \exp\left(\frac{-4E}{RT}\right)$$

The high-pressure p''_{O_2} is taken as a reference pressure so that the unknown pressure p'_{O_2} is readily determined. These equations are often seen in the form:

$$E = +\left(\frac{RT}{4F}\right) \ln\left(\frac{p''_{O_2}}{p'_{O_2}}\right)$$

$$E = +\left(\frac{RT}{4F}\right) \ln\left[\frac{\text{cathode pressure (high)}}{\text{anode pressure (low)}}\right]$$

where R is the gas constant, T the temperature (K), and the oxygen pressures are measured in atmospheres.

The same principles apply for oxygen in solutions as diverse as liquid metals or blood. Because the oxygen is dissolved, the voltage measured depends upon the activity of the oxygen in the solution. For low concentrations, the activity can be approximated to the concentration, hence:

$$E = -\left(\frac{RT}{4F}\right) \ln \left\{ \frac{[O_2(\text{solution})]}{p_{O_2}(\text{reference})} \right\}$$

where $[O_2 (\text{solution})]$ is the concentration of the oxygen molecules, which is assumed to be lower than the reference pressure. If p_{O_2} is taken as 1 atm, the equation becomes

$$E = -\left(\frac{RT}{4F}\right) \ln [O_2(\text{solution})]$$

These equations consider the oxygen present as molecules in solution. If the oxygen exists as atoms in solution, only two electrons are needed in the cell equation and the potential is given by:

$$E = -\left(\frac{RT}{2F}\right) \ln [O(\text{solution})]$$

The design of the sensor depends upon its ultimate use. For high-temperature applications a stabilized zirconia tube coated inside and outside with porous platinum forms a suitable design. The tube can be used directly as an oxygen meter if p_{O_2}'' is a standard pressure, such as 1 atm of oxygen or else the pressure of oxygen in air, which is approximately 0.21 atm. Such a system is utilized to monitor the exhaust oxygen content, and thus fuel efficiency, of a car engine. The coated zirconia tube is arranged to project into the exhaust stream of the engine (Fig. 6.16). Using air as the standard oxygen pressure, the output voltage of the sensor is directly related to the stoichiometry of the air–fuel mixture. The cell voltage may be used to alter the engine input fuel–air mix automatically to optimize engine efficiency.

6.8.4   Oxygen Pumps and Coulometric Titrations

If a current is passed through a cell using a stabilized zirconia electrolyte, oxygen is transferred from one side of the electrolyte to the other. Thus, these cells can be used as electrochemical pumps capable of delivering or removing very precisely monitored amounts of oxygen. The amount of oxygen transferred can be calculated by an application of Faraday's law of electrolysis. The number of moles of O^{2-} transferred during cell operation, δO, is

$$\delta O = \frac{It}{2F}$$

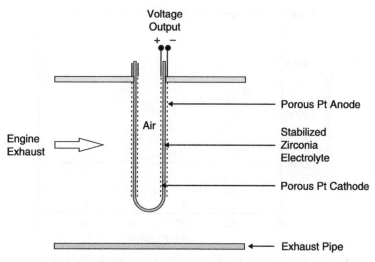

Figure 6.16 Car exhaust sensor (schematic) fitted with a stabilized zirconia ceramic tube as electrolyte.

where I is the current (A), t the time (s), and F is the Faraday constant. The factor 2 arises because two moles of electrons are involved in the formation of one mole of O^{2-} ions.

This technique, known as coulometric titration, allows the composition of a non-stoichiometric oxide to be varied in a precise way. The oxide in question is set as the anode of a cell with oxygen as the cathode:

$$\text{Pt, } (p'_{O_2}), MO_x \parallel \text{stab. zirconia} \parallel O_2(p''_{O_2}), \text{Pt}$$

In general, the practice is to use the coulometric titration to change the composition of the sample, then allow the cell to reach equilibrium and measure the cell voltage, which can be converted to the Gibbs energy or the equilibrium partial pressure over the sample (Fig. 6.17) (also see Section 7.4). Not all of the oxygen transferred will be incorporated or extracted from the oxide matrix, as the residual volume around the sample will take up some, creating an oxygen atmosphere. This effect can be allowed for during calibration.

6.9 PROTON (H⁺ ION) CONDUCTORS

Proton, that is, H⁺ ion, conductors are of importance as potential electrolytes in fuel cells. There are a number of hydroxides, zeolites, and other hydrated materials that conduct hydrogen ions, but these are not usually stable at moderate temperatures, when water or hydroxyl tends to be lost, and so have only limited applicability.

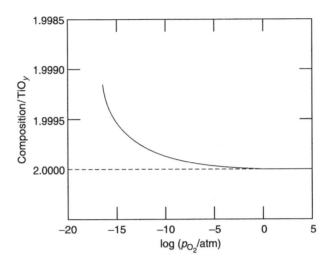

Figure 6.17 Composition of TiO_2 reduced by Coulombic titration to TiO_y plotted versus the corresponding equilibrium partial pressure of oxygen, determined by cell voltage, for a sample at 900°C. [Data taken from D-K. Lee, J-I. Jeon, W. Choi, and H-I. Yoo, *J. Solid State Chem.*, **178**, 185–193 (2005).]

This section will be limited to oxide materials that are inherently not H^+ ion conductors but can be made so by the introduction of suitable defects. These materials generally maintain H^+ conductivity to high temperatures.

The most important of these are perovskite structure solids with a formula $A^{2+}B^{4+}O_3$ that can be typified by $BaCeO_3$ and $BaZrO_3$. The way in which defects play a part in H^+ conductivity can be illustrated by reference to $BaCeO_3$. $BaCeO_3$ is an insulating oxide when prepared in air. This is converted to an oxygen-deficient phase by doping the Ce^{4+} sites with trivalent M^{3+} ions (Sections 8.2 and 8.6). The addition of the lower valence ions is balanced by a population of vacancies. A simple substitution reaction might be formulated:

$$2BaO + M_2O_3(BaCeO_3) \longrightarrow 2Ba_{Ba} + 2M'_{Ce} + 5O_O + V_O^{2\bullet}$$

and the formula of the resultant phase is $BaCe_{1-x}M_xO_{3-x/2}$. A typical dopant level of $x = 0.05$ is often used.

Note that this simple formalism disguises the fact that a considerable amount of chemical skill is involved in ensuring that the dopant M only occupies the Ce^{4+} sites. For example, when the apparently suitable dopant ion Nd^{3+} is used, it occupies both the Ce and Ba sites, thus suppressing vacancy formation:

$$Nd_2O_3(BaCeO_3) \longrightarrow Ba_{Ba} + Nd_{Ba}^\bullet + Ce_{Ce} + Nd'_{Ce} + 6O_O$$

Successfully doped materials are generally *p*-type semiconductors. They are heated in water vapor, and the oxygen vacancies are filled by OH^- that occupy O sites:

$$O_O + V_O^{2\bullet} + H_2O \longrightarrow 2OH_O^\bullet$$

The resulting phases are good proton conductors and operate up to high temperatures.

The mechanism of conduction is believed to be that of proton hopping. Under the influence of an electric field a proton can be transferred from one OH^- group to an adjacent O^{2-} ion. The electric field ensures that the H^+ ion rotates in the field direction, making the next jump easier.

Proton conductors are considered further in Section 8.8.

6.10 SOLID OXIDE FUEL CELLS

Fuel cells differ from batteries in that the supply of chemicals driving the cell reaction, the "fuel," is provided continuously from an external source (Fig. 6.8). There are several alternatives for the electrolyte in these cells (see Further Reading Section), but the most relevant from the point of view of defects in solids are those using a solid oxide as the electrolyte. Although oxides that are proton conductors are being widely explored, the term solid oxide fuel cells (often abbreviated to SOFCs) generally indicates that the working electrolyte is an oxygen ion conductor.

Among the oxide ion conductors, the stabilized zirconia family has been studied most as electrolyte materials as they have an oxide ion transport number, $t_a = 1$, at high temperatures. The commonest materials in use are calcia- or yttria-stabilized zirconia. Cells can be fabricated from stabilized zirconia tubes, but a flat-plate design is more suited to stacking into a battery (Fig. 6.18). The cells operate at temperatures of about 900°C, this high temperature being needed to maintain a high enough oxygen transport for useful cell output.

The high operating temperature of these cells imposes severe constraints upon the other components of the cell. They must be stable to high temperatures, have thermal expansion coefficients compatible with that of the electrolyte, and demanding electrical properties. The anode, for example, must be able to transport the fuel gas, O^{2-} ions from the electrolyte and electrons to the separator plates, all under reducing conditions. At present the favored anode is a porous cermet (a composite of ceramic and a metal) composed of the same stabilized zirconia as the electrolyte blended with nickel. However, much research is focused on the development of alternative oxide anode materials. The cathode materials are similarly constrained. The most widely used material at present is the perovskite structure $La_{1-x}Sr_xMnO_3$, typically $La_{0.8}Sr_{0.2}MnO_{3\pm\delta}$. The separators are also perovskite structure oxides, with the general formula $La_{1-x}Ca_xCrO_{3\pm\delta}$. These oxide materials support both electronic and ionic conductivity. The way in which oxides can be made to fulfill both these roles is discussed in more detail later (Section 8.8).

In operation, the fuel flows over the anode and reacts with oxide ions arriving from the electrolyte. Oxygen flows over the cathode where it is reduced and transported as

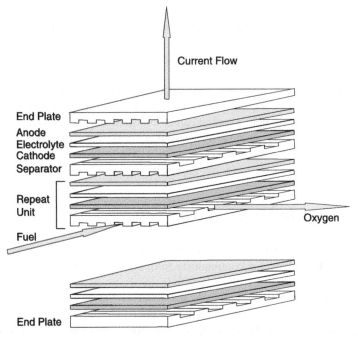

Figure 6.18   Schematic diagram of a fuel cell stack using a stabilized zirconia electrolyte.

ions from the high oxygen pressure cathode region to the low oxygen pressure anode region. The cell reactions depend upon the fuel. Typically these are

$$\text{Anode reaction:}\quad H_2(g) + O^{2-} \longrightarrow H_2O(g) + 2e^-$$

$$CO(g) + O^{2-} \longrightarrow CO_2(g) + 2e^-$$

$$CH_4(g) + 4O^{2-} \longrightarrow 2H_2O(g) + CO_2(g) + 8e^-$$

$$\text{Cathode reaction:}\quad O_2(g) + 4e^- \longrightarrow 2O^{2-}$$

$$\text{Typical cell reaction:}\quad 2H_2(g) + O_2(g) \longrightarrow 2H_2O(g)$$

The fact that these cells can operate with many different fuels is one advantage of high-temperature operation. However, there is at present much research directed toward lowering the operating temperature for cost reasons.

6.11   ANSWERS TO INTRODUCTORY QUESTIONS

Why are ionic conductors vital for battery operation?

A battery is a self-contained unit in which chemical reactants undergo a spontaneous chemical reaction and the resulting chemical energy is transformed into an external electric current. (Technically, a battery is a single galvanic cell or several connected together so as to increase power output.) In a battery, electrons generated as a result of the internal cell chemical reaction leave via the anode, cross an external circuit, doing useful work, such as powering a computer, and then reenter the battery via the cathode. The reaction is completed by a transfer of ions across the electrolyte from the anode region to the cathode region or vice versa.

In order for the transformation of chemical energy to external electrical energy to occur, it is essential that the electrons cannot traverse the battery internally. For this reason the electrolyte must be an insulator. However, for the reaction to continue, charge transport is needed, and for this to occur the electrolyte must be able to transfer ions. Thus, the electrolyte must consist of an insulating material with a high ionic conductivity for one ion taking part in the cell reaction.

How does an oxygen sensor work?

The principle of an oxygen sensor is similar to that of a battery. A battery uses a spontaneous chemical reaction to produce an external electric current. In an oxygen sensor the voltage generated by the reaction is used as a measure of oxygen concentration compared to a standard oxygen pressure, such as pure oxygen or oxygen in air. The sensor consists of an inert metal anode and cathode separated by a solid electrolyte that is a good O^{2-} ion conductor. For this calcia- or yttria-stabilized zirconia is generally used.

The voltage generated by a battery or fuel cell, the cell potential, is simply related to the Gibbs energy of the cell reaction, ΔG_r, by:

$$\Delta G_r = -nE_{cell}F$$

where E_{cell} is the cell potential, F is the Faraday constant, and n is the number of moles of electrons that migrate from anode to cathode in the cell reaction. The potential generated by a cell is dependent upon the concentration of the components present at the anode and cathode. For an oxygen sensor, this is

$$E_{cell} = -\left(\frac{RT}{4F}\right) \ln\left[\frac{p'_{O_2}}{p''_{O_2}}\right]$$

$$= -\left(\frac{RT}{4F}\right) \ln\left[\frac{\text{anode pressure (low)}}{\text{cathode pressure (high)}}\right]$$

where E_{cell} is the cell voltage, R is the gas constant, T the temperature (K), and F is the Faraday constant. Solving this equation for the oxygen partial pressure gives

$$p'_{O_2} = p''_{O_2} \exp\left(\frac{-4E}{RT}\right)$$

The high pressure p''_{O_2} is taken as a reference pressure, such as the oxygen partial pressure in air. A measurement of the cell voltage, together with knowledge of the temperature allows the unknown pressure p'_{O_2} to be determined.

What is a fuel cell?

A fuel cell is a form of battery. An ordinary battery consists of internal reactants that are converted into electrical energy, whereas in a fuel cell the chemical reactants are supplied from an external source. There are several designs of fuel cell, one of which is the solid oxide fuel cell (SOFC). These employ calcia- or yttria-stabilized zirconia. The cells operate at temperatures of about 900°C, this high temperature being needed to maintain a high enough oxygen transport for useful cell output.

In operation, the fuel flows over the anode and reacts with oxide ions arriving from the electrolyte. Oxygen flows over the cathode where it is reduced and transported as ions from the high oxygen pressure cathode region to the low oxygen pressure anode region. The cell reactions depend upon the fuel. Typically these are

$$\text{Anode reaction:} \quad H_2(g) + O^{2-} \longrightarrow H_2O(g) + 2e^-$$
$$CO(g) + O^{2-} \longrightarrow CO_2(g) + 2e^-$$
$$CH_4(g) + 4O^{2-} \longrightarrow 2H_2O(g) + CO_2(g) + 8e^-$$
$$\text{Cathode reaction:} \quad O_2(g) + 4e^- \longrightarrow 2O^{2-}$$
$$\text{Typical cell reaction:} \quad 2H_2(g) + O_2(g) \longrightarrow 2H_2O(g)$$

The fact that these cells can operate with many different fuels is one advantage of high-temperature operation. However, there is at present much research directed toward lowering the operating temperature as much as possible.

PROBLEMS AND EXERCISES

Quick Quiz

1. Assuming identical conductivity mechanisms, the conductivity of a solid will be greatest if the cation charge carriers are:
   (a) Monovalent
   (b) Divalent
   (c) Trivalent

2. One-fourth of the current is carried by anions across a solid conducting by way of both anions and cations. The transport number for anions is:
   (a) 0.25
   (b) −0.25
   (c) −0.75

3. The activation energy for ionic conductivity is derived from a plot of:

   (a) $\log \sigma$ vs. $1/T$

   (b) $\log(\sigma T)$ vs. $1/T$

   (c) $\log \sigma$ vs. $1/(\sigma T)$

4. The ratio of the tracer diffusion coefficient to the diffusion coefficient obtained from ionic conductivity measurements is called:

   (a) The Nernst–Einstein ratio

   (b) The Haven ratio

   (c) The correlation ratio

5. The bulk and grain boundary conductivities of a ceramic sample can be separated using the impedance, Z, by plotting:

   (a) Z' versus Z''

   (b) Z' versus ω

   (c) Z'' versus ω

6. A fast ion conductor with a molten sublattice contains:

   (a) A liquid layer of cations

   (b) A molten layer of metal atoms

   (c) Rapidly diffusing cations

7. The electrolyte β-alumina conducts using:

   (a) Al^{3+} ions

   (b) O^{2-} ions

   (c) Na^+ ions

8. The ZEBRA cell uses the reaction between:

   (a) Liquid sodium and sulfur

   (b) Liquid sodium and nickel chloride

   (c) Liquid sodium and β''-alumina

9. The electrolyte in an oxygen sensor is:

   (a) Stabilized zirconia

   (b) β-alumina

   (c) Doped $BaCeO_3$

10. A perovskite structure oxide such as $BaZrO_4$ can be made into a proton conductor by doping so as to introduce:

    (a) Cation vacancies

    (b) Anion vacancies

    (c) H^+ interstitials

Calculations and Questions

1. Estimate the activation energy for ionic conductivity in the $Lu_2Ti_2O_7$ phase illustrated in Figure 6.5b.

2. Estimate (a) the energy of migration and (b) the energy of aggregation of V_O defects in CeO_2 doped with 10 mol % Nd_2O_3 using the data in Figure 6.6.

3. The phase $EuNbO_4$ can be made nonstoichiometric by doping with Nb_2O_5. The ionic conductivity, $t_O \approx 1$, increases with dopant content, but the slope of the σT versus $1/T$ plots does not change. (a) Write a defect equation for the reaction of Nb_2O_5 with $EuNbO_4$, assuming that Nb does not occupy interstitial positions. (b) Explain why the conductivity increases with dopant concentration. (c) Estimate the activation energy for O^{2-} ion conductivity using the following data for $Eu_{0.77}NbO_{3.655}$.

log σT	Temperature/°C
−3.25	560
−3.62	500
−3.98	457
−4.3	412
−4.63	360
−5.19	315

Data adapted from K. Toda et al., *Solid State Ionics*, **136–137**, 25–30 (2000).

4. The values of the conductivity due to Na^+ ions in a glass are given in the following table. (a) Estimate the activation energy for conductivity. The conductivity at 700 K is 0.00316 S m^{-1}. (b) Estimate the value of the diffusion coefficient D_σ at this temperature if the concentration of Na^+ ions is 7.2×10^{27} m^{-3}. (c) Using the data in Chapter 5, Question 5, calculate the Haven ratio at this temperature.

Conductivity/S m^{-1}	Temperature/K
0.0316	803
0.0071	737
0.0011	653
0.00013	572
0.0000079	490

Data adapted from E. M. Tanguep Njiokop and H. Mehrer, *Solid State Ionics*, **177**, 2839–2844 (2006).

5. Why is it only possible to replace 50% of the Na ions in β-alumina by M^{2+} ions and by 33% M^{3+} ions? If more than these quantities were introduced, what would be the consequences for the defect structure of the phases?

6. The standard Gibbs energy of formation of NaCl is $-384\,kJ\,mol^{-1}$ and that of $NiCl_2$ is $-62\,kJ\,mol^{-1}$. Calculate the ideal voltage of a ZEBRA cell.

7. A conductivity cell is set up using an yttria-stabilized zirconia electrolyte. At 900°C the equilibrium pressure in the cell was 1.02×10^{-10} atm, and the reference pressure outside the cell was 7.94×10^{-18} atm. (a) What is the cell voltage? The temperature was dropped to 800°C and the reference pressure changed to 1.61×10^{-19} atm. The measured equilibrium voltage was 946 mV. (b) What is the equilibrium oxygen pressure in the cell? [Data adapted from D-K. Lee *et al.*, *J. Solid State Chem.*, **178**, 185–193 (2005).]

8. A conductivity cell is set up with an yttria-stabilized zirconia electrolyte to measure the nonstoichiometry of $BaTiO_3$ using oxygen gas as the reference. (a) Sketch the cell diagram. A current of 300 μA was passed through a pellet of $BaTiO_3$ weighing 16 g for 1500 s in such a direction as to remove oxygen by a Coulometric titration. (b) Calculate the composition of the reduced $BaTiO_{3-\delta}$ after this treatment. [Data adapted from D-K. Lee and H-I. Yu, *Solid State Ionics*, **144**, 87–97 (2001).]

9. Impedance plots for polycrystalline pellets of $SrTiO_3$ 10.4 mm diameter and 1.68 mm thick at 400, 500, and 600°C are shown in the figure. (a) What are the values of the bulk and grain boundary resistance for the samples? (b) What is the activation energy for bulk conductivity? (c) What is the activation energy for grain boundary conductivity? [Data adapted from S. K. Rout, S. Panigrahi, and J. Bera, *Bull. Mater. Sci.*, **28**, 275–279 (2005).]

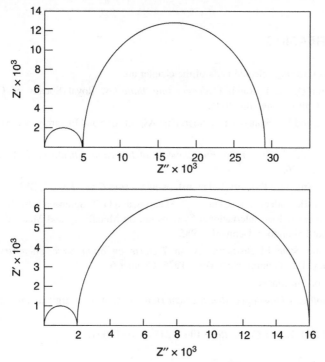

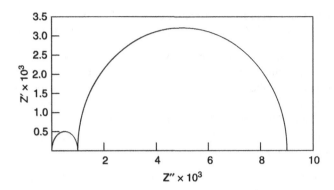

10. **(a)** Calculate the diffusion coefficient of Cu^+ ions in Cu_2O at 1000°C from the conductivity: $0.19\,S\,m^{-1}$. The measured Cu^+ tracer diffusion coefficient in Cu_2O at this temperature is $2.8 \times 10^{-12}\,m^2\,s^{-1}$. **(b)** What is the Haven ratio? The unit cell of Cu_2O is cubic, $a = 0.4267\,nm$, the unit cell contents, $Z = 2Cu_2O$.

11. The diffusion coefficient of Ca^{2+} in CaO is $3 \times 10^{-15}\,m^2\,s^{-1}$ at 1000°C. **(a)** Estimate the conductivity due to Ca^{2+} at this temperature. The total conductivity is measured as $8 \times 10^{-4}\,S\,m^{-1}$. **(b)** What does a comparison of these values suggest? The density of CaO is $3300\,kg\,m^{-3}$.

FURTHER READING

General sources covering selected parts of the chapter are:

R. M. Dell and D. A. J. Rand, *Understanding Batteries*, Royal Society of Chemistry Cambridge, United Kingodm, 2001.

N. E. W. de Reca and J. I. Franco, Crystallographic Aspects of Solid Electrolytes *Cryst. Rev.*, **2**, 241 (1992).

M. Greenblatt, Ionic Conductors, in *Encyclopedia of Inorganic Chemistry*, 2nd ed., Wiley, Hoboken, NJ, 2006.

R. E. Newnham Structure-Property Relationships in Sensors *Cryst. Rev.*, **1**, 253 (1988).

F. W. Poulsen, N. H. Andersen, K. Clausen, S. Skaarup, and O. T. Sørensen Eds., *Fast Ion and Mixed Conductors*, Risø international Symposium on Metallurgy and Materials Science, Risø National Laboratory, Denmark, 1985.

H. Rickert, Solid State Electrochemistry, in *Treatise on Solid State Chemistry* Vol. 4, N. B. Hannay, Ed., Plenum, New York, 1976, Chapter 6.

Measurement of Impedance:

Argilent Technologies Impedance Measurement Handbook, at www.argilent.com (2006).

Fuel cells:

M. Winter and R. J. Brodd *Chem. Rev.*, **104**, 4245–4269 (2004).

The first paper describing electrochemical use of β'-alumina is:

Y. F. Y. Yao and J. T. Kummer, *J. Inorg. Nucl. Chem.*, **29**, 2453 (1967).

For interesting sidelights on the Cole–Cole plot, see:

A. B. Pippard *Response and Stability*, Cambridge University Press, Cambridge, United Kingdom, 1985, Chapter 3, especially, pp. 57–60.

Fuel cells and batteries are surveyed by:

M. Winter and M. J. Brodd *Chem. Rev.*, **104**, 4245–4269 (2004).

Nonstoichiometry and Intrinsic Electronic Conductivity

How does composition variation change the electronic properties of a solid?

Why does the electronic conductivity of a nonstoichiometric solid change as the surrounding partial pressure changes?

What information does a Brouwer diagram display?

7.1 NONSTOICHIOMETRY AND ELECTRONIC DEFECTS IN OXIDES

7.1.1 Electronic and Ionic Compensation

Significant changes in electronic properties of a solid can result from composition variation. The examples chosen to illustrate this will mainly be drawn from oxides as these have been studied in most detail. In this chapter, pure (single-phase) solids will be described—intrinsic conductivity—while in the following chapter impurities and doping—that is, extrinsic conductivity—will be considered. Note that the principles described below apply equally well to doped crystals—the division into two chapters is a matter of convenience.

Chemical intuition can provide good guidance as to the electronic changes that nonstoichiometry brings about in oxides. Broadly speaking, materials in which all of the atoms have a fixed valence show little variation of composition at ordinary temperatures and are electrical insulators. These solids are typified by the oxides MgO, Al_2O_3, and $MgAl_2O_4$, which are stable to high temperatures in air. Any small variation in composition in these solids is compensated by the introduction of vacancies or interstitials. On the other hand, oxides in which the valence can change tend to show composition variation and associated electronic conductivity. For example, nickel oxide, NiO, a green insulating powder, or strontium titanate, $SrTiO_3$, a colorless ferroelectric solid, can be changed into electronic conductors by increasing or decreasing the amount of oxygen relative to the amount of metal present, to form a nonstoichiometric compound. Indeed, strontium titanate can even be made superconducting. The difference between the transition-metal

Defects in Solids, by Richard J. D. Tilley
Copyright © 2008 John Wiley & Sons, Inc.

compounds and oxides such as MgO is that the composition change is offset at a relatively low energy cost by a change of valence, that is, by electronic compensation. The change in composition can be conveniently divided into two types, composition change that results in metal excess solids and composition change that results in metal deficit phases.

7.1.2 Metal-Excess Phases

Taking as an example an ionic oxide MO, this material can be made into a metal-excess nonstoichiometric material by the loss of oxygen. As only neutral oxygen atoms are removed from the crystal, each anion removed will leave two electrons behind, which leads to electronic conductivity. The oxygen loss can be incorporated as oxygen vacancies to give a nonstoichiometric oxide with a formula MO_{1-x}, or the structure can assimilate the loss and compensate by the introduction of cation interstitials to give a formula $M_{1+x}O$.

The introduction of anion vacancies can be written:

$$x\, O_O(MO) \rightleftharpoons \tfrac{x}{2}O_2 + x\, V_O^{2\bullet} + 2e'$$

The electrons might be free or else may be associated with variable valence cations, chemically transforming M^{2+} ions (M_M) to M^+ ions (M'_M):

$$x\, MO \rightleftharpoons \tfrac{x}{2}O_2 + x\, V_O^{2\bullet} + xM'_M$$

The introduction of interstitial metal can be written:

$$x\, MO \rightleftharpoons \tfrac{x}{2}O_2 + x\, M_i^{2\bullet} + 2x\, e'$$

where the interstitial cation has two effective positive charges, $M_i^{2\bullet}$. As before, if the cations can accept a lower valence, interstitial M^+ ions can form:

$$2x\, MO \rightleftharpoons x\, O_2 + 2x\, M_i^{\bullet} + 2x\, e'$$

In both these materials, nonstoichiometry results in the introduction of electrons. The electronic properties will depend upon how strongly the additional electrons are bound to other species. Generally, though, if sufficient energy is supplied to make them move from one cation to another, the crystal will be able to conduct electricity. Often light will provide the energy needed to allow the electrons to migrate and the material shows photoconductivity. When thermal energy is sufficient to free the electrons, the electronic conductivity will increase with increasing temperature and the solids are *n*-type semiconductors. Both ZnO and CdO are believed to become metal excess via the introduction of metal interstitials, although the defect species present in both these oxides is more complex than the simple scheme presented above (Sections 1.11.5 and 4.3.5).

Exactly the same principles apply to oxides that are more complex from a compositional point of view. In these materials, however, there is a greater degree of

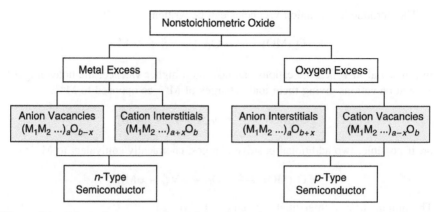

Figure 7.1 Schematic representation of the electronic consequences of nonstoichiometry in oxides.

flexibility to the electronic effects of stoichiometry variation. This can be illustrated by the phase $SrFeO_{2.75}$, which is derived from the parent cubic perovskite $SrFeO_3$, nominally composed of Sr^{2+} and Fe^{4+} ions. Normal preparations produce a material with a composition somewhere between $Sr_2Fe_2O_5$ ($SrFeO_{2.5}$) and $SrFeO_{2.75}$. The electronic consequences can be analyzed in the same way as the metal excess compounds described above. It is found that oxygen vacancies rather than cation interstitials form, and this can be represented by the following equation:

$$x\ O_O \longrightarrow \tfrac{x}{2}O_2 + x\ V_O^{2\bullet} + 2e'$$

The electrons are believed to be localized upon Fe^{4+} ions, converting them to Fe^{3+}:

$$2Fe_{Fe} + 2e' \longrightarrow 2Fe'_{Fe}$$

The material would be expected to be an n-type semiconductor. (Although the vacancies are ordered at lower temperatures to form brownmillerite, $Sr_2Fe_2O_5$ (Section 4.9), this does not change the analysis.)

The overall situation in metal-rich materials is summarized in the scheme in Figure 7.1.

7.1.3 Oxygen-Excess Phases

An ionic oxide MO can be made into an oxygen-excess nonstoichiometric material by the gain of oxygen. As only neutral oxygen atoms are added to the crystal, each atom added must capture two electrons to form anions, which leads to the production of holes and to electronic conductivity. The oxygen gain can be balanced by cation vacancies to give a nonstoichiometric oxide with a formula of $M_{1-x}O$, or the structure can compensate by the introduction of anion interstitials to give the formula of MO_{1+x}.

The introduction of anion vacancies can be written:

$$\tfrac{x}{2}O_2(MO) \rightleftharpoons x\,O_O + x\,V_M^{2\prime} + 2x\,h^{\bullet}$$

In a material in which the cations can take on a higher valence, the holes might be located on cations giving them ionic charges of M^{3+} as opposed to M^{2+}:

$$\tfrac{x}{2}O_2(MO) \rightleftharpoons x\,O_O + x\,V_M^{2\prime} + 2x\,M_M^{\bullet}$$

or, if possible, two additional positive charges, chemically equivalent to M^{4+}:

$$\tfrac{x}{2}O_2(MO) \rightleftharpoons x\,O_O + x\,V_M^{2\prime} + x\,M_M^{2\bullet}$$

The introduction of interstitial oxygen can be written:

$$\tfrac{x}{2}O_2(MO) \rightleftharpoons x\,O_i^{2\prime} + 2x\,h^{\bullet}$$

The holes can reside on normal cations converting them from M^{2+} to M^{3+} or M^{4+}:

$$\tfrac{x}{2}O_2(MO) \rightleftharpoons x\,O_i^{2\prime} + 2x\,M_M^{\bullet}$$

$$\tfrac{x}{2}O_2(MO) \rightleftharpoons x\,O_i^{2\prime} + x\,M_M^{2\bullet}$$

If the holes are able to gain enough energy to move from a cation when illuminated, the materials are photoconducting. Thermal energy may also be able to liberate the holes and the solids are p-type semiconductors. The transition-metal monoxides NiO and CoO represent this behavior (Sections 1.11.4 and 4.3.2).

The oxide La_2NiO_4, which adopts the K_2NiF_4 structure (Fig. 7.2), provides a typical example of an oxygen-excess ternary oxide. The oxide can take up oxygen to a composition of about $La_2NiO_{4.24}$. Although either metal vacancies or oxygen interstitials will provide the stoichiometric imbalance, in fact, the additional oxygen is in the form of interstitials, mainly distributed in the regions between the perovskite-like sheets of the structure. Assuming an ionic material, the charge balance is maintained by the generation of holes, according to the following equation:

$$\tfrac{x}{2}O_2(La_2NiO_4) \rightleftharpoons x\,O_i^{2\prime} + 2x\,h^{\bullet}$$

As in the preceding examples, the holes can reside on cations, converting them to a lower valence state. In the present example, the charge on the lanthanum is fixed as La^{3+}, and any holes must associate with the Ni^{2+} cations to convert them to Ni^{3+}:

$$\tfrac{x}{2}O_2 + 2Ni_{Ni}(La_2NiO_4) \rightleftharpoons O_i^{2\prime} + 2x\,Ni_{Ni}^{\bullet}$$

The material would be expected to be a hole (p-type) semiconductor. However, in this compound the interstitial oxygen ions can diffuse fairly quickly, and the oxygen diffusion coefficient is higher than normal, so that the compound shows both high oxygen diffusivity and electronic conductivity, a situation referred to as mixed conductivity (Section 8.7).

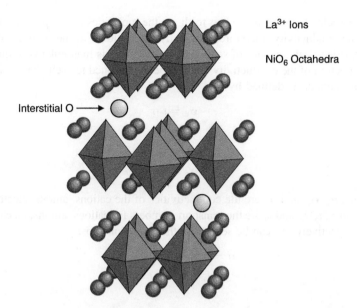

La^{3+} Ions

NiO$_6$ Octahedra

Interstitial O $\longrightarrow$

Figure 7.2 Location of interstitial oxygen ions in the K$_2$NiF$_4$ structure oxides La$_2$NiO$_{4+x}$ and La$_2$CuO$_{4+x}$.

The closely related phase La$_2$CuO$_4$ (Section 4.3.3) is similar. In both of these materials there is strong evidence to suggest that the interstitial oxygen atoms are not distributed at random but are partly or completely ordered, depending upon preparation conditions. This has significance for high-temperature superconductors (Section 8.6).

The overall situation in oxygen-rich materials is summarized in the scheme in Figure 7.1.

7.2 CONDUCTIVITY AND DEFECTS

7.2.1 Conductivity and Defect Concentrations

If the total conductivity of a material is made up of contributions from cations, anions, electrons, and holes, equations such as (6.1), for ionic conductivity, must be extended to include the electronic defects:

$$\sigma = \sigma_{cation} + \sigma_{anion} + \sigma_{electrons} + \sigma_{holes}$$
$$= c_a q_a \mu_a + c_c q_c \mu_c + c_e q_e \mu_e + c_h q_h \mu_h$$

where σ is the conductivity (Ω^{-1} m^{-1} = S m^{-1}),[1] c_a and so forth are the concentrations of mobile charge carriers (number m^{-3}), q_a and so forth are the charges

---

[1]Note: many published studies use centimeter rather than meter when reporting results. To convert conductivity in Ω^{-1} cm^{-1} to Ω^{-1} m^{-1}, multiply the value by 100. To convert conductivity in Ω^{-1} m^{-1} to Ω^{-1} cm^{-1}, divide the value by 100.

on the particles (C), and μ_a and so forth are the mobilities of the particles (m^2 V^{-1} s^{-1}). The conductivity, therefore, is directly proportional to the number of defects present. (Electrical behavior of semiconductor devices is manipulated in this way.)

The fraction of the conductivity that can be apportioned to each charge carrier, its transport number, is defined by

$$\sigma_c = t_c \sigma$$
$$\sigma_a = t_a \sigma$$
$$\sigma_e = t_e \sigma$$
$$\sigma_h = t_h \sigma$$

where σ_c, σ_a, σ_e, and σ_h are the conductivities of the cations, anions, electrons, and holes, and t_c, t_a, t_e, and t_h are the transport numbers for cations, anions, electrons, and holes, respectively. As can be seen from these relationships:

$$\sigma = \sigma(t_c + t_a + t_e + t_h)$$
$$t_c + t_a + t_e + t_h = 1$$

7.2.2 Holes, Electrons, and Valence

The viewpoint presented above suggests that electrons and holes in oxide semiconductors are located (usually) on cations that can take a variable valence, Ni^{2+}/Ni^{3+} or Cu^{2+}/Cu^{3+}, for instance. In such a model the electronic conductivity can take place by way of a hopping mechanism in which the charges jump from one site to another in a process akin to that of diffusion[2] (Chapters 5 and 6). However, it is by no means easy to state categorically whether such localization of charge states fits the experimental observations. Currently, the favored method of investigating this problem is via calculation, frequently using quantum mechanical methods. To illustrate the situation, two examples follow, the apparently simple oxides nickel oxide, NiO, and zinc oxide, ZnO.

Nickel oxide is a classical nonstoichiometric oxide that has been studied intensively over the last 30–40 years. Despite this, there is still uncertainty about the electronic nature of the defects present. It is well accepted that the material is an oxygen-excess phase, and the structural defects present are vacancies on cation sites. Although it is certain that the electronic conductivity is by way of holes, there is still hesitancy about the best description of the location of these charge carriers.

---

[2]There are a number of ways in which hopping conductivity can be treated theoretically. In semiconductor physics, the best known approach is Mott's variable range hopping model. This applies to disordered systems that have localized states within the band gap. The *low-temperature* conductivity in such a system is proportional to $\exp[-(T_0/T)^{1/4}]$, where T is the temperature and T_0 a constant. The Mott law has been found to account well for the conductivity of doped semiconductors at low temperatures. The same approach is not usually applicable when hopping involves valence state alternation, as this tends to occur at higher temperatures. In such cases considering the conductivity to be identical to atomic diffusion offers an alternative approach.

In the earliest models it was believed that at temperatures more or less below 1220 K the situation could be represented by the following equation:

$$\tfrac{x}{2}O_2(NiO) \rightleftharpoons x\,O_O + x\,V_{Ni}' + x\,h^\bullet$$

while above this temperature it is more correct to write

$$\tfrac{x}{2}O_2(NiO) \rightleftharpoons O_O + x\,V_{Ni}'' + 2x\,h^\bullet$$

It is now suggested that a more realistic possibility is given by

$$\tfrac{x}{2}O_2(NiO) \rightleftharpoons x\,O_O + xV_{Ni}^{\alpha'} + x\,\alpha h^\bullet$$

where the charge α can be 0, 1, or 2 depending upon temperature. Traditionally, the holes have been localized on Ni^{2+} ions to form Ni^{3+} ions. However, recent calculations suggest that the holes preferentially link with O^{2-} ions to form O^-. The resolution of the electronic structure of the material requires further calculation.

The situation with zinc oxide, ZnO, a material that has been investigated for a similar number of years, is comparable. Usually, nonstoichiometric ZnO is an n-type semiconductor. In the past it has been generally accepted that this is due to an excess of Zn in the form of Zn^+ interstitials:

$$2x\,ZnO \rightleftharpoons x\,O_2 + 2x\,Zn_i^\bullet + 2x\,e'$$

However, calculations show this to be true only over a limited range of conditions. Although the calculations differ in detail, it is now fairly certain that under zinc-rich conditions oxygen vacancies are the main defect:

$$x\,O_O(ZnO) \rightleftharpoons \tfrac{x}{2}O_2 + x\,V_O^{2\bullet} + 2e'$$

While zinc interstitials are possible, the formation energy for these defects is higher than that of oxygen vacancies. As in the case of NiO, continuing theoretical studies are needed to clarify the location of holes and electrons in these phases.

7.2.3 Localized Electrons and Polarons

The results of many conductivity measurements made on transition-metal oxide semiconductors are most easily explained if electronic conduction takes place by the diffusion type of process described in the previous section. For example, take an oxide containing iron in the two valence states Fe^{2+} and Fe^{3+}, Fe_3O_4, for instance. Electronic conductivity can be achieved (at least in principle) by an electron jumping from one Fe^{2+} ion to a neighboring Fe^{3+} ion. Similarly, slightly nonstoichiometric cobalt oxide, $Co_{1-x}O$ oxide can be thought of as containing both Co^{2+} and Co^{3+} ions. Electronic conductivity can be imagined to take place by way of hole diffusion from one Co^{3+} ion to a Co^{2+} ion.

This is essentially an ionic model of conductivity, which treats the electrons or holes as being localized at cations. The simple model must, however, be slightly modified to take into account other interactions. For example, an electron will attract positive ions and repel negative ions while a hole will have the opposite effect. This resulting polarization (distortion) of the structure is localized around the charged particle, and it is convenient to consider the particle and the deformation to jump together as a quasi-particle called a polaron.

When the interaction is weak, the polaron is called a large polaron. Large polarons have properties very similar to free particles and are considered to move in an energy band. Band theory equations are appropriate, with small modification, particularly in the effective mass of the electron or hole. When the interaction is strong, considerable deformation of the surrounding structure occurs and the term small polaron is appropriate. In such cases, the charged particle is more strongly trapped and movement of the charged particle is by a hopping mechanism. The two mechanisms can be separated (in ideal cases) by comparing the temperature variation of conductivity and mobility of the conducting entities (see below).

There are a number of ways in which an electron can become trapped by a structural distortion. Transition-metal ions Cr^{2+}, Mn^{3+}, Co^{2+}, Ni^{3+}, and Cu^{2+} in regular MO_6 octahedral coordination can gain stability by allowing the octahedron to flex so that there are two long and four short bonds or vice versa. This is known as the Jahn–Teller effect. These distortions can effectively trap either an electron or a hole at the cation position. In solids such as $BaBiO_3$, trapping can also result because of structural deformations. In this case the Bi ions are in two different charge states, Bi^{3+}, with a lone pair of electrons, and Bi^{5+} with no lone pair. The lone pair causes considerable local asymmetry in the structure, thus preventing easy charge migration.

7.2.4 Defects and Hopping Conductivity

Electron movement by way of discrete jumps is identical to that of atomic diffusion (Chapter 5) and ionic conductivity (Chapter 6), and Eq. (6.2) for the conductivity would be expected to hold provided that two simple changes are incorporated (also see footnote 2). The charge on the migrating particles is $\pm e$ rather than $\pm Ze$, of course. In addition, the geometrical factor that relates to crystal structure in ionic conductivity can be replaced by a factor that takes into account the composition of the nonstoichiometric phase. This is because a mobile particle can only jump to a site if it is occupied by a suitable ion. Thus, a hole on a Co^{2+} ion (Co^{3+}) can only jump if there is a neighboring Co^{2+} ion available. In general, it is possible to write the fraction of total sites that are occupied by mobile charges, either e' or h•, as φ, and the fraction of available unoccupied sites that the charge can jump to is $(1 - \varphi)$. The term $(1 - \varphi)$ now replaces the geometrical factor so that the form that Eq. (6.2) now adopts is

$$\sigma = (1 - \varphi)\left(\frac{c\,v\,a^2 e^2}{kT}\right)\exp\left(\frac{-\Delta h_m}{kT}\right) \tag{7.1}$$

where $(1 - \varphi)$ is the fraction of available unoccupied sites to which the charge can jump, c is the concentration of mobile electrons or holes present in the crystal, a is the jump distance, e the charge on the electrons or holes, v is the attempt frequency for a jump, k is the Boltzmann constant, T is the temperature (K), and Δh_m is the height of the barrier to be overcome during migration from one stable position to another.

Unlike a geometrical factor, the value of the factor φ will vary with composition in a predictable way. To illustrate this, suppose that stoichiometric MO_2 is heated in a vacuum so that it loses oxygen. Initially, all cations are in the M^{4+} state, and we expect the material to be an insulator. Removal of O^{2-} to the gas phase as oxygen causes electrons to be left in the crystal, which will be localized on cation sites to produce some M^{3+} cations. The oxide now has a few M^{3+} cations in the M^{4+} matrix, and thermal energy should allow electrons to hop from M^{3+} to M^{4+}. Thus, the oxide should be an n-type semiconductor. The conductivity increases until $\varphi = 0.5$ or $x = 1.75$, when there are equal numbers of M^{3+} and M^{4+} cations present. As reduction continues, eventually almost all the ions will be in the M^{3+} state and only a few M^{4+} cations will remain. In this condition it is convenient to imagine holes hopping from site to site and the material will be a p-type semiconductor. Eventually at $x = 1.5$, all cations will be in the M^{3+} state and M_2O_3 is an insulator (Fig. 7.3).

This argument can be repeated over the composition range from M_2O_3 to MO. Slight reduction of M_2O_3 will produce a few M^{2+} cations in the M^{3+} matrix, leading to n-type semiconductivity. This would persist in the composition range MO_x between $x = 1.5$ and $x = 1.25$, the conductivity passing through a maximum at the composition $MO_{1.25}$. Further reduction would lead to the situation where there are fewer M^{3+} cations than M^{2+} and p-type behavior is anticipated in the composition range between $x = 1.25$ and $x = 1.0$. The stoichiometric composition MO should be an insulator.

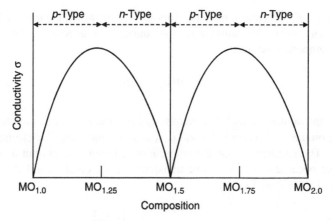

Figure 7.3 Theoretical variation of the conductivity as a function of composition for a hopping semiconducting oxide MO_x, where x can take values of between 1.0 and 2.0.

7.2.5 Band versus Hopping Conduction

The differentiation between whether delocalized (band theory) conductivity or diffu-sionlike hopping conductivity best explains experimental conductivity results is not always easy in practice but can be made by a comparison of the theoretical expressions for electrical conductivity and mobility of the charge carriers in a solid.

For a metal that obeys band theory well, and conducts by electrons, the conduc-tivity can be written:

$$\sigma = ne\mu_e$$

where n is the number of mobile electrons in the metal, which is more or less constant over small temperature ranges, e is the charge on the electron, and μ_e is the mobility of the electrons. The mobility of the electrons depends upon their mass and the number of times an electron collides with another electron or some other obstacle in the crystal. It can be expressed as

$$\mu_e = \frac{e\tau}{m^*}$$

where m^* is the effective mass of the electron in the crystal and τ is the mean lifetime between electron collisions. At normal temperatures μ_e is inversely proportional to a low power of temperature, typically:

$$\mu_e \propto \frac{1}{T^{3/2}}$$

For a metal, therefore, σ decreases with temperature, thus:

$$\sigma \propto \frac{ne}{T^{3/2}}$$

In the case of elemental semiconductors such as Si, which are also well described in band theory terms, the equation for the conductivity is composed of an electron and hole component so that:

$$\sigma = ne\mu_e + pe\mu_h$$

where n is the number of mobile electrons and p is the number of mobile holes, e is the charge on the electrons or holes and μ_e and μ_h are the respective electron and hole mobilities. The number of charge carriers, n or p, is not constant but increases with temperature because electrons can be transferred to the conduction band by thermal excitation. This is described by

$$n = p = n_0 T^{3/2} \exp\left[\frac{-E_g}{2kT}\right]$$

TABLE 7.1 Electrical Conductivity and Mobility of Charge Carriers in Metals, Band-like Semiconductors, and Hopping Semiconductors

	Metal	Band-like Semiconductor	Hopping Semiconductor
Conductivity, σ	$\propto T^{-m}$; falls slightly with T; no change with composition/ doping	$\propto \exp[-E/kT]$; increases with T; increases with amount of doping	$\propto \exp[-E/kT]$; increases with T; increases with degree of nonstoichiometry
Mobility, μ	$\propto T^{-m}$; falls slightly with T	$\propto T^{-m}$; falls slightly with T	$\propto \exp[-E/kT]$; increases with T

where E_g is the band gap and n_0 is a constant. Very similar equations hold for doped semiconductors in which the number of mobile charge carriers is also related to the dopant concentration. However, the mobility of the particles is still governed by the same processes as in a metal, so that the mobility is still given by

$$\mu = \frac{e\tau}{m^*} \propto \frac{1}{T^{3/2}}$$

(Note that the effective mass of holes and electrons can differ considerably.) The $1/T^{3/2}$ terms cancel in the expressions for n and p, hence:

$$\sigma \propto \exp\left[\frac{-E_g}{2kT}\right]$$

A hopping semiconductor such as an oxide is often best described by Eq. (7.1) rather than by classical band theory. In these materials the conductivity increases with temperature because of the exponential term, which is due to an increase in the successful number of jumps, that is, the mobility, as the temperature rises. Moreover, the concentration of charge carriers increases as the degree of nonstoichiometry increases.

A comparison of the relevant equations for metals, band theory semiconductors, and hopping semiconductors is given in Table 7.1. These equations can be used in a diagnostic fashion to separate one material type from another. In practice, it is not quite so easy to distinguish between the different conductivity mechanisms.

7.2.6 Seebeck Coefficient and Stoichiometry

Because the value of the Seebeck coefficient α depends upon the number of defects present, it should vary systematically with the composition. A common form of the dependence is

$$\alpha = \pm \left(\frac{k}{e}\right) \left[\ln\left(\frac{n_0}{n_d}\right)\right] \tag{7.2}$$

where n_0 is the number of sites in the sublattice containing defects and n_d is the number of defects giving rise to mobile electrons or holes (Section 1.6 and Supplementary Material S3). To consider how this relates to stoichiometric variation, consider a oxide MO_2, which is fairly readily reduced to form MO_{2-x} and that passes through the phases M_2O_3 and MO during the course of this reduction.

The sequence of events that occurs during the reduction has been described when the electrical conductivity of a nonstoichiometric oxide was discussed. Initial reduction will populate our MO_{2-x} crystal with a few M^{3+} ions, which will give rise to n-type semiconductivity. The value of α will therefore be large and negative. This value will fall as the number of defects increases. Turning to M_2O_3, a slight degree of oxidation will introduce into the M_2O_{3+x} phase a small number of M^{4+} ions in a matrix of M^{3+} ions. This will lead to p-type semiconductivity and a large positive value of α. Continued oxidation will cause this value to fall as the number of M^{4+} centers increases. Most significantly, there is a change from n-type to p-type behavior as the composition passes through a composition of $MO_{1.75}$ (Fig. 7.4).

A similar situation will hold as the composition range between M_2O_3 and MO is crossed, with α changing from positive to negative at a composition of $MO_{1.25}$. The value of α will change drastically from large and positive to large and negative as we pass through the stoichiometric position at M_2O_3 (Fig. 7.4).

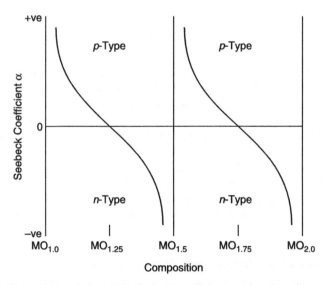

Figure 7.4 Theoretical variation of the Seebeck coefficient as a function of composition for a hopping semiconducting oxide MO_x, where x can take values of between 1.0 and 2.0.

The same analysis can be applied to compounds with a more complex formula. For example, the oxide $LaCoO_3$, which adopts the cubic perovskite structure, usually shows a large positive Seebeck coefficient, of the order of $+700 \, \mu V \, K^{-1}$, when prepared in air (Hébert *et al.*, 2007). This indicates that there are holes present in the material. The La ions have a fixed valence, La^{3+}, hence the presence of holes must be associated with the transition-metal ion present. Previous discussion suggests that $LaCoO_3$ has become slightly oxidized to $LaCoO_{3+\delta}$ and contains a population of Co^{4+} ions ($Co^{3+} + h^{\bullet}$ or $Co_{Co}^{\bullet}$). Each added oxygen ion will generate two holes, equivalent to two Co^{4+}:

$$\tfrac{1}{2}O_2 \longrightarrow O_O + 2 \, h^{\bullet} \quad \text{or} \quad \tfrac{1}{2}O_2 \longrightarrow O_O + 2Co_{Co}^{\bullet}$$

Using Eq. (7.2), the value of n_d, the number of holes (or Co^{4+} ions) present, is 3×10^{-4} of the total population, n_0, of Co^{3+}. The approximate formula of the oxide can be written $LaCoO_{3+x/2}$, where x is the number of holes. With $x = 3 \times 10^{-4}$, the material has a formula $LaCoO_{3.00015}$.

7.3 STOICHIOMETRY, DEFECT POPULATIONS AND PARTIAL PRESSURES

7.3.1 Equilibrium Partial Pressures

Nonstoichiometric phases can gain or loose components by reaction with the surrounding gaseous atmosphere. Broadly speaking, the composition of any pure phase will depend upon the equilibrium partial pressure of each of the components of the solid in the surrounding vapor. This implies that physical properties such as electronic conductivity will also vary as a function of the surrounding partial pressure of the components of the solid. To illustrate this, the relationship between the stability of an oxide and the surrounding oxygen partial pressure is described.

Silver oxide is a stoichiometric phase that decomposes at about 230°C to silver metal and oxygen gas when heated in air.

$$2Ag_2O \longrightarrow 4Ag + O_2$$

The change in the Gibbs free energy for this reaction, ΔG_r, can be related to the partial pressure of the oxygen gas, p_{O_2}, using the equation:

$$\Delta G_r = -RT \, \ln K_p$$

where K_p is the equilibrium constant of the reaction. In this example,

$$K_p = p_{O_2} \qquad \Delta G_r = -RT \, \ln p_{O_2}$$

At a fixed temperature, ΔG_r is constant, and so the equilibrium oxygen partial pressure, p_{O_2}, will also be constant. This oxygen pressure is called the decomposition pressure or dissociation pressure of the oxide and depends *only* upon the temperature of the system.

What does this imply? Suppose some silver metal and silver oxide is sealed in a closed silica ampoule, under a complete vacuum, and the ampoule is heated to a temperature somewhat below the decomposition temperature of 230°C. As there is no oxygen in the ampoule, some of the silver oxide will decompose and oxygen will be released. This will continue until the equilibrium decomposition pressure is reached. Provided that there is both silver and silver oxide in the tube, the oxygen pressure will be fixed (Fig. 7.5a). If the temperature is raised or lowered, either more silver

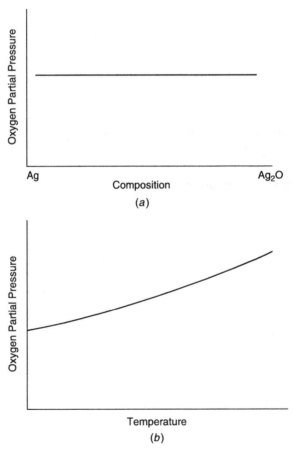

Figure 7.5 Variation of equilibrium oxygen partial pressure: (*a*) equilibrium between a metal, Ag, and its oxide, Ag$_2$O, generates a fixed partial pressure of oxygen irrespective of the amount of each compound present at a constant temperature; (*b*) the partial pressure increases with temperature; (*c*) a series of oxides will give a succession of constant partial pressures at a fixed temperature; and (*d*) the Mn–O system. [Data from T. B. Reed, *Free Energy of Formation of Binary Compounds: An Atlas of Charts for High-Temperature Chemical Calculations*, M.I.T. Press, Cambridge, MA, 1971.]

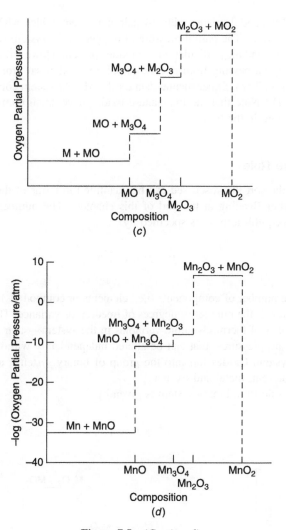

Figure 7.5 *(Continued)*.

oxide will decompose or some silver will oxidize until a new equilibrium decomposition pressure is reached, which is appropriate to the new temperature (Fig. 7.5b).

The same analysis will hold for any metal in contact with its single oxide. In order to determine the oxygen partial pressure over the pair, it is only necessary to know the value of the free energy of the reaction at the temperature required and insert the value into the following equation:

$$\Delta G_r = -RT \ln p_{O_2}$$

A multivalent metal M, will form several oxides, for example, MO_2, M_2O_3, M_3O_4, and MO. In this case the same analysis holds good for the metal M in contact with the lowest oxide, MO, and thereafter with any pair of adjacent oxides, MO/M_3O_4, M_3O_4/M_2O_3, M_2O_3 / MO_2. The partial pressure of oxygen in the system will depend only upon temperature and the particular pair of oxides present. However, the pressure will change abruptly on passing from one oxide pair to another say from MO/M_2O_3 to M_2O_3/MO_2 (Fig. 7.5c). Experimental data for the Mn–O system confirms this behavior (Fig. 7.5d). (Note that the manganese oxides have small composition ranges, which are ignored in the figure.)

7.3.2   Phase Rule

Equilibria of the sort just discussed can be quantified in terms of the (Gibbs) phase rule (see Further Reading at the end of this chapter). The number of phases that can coexist at equilibrium, P, is specified by

$$P + F = C + 2 \tag{7.3a}$$

where C is the number of components (i.e., elements or compounds) needed to form the system and F is the number of degrees of freedom or variance. The variance specifies the number of thermodynamic variables in the system—such as composition, temperature, and pressure—that can be altered independently without changing the state of the system. Oxides fall into the group of binary systems, as there are two components present, metal and oxygen.

The phase rule for a binary system is given by

$$P + F = 4 \tag{7.3b}$$

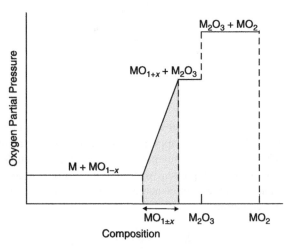

Figure 7.6   Variation of equilibrium oxygen partial pressure a series of oxides that includes one nonstoichiometric phase $MO_{1 \pm x}$.

When the system (e.g., a sealed ampoule containing Ag, Ag_2O, and O_2 gas) contains three phases, $P = 3$ and Eq. (7.3b) gives

$$F = 1$$

The system is said to have a variance of one; it is *univariant*. This means that under conditions in which three phases are in equilibrium, just one thermodynamic parameter needs to be given in order to specify the state of the system. In this case the most important parameters are temperature and oxygen partial pressure because in the Ag–O system the phases all have fixed compositions. Thus, knowledge of

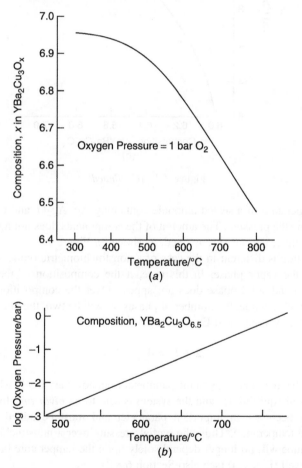

Figure 7.7   Equilibrium oxygen partial pressure for a nonstoichiometric oxide; $YBa_2Cu_3O_x$: (*a*) composition, *x*, versus temperature under an oxygen partial pressure of 1 bar; (*b*) oxygen partial pressure versus temperature for a composition of $YBa_2Cu_3O_{6.5}$; and (*c*) oxygen partial pressure versus composition, *x*, for a temperature of 600°C. [Adapted from data in P. Karen, *J. Solid State Chem.*, **179**, 3167–3183 (2006).]

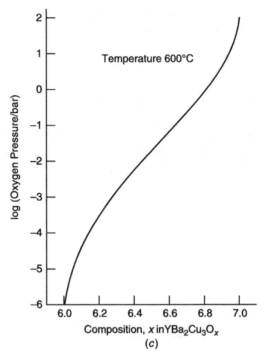

Figure 7.7   *(Continued).*

just the temperature of a sealed ampoule containing Ag, Ag_2O, and O_2 gas is sufficient to define the pressure. The amount of the components does not figure. At a constant temperature, the pressure is fixed, as shown in Figure 7.5.

The situation is different in the case of a nonstoichiometric oxide MO_x in equilibrium with the vapor phase. In this case, if the composition of the oxide varies
slightly, a second solid phase does not appear. Over the composition range of the
nonstoichiometric oxide, the number of phases, P, will be two, the nonstoichiometric
oxide MO_x and O_2, gas, and Eq. (7.3b) gives

$$2 + F = 4 \qquad F = 2$$

This means that two thermodynamic parameters can now vary freely while the system
still remains in equilibrium, and the system is said to be *bivariant*. In this case, the
composition becomes an important parameter and must be added to the partial
pressure and temperature. Thus, the partial pressure over a nonstoichiometric oxide
in a sealed tube will no longer depend solely upon the temperature but on the composition as well (Fig. 7.6) (see also Section 6.8.2).

This behavior can be illustrated by the nonstoichiometric oxide $YBa_2Cu_3O_x$,
where x can take values between 6.0 and 7.0. Suppose that some oxide with a composition within the stoichiometry range $YBa_2Cu_3O_{6.0}$–$YBa_2Cu_3O_{7.0}$ and O_2 gas are

present in a sealed tube. Two out of the three parameters—temperature, composition, and oxygen partial pressure—are now required to specify equilibrium. For instance, if the temperature is 500°C and the pressure is 1 bar of oxygen, at equilibrium the composition is $YBa_2Cu_3O_{6.88}$ (Fig. 7.7a). Similarly, if the composition is $YBa_2Cu_3O_{6.5}$ and the temperature is 650°C, the equilibrium oxygen partial pressure will be approximately 0.78 bar ($\log p_{O_2} \approx -1.1 \times 10^{-1}$) (Fig. 7.7b). Finally, the oxygen partial pressure will not be fixed at a definite temperature but will vary with the composition of the oxide (Fig. 7.7c).

7.4 VARIATION OF DEFECT POPULATIONS WITH PARTIAL PRESSURE

The previous sections imply that in the case of nonstoichiometric oxides that favor electronic compensation of composition change a relationship exists between the variation in partial pressure and the electronic conductivity. The form that this relationship takes depends upon the way in which the oxide accommodates composition variation.

7.4.1 Metal-Excess Oxides

Metal-excess oxides can change composition by way of metal interstitials or oxygen vacancies. The formation of cation interstitials in a nonstoichiometric oxide MO can be represented by

$$MO \rightleftharpoons \tfrac{1}{2}O_2 + M_i^{2\bullet} + 2e' \tag{7.4}$$

The assumption that the defects are noninteracting allows the law of mass action in its simplest form, with concentrations instead of activities, to be used for this purpose. In this case, the equilibrium constant K for this reaction is

$$K = \frac{[M_i^{2\bullet}][e']^2 p_{O_2}^{1/2}}{[MO]}$$

where the brackets [ ] denote concentrations. The concentration of the parent oxide, MO, can be taken as more or less static and can be included in a new constant K_1, giving

$$K_1 = [M_i^{2\bullet}][e']^2 p_{O_2}^{1/2}$$

Equation (7.4) shows that the concentrations of interstitials and electrons are related thus:

$$2[M_i^{2\bullet}] = [e']$$

hence

$$K_1 = \tfrac{1}{2}[e']^3 {p_{O_2}}^{1/2}$$

so that the concentration of electrons $[e']$ is given by:

$$[e'] = 2K_1^{1/3} {p_{O_2}}^{-1/6}$$

The conductivity of the solid, σ, will be proportional to the electron concentration so that

$$\sigma \propto [e'] \propto {p_{O_2}}^{-1/6}$$

The electronic conductivity will *fall* as the oxygen pressure increases.

A similar analysis can be carried out for the case of oxygen vacancies:

$$O_O(MO) \rightleftharpoons \tfrac{1}{2}O_2 + V_O^{2\bullet} + 2e'$$

Repeating the steps above gives exactly the same pressure dependence, namely

$$\sigma \propto [e'] \propto {p_{O_2}}^{-1/6}$$

The variation of conductivity with partial pressure is found to be sensitive to the number of electrons present. To illustrate this, suppose that the interstitial cation loses only one electron to form $M_i^{\bullet}$ defects:

$$MO \rightleftharpoons \tfrac{1}{2}O_2 + M_i^{\bullet} + e' \tag{7.5}$$

The equilibrium constant K for this reaction is

$$K = \frac{[M_i^{\bullet}][e']{p_{O_2}}^{1/2}}{[MO]}$$

including the term $[MO]$ in a revised constant K_1 gives

$$K_1 = [M_i^{\bullet}][e']{p_{O_2}}^{1/2}$$

Equation (7.5) shows that the concentration of interstitials and electrons are related; thus,

$$[M_i^{\bullet}] = [e']$$

Hence:

$$K_1 = [e']^2 {p_{O_2}}^{1/2}$$

so that the concentration of electrons $[e']$ is given by:

$$[e'] = K_1^{1/2} {p_{O_2}}^{-1/4}$$

that is,

$$\sigma \propto [e'] \propto p_{O_2}{}^{-1/4}$$

The electronic conductivity is now proportional to the $-\frac{1}{4}$ power of the oxygen partial pressure. In all cases described the conductivity falls as the partial pressure of oxygen rises, but the exact power relationship, $-\frac{1}{4}$ or $-\frac{1}{6}$, depends upon the charge state of the interstitial cation.

7.4.2 Oxygen-Excess Oxides

The analysis of oxygen-excess oxides is similar to that for metal-rich phases just given. For example, the creation of oxygen excess by cation vacancies can be written:

$$\tfrac{1}{2}O_2(MO) \rightleftharpoons O_O + V_M^{2\prime} + 2h^\bullet \tag{7.6}$$

The equilibrium constant, K, of Eq. (7.6) is

$$K = \frac{[h^\bullet]^2[V_M^{2\prime}][O_O]}{p_{O_2}{}^{1/2}}$$

where [ ] represent concentrations and p_{O_2} the oxygen partial pressure. Now the value of $[O_O]$ is essentially constant, as the change in stoichiometry is small. Hence we can assimilate it into a new constant, K_1, and write

$$K_1 = \frac{[h^\bullet]^2[V_M^{2\prime}]}{p_{O_2}{}^{1/2}}$$

For every vacancy in the crystal there are two holes generated [Eq. (7.6)]: thus:

$$[V_M^{2\prime}] = \tfrac{1}{2}[h^\bullet]$$

and substituting:

$$K_1 = \frac{[h^\bullet]^3}{2p_{O_2}{}^{1/2}}$$

so that the concentration of holes is given by

$$[h^\bullet] = (2K_1)^{1/3}p_{O_2}{}^{1/6}$$

That is, the concentration of holes, $[h^\bullet]$, and the electronic conductivity is proportional to the $\frac{1}{6}$ power of the oxygen partial pressure:

$$\sigma \propto [h^\bullet] \propto p_{O_2}{}^{1/6}$$

Thus, the conductivity increases as the partial pressure of oxygen increases, which is the opposite behavior to that of the metal-excess oxides.

Interstitial formation

$$\tfrac{1}{2}O_2(MO) \rightleftharpoons O_i^{2'} + 2h^{\bullet}$$

will lead to precisely the same oxygen dependence.

7.4.3 $Ba_2In_2O_5$

More complex oxides are treated in exactly the same way as binary oxides. The oxide $Ba_2In_2O_5$ adopts the brownmillerite structure (see Section 4.9) below approximately 925°C. The main intrinsic defects present are believed to be Frenkel defects consisting of interstitial oxygen, $O_i^{2'}$, and oxygen vacancies, $V_O^{2\bullet}$. The conductivity at temperatures of the order of 700°C is p-type when the oxygen partial pressure approaches 1 atm and n-type when the oxygen partial pressure is in the region of 10^{-4} atm (Fig. 7.8). This variation can be explained in the following way (see Further Reading section at the end of this chapter).

At higher oxygen partial pressures, assume that oxygen is incorporated into the structure and adds to the oxygen interstitials already present as part of the Frenkel equilibrium. The incorporation of neutral oxygen requires the abstraction of two electrons to form oxide ions, thus generating an equal number of holes:

$$\tfrac{1}{2}O_2 \rightleftharpoons O_i^{2'} + 2h^{\bullet}$$

The equilibrium constant is given by

$$K = \frac{[h^{\bullet}]^2[O_i^{2'}]}{p_{O_2}^{1/2}}$$

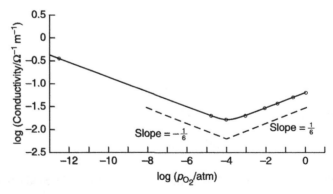

Figure 7.8   Variation of the conductivity of $Ba_2In_2O_5$ as a function of oxygen partial pressure at 700°C. [Redrawn from data in G. B. Zhang and D. M. Smyth, *Solid State Ionics*, **82**, 161–172 (1995).]

For every interstitial introduced into the crystal, two holes are generated:

$$[O_i^{2'}] = \tfrac{1}{2}[h^\bullet]$$

that is,

$$K = \frac{[h^\bullet]^3}{2\,p_{O_2}^{1/2}}$$

Rearrangement gives

$$[h^\bullet] = (2K)^{1/3}p_{O_2}^{1/6}$$

The conductivity is proportional to the hole concentration:

$$\sigma \propto [h^\bullet] \propto p_{O_2}^{1/6}$$

Under oxidizing conditions the conductivity increases with oxygen partial pressure to the power $+\tfrac{1}{6}$, in agreement with the experimental results (Fig. 7.8).

At low oxygen partial pressures, that is, reducing conditions, assume that oxygen is lost to the surrounding atmosphere to create oxygen vacancies to add to those present due to the Frenkel defects. The charged oxide ion is extracted as a neutral atom, leaving behind two electrons:

$$O_O \rightleftharpoons \tfrac{1}{2}O_2 + V_O^{2\bullet} + 2e'$$

The equilibrium constant is given by

$$K = \frac{[e']^2[V_O^{2\bullet}]p_{O_2}^{1/2}}{[O_O]}$$

For every vacancy introduced into the crystal, two electrons are generated:

$$[V_O^{2\bullet}] = \tfrac{1}{2}[e']$$

that is,

$$K = \frac{[e']^3}{2[O_O]}\,p_{O_2}^{1/2}$$

Regarding the amount of structural oxygen, $[O_O]$, as constant and incorporating it into a constant K_1, followed by rearrangement gives

$$[e'] = (2K_1)^{1/3}p_{O_2}^{-1/6}$$

The conductivity is proportional to the electron concentration:

$$\sigma \propto [e'] \propto p_{O_2}^{-1/6}$$

Under reducing conditions the conductivity is n type and decreases with oxygen partial pressure, in agreement with the experimental results (Fig. 7.8).

7.5   BROUWER DIAGRAMS

Defect populations and physical properties such as electronic conductivity can be altered and controlled by manipulation of the surrounding atmosphere. To specify the exact electronic conductivity of such a material, it is necessary to specify its chemical composition, the defect types and populations present, the temperature of the crystal, and the surrounding partial pressures of all the constituents. In this complex situation it is extremely helpful if the defect concentrations (and indirectly the associated electrical conductivity) can be displayed graphically as a function of some variable such as partial pressure of one of the components. A procedure introduced by Brouwer (1954) allows such figures to be compiled. These diagrams, Brouwer or Kröger–Vink diagrams, are widely used, especially for solids with narrow composition ranges.

To construct such a diagram, a set of defect reaction equations is formulated and expressions for the equilibrium constants of each are obtained. The assumption that the defects are noninteracting allows the law of mass action in its simplest form, with concentrations instead of activities, to be used for this purpose. To simplify matters, only one defect reaction is considered to be dominant in any particular composition region, this being chosen from knowledge of the chemical attributes of the system under consideration. The simplified equilibrium expressions are then used to construct plots of the logarithm of defect concentration against an experimental variable such as the log (partial pressure) of the components. The procedure is best illustrated by an example.

7.5.1   Initial Assumptions

The first step is to set out the assumptions concerning the defects that are likely to occur, using physical and chemical intuition about the system in mind. For illustrative purposes, take a nonstoichiometric phase of composition MX, nominally containing M^{2+} and X^{2-} ions, with a stoichiometric composition, $MX_{1.0}$. In this example the following is assumed:

1. The only ionic defects of importance are vacancies on metal sites, $V_M^{2'}$, and vacancies on anion sites, $V_X^{2\bullet}$; interstitial defects are ignored. The stoichiometric composition $MX_{1.0}$ will occur when the number of vacancies on the cation sublattice is exactly equal to the number of vacancies on the anion sublattice, which is, therefore, a population of Schottky defects.

2. MX can have an existence range that spans both sides of the stoichiometric composition, $MX_{1.0}$, which is due to an imbalance in the population of vacancies.

3. As well as vacancies, MX contains a varying population of electrons e', or holes, $h^\bullet$, which can act to maintain electroneutrality when the solid becomes nonstoichiometric. At the stoichiometric composition of $MX_{1.0}$, the number of electrons is equal to the number of holes, which are simply the normally present intrinsic electrons and holes.

4. The most important gaseous component is X_2, as is the case in most oxides, halides, and sulfides. The stoichiometric variation will be linked to the partial pressure of the surrounding nonmetal atmosphere. The nonmetal component will be gained at high pressures and lost at low pressures. These options correspond to oxidation and reduction.

These assumptions mean that there are only four defects to consider, electrons, e', holes, $h^\bullet$, vacancies on metal sites $V_M^{2\prime}$ and vacancies on anion sites, $V_X^{2\bullet}$.

7.5.2 Defect Equilibria

The equilibrium between the defects present can be described in terms of chemical equations:

1. *Creation and Elimination of Schottky Defects*   These defects can form at the crystal surface or vanish by diffusing to the surface. The equation describing this and the associated equilibrium constant is

$$\text{nul} \rightleftharpoons V_M^{2\prime} + V_X^{2\bullet} \qquad K_S = [V_M^{2\prime}][V_X^{2\bullet}] \tag{7.7}$$

2. *Creation and Elimination of Electronic Defects*   These are the normal intrinsic electrons and holes present in a semiconductor. Electrons can combine with holes to be eliminated from the crystal thus:

$$\text{nul} \rightleftharpoons e' + h^\bullet \qquad K_e = [e'][h^\bullet] = np \tag{7.8}$$

(Electron and hole concentrations are often written as n and p, but this alternative will not be followed here to avoid confusion with pressure, also denoted as p.)

3. *Composition Change*   This comes about by interaction with the gas phase to produce cation or anion vacancies. For oxidation the equation is

$$\tfrac{1}{2}X_2 \rightleftharpoons X_X + V_M^{2\prime} + 2h^\bullet \qquad K_o = [V_M^{2\prime}][h^\bullet]^2 p_{X_2}^{-1/2} \tag{7.9}$$

A change in composition to produce reduction is written:

$$X_X \rightleftharpoons \tfrac{1}{2}X_2(g) + V_X^{2\bullet} + 2e' \qquad K_r = [V_X^{2\bullet}][e']^2 p_{X_2}{}^{1/2} \tag{7.10}$$

The equilibrium constant K_r is redundant, being equal to

$$K_r = \frac{K_S K_e{}^2}{K_o} \tag{7.11}$$

but it simplifies some of the following equations if it is retained.

4. *Electroneutrality*   At all times the crystal must remain electrically neutral. Equations (7.9) and (7.10) define the formation of charged species, so that the appropriate electroneutrality equation is

$$2[V_M^{2\prime}] + [e'] = 2[V_X^{2\bullet}] + [h^\bullet] \tag{7.12}$$

The four equations [(7.9)–(7.12)] are simplified using chemical and physical intuition. Two examples are given. In the first (Sections 7.6.3–7.6.6) the case where ionic point defects are more important than electrons and holes is considered, and in the following sections (Sections 7.7.1–7.7.5) the reverse case, where electronic defects are more important than vacancies, is described.

7.5.3   Stoichiometric Point: Ionic Defects

The compound will be stoichiometric, with an exact composition of $MX_{1.0000}$ when the number of metal vacancies is equal to the number of nonmetal vacancies. At the same time, the number of electrons and holes will be equal. In an inorganic compound, which is an insulator or poor semiconductor with a fairly large band-gap, the number of point defects is greater than the number of intrinsic electrons or holes. To illustrate the procedure, suppose that the values for the equilibrium constants describing Schottky disorder, K_S, and intrinsic electron and hole numbers, K_e, are

$$K_S = [V_M^{2\prime}][V_X^{2\bullet}] = 10^{40} \text{ (defects m}^{-3})^2$$

$$[V_M^{2\prime}] = [V_X^{2\bullet}] = 10^{20} \text{ defects m}^{-3} \qquad \log[V_M^{2\prime}] = \log[V_X^{2\bullet}] = 20$$

$$K_e = [e'][h^\bullet] = 10^{32} \text{ (defects m}^{-3})^2$$

$$[e'] = [h^\bullet] = 10^{16} \text{ defects m}^{-3} \qquad \log[e'] = \log[h^\bullet] = 16$$

Equilibrium partial pressure $= 1$ atm of X_2 \qquad $\log p_{X_2} = 0$

These are plotted in Figure 7.9a.

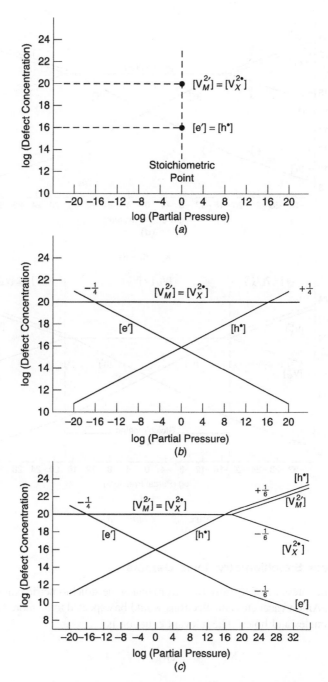

Figure 7.9 Brouwer diagram for a phase ~MX in which Schottky defects are the main point defect type: (*a*) initial points on the diagram, (*b*) variation of defect concentrations in the near-stoichiometric region, (*c*) extension to show variation of defect concentrations in the high partial pressure region, (*d*) extension to show variation of defect concentrations in the low partial pressure region, and (*e*) complete diagram.

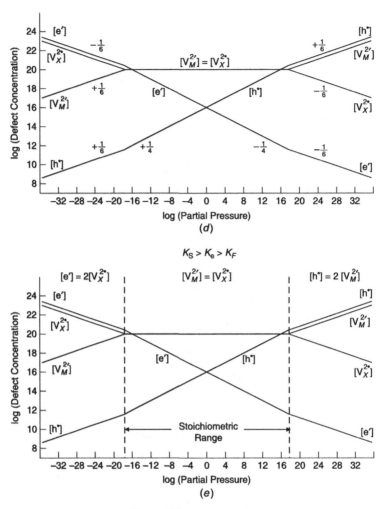

Figure 7.9   *(Continued).*

7.5.4   Near Stoichiometry: Ionic Defects

As the partial pressure of X_2 falls away from that at the stoichiometric point, reduction will occur. Anion vacancies and electrons would be expected to be more favored than cation vacancies and holes. The relevant equation is

$$X_X \rightleftharpoons \tfrac{1}{2}X_2(g) + V_X^{2\bullet} + 2e' \qquad (7.10)$$

The effects of this change will depend upon the equilibrium constants. Suppose that 10^{17} m^{-3} of vacancies, and double that number of electrons, is produced by the change in partial pressure. In the present example, the concentration of electrons will change considerably, from $1 \times 10^{16}\,\text{m}^{-3}$ to $20 \times 10^{16}\,\text{m}^{-3}$. However, the concentration of vacancies will change only slightly, from $1 \times 10^{20}\,\text{m}^{-3}$ to $1.001 \times 10^{20}\,\text{m}^{-3}$.

As the partial pressure of X_2 increases beyond that at the stoichiometric point, oxidation will occur. Cation vacancies would be expected to be more favored than anion vacancies. The relevant equation is

$$\tfrac{1}{2}X_2 \rightleftharpoons X_X + V_M^{2\prime} + 2h^\bullet \tag{7.9}$$

Under these conditions, cation vacancies and holes will be formed. As before, the effect of this change will depend upon the equilibrium constants. Suppose $10^{17}\,\text{m}^{-3}$ of vacancies, and double that number of holes, is produced by the change in partial pressure. The concentration of holes will change considerably, from $1 \times 10^{16}\,\text{m}^{-3}$ to $20 \times 10^{16}\,\text{m}^{-3}$. The concentration of vacancies will change only slightly, from $1 \times 10^{20}\,\text{m}^{-3}$ to $1.001 \times 10^{20}\,\text{m}^{-3}$.

Because it is assumed that K_S is a lot greater than K_e it is reasonable to ignore the minority electronic defects and approximate the electroneutrality Eq. (7.12) by the relation:

$$[V_M^{2\prime}] = [V_X^{2\bullet}] \tag{7.13}$$

This equation is now substituted into Eqs. (7.7), (7.9), and (7.10) to obtain a new set of equations describing the appropriate defect concentrations:
From Eq. (7.7):

$$[V_M^{2\prime}] = [V_X^{2\bullet}] = K_S^{1/2}$$
$$\log[V_M^{2\prime}] = \log[V_X^{2\bullet}] = \tfrac{1}{2}\log K_S \tag{7.7}$$

From Eq. (7.9):

$$[h^\bullet]^2 = \left(\frac{K_o}{[V_M^{2\prime}]}\right)p_{X_2}^{1/2} = \left(\frac{K_o}{K_S^{1/2}}\right)p_{X_2}^{1/2}$$
$$[h^\bullet] = K_o^{1/2}K_S^{-1/4}p_{X_2}^{1/4} \tag{7.14}$$
$$\log[h^\bullet] = \tfrac{1}{2}\log K_o - \tfrac{1}{4}\log K_S + \tfrac{1}{4}\log p_{X_2}$$

From Eq. (7.10):

$$[e^\prime]^2 = \frac{K_r}{[V_X^{2\bullet}]p_{X_2}^{1/2}} = \frac{K_r}{K_S^{1/2}p_{X_2}^{1/2}}$$
$$[e^\prime] = K_r^{1/2}K_S^{-1/4}p_{X_2}^{-1/4} \tag{7.15}$$
$$\log[e^\prime] = \tfrac{1}{2}\log K_r - \tfrac{1}{4}\log K_S - \tfrac{1}{4}\log p_{X_2}$$

These equations apply to both sides of the stoichiometric composition. They can be plotted as log(defect concentration) versus log p_{X_2}. A plot of log [$h^\bullet$] versus log p_{X_2} has a slope of $\frac{1}{4}$ and passes through the (electronic) stoichiometric point. A plot of log [e'] versus log p_{X_2} has a slope of $-\frac{1}{4}$ and passes through the (electronic) stoichiometric point (Fig. 7.9b).

7.5.5   High X$_2$ Partial Pressures: Ionic Defects

At ever higher partial pressures the number of nonmetal vacancies will diminish more and more. In addition, Figure 7.9b suggests that the number of holes will eventually exceed the number of vacancies in this regime. Thus, the approximate electroneutrality Eq. (7.13) will no longer be representative. A more appropriate form of the electroneutrality equation for the high-pressure region is obtained by ignoring the minority species, nonmetal vacancies and electrons, to give

$$2[V_M^{2\prime}] = [h^\bullet] \tag{7.16}$$

This can now be substituted into Eqs. (7.7)–(7.10) to obtain relationships between the partial pressure of X$_2$ and the defect concentrations present in the material relevant to the high partial pressure region.

From Eq. (7.9):

$$[V_M^{2\prime}] = \frac{K_o p_{X_2}{}^{1/2}}{[h^\bullet]^2} = \frac{K_o p_{X_2}{}^{1/2}}{(2[V_M^{2\prime}])^2}$$

$$4[V_M^{2\prime}]^3 = K_o p_{X_2}{}^{1/2}$$

$$8[V_M^{2\prime}]^3 = 2K_o p_{X_2}{}^{1/2}$$

$$[V_M^{2\prime}] = \tfrac{1}{2}(2K_o)^{1/3} p_{X_2}{}^{1/6} \tag{7.17}$$

$$\log [V_M^{2\prime}] = \log\tfrac{1}{2} + \tfrac{1}{3}\log (2K_o) + \tfrac{1}{6}\log p_{X_2}$$

From Eq. (7.16):

$$[h^\bullet] = (2K_o)^{1/3} p_{X_2}{}^{1/6} \tag{7.18}$$

$$\log [h^\bullet] = \tfrac{1}{3}\log (2K_o) + \tfrac{1}{6}\log p_{X_2}$$

From Eq. (7.7):

$$[V_X^{2\bullet}] = \frac{K_S}{[V_M^{2\prime}]} = 2K_S(2K_o)^{-1/3} p_{X_2}{}^{-1/6} \tag{7.19}$$

$$\log [V_X^{2\bullet}] = \log 2 + \log K_S - \tfrac{1}{3}\log (2K_o) - \tfrac{1}{6}\log p_{X_2}$$

From Eq. (7.8):

$$[e'] = \frac{K_e}{[h^\bullet]} = K_e(2K_o)^{-1/3}p_{X_2}^{-1/6} \tag{7.20}$$

$$\log[e'] = \log K_e - \tfrac{1}{3}\log(2K_o) - \tfrac{1}{6}\log p_{X_2}$$

A plot of $\log[h^\bullet]$ versus $\log p_{X_2}$ has a slope of $\tfrac{1}{6}$, a plot of $\log[e']$ versus $\log p_{X_2}$ has a slope of $-\tfrac{1}{6}$, a plot of $\log[V_M^{2\prime}]$ versus $\log p_{X_2}$ has a slope of $\tfrac{1}{6}$, and a plot of $\log[V_X^{2\bullet}]$ versus $\log p_{X_2}$ has a slope of $-\tfrac{1}{6}$.

The point at which the slopes change is chosen to coincide with the point at which the region obeying the electroneutrality condition given by Eq. (7.13) is replaced by Eq. (7.16). Thus, the change of slope occurs when $2[V_M^{2\prime}] = [h^\bullet]$, and using the values of $[V_M^{2\prime}]$ as 10^{20} defects m^{-3}, the slope of the line displaying the concentration of holes changes slope at a value of $[h^\bullet] = 2 \times 10^{20}$ defects m^{-3}. This is close to the intersection of the line representing the concentration of holes with the line representing the concentration of vacancies. The results can be used to continue Figure 7.9*b* to the right (Fig. 7.9*c*).

7.5.6 Low X$_2$ Partial Pressures: Ionic Defects

As the pressure diminishes far below that of the stoichiometric point, the number of metal vacancies and holes will continue to fall and the electroneutrality equation chosen, Eq. (7.16), will no longer be representative. A more appropriate form of the electroneutrality equation for the low-pressure region ignores the minority defects, which are now metal vacancies and holes, to give

$$[e'] = 2[V_X^{2\bullet}] \tag{7.21}$$

This is substituted into Eqs. (7.7)–(7.10) to derive the defect concentrations relevant to the low partial pressure region.

From Eq. (7.10):

$$[V_X^{2\bullet}] = \frac{K_r}{[e']^2 p_{X_2}^{1/2}} = \frac{K_r}{(2[V_X^{2\bullet}])^2 p_{X_2}^{1/2}}$$

$$4[V_X^{2\bullet}]^3 = \frac{K_r}{p_{X_2}^{1/2}}$$

$$8[V_X^{2\bullet}]^3 = \frac{2K_r}{p_{X_2}^{1/2}}$$

$$[V_X^{2\bullet}] = \tfrac{1}{2}(2K_r)^{1/3}p_{X_2}^{-1/6} \tag{7.22}$$

$$\log[V_X^{2\bullet}] = \log\tfrac{1}{2} + \tfrac{1}{3}\log(2K_r) - \tfrac{1}{6}\log p_{X_2}$$

From Eq. (7.21):

$$[e'] = (2K_r)^{1/3} p_{X_2}{}^{-1/6}$$

$$\log [e'] = \tfrac{1}{3} \log (2K_r) - \tfrac{1}{6} \log p_{X_2} \tag{7.23}$$

From Eq. (7.7):

$$[V_M^{2\prime}] = \frac{K_S}{[V_X^{2\bullet}]} = 2K_S (2K_r)^{-1/3} p_{X_2}{}^{1/6} \tag{7.24}$$

$$\log [V_M^{2\prime}] = \log 2 + \log K_S - \tfrac{1}{3} \log (2K_r) + \tfrac{1}{6} \log p_{X_2}$$

From Eq. (7.8):

$$[h^\bullet] = \frac{K_e}{[e']} = K_e (2K_r)^{-1/3} p_{X_2}{}^{1/6} \tag{7.25}$$

$$\log [h^\bullet] = \log K_e - \tfrac{1}{3} \log (2K_r) + \log p_{X_2}{}^{1/6}$$

A plot of $\log [h^\bullet]$ versus $\log p_{X_2}$ has a slope of $\tfrac{1}{6}$, a plot of $\log [e']$ versus $\log p_{X_2}$ has a slope of $-\tfrac{1}{6}$, a plot of $\log [V_M^{2\prime}]$ versus $\log p_{X_2}$ has a slope of $\tfrac{1}{6}$, and a plot of $\log [V_X^{2\bullet}]$ versus $\log p_{X_2}$ has a slope of $-\tfrac{1}{6}$, exactly as in the high-pressure region.

The point at which the slopes change is chosen to coincide with the point at which the region obeying the electroneutrality condition given by Eq. (7.13) is replaced by Eq. (7.21). Thus, the change of slope occurs when $[e'] = 2[V_X^{2\bullet}]$, and using the values of $[V_X^{2\bullet}]$ as 10^{20} defects m^{-3}, the slope of the line displaying the concentration of holes changes slope at a value of $[e'] = 2 \times 10^{20}$ defects m^{-3}. This is close to the intersection of the line representing the concentration of holes with the line representing the concentration of vacancies. The results can be used to continue Figure 7.9b to the left (Fig. 7.9d).

7.5.7 Complete Diagram: Ionic Defects

It is important that the complete diagram displays prominently information about the assumptions made. Thus, the assumption that Schottky defect formation was preferred to the formation of electronic defects is explicitly stated in the form $K_S > K_e$ (Fig. 7.9e). As Frenkel defect formation has been ignored altogether, it is also possible to write $K_S > K_e >> K_F$, where K_F represents the equilibrium constant for the formation of Frenkel defects in MX.

This central, medium pressure, region of the diagram displays the electroneutrality equation approximation $[V_M^{2\prime}] = [V_X^{2\bullet}]$ at the top. In this region the compound remains stoichiometric across this range, as the number of cation and anion vacancies remains the same. The number of holes and electrons will be below the number of cation and anion vacancies for most of this region, and the material will be a stoichiometric insulator with a composition $MX_{1.000}$ containing Schottky defects.

The high partial pressure region displays the electroneutrality equation approximation $[h^\bullet] = 2[V_M^{2/}]$. The defects with the highest concentration are holes, so that the material is a p-type semiconductor in this regime. In addition, the conductivity will increase as the partial pressure of the gaseous X_2 component increases. The number of metal vacancies will increase as the partial pressure of the gaseous X_2 component increases and, because of the corresponding fall in the concentration of anion vacancies, the phase will be nonstoichiometric and rich in nonmetal. In this region the high concentration of cation vacancies would be expected to enhance cation diffusion.

The low-pressure region displays the electroneutrality equation approximation $[e'] = 2[V_X^{2\bullet}]$. Electrons predominate so that the material is an n-type semiconductor in this regime. In addition, the conductivity will increase as the partial pressure of the gaseous X_2 component decreases. The number of nonmetal vacancies will increase as the partial pressure of the gaseous X_2 component decreases, and the phase will display a metal-rich nonstoichiometry opposite to that in the high-pressure domain. Because there is a high concentration of anion vacancies, easy diffusion of anions is to be expected.

The form of this diagram in which abrupt changes occur between the three regions is clearly too extreme. The diagram can easily be modified to smooth out these discontinuities by inserting additional segments, employing additional electroneutrality equations. For example, between the low-pressure and medium-pressure regions the electroneutrality equation

$$2[V_M^{2/}] = 2[V_X^{2\bullet}] + [h^\bullet]$$

is appropriate. Between the medium-pressure and high-pressure regions

$$2[V_M^{2/}] + [e'] = [V_X^{2\bullet}]$$

can be used. The value of these extensions depends upon the precision of available experimental data and the intended use of the diagram.

7.6 BROUWER DIAGRAMS: ELECTRONIC DEFECTS

7.6.1 Electronic Defects

In many compound semiconductors the band gap is rather small, leading to a significant concentration of intrinsic electrons and holes while the population of interstitial or vacancy point defects is lower. To illustrate differences that this changed state of affairs brings about, the Brouwer diagram for a nonstoichiometric phase of composition MX, nominally containing M^{2+} and X^{2-} ions, but with a sizable concentration of electrons and holes will be derived. Exactly the same procedures as those described in Sections 7.5.1–7.5.7 will be followed for this changed situation. To summarize, the nonstoichiometric phase of composition MX, nominally containing M^{2+} and X^{2-} ions, has a stoichiometric composition, $MX_{1.0}$. The only ionic defects of importance are vacancies on metal sites, $V_M^{2\prime}$, and vacancies on anion sites, $V_X^{2\bullet}$. The electronic defects present are free electrons e' and holes $h^\bullet$. At the stoichiometric composition of $MX_{1.0}$, the numbers of cation and anion vacancies are equal (as Schottky defects) as are the numbers of (intrinsic) electrons and holes. These assumptions mean that there are only four defects to consider: electrons e', holes $h^\bullet$, vacancies on metal sites $V_M^{2\prime}$, and vacancies on anion sites $V_X^{2\bullet}$. As the electronic defects are of greatest importance, the value of K_e is greater than K_S.

The equations involving defect formation will be identical to Eqs. (7.7–7.10). However, the values for the equilibrium constants chosen in Section 7.5.3 are now reversed to give greater concentrations of electronic defects at the stoichiometric point:

$$K_S = [V_M^{2\prime}][V_X^{2\bullet}] = 10^{32} \text{ (defects m}^{-3})^2$$

$$[V_M^{2\prime}] = [V_X^{2\bullet}] = 10^{16} \text{ defects m}^{-3} \qquad \log[V_M^{2\prime}] = \log[V_X^{2\bullet}] = 16$$

$$K_e = [e'][h^\bullet] = 10^{40} \text{ (defects m}^{-3})^2$$

$$[e'] = [h^\bullet] = 10^{20} \text{ defects m}^{-3} \qquad \log[e'] = \log[h^\bullet] = 20$$

$$\text{Equilibrium partial pressure} = 1 \text{ atm of } X_2 \qquad \log p_{X_2} = 0$$

These values are plotted on Figure 7.10a.

7.6.2   Near Stoichiometry: Electronic Defects

As the partial pressure of X_2 falls away from that at the stoichiometric point, anion vacancies and electrons would be expected to be more favored than cation vacancies and holes, as before:

$$X_X \rightleftharpoons \tfrac{1}{2}X_2(g) + V_X^{2\bullet} + 2e' \qquad (7.10)$$

Suppose that 10^{17} m^{-3} of vacancies, and double that number of electrons, is produced by the change in partial pressure. The change in the concentration of electrons will be negligible, from $1 \times 10^{20} \text{ m}^{-3}$ to $1.002 \times 10^{20} \text{ m}^{-3}$. The concentration of vacancies will change appreciably, from $1 \times 10^{16} \text{ m}^{-3}$ to $1 \times 10^{17} \text{ m}^{-3}$.

As the partial pressure of X_2 increases beyond that at the stoichiometric point, oxidation will occur and cation vacancies and holes would be expected to be more favored than anion vacancies and electrons:

$$\tfrac{1}{2}X_2 \; \rightleftharpoons \; X_X + V_M^{2\prime} + 2\,h^{\bullet} \tag{7.9}$$

As above, suppose that 10^{17} m^{-3} of vacancies, and double that number of holes, is produced by the change in partial pressure. The concentration of vacancies will

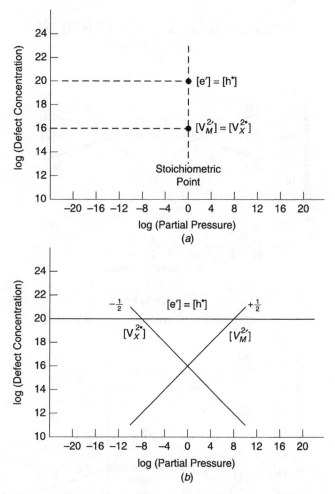

Figure 7.10 Brouwer diagram for a phase $\sim$MX in which electronic defects are the main point defect type: (*a*) initial points on the diagram, (*b*) variation of defect concentrations in the near-stoichiometric region, (*c*) extension to show variation of defect concentrations in the high partial pressure region, (*d*) extension to show variation of defect concentrations in the low partial pressure region, and (*e*) the complete diagram.

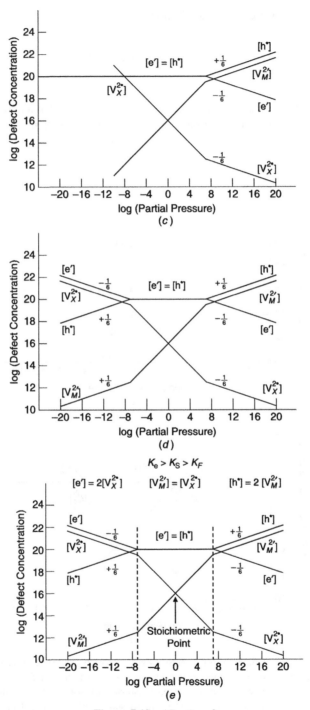

Figure 7.10 (*Continued*).

change considerably, from 1×10^{16} m^{-3} to 1×10^{17} m^{-3}. The concentration of holes will change only slightly, from 1×10^{20} m^{-3} to 1.002×10^{20} m^{-3}.

In both of these regimes it is more reasonable to approximate the electroneutrality Eq. (7.12) by

$$[e'] = [h^{\bullet}] \qquad (7.26)$$

This is equivalent to the assumption that K_e is a lot greater than K_S, as stated.

This equation is now substituted into Eqs. (7.8) and (7.9) to obtain a new set of equations for the defect concentrations:

$$[h^{\bullet}] = [e'] = K_e^{1/2} \qquad (7.27)$$
$$\log[h^{\bullet}] = \log[e'] = \tfrac{1}{2}\log K_e$$

$$[V_X^{2\bullet}] = \left(\frac{K_r}{K_e}\right) p_{X_2}^{-1/2} \qquad (7.28)$$

$$\log[V_X^{2\bullet}] = \log K_r - \log K_e - \tfrac{1}{2}\log p_{X_2}$$

$$[V_M^{2\prime}] = \left(\frac{K_o}{K_e}\right) p_{X_2}^{1/2} \qquad (7.29)$$

$$\log[V_M^{2\prime}] = \log K_o - \log K_e + \tfrac{1}{2}\log p_{X_2}$$

As expected, these equations are similar in form to Eqs. (7.13)–(7.15) and apply to both sides of the stoichiometric composition. They can be plotted as log(defect concentration) versus $\log p_{X_2}$ graph. A plot of $\log[V_X^{2\bullet}]$ versus $\log p_{X_2}$ has a slope of $-\tfrac{1}{2}$ and passes through the (ionic) stoichiometric point. A plot of $\log[V_M^{2\prime}]$ versus $\log p_{X_2}$ has a slope of $+\tfrac{1}{2}$ and passes through the (ionic) stoichiometric point (Fig. 7.10b).

7.6.3 High X$_2$ Partial Pressures: Electronic Defects

At ever higher partial pressures the number of nonmetal vacancies will diminish more and more and the number of holes will exceed the number of vacancies in this regime. Thus the approximate electroneutrality Eq. (7.13) will no longer be representative. A more appropriate form of the electroneutrality equation for the high-pressure region is obtained by ignoring the minority species, nonmetal vacancies and electrons, to give

$$[h^{\bullet}] = 2[V_M^{2\prime}] \qquad (7.16)$$

This can now be substituted into Eqs. (7.7)–(7.10) to obtain relationships between the partial pressure of X$_2$ and the defect concentrations present in the material relevant to the high partial pressure region. This is identical to the situation described for ionic defects, and so some detail will be condensed in the following.

From Eq. (7.9):

$$[V_M^{2\prime}] = \tfrac{1}{2}(2K_o)^{1/3}p_{X_2}{}^{1/6} \tag{7.17}$$

$$\log [V_M^{2\prime}] = \log \tfrac{1}{2} + \tfrac{1}{3}\log (2K_o) + \tfrac{1}{6}\log p_{X_2}$$

From Eq. (7.16):

$$[h^{\bullet}] = (2K_o)^{1/3}p_{X_2}{}^{1/6} \tag{7.18}$$

$$\log [h^{\bullet}] = \tfrac{1}{3}\log (2K_o) + \tfrac{1}{6}\log p_{X_2}$$

From Eq. (7.7):

$$[V_X^{2\bullet}] = 2K_S(2K_o)^{-1/3}p_{X_2}{}^{-1/6} \tag{7.19}$$

$$\log [V_X^{2\bullet}] = \log 2 + \log K_S - \log (2K_o) - \tfrac{1}{6}\log p_{X_2}$$

From Eq. (7.8):

$$[e^{\prime}] = K_e(2K_o)^{-1/3}p_{X_2}{}^{-1/6}$$
$$\log [e^{\prime}] = \log K_e - \tfrac{1}{3}\log (2K_o) - \tfrac{1}{6}\log p_{X_2} \tag{7.20}$$

A plot of $\log [h^{\bullet}]$ versus $\log p_{X_2}$ has a slope of $\tfrac{1}{6}$, a plot of $\log [e^{\prime}]$ versus $\log p_{X_2}$ has a slope of $-\tfrac{1}{6}$, a plot of $\log [V_M^{2\prime}]$ versus $\log p_{X_2}$ has a slope of $\tfrac{1}{6}$, and a plot of $\log [V_X^{2\bullet}]$ versus $\log p_{X_2}$ has a slope of $-\tfrac{1}{6}$.

The point at which the slopes change is chosen to coincide with the point at which the region obeying the electroneutrality condition given by Eq. (7.13) is replaced by Eq. (7.16). Thus the change of slope occurs when $2[V_M^{2\prime}] = [h^{\bullet}]$, and using the value of $[h^{\bullet}]$ as 10^{20} defects m^{-3}, the slope of the line displaying the concentration of metal vacancies changes slope at a value of $[V_M^{2\prime}] = 5 \times 10^{19}$ defects m^{-3}. The results can be used to continue Figure 7.10b to the right (Fig. 7.10c).

7.6.4   Low X_2 Partial Pressures: Electronic Defects

As the pressure diminishes far below that of the stoichiometric point, the number of metal vacancies and holes will continue to fall, and, as before, the electroneutrality equation chosen, Eq. (7.16), will no longer be representative. A more appropriate form of the electroneutrality equation for the low-pressure region ignores the minority defects, which are now metal vacancies and holes, to give

$$[e^{\prime}] = 2[V_X^{2\bullet}] \tag{7.21}$$

This equation is substituted into Eqs. (7.7)–(7.10) to derive the defect concentrations relevant to the low partial pressure region. Once again these will be similar to those derived for ionic defects.

From Eq. (7.10):

$$[V_X^{2\bullet}] = \tfrac{1}{2}(2K_r)^{1/3} p_{X_2}{}^{-1/6} \tag{7.22}$$

$$\log[V_X^{2\bullet}] = \log\tfrac{1}{2} + \tfrac{1}{3}\log(2K_r) - \tfrac{1}{6}\log p_{X_2}$$

From Eq. (7.21):

$$[e'] = (2K_r)^{1/3} p_{X_2}{}^{-1/6} \tag{7.23}$$

$$\log[e'] = \tfrac{1}{3}\log(2K_r) - \tfrac{1}{6}\log p_{X_2}$$

From Eq. (7.7):

$$[V_M^{2\prime}] = 2K_S(2K_r)^{-1/3} p_{X_2}{}^{1/6} \tag{7.24}$$

$$\log[V_M^{2\prime}] = \log 2 + \log K_S - \tfrac{1}{3}\log(2K_r) + \tfrac{1}{6}\log p_{X_2}$$

From Eq. (7.8):

$$[h^\bullet] = K_e(2K_r)^{-1/3} p_{X_2}{}^{1/6} \tag{7.25}$$

$$\log[h^\bullet] = \log K_e - \tfrac{1}{3}\log(2K_r) + \log p_{X_2}{}^{1/6}$$

A plot of $\log[h^\bullet]$ versus $\log p_{X_2}$ has a slope of $\tfrac{1}{6}$, a plot of $\log[e']$ versus $\log p_{X_2}$ has a slope of $-\tfrac{1}{6}$, a plot of $\log[V_M^{2\prime}]$ versus $\log p_{X_2}$ has a slope of $\tfrac{1}{6}$, and a plot of $\log[V_X^{2\bullet}]$ versus $\log p_{X_2}$ has a slope of $-\tfrac{1}{6}$, exactly as in the high-pressure region.

The point at which the slopes change is chosen to coincide with the point at which the region obeying the electroneutrality condition given by Eq. (7.13) is replaced by Eq. (7.21). Thus the change of slope occurs when $[e'] = 2[V_X^{2\bullet}]$, and using the values of $[e']$ as 10^{20} defects m^{-3}, the slope of the line displaying the concentration of nonmetal vacancies changes slope at a value of $[V_X^{2\bullet}] = 5 \times 10^{19}$ defects m^{-3}. The results can be used to continue Figure 7.10b to the left (Fig. 7.10d).

7.6.5 Complete Diagram: Electronic Defects

The original assumption that electronic defect formation was preferred to the formation of Schottky defects is explicitly stated at the top of Figure 7.10e. As Frenkel defect formation has been ignored altogether, it is also possible to write $K_e > K_S \gg K_F$, where K_F represents the equilibrium constant for the formation of Frenkel defects in MX.

Although Fig. 7.10e is similar to that for ionic defects (Fig. 7.9e), there are a number of significant differences. In particular, the stoichiometric range is now far narrower. This central region of the diagram displays the electroneutrality equation chosen, $[e'] = [h^\bullet]$, prominently. In this region the compound remains an intrinsic semiconductor. However, the number of anion and cation vacancies change unsymmetrically away from the stoichiometric point and the composition will vary from $MX_{1.000}$. However, this variation will be extremely small. For example, at the high partial pressure limit ($\log p_{X_2} = +7$) side the metal vacancy concentration is 5×10^{19} metal vacancies m^{-3}; then from Figure 7.10e it is found that there are 5×10^{13} nonmetal vacancies m^{-3}. Taking a NaCl crystal as typical, there are of the order of 10^{28} cations and anions m^{-3}, so that the composition change is minute.

The high-pressure region is associated with the electroneutrality equation $[h^\bullet] = 2[V_M^{2\prime}]$. Holes predominate, so that the material is a p-type semiconductor in this regime. In addition, the conductivity will increase as the $\frac{1}{6}$ power of the partial pressure of the gaseous X_2 component increases. The number of metal vacancies (and nonmetal excess) will increase as the partial pressure of the gaseous X_2 component increases and the phase will be distinctly nonstoichiometric. There is a high concentration of cation vacancies that would be expected to enhance cation diffusion.

The low-pressure region is associated with the electroneutrality equation $[e'] = 2[V_X^{2\bullet}]$. Electrons predominate so that the material is an n-type semiconductor in this regime. In addition, the conductivity will increase as the $-\frac{1}{6}$ power of the partial pressure of the gaseous X_2 component increases. The number of nonmetal vacancies (and metal excess) will increase as the partial pressure of the gaseous X_2 component decreases and the phase will display a nonstoichiometry opposite to that in region III. Because there is high concentration of anion vacancies, easy diffusion of anions is to be expected.

As in the case of ionic defects, the form of this diagram can easily be modified to smooth out the abrupt changes between the three regions by including intermediate electroneutrality equations.

7.7 BROUWER DIAGRAMS: MORE COMPLEX EXAMPLES

In general, there is little problem in extending the concepts just outlined to more complex materials. The procedure is to write down the equations specifying the various equilibria: point defect formation, electronic defect formation, the oxidation reaction, and the reduction reaction. These four equations, only three of which are independent, are augmented by the electroneutrality equation. Two examples will be sketched for the oxides Cr_2O_3 and $Ba_2In_2O_5$.

7.7.1 Cr_2O_3

The presence of relatively small Cr^{3+} cations and the close-packed corundum structure anions suggest that Schottky defects will be preferred. Oxidation will add anions and reduction will remove anions. The four equilibria can now be written.

Point defect formation; Schottky defects:

$$\text{nul} \rightleftharpoons 2V_{Cr}^{3\prime} + 3V_O^{2\bullet} \qquad K_S = [V_{Cr}^{3\prime}]^2[V_O^{2\bullet}]^3$$

Electronic defect formation:

$$\text{nul} \rightleftharpoons e^{\prime} + h^{\bullet} \qquad K_e = [e^{\prime}][h^{\bullet}]$$

Oxidation:

$$\tfrac{3}{2}O_2 \rightleftharpoons 3O_O + 2V_{Cr}^{3\prime} + 6h^{\bullet} \qquad K_o = \frac{[V_{Cr}^{3\prime}]^2[h^{\bullet}]}{p_{O_2}^{3/2}}$$

Reduction:

$$O_O \rightleftharpoons V_O^{2\bullet} + 2e^{\prime} + \tfrac{1}{2}O_2 \qquad K_r = [V_O^{2\bullet}][e^{\prime}]^2 p_{O_2}^{1/2}$$

Electroneutrality:

$$3[V_{Cr}^{3\prime}] + [e^{\prime}] = [h^{\bullet}] + 2[V_O^{2\bullet}]$$

These equations can now be simplified in the manner described above. The electroneutrality point is given by

$$[e^{\prime}] = [h^{\bullet}]$$

and the point at which the compound is strictly stoichiometric by

$$3[V_{Cr}^{3\prime}] = 2[V_O^{2\bullet}]$$

The approximations to use depend upon the pressure regime and the values of the equilibrium constants. This oxide is an insulator under normal conditions, and so, in the middle region of the diagram, Schottky equilibrium is dominant, that is, $K_S > K_e$ and the electroneutrality equation is approximated by

$$3[V_{Cr}^{3\prime}] = 2[V_O^{2\bullet}]$$

In the oxidized region, holes and cation vacancies are preferred, so the electroneutrality equation is approximated by

$$3[V_{Cr}^{3\prime}] = [h^{\bullet}]$$

In the reduced region, electrons and anion vacancies are to be preferred, and the electroneutrality equation is approximated by

$$[e'] = 2[V_O^{2\bullet}]$$

Insertion of the relevant equilibrium constants allows the diagram to be drawn.

7.7.2   Ba$_2$In$_2$O$_5$

Although the formula of this phase is more complex than the binary compounds described above, the procedure is exactly the same. It is only necessary to identify the likely defect equilibria that pertain to the experimental situation.

The structure of the oxide Ba$_2$In$_2$O$_5$ suggests that anion Frenkel defects will be preferred. Oxidation will add anions and reduction will remove anions. The four equilibria can now be written.

Point defect formation; Frenkel defects:

$$\text{nul} \rightleftharpoons O_i^{2'} + V_O^{2\bullet} \qquad K_F = [O_i^{2'}][V_O^{2\bullet}]$$

Electronic defect formation:

$$\text{nul} \rightleftharpoons e' + h^\bullet \qquad K_e = [e'][h^\bullet]$$

Oxidation:

$$\tfrac{1}{2}O_2 \rightleftharpoons O_i^{2'} + 2h^\bullet \qquad K_o = \frac{[O_i^{2'}][h^\bullet]^2}{p_{O_2}^{1/2}}$$

Reduction:

$$O_O \rightleftharpoons V_O^{2\bullet} + 2e' + \tfrac{1}{2}O_2 \qquad K_r = [V_O^{2\bullet}][e']^2 p_{O_2}^{1/2}$$

Electroneutrality:

$$2[O_i^{2'}] + [e'] = [h^\bullet] + 2[V_O^{2\bullet}]$$

These equations can now be simplified in the manner described above. The electroneutrality point is given by

$$[e'] = [h^\bullet]$$

and the point at which the compound is strictly stoichiometric by

$$[O_i^{2\prime}] = [V_O^{2\bullet}]$$

The approximations to use depend upon the pressure regime and the values of the equilibrium constants. This oxide is an insulator under normal conditions, and so, in the middle region of the diagram, Frenkel equilibrium is dominant, that is, $K_F > K_e$ and the electroneutrality equation is approximated by

$$[O_i^{2\prime}] = [V_O^{2\bullet}]$$

In the oxidized region, holes and anion interstitials are preferred, so the electroneutrality equation is approximated by

$$2[O_i^{2\prime}] = [h^\bullet]$$

In the reduced region, electrons and anion vacancies are to be preferred, and the electroneutrality equation is approximated by

$$[e'] = 2[V_O^{2\bullet}]$$

The diagram can now be constructed.

For a fuller discussion of this material, see the Further Reading section at the end of this chapter.

7.8 BROUWER DIAGRAMS: EFFECTS OF TEMPERATURE

A Brouwer diagram is drawn up for one temperature and is an isothermal representation of the situation in the phase. A number of parameters determine the way in which any diagram changes with temperature.

Frenkel and Schottky defect equilibria are temperature sensitive and at higher temperatures defect concentrations rise, so that values of K_S and K_F increase with temperature. The same is true of the intrinsic electrons and holes present, and K_e also increases with temperature. This implies that the defect concentrations in the central part of a Brouwer diagram will move upward at higher temperatures with respect to that at lower temperatures, and the whole diagram will be shifted vertically.

The temperature dependence of the other equilibrium constants is more difficult to generalize. The partial pressure of a gaseous component will increase with temperature, and so equilibrium constants with partial pressure in the numerator will be likely to increase with temperature. Similarly, equilibrium constants with the partial pressure in the denominator may well decrease with temperature.

If a species tends to move from an oxidized to a reduced form at higher temperatures, say A^{2+} moving toward A^+, the overall diagram will tend to shift toward the low pressure (left hand) side of the diagram. Thus the diagram will tend to shift upward and to the left at high temperatures for reducible ions. The stoichiometric point will tend to move to the left at the same time. The reverse is true if the main species tends to be oxidized at higher temperatures, such as A^{3+} changing to A^{4+}. Here the diagram will tend to move upward and to the right at higher temperatures, and the stoichiometric point will tend to move to the right.

Finally, it is apparent that the principal defect type may change as the temperature increases, so that, for example, electronic defects may become more important than ionic defects. In such cases the diagram will change appreciably.

These rule-of-thumb generalizations must be treated cautiously. For precise work it is necessary to recalculate values of the equilibrium constants at the new temperature and then replot the diagram. The equilibrium constants have a general form (Chapter 2):

$$K = \exp\left(\frac{-\Delta H}{RT}\right)$$

where ΔH is the enthalpy of the defect formation reaction (often substituted for ΔG), R is the gas constant, and T the temperature (K). Thus, knowing a value of the enthalpy change allows the value of K to be calculated at other temperatures.

7.9   POLYNOMIAL FORMS FOR BROUWER DIAGRAMS

The approximations inherent in Brouwer diagrams can be bypassed by writing the appropriate electroneutrality equation as a polynomial equation and then solving this numerically using a computer. (This is not always a computationally trivial task.) To illustrate this method, the examples given in Sections 7.5 and 7.6, the MX system, will be rewritten in this form.

7.9.1   Ionic Defects

The five important equilibrium equations are:
    Point defect formation:

$$K_S = [V_M^{2\prime}][V_X^{2\bullet}] = 10^{40} \text{ (defects m}^{-3})^2 \tag{7.30}$$

Electronic defect formation:

$$K_e = [e^\prime][h^\bullet] = 10^{32} \text{ (defects m}^{-3})^2 \tag{7.31}$$

Oxidation:

$$K_o = [V_M^{2\prime}][h^\bullet]^2 p_{X_2}^{-1/2} = 10^{52} p_{X_2}^{-1/2} \text{ (defects m}^{-3})^3 \tag{7.32}$$

Reduction:

$$K_r = [V_X^{2\bullet}][e']^2 p_{X_2}{}^{1/2} = 10^{52} p_{X_2}{}^{1/2} \text{ (defects m}^{-3})^3 \tag{7.33}$$

Electroneutrality:

$$2[V_M^{2\prime}] + [e'] = 2[V_X^{2\bullet}] + [h^\bullet] \tag{7.34}$$

Equilibrium partial pressure $= 1$ atm of X_2; $\log p_{X_2} = 0$.
Note that the values for K_o and K_r are obtained by using:

$$[V_M^{2\prime}] = [V_X^{2\bullet}] = 10^{20} \text{ (defects m}^{-3})$$

$$[e'] = [h^\bullet] = 10^{16} \text{ (defects m}^{-3})$$

derived from Eqs. (7.30) and (7.31).

It is necessary to write the electroneutrality equation in terms of just two variables, a defect type and the partial pressure, to obtain a polynomial capable of solution. For example, the equation for the concentration of holes, $[h^\bullet]$, is obtained thus:

Electroneutrality:

$$2[V_M^{2\prime}] + [e'] = 2[V_X^{2\bullet}] + [h^\bullet]$$

From Eq. (7.31):

$$[e'] = \frac{K_e}{[h^\bullet]}$$

From Eq. (7.32):

$$[V_M^{2\prime}] = \frac{K_o \, p_{X_2}{}^{1/2}}{[h^\bullet]^2}$$

From Eq. (7.33):

$$[V_X^{2\bullet}] = \frac{K_r}{[e']^2 p_{X_2}{}^{1/2}} = \frac{K_r[h^\bullet]^2}{K_e{}^2 p_{X_2}{}^{1/2}}$$

Substituting:

$$\frac{2K_o \, p_{X_2}{}^{1/2}}{[h^\bullet]^2} + \frac{K_e}{[h^\bullet]} = \frac{2K_r[h^\bullet]^2}{K_e{}^2 p_{X_2}{}^{1/2}} + [h^\bullet]$$

Simplifying:

$$2K_o \, K_e^2 p_{X_2} + K_e^3[h^\bullet] p_{X_2}{}^{1/2} - 2K_r \, [h^\bullet]^4 - K_e{}^2 \, [h^\bullet]^3 p_{X_2}{}^{1/2} = 0 \tag{7.35}$$

Substituting the numerical values for the constants:

$$2 \times 10^{116}\, p_{X_2} + 10^{96}[\text{h}^{\bullet}]p_{X_2}{}^{1/2} - 2 \times 10^{52}[\text{h}^{\bullet}]^4 - 10^{64}[\text{h}^{\bullet}]^3 p_{X_2}{}^{1/2} = 0 \quad (7.36)$$

This equation can be solved to obtain the relationship between p and $[\text{h}^{\bullet}]$ and the results plotted as log $[\text{h}^{\bullet}]$ versus log p_{X_2} (Fig. 7.11a). Polynomial equations can also be written for the other defects as a function of the partial pressure and solved in the same way. Note, however, that these other equations are not strictly necessary

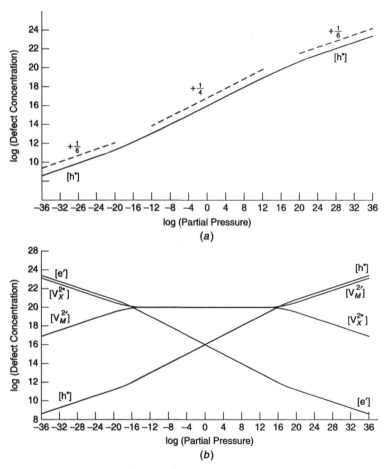

Figure 7.11 Computed polynomial plot for defect concentrations in a compound MX for the situation where Schottky defects are dominant, $K_S = 10^{40}$ defects m^{-3}, $K_e = 10^{32}$ defects m^{-3} [Eq. (7.36)]: (a) hole concentration in MX as a function of partial pressure and (b) the complete diagram.

because of the relations between the constants, as set out above:

$$[e'] = \frac{K_e}{[h^\bullet]}$$

$$[V_M^{2\prime}] = \frac{K_o \, p_{X_2}^{1/2}}{[h^\bullet]^2}$$

$$[V_X^{2\bullet}] = \frac{K_S}{[V_M^{2\prime}]}$$

The results, plotted as log (defect concentration) versus log p_{X_2}, give the complete diagram (Fig. 7.11b). It is seen that the Brouwer approximation bears a good correspondence to this diagram and indicates that the simplification procedure may be adequate in most cases.

7.9.2 Electronic Defects

Exactly the same equations apply for cases in which the electronic defects dominate the equilibrium, so that Eq. (7.35) still represents the relationship between the hole concentration and partial pressure. The values of the equilibrium constants, however, are now changed. The revised values are:

$$K_S = 10^{32} \text{ (defects m}^{-3})^2$$

$$K_e = 10^{40} \text{ (defects m}^{-3})^2$$

$$K_o = 10^{56} p_{X_2}^{-1/2} \text{ (defects m}^{-3})^3$$

$$K_r = 10^{56} p_{X_2}^{-1/2} \text{ (defects m}^{-3})^3$$

Substituting these values into Eq. (7.35) gives

$$2 \times 10^{136} p_{X_2} + 10^{120} [h^\bullet] p_{X_2}^{1/2} - 2 \times 10^{56} [h^\bullet]^4 - 10^{80} [h^\bullet]^3 p_{X_2}^{1/2} = 0 \qquad (7.37)$$

This equation can be solved to obtain the relationship between p and $[h^\bullet]$. Similarly, other equations can be set up, but, as mentioned, these other polynomials are not strictly necessary because of the relations between the constants, as set out above:

$$[e'] = \frac{K_e}{[h^\bullet]}$$

$$[V_M^{2\prime}] = \frac{K_o \, p_{X_2}^{1/2}}{[h^\bullet]^2}$$

$$[V_X^{2\bullet}] = \frac{K_S}{[V_M^{2\prime}]}$$

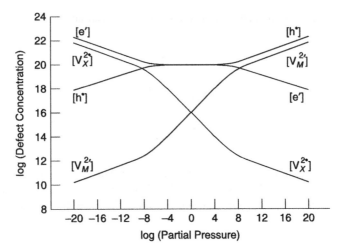

Figure 7.12   Computed polynomial plots for the defect concentrations in MX as a function of partial pressure for the situation where electronic defects are dominant, $K_S = 10^{32}$ defects m^{-3}, $K_e = 10^{40}$ defects m^{-3} [Eqs. (7.37) and (7.38)].

The solution of these equations allows the full diagram to be constructed (Fig. 7.12.) This bares a strong resemblance to the diagram constructed using approximations, and for many applications the simpler form may suffice. This topic is expanded in Section 8.4.

7.10   ANSWERS TO INTRODUCTORY QUESTIONS

How does composition variation change the electronic properties of a solid?

Changes in electronic properties of a solid are a result of the changing point defect population brought about by composition change. Chemical intuition can provide good guidance as to the electronic changes that nonstoichiometry brings about in oxides. Oxides in which the valence can change tend to become electronic conductors when the composition varies. The resulting solid has a cation in two valence states. The cation in the new valence state, induced by the change in stoichiometry, can be regarded as an unchanged cation plus an electron or a hole. Thus Ni^{3+} in NiO can be regarded as Ni^{2+} plus a hole, and Fe^{2+} in Fe_2O_3 can be regarded as Fe^{3+} plus an electron. Electronic conductivity can then take place by electron or a hole hopping from one ion to another, as appropriate.

The change in composition in oxides can be conveniently divided into two types, composition change which results in metal-excess solids, and composition change which results in metal-deficit phases. Nonstoichiometric metal-excess phases are generally regarded as containing extra electrons and the solids are *n*-type semiconductors. Nonstoichiometric oxygen-excess phases are generally regarded as containing extra holes and the solids are *p*-type semiconductors.

Why does the electronic conductivity of a nonstoichiometric solid change as the surrounding partial pressure changes?

A solid in a perfect vacuum will loose material to the vapor phase until equilibrium is reached. In the case of a solid with no composition range, this results in the production of a new phase. In a nonstoichiometric solid it results in a change of composition. When a nonstoichiometric solid is surrounded by an atmosphere with a greater pressure of these components than the equilibrium pressure, then the solid will tend to gain material, and when surrounded by an atmosphere with a lower partial pressure, then it will tend to loose material. In either case the change in composition is accomplished by a change in the defect structure of the material. The change in defect structure is responsible for the change in electronic properties. Because the composition change is a function of the surrounding partial pressure, the electronic properties are a function of the surrounding partial pressure.

Models describing the point defect population can be used to determine how the electronic conductivity will vary with changes in the surrounding partial pressure.

What information does a Brouwer diagram display?

Defect populations and physical properties such as electronic conductivity can be altered and controlled by manipulation of the surrounding atmosphere. To specify the exact electronic conductivity of such a material, it is necessary to specify its chemical composition, the defect types and populations present, the temperature of the crystal, and the surrounding partial pressures of all the constituents. Brouwer diagrams display the defect concentrations present in a solid as a function of the partial pressure of one of the components. Because the defect populations control such properties as electronic and ionic conductivity, it is generally easy to determine how these vary as the partial pressure varies.

To construct such a diagram, a set of defect reaction equations is written down, and expressions for the number of defects present as a function of the partial pressure are derived. The diagram can then be drawn, either by using the simplification that only one defect reaction is considered to be dominant in any particular composition region, or else the polynomial describing the reaction can be solved across the whole partial pressure range using a computer.

PROBLEMS AND EXERCISES

Quick Quiz

1. Which of the following oxides, (a), (b), or (c), is likely to employ electronic compensation to balance oxygen loss:
   (a) BaO
   (b) WO_3
   (c) $SrAl_2O_4$

2. A hole localized on an Mn^{3+} cation can be written as:

  (a) Mn^{2+}

  (b) Mn^{4+}

  (c) Mn^{6+}

3. A polaron is a description of:

  (a) A mobile electron in an energy band

  (b) A mobile electron at a defect

  (c) A mobile electron plus a lattice distortion

4. The reduction of an oxide M_2O_{3+x} to M_2O_{3-x} will show a change in conductivity from:

  (a) p-type to n-type

  (b) n-type to p-type

  (c) Insulator to p-type

5. The Seebeck coefficient of a slightly nonstoichiometric $\sim La^{3+}B^{3+}O_3$ perovskite structure oxide is negative. The defects present are:

  (a) B^{2+}

  (b) B^{3+}

  (c) B^{4+}

6. An increase of oxygen partial pressure will cause the conductivity of a metal-excess nonstoichiometric oxide to:

  (a) Increase

  (b) Decrease

  (c) Remain unchanged

7. The slope of the log (conductivity) versus log(oxygen pressure) over a nonstoichiometric oxide has a slope of $+\frac{1}{4}$. The material is:

  (a) A metallic conductor

  (b) An n-type semiconductor

  (c) A p-type semiconductor

8. A Brouwer diagram displays:

  (a) Defect concentrations versus partial pressure

  (b) Defect concentrations versus conductivity

  (c) Conductivity versus partial pressure

9. The electroneutrality equation defines the condition at which:

  (a) The number of holes equals the number of electrons

  (b) The charges on the ionic defects are equal

  (c) The total of the positive and negative charges are equal

10. An increase in temperature will cause the general form of a Brouwer diagram to:

   (a) Move upward

   (b) Move downward

   (c) Move laterally

Calculations and Questions

1. The composition of a slightly nonstoichiometric cobalt oxide is $Co_{0.999}O$. (a) Write a defect formation equation for this phase assuming that electronic compensation occurs. (b) If conductivity takes place by hopping, what is the value of the factor φ in Eq. (7.1)?

2. A metal–metal oxide mixture A with a low equilibrium partial pressure of oxygen is sealed in a vacuum and heated to a temperature T. A metal–metal oxide mixture B with a high equilibrium partial pressure of oxygen is treated similarly. If the two containers are linked so that there is a free interchange of gas, what will be the final solids present?

3. The conductivity of high-purity single crystals of CoO at 1300°C was found to depend upon the oxygen partial pressure as in the following table. What conclusions can be drawn concerning the nonstoichiometry of CoO under these circumstances?

Conductivity/S m^{-1}	Oxygen Partial Pressure/atm
0.490	0.0001
0.759	0.001
1.202	0.01
1.906	0.1
2.884	1

4. The conductivity of nickel oxide at 1000°C was found to depend upon the oxygen partial pressure as in the following table. What conclusions can be drawn concerning the nonstoichiometry of NiO under these circumstances?

Conductivity/S m^{-1}	Oxygen Partial Pressure/atm
23.44	1.0
20.42	0.355
15.49	0.079
7.94	0.0126
6.61	8.91×10^{-4}
5.75	2.24×10^{-4}
5.01	8.91×10^{-5}

Data adapted from I. Rom, W. Jantscher, and W. Sitte, *Solid State Ionics*, **135**, 731–736 (2000).

5. The conductivity of titanium dioxide, TiO_2, at 1166 K was found to depend upon the oxygen partial pressure as in the following table. Assuming that oxygen vacancies are introduced into the oxide as the main point defect, is the vacancy more likely to be doubly or singly charged?

Conductivity/S m^{-1}	Oxygen Partial Pressure/atm
1.047	1.0×10^{-2}
5.012	1.58×10^{-6}
7.94	1.0×10^{-7}
15.85	1.58×10^{-9}
39.81	5.01×10^{-12}
83.18	7.94×10^{-14}

Data adapted from J. Novotny, M. Radecka, and M. Rekas, *J. Phys. Chem. Solids*, **58**, 927–937 (1997).

6. Using the information in Section 7.7.1, sketch the Brouwer diagram for Cr_2O_3. The oxide is an insulator under normal conditions, so assume that Schottky defects dominate.

7. Using the information in Section 7.7.2 derive the dependence of conductivity on oxygen partial pressure for $Ba_2In_2O_5$. Assume that (a) at higher oxygen partial pressures, oxygen is incorporated into the structure and fills vacancies created by the Frenkel equilibrium and (b) at lower oxygen partial pressures, oxygen interstitials created by the Frenkel equilibrium are lost to the surrounding atmosphere.

8. Derive the polynomial expressions for the concentration dependence of electrons, metal vacancies, and anion vacancies for a compound MX showing Schottky equilibrium (Section 7.9.4). Insert values of the equilibrium constants to obtain polynomial expressions for the cases where electronic defects dominate.

9. The Seebeck coefficient of the layered structure $Bi_2Sr_2CoO_{6+\delta}$ is -33 μV K^{-1} at 250 K. The defects are confined to the CoO_2 planes in the structure. (a) Is the conductivity by way of holes or electrons? (b) What are the ionic states of the Co ions in the phase? (c) What is the ionic formula of the compound. (d) What is the value of δ? [Note the structure is very similar to that of $Bi_2Sr_2CuO_6$, Section 8.6. Data adapted from Y. Nagao and I. Terasaki, *Phys. Rev.*, **B76**, 144203-1–144203-4 (2007).]

10. The equilibrium oxygen pressure over the series of iron oxides at 1100°C is shown in the figure. (a) What phases are involved? (b) What does the sloping part of the plot indicate? (c) Write equations for each of the formation reactions taking place. (d) Estimate the value of the reaction Gibbs energy, ΔG_r per mole of product, at 1100°C for each reaction. The value of ΔG_r for the formation of

$FeO_{1.084}$ is $162.5 \, kJ \, mol^{-1}$ of $FeO_{1.1084}$. **(e)** What is the equilibrium partial pressure over a mixture of $FeO_{1.084}$ and Fe at 1100°C? [Data adapted from K. Kitayama, M. Sakaguchi, Y. Takahara, H. Endo, and H. Ueki, *J. Solid State Chem.*, **177**, 1933–1938 (2004).]

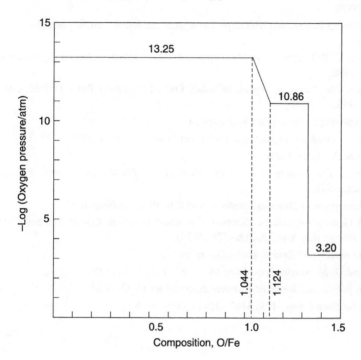

REFERENCES

G. Brouwer, A General Asymmetric Solution of Reaction Equations Common in Solid State Chemistry, *Philips Res. Rep.*, **9**, 366–376 (1954).

S. Hébert, D. Flahaut, C. Martin, A. Lemonnier, J. Noudem, C. Goupil, A. Maignan, and J. Hejtmanek, *Prog. Solid State Chem.*, **35**, 457–467 (2007).

FURTHER READING

The thermodynamics of solid equilibrium with gas atmospheres is considered in many textbooks of thermodynamics and, often, with useful examples, in textbooks of metallurgy and geology. The thermodynamics of many metal–oxygen systems are especially well characterized in view of their industrial importance. Two old but extremely useful descriptions of gas–solid equilibria are:

A. Muan, The Effect of Oxygen Pressure on Phase Relations in Oxide Systems, *Am. J. Sci.*, **256**, 171–207 (1958).

T. B. Reed, *Free Energy of Formation of Binary Compounds: An Atlas of Charts for High-Temperature Chemical Calculations*, M.I.T. Press, Cambridge, MA, 1971.

Much of the material covered here is discussed in:

D. M. Smyth, *The Defect Chemistry of Metal Oxides*, Oxford University Press, Oxford, United Kingdom, 2000.

Expansion and further explanation with respect to defects and defect chemistry and physics will be found in:

F. Agullo-Lopez, C. R. Catlow, and P. D. Townsend, *Point Defects in Materials*, Academic, New York, 1988.

A. M. Stoneham, *The Theory of Defects in Solids*, Oxford University Press, Oxford, United Kingdom, 1985.

The Mott variable-range hopping law is derived in:

N. F. Mott, *Metal–Insulator Transitions*, Taylor and Francis, London, 1974, p. 35 ff.

The phase rule is clearly explained in:

E. G. Ehlers, *The Interpretation of Geological Phase Diagrams*, Freeman, San Francisco, 1972.

The original description of Brouwer diagrams is well worth consulting. It is:

G. Brouwer, A General Asymmetric Solution of Reaction Equations Common in Solid State Chemistry, *Philips Res. Rept.*, **9**, 366–376 (1954).

Details of point defect equilibria in $Ba_2In_2O_5$ are in:

G. B. Zhang and D. M. Smyth, *Solid State Ionics*, **82**, 161–172 (1995).

Information on polynomial forms of Brouwer diagrams for Cr_2O_3 is in:

P. Karen, *J. Solid State Chem.*, **179**, 3167–3183 (2006).

Nonstoichiometry and Extrinsic Electronic Conductivity

What are donor and acceptor dopants?

How do negative temperature coefficient (NTC) thermistors respond to temperature changes?

What are mixed conductors?

8.1 EFFECT OF IMPURITY ATOMS

The composition variation described in the previous chapter has a considerable impact upon the electronic properties of the solid. However, it is often difficult to alter the composition of a phase to order, and stoichiometry ranges are frequently too narrow to allow desired electronic properties to be achieved. Traditionally, the problem has been circumvented by using selective doping by aliovalent impurities, that is, impurities with a different nominal valence to those present in the parent material. However, it is important to remember that all the effects described in the previous chapter still apply to the materials below. The division into two chapters is a matter of convenience only.

Depending upon the electronic consequences of the doping, aliovalent additives are often called donor impurities or donor dopants when they tend to provide electrons and enhance intrinsic *n*-type semiconducting behavior, or acceptor impurities or acceptor dopants when they tend to give a population of mobile holes and enhance *p*-type semiconducting behavior. Alternatively, the process is simply referred to as electron doping or hole doping. The process of creating electronic defects in a crystal in this way is called valence induction. Extreme changes in properties can be brought about by use of this technique. Examples given below include cases where insulating oxides can be made metallic, representing examples of metal–insulator transitions. In the case of some oxides, especially well represented by the cuprates, insulating parent phases can be made superconducting. Moreover, these transformations can often be controlled with great precision, as the introduction of small amounts of impurity poses no inherent problems.

Defects in Solids, by Richard J. D. Tilley
Copyright © 2008 John Wiley & Sons, Inc.

As in the previous chapter, most work has been carried out on oxides, and these figure prominently here. As the literature on oxides alone is not only vast but is also rapidly increasing, this chapter focuses upon a number of representative structure types to explain the broad principles upon which the defect chemistry depends. However, despite considerable research, the defect chemistry and physics of doped crystals is still open to considerable uncertainty, and even well-investigated "simple" oxides such as lithium-doped nickel oxide, $Li_xNi_{1-x}O$, appear to have more complex defect structures than thought some years ago.

8.2 IMPURITIES IN OXIDES

8.2.1 Donor Doping

Donor dopants are impurity ions of a *higher* valence than that of the parent ions, as when small amounts of Nb_2O_5 are incorporated into TiO_2, so that Nb^{5+} substitutes for Ti^{4+}. The donor species has an effective *positive* charge, $Nb^{\bullet}_{Ti}$, in this example, and the introduction of donor species tends to introduce counterbalancing *electrons* into the structure. The inclusion of higher valence impurities will have an effect upon the stoichiometry of the phase and the charge balance of the host, which need to be compensated by a set of balancing defects. There are three ways in which this can be achieved for donor doping:

1. Cation vacancies
2. Anion interstitials
3. Electrons

8.2.2 Donor Doping of Cr_2O_3

The doping of chromium(III) oxide, Cr_2O_3, doped with titanium dioxide, TiO_2, gives a material that is of use in sensors that detect reducing gases. The general principle underlying the operation of this and similar sensors is that atmospheric oxygen is chemisorbed onto the surface. This process removes electrons from the bulk, enhancing bulk *p*-type semiconductivity or reducing bulk *n*-type semiconductivity. A reducing gas, such as carbon monoxide, CO, will remove chemisorbed oxygen, liberating electrons back to the bulk, so that *n*-type conductivity increases and *p*-type semiconductivity decreases. For successful operation, a low concentration of bulk carriers is needed, so that the weak absorption effects make a measurable difference in electronic properties.

Pure Cr_2O_3 is an intrinsic semiconductor with a band-gap of approximately 3.3 eV. Generally, Cr_2O_3 shows little stoichiometric variation. On doping with TiO_2, Ti^{4+} ions substitute on Cr^{3+} sites in the structure. The conductivity of the doped solid is *n*-type and has a $-\frac{1}{8}$ dependence upon oxygen partial pressure. The three alternative compensation mechanisms can be written as follows:

Cation vacancies: $3TiO_2(Cr_2O_3) \rightleftharpoons 3Ti_{Cr}^{\bullet} + V_{Cr}^{3\prime} + 6O_O$ (8.1a)

Anion interstitials: $2TiO_2(Cr_2O_3) \rightleftharpoons 2Ti_{Cr}^{\bullet} + 3O_O + O_i^{2\prime}$ (8.1b)

Electrons: $2TiO_2(Cr_2O_3) \rightleftharpoons 2Ti_{Cr}^{\bullet} + 2e' + 3O_O + \frac{1}{2}O_2$ (8.1c)

Equation (8.1c) is derived from Eq. (8.1b) by supposing that the oxygen interstitials are lost to the atmosphere:

$$O_i^{2\prime} \rightleftharpoons \frac{1}{2}O_2 + 2e'$$

The partial pressure dependence is readily explained using Eq. (8.1c). The equilibrium constant, K, is

$$K = [Ti_{Cr}^{\bullet}]^2 \, [e']^2 \, [O_O]^3 \, p_{O_2}^{1/2}$$

As the incorporation of one substitutional cation results in the production of one electron:

$$[Ti_{Cr}^{\bullet}] = [e']$$

and including the term $[O_O]$ into a new constant, K_1,

$$K_1 = [e']^4 \, p_{O_2}^{1/2}$$

$$[e']^4 = K_1 \, p_{O_2}^{-1/2}$$

$$\sigma \propto [e'] \propto p_{O_2}^{-1/8}$$

However, density measurements suggest that the formation of chromium vacancies occurs, and atomistic simulations indicate that the vacancies are likely to cluster near to titanium substituents to form neutral entities:

$$3TiO_2(Cr_2O_3) \rightleftharpoons (3Ti_{Cr}^{\bullet} + V_{Cr}^{3\prime})^x + 6O_O$$

This indicates that the defect structure is complex and may vary with degree of doping. Further studies are needed to clarify the defect structure of this notionally simple solid.

8.2.3 Acceptor Doping

Acceptor dopants are impurity ions of a *lower* valence than that of the parent ions, as when small amounts of Al_2O_3 are incorporated into TiO_2, so that the Al^{3+} ion substitutes for Ti^{4+}. The acceptor species has an effective *negative* charge, Al_{Ti}' in this example, and the introduction of acceptor species tends to introduce counterbalancing *holes* into the structure. The inclusion of lower valence impurities will have an effect upon the stoichiometry of the phase and the charge balance of the host, which need to be compensated by a set of balancing defects. There are three ways in which this can be achieved for acceptor doping:

1. Cation interstitials
2. Anion vacancies
3. Holes

8.2.4   Acceptor Doping of NiO

Lithium-doped nickel oxide provides a good example of acceptor doping. This is an extremely well-studied system but (as in the case of TiO_2 doped Cr_2O_3), there is still uncertainty about the real defect structure of the material. Nonstoichiometry in nickel oxide has been mentioned previously (Sections 1.11.4 and 7.2.2). Although the electronic properties of the pure oxide are greatly influenced by degree of composition variation, this is too small to be exploited for device application. The limitation in stoichiometric variation can be overcome by aliovalent acceptor doping, most often by the incorporation of lithium oxide, Li_2O. The process is accomplished by heating colorless Li_2O and green NiO together at high temperatures. The resulting compound is black in color with a formula $Li_xNi_{1-x}O$. At the same time, the electronic conductivity changes from approximately 10^{-14} to approximately $10^{-1}\ \Omega^{-1}\ cm^{-1}$, depending upon dopant concentration and oxygen partial pressure. The material is a *p*-type semiconductor, and shows high (quasi-metallic) conductivity, which is still the subject of investigation. The electronic conductivity is also very temperature dependent, and lithium-doped nickel oxide and related materials are widely used in thermistor temperature sensors (Section 8.3).

Assuming that the material is ionic and the ions have their normal charges, Li^+, Ni^{2+}, O^{2-}, the three compensation mechanisms possible for donor doping are:

$$\text{Cation interstitials:} \qquad Li_2O(NiO) \rightleftharpoons Li'_{Ni} + Li_i^\bullet + O_O \qquad (8.2a)$$

$$\text{Anion vacancies:} \qquad Li_2O(NiO) \rightleftharpoons 2Li'_{Ni} + V_O^{2\bullet} + O_O \qquad (8.2b)$$

$$\text{Holes:} \qquad Li_2O(NiO) + \tfrac{1}{2}O_2 \rightleftharpoons 2Li'_{Ni} + 2O_O + 2h^\bullet \qquad (8.2c)$$

Equation (8.2c) is derived from Eq. (8.2b) by assuming that the anion vacancy is filled by reaction with atmospheric oxygen:

$$\tfrac{1}{2}O_2(Li_2O/NiO) + V_O^{2\bullet} \rightleftharpoons O_O + 2h^\bullet$$

The variation expected in the electronic conductivity is readily derived from Eq. (8.2c):

$$K = \frac{[Li'_{Ni}]^2\,[O_O]^2\,[h^\bullet]^2}{p_{O_2}^{1/2}}$$

The incorporation of one Li^+ ion results in the generation of one hole, so:

$$[Li'_{Ni}] = [h^\bullet]$$

and incorporating the concentration of normal oxygen ions, $[O_O]$, into a new equilibrium constant, K_1, it is possible to write:

$$K_1 = \frac{[h^\bullet]^4}{p_{O_2}^{1/2}}$$

$$[h^\bullet]^4 = K_1 p_{O_2}^{1/2}$$

$$\sigma \propto [h^\bullet] \propto p_{O_2}^{1/8}$$

Despite the many investigations of the defect chemistry of lithium-oxide-doped nickel oxide, the real nature of the defect structure still remains uncertain. For many years the holes were regarded as being localized on Ni^{2+} ions to form Ni^{3+}, written $Ni^\bullet_{Ni}$:

$$Li_2O + 2NiO + \tfrac{1}{2}O_2 \longrightarrow 2Li'_{Ni} + 4O_O + 2Ni^\bullet_{Ni}$$

Thus, every Li^+ on a Ni^{2+} site in the lattice results in the formation of a Ni^{3+} ion elsewhere. However, an increasing number of studies suggest that the hole may be preferentially trapped on an oxygen ion, thus:

$$Li_2O + 2NiO + \tfrac{1}{2}O_2 \longrightarrow 2Li'_{Ni} + 2O_O + 2O^\bullet_O + 2Ni_{Ni}$$

where $O^\bullet_O$ represents an oxygen ion carrying a single negative charge, O^-. Similarly, there is still some debate about whether the conductivity takes place by a genuine hopping mechanism or through an energy band.

8.3 NEGATIVE TEMPERATURE COEFFICIENT (NTC) THERMISTORS

The resistivity of doped transition-metal oxides such as NiO doped with Li_2O, Fe_2O_3 doped with TiO_2, and Mn_3O_4 doped with NiO generally decreases exponentially with temperature (Fig. 8.1a). This makes these materials suitable for temperature measurement and sensing. [Note that the change in resistance is continuous and does not occur over a narrow temperature range as in the case of PCT thermistors (Section 3.14.4)]. The transition-metal oxide pellet that makes up a thermistor bead is small, strong, and rugged (Fig. 8.1b), with a fast response time. These characteristics make thermistors ideal components of electronic equipment, from spacecraft to domestic ovens.

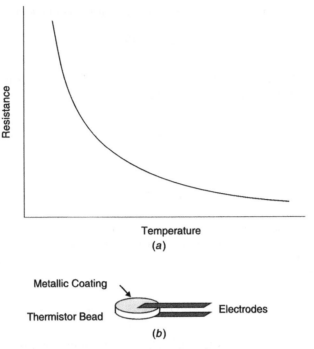

Figure 8.1 (*a*) Resistivity of a doped transition-metal oxide (schematic) and (*b*) an oxide thermistor bead (schematic).

The mechanism of generation of the mobile charge carriers follows that described in the previous two sections. Donor doping is expected to result in *n*-type thermistors. The situation in which Fe_2O_3 is doped with TiO_2, analogous to the situation outlined above for TiO_2-doped Cr_2O_3, provides an example. The favored mechanism is the formation of electrons, comparable to Eq. (8.1):

$$2TiO_2(Fe_2O_3) \longrightarrow 2Ti^{\bullet}_{Fe} + 2e' + 3O_O + \tfrac{1}{2}O_2$$

or

$$2TiO_2(Fe_2O_3) \longrightarrow 2Ti^{\bullet}_{Fe} + 2Fe'_{Fe} + 3O_O + \tfrac{1}{2}O_2$$

where Fe'_{Fe} is equivalent to Fe^{2+}. Each added Ti^{4+} ion induces the formation of a counterbalancing Fe^{2+} ion, so that the formula of the doped phase can be written $Fe^{3+}_{2-2x} Fe^{2+}_{x} Ti^{4+}_{x}O_3$.

Acceptor doping, as in lithium oxide doping of nickel oxide, produces *p*-type thermistors. The situation in nickel-oxide-doped Mn_3O_4 is similar but slightly more complex. This oxide has a distorted spinel structure (Supplementary Material S1), with Mn^{2+} occupying tetrahedral sites and Mn^{3+} occupying octahedral sites in the crystal, to give a formula $Mn^{2+}[Mn^{3+}]_2O_4$, where the square parentheses enclose the ions in octahedral sites. The dopant Ni^{2+} ions preferentially occupy octahedral sites. Because the tetrahedral sites and the Mn^{2+} ions do not figure in the reaction, the spinel can be written $MnO \cdot Mn_2O_3$ and the MnO part ignored. Thus, the defect chemistry can be written in terms of NiO doping into Mn_2O_3. The defect equations are then equivalent to acceptor doping into Mn_2O_3:

$$2NiO(Mn_2O_3) \longrightarrow 2Ni'_{Mn} + 2O_O + V_O^{2\bullet}$$

$$2NiO(Mn_2O_3) + \tfrac{1}{2}O_2 \longrightarrow 2Ni'_{Mn} + 3O_O + 2h^\bullet$$

The holes can be regarded a localized on Mn^{3+} ions to give Mn^{4+}, and the doped oxide has a nominal formula $Mn^{2+}[Mn^{3+}_{2-x} Mn^{4+}_x Ni^{2+}_x]O_4$. The material is a p-type semiconductor.

Impurity substitution that is effectively neutral, that is, neither donor nor acceptor, can also lead to significant changes in properties that are utilized in NTC thermistors. For example, the replacement of Ga^{3+} in the spinel $MgGa_2O_4$ by Mn^{3+} involves no apparent donor or acceptor action. The conductivity in the system $MgGa_{2-x}Mn_xO_4$ evolves from insulating (conductivity about $10^{-9}\ \Omega^{-1}\,m^{-1}$) for the parent phase with $x = 0$, to a conductivity approximately equal to that of germanium ($10^{-1}\,\Omega^{-1}$ m^{-1}) in the compound $MgGaMnO_4$, in which $x = 1$. The resistivity decreases markedly with temperature and the compounds display typical NCT behavior.

The parent phases for this series can be regarded as $MgGa_2O_4$ and $MgMn_2O_4$. Both are mixed spinels, but it is convenient to regard them as normal spinels initially. The replacement of Ga^{3+} by Mn^{3+} then only concerns the octahedral sites in the structure. Ignoring the MgO component, this is written:

$$Mn_2O_3(Ga_2O_3) \longrightarrow 2Mn_{Ga} + 3O_O$$

The conductivity comes about because of electron hopping between a pair of Mn^{3+} ions. This is equivalent to disproportionation:

$$Mn_{Ga} + Mn_{Ga} \longrightarrow Mn^\bullet_{Ga} + Mn'_{Ga}$$

or in ionic terms:

$$Mn^{3+} + Mn^{3+} \longrightarrow Mn^{4+} + Mn^{2+}$$

In mixed spinels, electron transfer between tetrahedral and octahedral Mn^{3+} ions is hampered. As the cation distribution is temperature sensitive, preparation of materials for NTC use must optimize the amount of Mn in octahedral sites.

Although these materials are often best described as hopping semiconductors in which the mobility is proportional to the exponential of the energy required to liberate the charge carriers, the conduction mechanism in devices is often complex. For application, mechanisms are sometimes neglected in favor of empirical relationships.

Thus, the thermistor resistance R is frequently written as:

$$R_T = R_{T0} \exp\left[\frac{\beta(T_0 - T)}{T_0 T}\right]$$

where R_T is the resistance at a temperature T (K), β is the thermistor constant, and R_{T0} is the resistance at a specified reference temperature T_0. The thermistor constant, β, is given by

$$\beta = \frac{E_a}{k}$$

where E_a is the activation energy associated with electron movement, usually considered to be by a hopping mechanism, and k is the Boltzmann constant. Thermistors are also frequently characterized by the temperature coefficient of resistance, α, which is the percentage change in R with temperature. The parameter β generally lies in the range 2000–6000 and devices have values of α of the order of 4% per degree.

8.4 BROUWER DIAGRAMS FOR DOPED SYSTEMS

8.4.1 Construction

Brouwer diagrams plot the defect concentrations in a solid as a function of the partial pressure of the components of the material and are a convenient way of displaying electronic properties (Sections 7.6–7.9). These can be readily extended to include the effects of doping by acceptors or donors.

It is important to know that the defect equilibria that apply to the pure material, and the associated equilibrium constants, also apply to the doped material. The only additional information required is the nature and concentration of the dopant. To illustrate the construction of a diagram, an example similar to that given in Chapter 7 will be presented, for a nonstoichiometric phase of composition MX, nominally containing M^{2+} and X^{2-} ions, with a stoichiometric composition $MX_{1.0}$. In this example, it is assumed that the relevant defect formation equations are the same as those given in Chapter 7:

1. The creation and elimination of Schottky defects:

$$\text{nul} \rightleftharpoons V_M'' + V_X^{\bullet\bullet} \qquad K_S = [V_M''][V_X^{\bullet\bullet}] \qquad (8.3)$$

2. The creation and elimination of electronic defects:

$$\text{nul} \rightleftharpoons e' + h^{\bullet} \qquad K_e = [e'][h^{\bullet}] \qquad (8.4)$$

3. Oxidation:

$$\tfrac{1}{2}X_2 \rightleftharpoons X_X + V_M^{2\prime} + 2h^{\bullet} \qquad K_o = [V_M^{2\prime}][h^{\bullet}]^2 \, p_{X_2}^{-1/2} \tag{8.5}$$

4. Reduction:

$$X_X \rightleftharpoons \tfrac{1}{2}X_2(g) + V_X^{2\bullet} + 2e' \qquad K_r = [V_X^{2\bullet}][e']^2 \, p_{X_2}^{1/2} \tag{8.6}$$

The addition of either donors or acceptors will, however, upset the charge balance, and these must be included in the electroneutrality equation. Consider donor doping by a trivalent ion D^{3+} due to reaction with D_2X_3 to introduce $D_M^{\bullet}$ defects, once again assuming that Frenkel defects are not important. The original electroneutrality Eq. (7.12):

$$2[V_M^{2\prime}] + [e'] = 2[V_X^{2\bullet}] + [h^{\bullet}]$$

now becomes:

$$2[V_M^{2\prime}] + [e'] = 2[V_X^{2\bullet}] + [h^{\bullet}] + [D_M^{\bullet}] \tag{8.7}$$

The donor defects can be compensated by cation vacancies or electrons (Section 8.2.1). That is, the addition of donors to the right-hand side of Eq. (8.7) requires an increase in the number of vacancies:

$$D_2X_3(MX) \longrightarrow 2D_M^{\bullet} + 3X_X + V_X^{2\prime} \tag{8.8}$$

or electrons:

$$D_2X_3(MX) \longrightarrow 2D_M^{\bullet} + 2X_X + \tfrac{1}{2}X_2 + 2e' \tag{8.9}$$

In general only one of these options will be preferred in a particular partial pressure regime depending upon the values of the various equilibrium constants.

In the case of acceptor doping, a similar change in the electroneutrality equation is required. Consider acceptor doping by a monovalent ion A^+ due to reaction with A_2X to introduce A_M' defects, once again assuming that Frenkel defects are not important. The original electroneutrality Eq. (7.12):

$$2[V_M^{2\prime}] + [e'] = 2[V_X^{2\bullet}] + [h^{\bullet}]$$

now becomes

$$2[V_M^{2\prime}] + [e'] + [A_M'] = 2[V_X^{2\bullet}] + [h^{\bullet}] \tag{8.10}$$

The acceptor defects can be compensated by anion vacancies or holes (Section 8.2.3). That is, the addition of acceptors to the left-hand side of Eq. (8.10) requires an increase in the number of anion vacancies:

$$A_2X(MX) \longrightarrow 2A_M' + V_X^{2\bullet} + X_X \tag{8.11}$$

or holes:

$$A_2X(MX) + \tfrac{1}{2}X_2 \longrightarrow 2A'_M + 2X_X + 2h^{\bullet} \qquad (8.12)$$

As in the case of donor doping, only one of these options will be preferred in a particular partial pressure regime, depending upon the values of the various equilibrium constants.

The four Eqs. [(8.3)–(8.6)] are simplified using chemical and physical intuition and appropriate approximations to the electroneutrality Eqs. (8.7) and (8.10). Brouwer diagrams similar to those given in the previous chapter can then be constructed. However, by far the simplest way to describe these equilibria is by way of polynomials. This is because the polynomial appropriate for the doped system is simply the polynomial equation for the undoped system, together with one extra term, to account for the donors or acceptors present. For example, following the procedure described in Section 7.9, and using the electroneutrality equation for donors, Eq. (8.9), the polynomial appropriate to donor doping is:

$$2K_oK_e^2\, p_{X_2} + K_e^3\, [h^{\bullet}]\, p_{X_2}^{1/2} - 2K_r\, [h^{\bullet}]^4 - K_e^2\, [h^{\bullet}]^3\, p_{X_2}^{1/2}$$
$$- K_e^2 D_M\, [h^{\bullet}]^2\, p_{X_2}^{1/2} = 0 \qquad (8.13)$$

where D_M represents the concentration of donor atoms (with a single effective charge) on metal sites. This can be compared to the almost identical Eq. (7.35). In the same way, the polynomial appropriate to acceptor doping, derived from the acceptor electroneutrality Eq. (8.10), is

$$2K_oK_e^2\, p_{X_2} + K_e^3\, [h^{\bullet}]\, p_{X_2}^{1/2} + K_e^2 A_M\, [h^{\bullet}]^2\, p_{X_2}^{1/2} - 2K_r\, [h^{\bullet}]^4$$
$$- K_e^2\, [h^{\bullet}]^3\, p_{X2}^{1/2} = 0 \qquad (8.14)$$

where A_M represents the concentration of acceptor atoms (with a single effective charge) on metal sites.

Insertion of the equilibrium constants will result in equations that can be solved for $[h^{\bullet}]$ as a function of p_{X_2}, which, in turn, can be used to determine the values of the other defects as a function of p_{X_2} (Section 7.10.2). The diagrams for donor and acceptor doping of MX in which electronic defects dominate over Schottky equilibrium (Fig. 8.2a and 8.2c) can be compared to that for undoped material (Fig. 7.11), redrawn here (Fig. 8.2b).

The effect of changes in the numerical values of the equilibrium constants or the degree of doping are easily discovered, as there is almost no change in the polynomials involved.

8.4.2 General Trends: MX

The construction of diagrams over a range of concentrations and equilibrium constants allows a number of general conclusions to be drawn.

8.4.2.1 $K_e > K_S$ In the undoped material a plateau is found over the middle part of the diagram where the dominant defects, $[e']$ and $[h^\bullet]$, have equal concentrations. The point at which the solid is electronically neutral, when $[e'] = [h^\bullet]$, occurs at the same partial pressure as when the solid is stoichiometric, $[V_M^{2\prime}] = [V_X^{2\bullet}]$.

In acceptor-doped material, the diagram is virtually unchanged for dopant levels below that of the intrinsic defects, that is, less than 10^{20} defects m^{-3} in the examples given above. When the concentration of acceptors passes this quantity, the $[e'] = [h^\bullet]$ plateau lengthens and divides with $[h^\bullet]$ greater than $[e']$. The separation of these

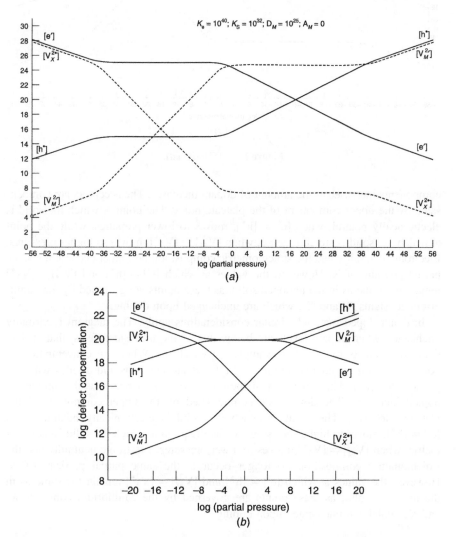

Figure 8.2   Computed polynomial plots for the defect concentrations for MX as a function of partial pressure for the situation where electronic defects are dominant, $K_S = 10^{32}$ defects m^{-3}, $K_e = 10^{40}$ defects m^{-3}: (a) doped with 10^{25} donor atoms m^{-3}, (b) undoped, and (c) doped with 10^{25} acceptor atoms m^{-3}.

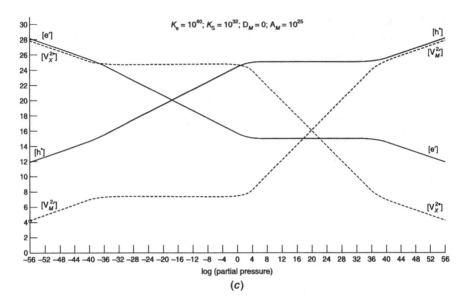

Figure 8.2 (*Continued*).

components increases as the amount of dopant increases. The acceptors are compensated by the upper component of the plateau, holes. The point at which the solid is electronically neutral, when $[e'] = [h^\bullet]$, moves to lower pressures, while the point at which the solid is stoichiometric, when $[V_M^{2\prime}] = [V_X^{2\bullet}]$, moves to higher pressures. Electronic neutrality and the stoichiometric composition no longer occur at the same partial pressure of X. However, the values at which $[e'] = [h^\bullet]$ and $[V_M^{2\prime}] = [V_X^{2\bullet}]$ remain the same as in the undoped solid, as these points are governed by the equilibrium constants K_e and K_S, which are unchanged upon doping.

In donor-doped material, similar considerations apply. The diagram is virtually unchanged when dopant levels are below that of the intrinsic defects, that is, less than 10^{20} defects m^{-3} in the examples given above. When the concentration of donors passes this quantity, the $[e'] = [h^\bullet]$ plateau lengthens and divides with $[e']$ greater than $[h^\bullet]$. The separation of these components increases as the amount of dopant increases. The donors are compensated by the upper component of the plateau, electrons. The point at which the solid is electronically neutral, when $[e'] = [h^\bullet]$, moves to higher pressures, while the point at which the solid is stoichiometric, when $[V_M^{2\prime}] = [V_X^{2\bullet}]$, moves to lower pressures. Electronic neutrality and the stoichiometric composition no longer occur at the same partial pressure of X. However, the values at which $[e'] = [h^\bullet]$ and $[V_M^{2\prime}] = [V_X^{2\bullet}]$ remain the same as in the undoped solid, as these points are governed by the equilibrium constants K_e and K_S, which are unchanged upon doping.

8.4.2.2 $K_S > K_e$ In the undoped material a plateau is found over the middle part of the diagram where the dominant defects, $[V_M^{2\prime}]$ and $[V_X^{2\bullet}]$, have equal concentrations. The point at which the solid is electronically neutral, when $[e'] = [h^\bullet]$, occurs at the same partial pressure as when the solid is stoichiometric, $[V_M^{2/}] = [V_X^{2\bullet}]$.

In acceptor-doped material, the diagram is virtually unchanged for dopant levels below that of the intrinsic defects, that is, less than 10^{20} defects m^{-3} in the examples given above. When the concentration of acceptors passes this quantity, the $[V_M^{2/}] = [V_X^{2\bullet}]$ plateau lengthens and divides with $[V_X^{2\bullet}]$ greater than $[V_M^{2/}]$. The separation of these components increases as the amount of dopant increases. The acceptors are compensated by the upper component of the plateau, anion vacancies. The point at which the solid is electronically neutral, when $[e'] = [h^\bullet]$, moves to lower pressures, while the point at which the solid is stoichiometric, when $[V_M^{2/}] = [V_X^{2\bullet}]$, moves to higher pressures. Electronic neutrality and the stoichiometric composition no longer occur at the same partial pressure of X. However, the values at which $[e'] = [h^\bullet]$ and $[V_M^{2/}] = [V_X^{2\bullet}]$ remain the same as in the undoped solid, as these points are governed by the equilibrium constants K_e and K_S, which are unchanged upon doping.

In donor-doped material, similar considerations apply. The diagram is virtually unchanged when dopant levels are below that of the intrinsic defects, that is, less than 10^{20} defects m^{-3} in the examples given above. When the concentration of donors passes this quantity, the $[V_M^{2/}] = [V_X^{2\bullet}]$ plateau lengthens and divides with $[V_M^{2/}]$ greater than $[V_X^{2\bullet}]$. The separation of these components increases as the amount of dopant increases. The donors are compensated by the upper component of the plateau, cation vacancies. The point at which the solid is electronically neutral, when $[e'] = [h^\bullet]$, moves to higher pressures, while the point at which the solid is stoichiometric, when $[V_M^{2/}] = [V_X^{2\bullet}]$, moves to lower pressures. Electronic neutrality and the stoichiometric composition no longer occur at the same partial pressure of X. However, the values at which $[e'] = [h^\bullet]$ and $[V_M^{2/}] = [V_X^{2\bullet}]$ remain the same as in the undoped solid, as these points are governed by the equilibrium constants K_e and K_S, which are unchanged upon doping.

8.5 METALS AND INSULATORS

Doping can have profound effects upon the electronic properties of a crystal. In this section three examples of changes from insulating to metallic properties are described.

8.5.1 Acceptor Doping into La$_2$CuO$_4$

The compound La$_2$CuO$_4$ adopts a slightly distorted form of the K$_2$NiF$_4$ structure (Figs. 4.5 and 4.28a), which is often considered to be made up of slabs of perovskite structure one octahedron thick. (Note that in the superconductor literature this structure is often called the T or T/O structure.) The stoichiometric oxide is an insulator. However, acceptor doping can lead to remarkable changes in electronic conductivity resulting in metallic and superconducting states. These come about as copper is able to change its valence state from the normally stable Cu^{2+}.

Acceptor doping of La$_2$CuO$_4$ is simply carried out by replacing the La^{3+} ion with a similar sized divalent ion, typically an alkaline earth ion such as Ca^{2+}, Sr^{2+}, or

Ba^{2+}. In this process, small changes in the oxygen content also occur, which will be ignored in the following discussion.

Following substitution, charge neutrality can be maintained in one of the three ways, as described above, that is, by the introduction of cation interstitials, anion vacancies, or holes. Replacement of some of the La^{3+} by Sr^{2+} can be written in terms of La_2O_3 alone (Section 1.11.7):

Cation interstitials: $3SrO(La_2O_3) \longrightarrow 2Sr'_{La} + 3O_O + Sr_i^{2\bullet}$ (a)

Anion vacancies: $2SrO(La_2O_3) \longrightarrow 2Sr'_{La} + 2O_O + V_O^{2\bullet}$ (b)

Holes: $2SrO(La_2O_3) + \frac{1}{2}O_2 \longrightarrow 2Sr'_{La} + 3O_O + 2h^\bullet$ (c)

This last equation is equivalent to filling of the anion vacancies by atmospheric oxygen:

$$V_O^{2\bullet} + \tfrac{1}{2}O_2 \rightleftharpoons O_O + 2h^\bullet$$

Including the component, CuO gives:

Cation interstitials: $3SrO(La_2CuO_4) \longrightarrow 2Sr'_{La} + 4O_O + Sr_i^{2\bullet} + Cu_{Cu}$ (a)

Anion vacancies: $2SrO(La_2CuO_4) \longrightarrow 2Sr'_{La} + 3O_O + V_O^{2\bullet} + Cu_{Cu}$ (b)

Holes: $2SrO(La_2CuO_4) + \frac{1}{2}O_2 \longrightarrow 2Sr'_{La} + 4O_O + Cu_{Cu} + 2h^\bullet$ (c)

The defect structure that occurs on acceptor doping is a function of dopant concentration. At low concentrations, hole generation is preferred—Eq. (c). The incorporation of one acceptor ion generates one hole in the structure. In terms of superconductivity, these are regarded as being generated in the LaO layers before migrating into the CuO sheets. Within the CuO layers the holes are generally regarded as located on some of the Cu^{2+} ions to form Cu^{3+} defects, although there is evidence that holes may also reside on O^{2-} ions to generate O^- species. Insulating La_2CuO_4 becomes a semiconductor when the hole concentration reaches about 0.05 holes per Cu^{2+}. As each Sr^{2+} dopant ion generates 1 hole, the composition is approximately $La_{1.95}Sr_{0.05}Cu^{2+}_{0.95}Cu^{3+}_{0.05}O_4$. Continued doping continues the transformation, and the solid shows superconductivity when the hole concentration reaches about 0.1 holes per Cu^{2+}, that is, at a composition $La_{1.9}Sr_{0.1}Cu^{2+}_{0.9}Cu^{3+}_{0.1}O_4$. Further doping transforms the phase into a metallic conductor. The material that has a maximum superconducting transition temperature of 37 K has a composition of $La_{1.85}Sr_{0.15}CuO_4$ and is referred to as a "hole" superconductor. As doping increases past this amount, the compensation mechanism changes from hole generation to vacancy generation—Eq. (b). One oxygen vacancy is generated for every two Sr^{2+} dopant ions to give a formula $La_{2-x}Sr_xCu^{2+}O_{4-x/2}$. Holes are no longer generated and superconductivity is lost. Preparations made under high oxygen pressures can reverse this trend, and vacancy-free materials can be prepared with higher dopant concentrations. Under these conditions the material loses superconductor properties when the concentration of Sr^{2+} reaches approximately 0.32, although it remains a metal.

Although the superconducting state of this material has attracted considerable attention, the metallic state is also of interest. For instance, metallic samples are being explored for use as cathode materials in solid oxide fuel cells (Section 6.10).

8.5.2 Donor Doping into Nd_2CuO_4

The structure of Nd_2CuO_4 is very similar to La_2CuO_4, the principal difference lies in the disposition of the oxygen atoms, as the cations in the two structures are in almost identical positions (Fig. 8.3). However, in Nd_2CuO_4 the Nd^{3+} ions are in the centers of oxygen cubes, and so this region of the structure can be likened to slabs of the fluorite type, and the structure may be described as an intergrowth of perovskite and fluorite structures. The Cu atoms lie at the centers of square coordination groups between these fluorite slabs rather than octahedra, as in La_2CuO_4. (Note that in the superconductor literature this structure is often called the T' structure.)

The structural and chemical similarity of La_2CuO_4 and Nd_2CuO_4 suggest that Nd_2CuO_4 should become metallic following acceptor (hole) doping. This turns out not to occur, and hole superconductivity does not arise in this phase when Nd^{3+} is substituted by Ca^{2+}, Sr^{2+}, or Ba^{2+}. However, Nd_2CuO_4 can be made metallic and superconducting by donor doping involving the substitution of the lanthanide by a higher valence cation such as Ce^{4+} to form, for example, $Nd_{2-x}Ce_xCuO_4$. As in the case of La_2CuO_4, doping is usually accompanied by small changes in oxygen content, which will be ignored in the present discussion.

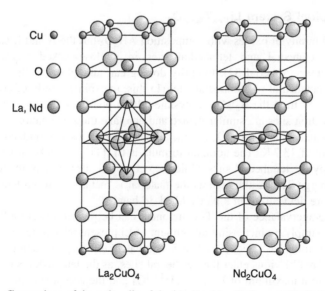

Cu ○

○ ◗

La, Nd ◔

La_2CuO_4 Nd_2CuO_4

Figure 8.3 Comparison of the unit cells of the idealized La_2CuO_4 and Nd_2CuO_4 structures.

Using the same formalism as above, and writing Nd_2CuO_4 as $Nd_2O_3 + CuO$, and focusing upon donor doping into Nd_2O_3, charge compensation can be by the following:

Cation vacancies: $\quad 3CeO_2(Nd_2O_3) \longrightarrow 3Ce_{Nd}^{\bullet} + V_{Nd}^{3\prime} + 6O_O \qquad$ (a)

Anion interstitials: $\quad 2CeO_2(Nd_2O_3) \longrightarrow 2Ce_{Nd}^{\bullet} + 3O_O + O_i^{2\prime} \qquad$ (b)

Electrons: $\quad 2CeO_2(Nd_2O_3) \longrightarrow 2Ce_{Nd}^{\bullet} + 2e' + 3O_O + \frac{1}{2}O_2 \qquad$ (c)

The oxygen interstitials in Eq. (b) are regarded as lost to the atmosphere:

$$O_i^{2\prime} \rightleftharpoons \frac{1}{2}O_2 + 2e'$$

to give Eq. (c). The component CuO as Cu_{Cu} and O_O can be added to each of these equations to make them represent stoichiometric Nd_2CuO_4.

As in the case of La_2CuO_4, initial doping turns the insulator into a metal. Further doping creates the superconducting state. Chemically, the donor doping can be written in terms of the generation of Cu^+ in place of Cu^{2+}. The Cu^+ ion is equivalent to a Cu^{2+} ion together with a trapped electron, and the formula is $Nd_{2-x}Ce_xCu_{1-x}^{2+}Cu_x^+O_4$ where one Cu^+ ion is created for every donor Ce^{4+} in the crystal. The compounds are considered to be electron superconductors rather than hole superconductors as in the La_2CuO_4-related phases. The superconducting state occurs over a rather narrow range of substitutions with x taking values of approximately 0.12–0.18. The maximum value of the superconducting transition temperature, T_c, is approximately 24 K, reached in the compound $Nd_{1.85}Ce_{0.15}CuO_4$.

8.5.3 Spinel System $Li_{1+x}Ti_{2-x}O_4$

The spinel family of oxides with composition AB_2O_4 has the A and B cations distributed in octahedral and tetrahedral sites in a close-packed oxygen structure (Supplementary Material S1). Impurity doping can take place by the addition of a dopant to octahedral or tetrahedral sites. In this, the spinel family of compounds is quite different from the A_2BO_4 perovskite-related phases of the previous section in that both cation sites are similar in size and can take the same cations.

The normal spinel $Li[Ti_2]O_4$ is a metallic oxide with a superconducting transition temperature of 13.7 K. The nominal formula is $Li^+[Ti^{3+}Ti^{4+}]O_4$, in which the Li^+ ions occupy the tetrahedral sites while the octahedral sites contain titanium with an average charge $Ti^{3.5+}$, although as the material is metallic at room temperature the electrons are delocalized in a partly filled $3d$ band.

The average valance of the Ti component can be pushed toward 4.0 by valence induction, simply by increasing the amount of Li present relative to that of titanium, to give a formula $Li_{1+x}Ti_{2-x}O_4$. This is equivalent to acceptor doping of the phase. The additional Li^+ ions enter the octahedral sites, as the tetrahedral sites are already occupied, to a maximum of $Li^+[Li_{1/3}^+Ti_{5/3}^{4+}]O_4$ and the acceptor doping is balanced by holes. In order to write formal defect equations for this situation, it is convenient to regard the doping as affecting only the Ti^{3+} component. Writing the spinel as $Li_2O + Ti_2O_3 + 2TiO_2$ (to give $2LiTi_2O_4$), the defect reaction of importance applies only to the Ti_2O_3 component:

$$Li_2O(Ti_2O_3) + O_2 \longrightarrow 2Li_{Ti3}^{2\prime} + 3O_O + 4h^{\bullet}$$

where only the Ti^{3+} sites are involved and are written as Ti3. Clearly a Ti^{3+} ion on its normal site shows no effective charge. The other components can be added to these equations to give

$$Li_2O(2LiTi_2O_4) \longrightarrow 2Li_{Li} + 2Li_{Ti3}^{2\prime} + 2Ti_{Ti4} + 8O_O + 4h^{\bullet}$$

where the Ti^{4+} sites involved are written as Ti4.

The holes can be regarded as transferred to Ti^{3+} ions converting them to Ti^{4+} by adding in the requisite number of extra ions:

$$Li_2O(6LiTi_2O_4) + O_2 \longrightarrow 6Li_{Li} + 2Li_{Ti3}^{2\prime} + 4Ti_{Ti3}^{\bullet} + 6Ti_{Ti4} + 24O_O$$

where $Ti_{Ti3}^{\bullet}$ represents a Ti^{3+} ion plus a hole on a Ti^{3+} site. In terms of an ionic picture each extra Li^+ present forces a Ti^{3+} ion to become Ti^{4+}. The limit occurs when $x = \frac{1}{3}$, to give $Li_{4/3}Ti_{5/3}O_4$. The $3d$ band is now completely empty. As would be expected, this material is a colorless insulator.

The conductivity changes in a discontinuous fashion at a composition of approximately $x = 0.12$ at which composition the metal becomes a poor semiconductor. The formula is now $Li^+[Li_{0.12}^+Ti_{0.64}^{3+}Ti_{1.24}^{4+}]O_4$. This transition is reversible. Donor doping of insulating $Li_{4/3}Ti_{5/3}O_4$ will return the compound to the metallic state.

8.6 CUPRATE HIGH-TEMPERATURE SUPERCONDUCTORS

8.6.1 Perovskite-Related Structures and Series

The important and widely studied copper-oxide-derived high-temperature super-conductors, known as cuprate superconductors, are basically insulators. Doping converts these into metallic materials, many of which are superconductors over rather more restricted composition ranges. Several of these materials have already been discussed: La_2CuO_4 and $Sr_2CuO_2F_2$ (Section 4.3.3), $La_{2-x}Sr_xCuO_4$ (Section 8.5.1), and $Nd_{2-x}Ce_xCuO_4$ (Section 8.5.2).

In the materials that follow, the structures are all layered. This structural feature has lead to a description of the doping in terms of charge reservoirs, a different approach to that described previously, and which is detailed below. Structurally the phases are all related to the perovskite-layered structures (Figs. 4.27, 4.28, 4.29, and 4.30). The similarity can be appreciated by comparison of the *idealized* structures and formulas of some of these materials, $Bi_2Sr_2CuO_6$

($=Tl_2Ba_2CuO_6$), $Bi_2CaSr_2Cu_2O_8$ ($= Tl_2CaBa_2Cu_2O_8$), and $Bi_2Ca_2Sr_2Cu_3O_{10}$ ($=Tl_2Ca_2Ba_2Cu_3O_{10}$) (Fig. 8.4). The single perovskite sheets in the idealized structure of $Bi_2Sr_2CuO_6$ are complete (Fig. 8.4a). These are separated by Bi_2O_2 (or Tl_2O_2) layers similar, but not identical, to those in the Aurivillius phases. In the other compounds, the oxygen structure needed to form the perovskite framework is

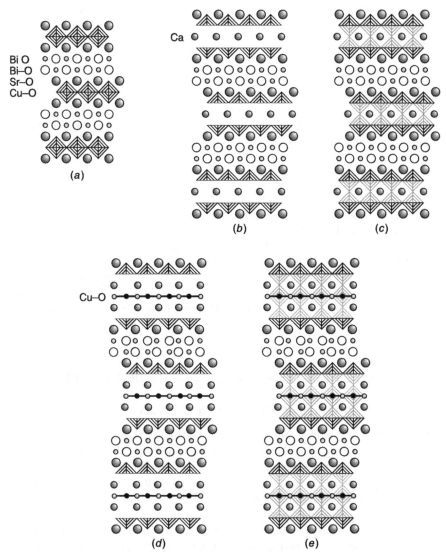

Figure 8.4 Idealized structures of some high-temperature superconductors: (a) $Bi_2Sr_2O_6$ ($=Tl_2Ba_2O_6$); (b) $Bi_2CaSr_2Cu_2O_8$ ($=Tl_2CaBa_2Cu_2O_8$); (c) as (b) but with the nominal CuO_6 octahedra completed in faint outline; (d) $Bi_2Ca_2Sr_2Cu_3O_{10}$ ($=Tl_2Ca_2Ba_2Cu_3O_{10}$); (e) as (d) but with the nominal CuO_6 octahedra completed in faint outline.

incomplete. The nominal double layer of CuO_6 octahedra needed to form a sheet of the idealized perovskite structure is replaced by square pyramids in $Bi_2CaSr_2Cu_2O_8$ (Fig. 8.4b). To make the relationship clearer, the octahedra are completed in faint outline in Figure 8.4c. In the case of $Ba_2Ca_2Sr_2Cu_3O_{10}$ the three CuO_6 octahedral perovskite layers have been replaced by two sheets of square pyramids and middle layer by a sheet of CuO_4 squares (Fig. 8.4d). The octahedra are completed in faint outline in Figure 8.4e.

In these and the other cuprate superconductors, the part of the structure that leads to superconductivity is the slab of CuO_2 sheets. When more than one sheet is present, they are separated by cation layers, Q (usually Ca or Y) to give a sequence CuO_2- $(Q-CuO_2)_{n-1}$, which forms the superconducting layer in the material. The index n is the total number of CuO_2 layers in the phase, which is equal to the formula number of Cu atoms present (Fig. 8.5).

The CuO_2 layers themselves are not, as such, superconducting and have to be doped with (usually) holes. These are provided by the charge reservoir layers that separate each $CuO_2-(Q-CuO_2)_{n-1}$ slab. The general structure of a charge reservoir sheet is $AO-[MO_x]_m-AO$, where A is a lanthanide such as La, or alkaline earth, typically Sr or Ba, M is a metal, typically Bi, Pb, Tl, or Hg, present as a nonstoichiometric oxide layer with x usually close to 0 or 1.0, and m takes values of 0, 1, 2, and so on. Doping the charge reservoir layer with acceptors, mainly oxygen, favors the creation of holes, which are subsequently transferred to the CuO_2 slab for metallic behavior and superconductivity to occur (Fig. 8.6).

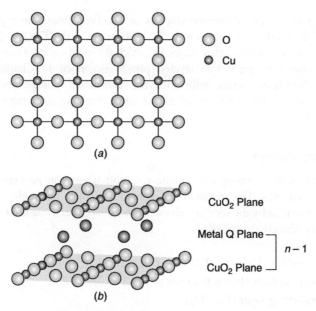

Figure 8.5 Superconducting planes found in cuprate superconductors: (a) a single CuO_2 sheet and (b) a CuO_2 $(Q-CuO_2)_{n-1}$ superconducting layer.

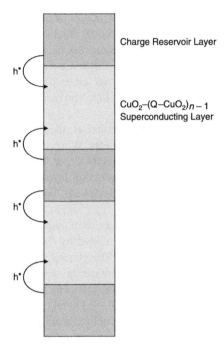

Figure 8.6 Schematic depiction of a cuprate superconductor. Doping into the charge reservoir layers results in the transfer of holes to the superconducting layers.

Homologous series of superconductors are derived from one type of charge reservoir slab interleaved with $n = 1$, 2, 3, 4, and so on—$CuO_2-(Q-CuO_2)_{n-1}$ slabs (Table 8.1). Clearly, other series in which different ordered slab thicknesses n, $n + 1$, similar to perovskite layered phases described in Chapter 4 can be envisaged. Similarly, series with different alternating charge reservoir slabs can be proposed, but synthesis of all of these complex structures might well pose problems.

8.6.2 Hole Doping

The concept of hole doping into charge reservoir layers can be explained with a number of examples. The materials are treated as if they are ionic. The simplest example is provided by the series of phases with a charge reservoir (AO AO), typified by La_2CuO_4, already described.

Charge reservoir $(La^{3+}O^{2-}\ La^{3+}O^{2-})$
The charge on each charge reservoir $= 2(+3 - 2) = +2$
Superconducting layer $(Cu^{2+}O_2)$
Charge on a superconducting layer $= (+2 - 2 \times 2) = -2$

TABLE 8.1 Homologous Series of Cuprate Superconductors

Charge Reservoir[a]	Charge Reservoir Formula	Superconducting Slab Formula	Idealized Series Formula[a]	Examples[b]
AO AO	LaO	$CuO_2\,(CuO_2)_{n-1}$	$La_2Cu_nO_{2n+2}$	$La_2CuO_{4+\delta}$ ($n=1$) $La_{2-x}Sr_xCuO_4$ ($n=1$)
	SrF	$CuO_2\,(CuO_2)_{n-1}$	$Sr_2Cu_nO_{2n}F_2$	$Sr_2CuO_2F_{2+\delta}$ ($n=1$)
AO M AO	BaO Cu BaO	$CuO_2\,(Y\,CuO_2)_{n-1}$	$Ba_2Y_{n-1}Cu_{n+1}O_{2n+\delta}$	$CuBa_2YCu_2O_{6+\delta} = YBa_2Cu_3O_{6+\delta}$ ($n=2$)
	BaO Hg BaO	$CuO_2\,(Ca\,CuO_2)_{n-1}$	$CuBa_2Ca_{n-1}Cu_nO_{2n+2}$	$CuBa_2Ca_2Cu_3O_{8+\delta}$ ($n=3$) $CuBa_2Ca_3Cu_4O_{10+\delta}$ ($n=4$)
		$CuO_2\,(Ca\,CuO_2)_{n-1}$	$HgBa_2Ca_{n-1}Cu_nO_{2n+2}$	$HgBa_2CuO_{4+\delta}$ ($n=1$) $HgBa_2CaCu_2O_{6+\delta}$ ($n=2$) $HgBa_2Ca_2Cu_3O_{8+\delta}$ ($n=3$)
AO MO AO	BaO TlO BaO	$CuO_2\,(Ca\,CuO_2)_{n-1}$	$TlBa_2Ca_{n-1}Cu_nO_{2n+3}$	$TlBa_2CuO_5$ ($n=1$) $TlBa_2CaCu_2O_{7+\delta}$ ($n=2$) $TlBa_2Ca_2Cu_3O_{9+\delta}$ ($n=3$)
	BaO BiO BaO	$CuO_2\,(Ca\,CuO_2)_{n-1}$	$BiBa_2Ca_{n-1}Cu_nO_{2n+3}$	$BiBa_2CuO_{5+\delta}$ ($n=1$) $BiBa_2CaCu_2O_{7+\delta}$ ($n=2$) $BiBa_2Ca_2Cu_3O_{9+\delta}$ ($n=3$)
	SrO GaO SrO	$CuO_2\,(Y, Ca\,CuO_2)_{n-1}$	$GaSr_2(Y,\,Ca)_{n-1}Cu_nO_{2n+3}$	$GaSr_2(Y,\,Ca)Cu_2O_{7+\delta}$ ($n=2$) $GaSr_2(Y,\,Ca)_2Cu_3O_{9+\delta}$ ($n=3$)
AO MO MO AO	SrO BiO BiO SrO	$CuO_2\,(Ca\,CuO_2)_{n-1}$	$Bi_2CaSr_{2n-1}Cu_nO_{2n+4}$	$Bi_2Sr_2CuO_{6+\delta}$ ($n=1$) $Bi_2CaSr_2Cu_2O_{8+\delta}$ ($n=2$) $Bi_2Ca_2Sr_2Cu_3O_{10+\delta}$ ($n=3$)
	BaO TlO TlO BaO	$CuO_2\,(Ca\,CuO_2)_{n-1}$	$Tl_2Ba_2Ca_{n-1}Cu_nO_{2n+4}$	$Tl_2Sr_2CuO_{6+\delta}$ ($n=1$) $Tl_2CaSr_2Cu_2O_{8+\delta}$ ($n=2$) $Tl_2Ca_2Sr_2Cu_3O_{10+\delta}$ ($n=3$)
AO MO M' MO AO	SrO PbO Cu PbO SrO	$CuO_2\,(Y, Ca\,CuO_2)_{n-1}$	$Pb_2Sr_2(Y,\,Ca)_{n-1}Cu_{n+1}O_8$	$Pb_2Sr_2Y_{0.5}Ca_{0.5}Cu_3O_{8+\delta}$ ($n=2$)

[a]The charge reservoir layer is nonstoichiometric. This is not indicated in the idealized formulas given.

[b]The composition is variable due to the nonstoichiometric nature of the charge reservoir slabs; δ indicates this and may be either positive or negative.

371

These charges exactly balance, and the material would be an insulator, a normal ionic compound. Hole doping can be achieved by adding lower valence cation such as Sr^{2+} or oxygen interstitials to the charge reservoir. For Sr^{2+} aceptor dopants:

Charge reservoir $(La_{1-x/2}Sr_{x/2}O)(La_{1-x/2}Sr_{x/2}O)$
The charge on each charge reservoir $= 2[+3(1-x/2)+2(x/2)-2] = +2-x$
Superconducting layer (CuO_2)
Charge on a superconducting layer $= (+2-2\times2) = -2$
Charge difference $= -x$

The charge difference between the charge reservoir and the superconducting layers must be achieved by the addition of balancing charges, $x\,h^{\bullet}$ in this case. For oxygen interstitials:

Charge reservoir $(LaO_{1+\delta/2})(LaO_{1+\delta/2})$
The charge on each charge reservoir $= 2[+3-2(1+\delta/2)] = +2-2\delta$
Superconducting layer (CuO_2)
Charge on a superconducting layer $= (+2-2\times2) = -2$
Charge difference $= -2\delta$

The charge difference between the charge reservoir and the superconducting layers must be balanced by the addition of $2\delta\,h^{\bullet}$, which are introduced into the superconducting layers.

The formal defect chemistry description of these processes has already described (Sections 4.3.3 and 8.5.1).

The high-temperature superconductor (with nominal composition) $YBa_2Cu_3O_7$, the first cuprate discovered with a superconducting transition temperature, T_c, above the boiling point of liquid nitrogen, is best discussed in terms of the nonstoichiometric phase $YBa_2Cu_3O_{6+\delta}$, which is able to take any composition over the range $YBa_2Cu_3O_6$ to $YBa_2Cu_3O_7$. The charge reservoir lamellae in $YBa_2Cu_3O_6$ are made up of AO M AO slices comprising BaO Cu BaO and the superconducting layers have the composition CuO_2 Y CuO_2 (Fig. 8.7a). Regarding these species as ions, and assuming that the charge reservoir layers contain Cu^+ ions:

Charge reservoir $(Ba^{2+}O^{2-}\ Cu^+\ Ba^{2+}O^{2-})$
The charge on each charge reservoir $= (+5-4) = +1$
Superconducting layer $(Cu^{2+}O_2^{2-}\ Y^{3+}\ Cu^{2+}O_2^{2-})$
Charge on a superconducting layer $= (+4+3-8) = -1$

These charges exactly balance, and the material is an insulator. Hole doping is achieved by adding oxygen interstitials to the charge reservoir to give a nonstoichiometric CuO_δ composition:

Charge reservoir $(BaO\ CuO_\delta\ BaO)$
The charge on each charge reservoir $= (+5-4-2\delta) = +1-2\delta$

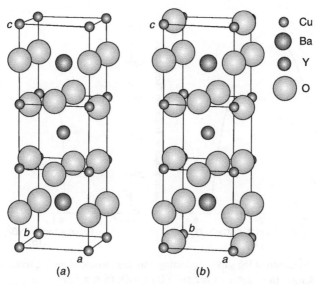

Figure 8.7 Structures of (a) $YBa_2Cu_3O_6$, tetragonal, $a = b$; and (b) $YBa_2Cu_3O_7$, orthorhombic. Oxygen atoms added to $YBa_2Cu_3O_6$ lie at positions along the **b** axes and can order at several intermediate compositions.

Superconducting layer (CuO_2 Y CuO_2)

Charge on a superconducting layer $= (+4 + 3 - 8) = -1$

Charge difference $= -2\delta$

The charge difference between the charge reservoir and the superconducting layers must be balanced by the addition of $2\delta h^\bullet$. The oxygen content can be simply manipulated by heating the sample in air at a suitable temperature. The reaction can be formally written as:

$$\delta/2\,O_2 \longrightarrow \delta\,O_i^{2\prime} + 2\delta h^\bullet$$

In the real structures the oxygen atoms are added at specific sites aligned along the **b** axis of the structure (Fig. 8.7b). A number of ordered superstructures over this composition range form when samples are carefully annealed.

The idea of charge reservoirs can be applied to the other members of the series.

8.6.3 Defect Structures

Although the high-temperature superconducting phases are formed from insulating materials by the introduction of defects, the precise relationship between dopant, structure, and properties is not fully understood yet. For example, in most of the cuprate phases it is extremely difficult to be exactly sure of the charges on the individual ions, and because of this the real defect structures are still uncertain.

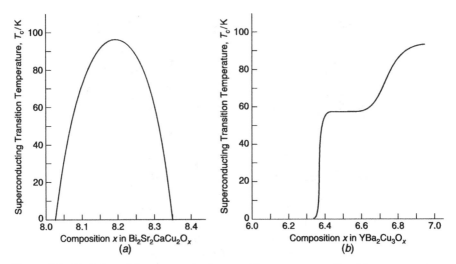

Figure 8.8   Variation of the superconducting transition temperature, T_c, with oxygen content: (a) $Bi_2Sr_2CaCu_2O_x$ ($8 < x < 8.4$) and (b) $YBa_2Cu_3O_x$ ($6 < x < 7$).

One feature of interest is the strong dependence of the appearance of superconductivity and the superconducting transition temperature T_c, upon the oxygen composition (and thus upon the defect content) of the phase. Typically, the superconducting transition temperature rises to a maximum and then falls as the oxygen content changes. For example, the maximum value of T_c for the superconductor $Bi_2CaSr_2Cu_2O_{8+\delta}$ reaches a maximum of 95 K at a composition close to $Bi_2CaSr_2Cu_2O_{8.19}$ (Fig. 8.8a).

The high-temperature superconductor $YBa_2Cu_3O_7$ shows more complex behavior. The maximum value of T_c, close to 94 K, is found at the composition of $YBa_2Cu_3O_{6.95}$. As more oxygen is removed, the value of T_c falls to a plateau of approximately 60 K when the composition lies between the approximate limits of $YBa_2Cu_3O_{6.7}$ to $YBa_2Cu_3O_{6.5}$. Continued oxygen removal down to the phase limit of $YBa_2Cu_3O_{6.0}$ rapidly leads to the formation of a semiconducting and then an insulating phase (Fig. 8.8b). The form of this curve suggests that there may be two superconducting phases forming, each of which contributes a curve similar to that in Fig. 8.8a, one cenetred near to $YBa_2Cu_3O_{6.5}$ and the other near to $YBa_2Cu_3O_{6.95}$.

A simplified "generic" phase diagram of these materials shows that at low doping concentrations, the solids are antiferromagnetic insulators. At greater concentrations they become semiconductors and then metals.[1] The superconducting range is rather narrow, and the value of the superconducting transition temperature, T_c, peaks at a doping level of about 0.15–0.20 holes per CuO_2 sheet (Fig. 8.9).

---

[1]There are many puzzling features of the physics of this phase diagram that have yet to be explained.

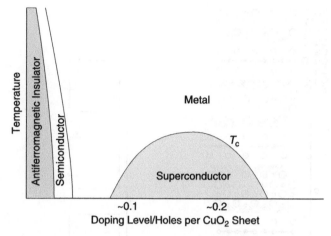

Figure 8.9 Simplified "generic" phase diagram for cuprate superconductors.

Doping, particularly oxygen gain and loss, introduces a wide range of defects into these structures. One of the simpler consequences is the appearance of new structures due to the ordering of oxygen interstitials or vacancies, such as those reported in $La_2CuO_{4+\delta}$ or $YBa_2Cu_3O_{6+\delta}$. A large number of such ordered structures can be postulated, and several ordering patterns have been observed, especially in annealed samples of $YBa_2Cu_3O_{6+\delta}$. The rather complex composition of many superconducting phases also means that antisite defects and substitutional defects arise, although these are not easy to characterize.

Stacking defects, consisting of different thicknesses of the CuO_2-containing slabs, commonly occur in all members of the superconducting cuprate series. Long-range ordering of these intergrowths, commonly observed in all of the layered structures derived from perovskite lamellae (Section 4.8.4) has also been recorded.

A defect structure mostly found in the $Bi_2Sr_2Ca_{n-1}Cu_nO_{2n+4}$ series consists of a modulation along or close to one of the short axes, usually taken to be the **b** axis of the structure (Fig. 8.10a). The modulation is associated with the incorporation of extra oxygen into the Bi–O layers of the structure and can be represented as an approximately sinusoidal buckling of these layers in order to integrate the additional atoms (Fig. 8.10b and 8.10c). This distortion is transmitted to the perovskite-like blocks so that the whole structure takes on a wavelike structure. The complexity of the real structures is considerable. The modulation wavelength is rarely commensurate with the b-lattice parameter. The modulation waves in one set of Bi–O planes can correspond with those in the adjacent layers (as in Fig. 8.10c) to give an approximately orthorhombic unit cell or be displaced by a small amount to give an approximately monoclinic cell (Fig. 8.10d). Moreover, the modulation waves may also run at an angle to the **b** axis in some or all of the layers.

Twin boundaries are frequently encountered in cuprate superconductors. There are a number of ways in which twins might form in a crystal, one of which is

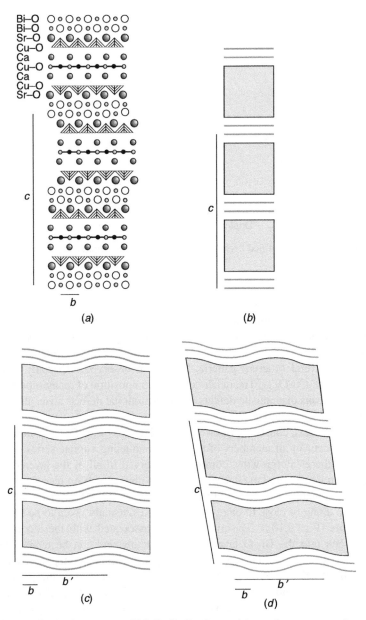

Figure 8.10   Idealized structures of $Bi_2Ca_2Sr_2Cu_3O_{10+\delta}$: (*a*) atomic structure projected down [100]; (*b*) structure as Bi–O and perovskite lamellae; (*c*) incommensurate modulation (exaggerated) along the **b** axis with a period $b' \approx 5.8\ b$ (orthorhombic); and (*d*) incommensurate modulation (exaggerated) along the **b** axis with a period $b' \approx 5.8\ b$ (monoclinic).

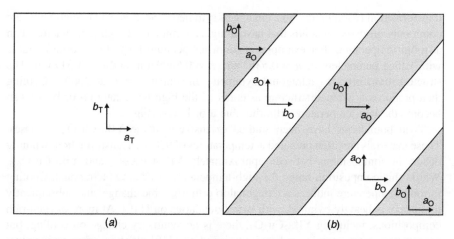

Figure 8.11 Transformation of a tetragonal crystal, with lattice parameters $a_T = b_T$, into a multiply twinned orthorhombic crystal with lattice parameters a_O, b_O. The twinned regions are often called domains and the boundaries may occur on a variety of crystallographic planes.

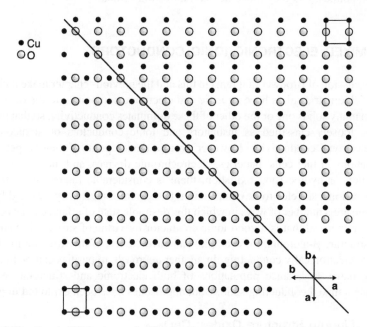

Figure 8.12 Basal (001) plane of YBa$_2$Cu$_3$O$_7$ containing a [110] twin boundary. The projection of the unit cell is outlined. The dopant oxygen atoms run along the **b** axis in each twin.

transformation twinning due to a change of symmetry (Section 3.11). Such faults are commonly observed in solids that have a higher symmetry at high temperatures than at room temperature. For example, at room temperature La_2CuO_4 is orthorhombic with lattice parameters of $a = 0.5363$ nm, $b = 0.5409$ nm, and $c = 1.317$ nm. The structure transforms to tetragonal symmetry in which $a = b$, at 260°C. Cooling then produces a twinned structure as either of the high temperature **a** or **b** axis can become the low-temperature orthorhombic **a** or **b** axis (Fig. 8.11).

Twin boundaries have been studied extensively in the $YBa_2Cu_3O_{6+\delta}$ phases. These materials are often prepared at temperatures of 950°C. Transformation twinning occurs in compositions between approximately $YBa_2Cu_3O_{6.25}$ and $YBa_2Cu_3O_{6.6}$. Within this composition range the high-temperature structure is tetragonal. Cooling to room temperature induces a tetragonal to orthorhombic change, and subsequently the matrix is heavily twinned, with twin boundaries on [110]. At more oxygen-rich compositions, including $YBa_2Cu_3O_7$, there is no symmetry change on cooling, but nevertheless most crystals are heavily twinned on [110] planes at room temperature (Fig. 8.12). In this case transformation twinning is not possible and must arise via a different mechanism. One possibility is that these faults are growth twins in which the chains of dopant oxygen atoms line up along either of the possible **a/b** axial directions in different crystal nuclei as the formation reactions proceed.

Twin planes play an important role in decreasing the current-carrying capability of the compounds, and they make device fabrication difficult.

8.7 MIXED ELECTRONIC/IONIC CONDUCTORS

The discussion of Brouwer diagrams in this and the previous chapter make it clear that nonstoichiometric solids have an ionic and electronic component to the defect structure. In many solids one or the other of these dominates conductivity, so that materials can be loosely classified as insulators and ionic conductors or semiconductors with electronic conductivity. However, from a device point of view, especially for applications in fuel cells, batteries, electrochromic devices, and membranes for gas separation or hydrocarbon oxidation, there is considerable interest in materials in which the ionic and electronic contributions to the total conductivity are roughly equal.

There are a number of ways in which this desirable state of affairs can be achieved. In one, a material that is a good ionic conductor by virtue of structural features (the layer structure β-alumina, for example) can have the rest of the structure modified to become electronically conducting. In another approach, impurities can be introduced into a matrix to balance populations of both electronic and structural defects to generate a mixed conducting solid. Both approaches have been exploited in practice.

8.7.1 Fluorite Structure Oxides: $CeO_{2-\delta}$

There has been considerable interest in CeO_2 as a component of solid oxide fuel cells, especially as an anode material, and also for use in oxygen separation membranes. The material shows a wide nonstoichiometry range, with oxygen vacancies as the major point defect. [Note that defect clusters certainly occur over part of the composition range (see Section 4.4.5)]. Under oxidizing conditions, in air, for instance, the oxygen deficit δ is close to zero. Oxygen vacancies can be introduced by slightly reducing the oxygen partial pressure. The loss of oxygen is balanced by the introduction of electrons:

$$O_O \rightleftharpoons V_O^{2\bullet} + \tfrac{1}{2}O_2 + 2e'$$

It is believed that the electrons are located on Ce^{4+} ions to form Ce^{3+} (Ce'_{Ce}):

$$Ce_{Ce} + e' \rightleftharpoons Ce'_{Ce}$$

or

$$2CeO_2 \rightleftharpoons \tfrac{1}{2}O_2 + 2Ce'_{Ce} + 3O_O + V_O^{2\bullet}$$

There is little electron transfer between the Ce^{4+} and Ce^{3+} ions in oxidizing conditions, and the material shows predominantly O^{2-} ion conductivity. As the reducing conditions become more severe, more of the Ce^{4+} cations are reduced to Ce^{3+} cations, leading to mixed ionic and electronic n-type conductivity. For oxides to be used as anodes under quite severe reducing conditions, this is acceptable, but if mixed conductivity is required at higher oxygen partial pressures, for example, in a membrane designed to separate oxygen gas from air, the electronic conductivity must be improved.

One successful technique involves the substitution of Ce^{4+} by a more readily reduced cation such as Pr^{4+} to form the oxide $Ce_{1-x}Pr_xO_2$. The Pr^{4+} ion is not as stable as the Ce^{4+} ion, and the oxide $Ce_{1-x}Pr_xO_2$ is reduced when the oxygen partial pressure falls below that in air. Reduction affects the Pr^{4+} ions almost exclusively, with the generation of oxygen vacancies and electrons:

$$O_O \rightleftharpoons V_O^{2\bullet} + \tfrac{1}{2}O_2 + 2e'$$

and if the electrons are presumed to sit on Pr^{4+} ions to form Pr^{3+} (Pr'_{Ce}):

$$Pr_{Ce} + e' \rightleftharpoons Pr'_{Ce}$$

The resulting materials have approximately equal ionic and electronic contributions to the total conductivity at doping levels between $Ce_{0.8}Pr_{0.2}O_{2-\delta}$ and $Ce_{0.75}Pr_{0.25}O_{2-\delta}$. The electronic conductivity mechanism in these oxides is believed to be by way of electron hopping between Pr^{4+} and Pr^{3+}.

At greater degrees of reduction, all of the Pr ions are in the trivalent state, and the oxide is in essence an acceptor-doped oxide with oxygen vacancy compensation. Any further reduction must then be accomplished by the transformation of Ce^{4+} to Ce^{3+}, repeating the previous cycle:

$$O_O \rightleftharpoons V_O^{2\bullet} + \tfrac{1}{2}O_2 + 2e'$$

and if the electrons are presumed to sit on Ce^{4+} ions to form Ce^{3+} (Ce'_{Ce}):

$$Ce_{Ce} + e' \rightleftharpoons Ce'_{Ce}$$

8.7.2   Layered Structures: Li_xMX_2

The cathode of a battery or fuel cell must allow good ionic conductivity for the ions arriving from the electrolyte and allow for electron conduction to any interconnects between cells and to external leads. In addition these properties must persist under oxidizing conditions. An important strategy has been to employ layered structure solids in which rapid ionic motion occurs between the layers while electronic conductivity is mainly a function of the layers themselves.

One of the first mixed conductors to be utilized as a battery cathode was lithium intercalated Li_xTiS_2. This material is composed of TiS_2 sheets (Section 4.7). Lithium can be inserted between the layers reversibly, making Li_xTiS_2 suitable for use as a cathode material in batteries that employ Li^+ ion transport from the anode across the electrolyte (Fig. 8.13). The parent phase, TiS_2, is a semiconductor. As Li enters the structure, the electrons on the Li atoms delocalize to give a significant electronic conductivity. The reactions that take place when the battery is used (discharged) or recharged are:

$$
\begin{aligned}
\text{Discharge:} \quad & x\,Li + TiS_2 \longrightarrow Li_xTiS_2 \\
\text{Recharge:} \quad & Li_xTiS_2 \longrightarrow x\,Li + TiS_2
\end{aligned}
$$

and the cell reaction is

$$x\,Li + TiS_2 \longrightarrow Li_xTiS_2$$

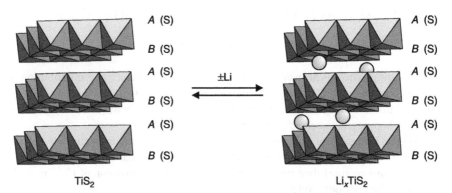

Figure 8.13   Structure of TiS_2 and Li_xTiS_2. The packing of the sulfur atoms remains in hexagonal ...$ABAB$... sequence across the composition range from TiS_2 to $Li_{1.0}TiS_2$.

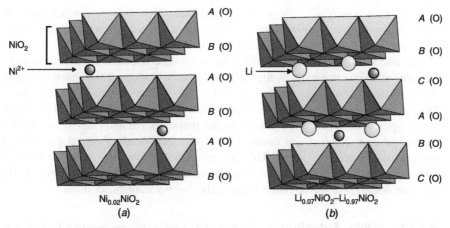

Figure 8.14 Layer structure of oxides with composition $Li_{1-y}Ni_{1+y}O_2$: (a) $Ni_{1.02}O_2$, with the oxygen atoms in hexagonal ... $ABAB$ ... sequence; and (b) nominal compositions between $Li_{0.07}NiO_2$–$Li_{1.0}NiO_2$ have the oxygen atoms in cubic ... $ABCABC$ ... sequence.

The $Li–TiS_2$ battery was superceded by the "Sony cell," which uses a cathode Li_xCoO_2. This has a similar structure to TiS_2 and intercalation of Li takes place in a similar fashion (Section 4.6). The potential cathode material Li_xNiO_2 behaves in a similar way. Under normal preparation conditions the material is always nickel rich, with a formula $Li_{1-y}Ni_{1+y}O_2$. In this phase the extra nickel ions are located as defects in the Li layer. The nominal ionic charges are $(Li_{1-y}Ni_y^{2+})(Ni_y^{2+}Ni_{1-y}^{3+})O_2$ and the defect species are nominally $Ni_{Li}^{\bullet}$ and Ni_{Ni}'. The close to stoichiometric $Ni_{1.02}O_2$ has the same CdI_2 structure as TiS_2 and CoO_2, (Fig. 8.14a), although preparations usually produce a material in which the ... $ABAB$ ... packing of the anions is interrupted by stacking faults to give occasional lamellae with an ... ABC ... packing. The stacking of the NiO_2 layers changes completely to ... $ABCABC$ ... packing for compositions in the range $Li_{0.07}NiO_2$–$Li_{1.0}NiO_2$ (Fig. 8.14b). Ionic conductivity is due to the easy movement of Li^+ ions between the close-packed NiO_2 layers. Electronic conductivity appears to be mainly due to electron hopping from Ni^{2+} to Ni^{3+} within the layers.

In practice, the defect structure of the materials $Li_x(Co, M)O_2$ and $Li_x(Ni, M)O_2$ under oxidizing conditions found at cathodes, is complex. For example, oxidation of Fe^{3+} substituted lithium nickelate, $Li_x(Ni, Fe)O_2$, under cathodic conditions leads to the formation of Fe^{4+} and Ni^{4+}. Conductivity can then take place by means of rapid charge hopping between Fe^{3+}, Ni^{3+}, Fe^{4+}, and Ni^{4+}, giving average charges of $Fe^{3+\delta}$ and $Ni^{3+\delta}$. These solids are the subject of ongoing research.

8.7.3 Acceptor Doping in Perovskite Structure Oxides

The introduction of an impurity cation onto one sublattice of perovskite structure oxides can change the defects on the other cation sublattice, on the oxygen sublattice, or both, in a controlled fashion. These materials often show a conflict between electronic and ionic compensation and doping often results in materials that show both appreciable ionic and electronic conductivity.

Because of this, oxides that adopt the perovskite structure have been widely explored as anode materials in fuel cells and related devices. These must be stable under reducing conditions as well as being able to transport electrons and ions. Because perovskites tend to show n-type conduction at low oxygen partial pressures and p-type conduction at high oxygen partial pressures, the main task is to improve the oxygen ion conductivity component of the total conductivity in the desired oxygen partial pressure range. This can often be achieved by selective doping; donor impurities will create cation vacancies, anion interstitials, or electrons, while acceptor doping will create cation interstitials, anion vacancies, or holes.

8.7.3.1 A-Site Substitution

The A-site doping of the perovskite $LaCoO_3$ has been widely studied. When prepared in air, the phase is virtually stoichiometric, corresponding to a nominal ionic distribution $La^{3+}Co^{3+}O_3$, although oxygen loss can occur under changes of oxygen partial pressure and temperature (Section 7.3). A-site acceptor doping is readily achieved by replacement of La^{3+} by an alkaline earth cation, typically Sr^{2+}. This acceptor substitution can be compensated in the following ways: (a) cation interstitials, (b) anion vacancies, or (c) holes.

As the introduced cations are large, interstitial formation can be regarded as less likely, leaving anion vacancy formation:

$$2SrO(LaCoO_3) \longrightarrow 2Sr'_{La} + 2Co_{Co} + 5O_O + V_O^{2\bullet}$$

The concentration of oxygen vacancies is half that of the concentration of Sr dopant:

$$[Sr'_{La}] = 2[V_O^{2\bullet}]$$

The composition of the phase is then $La_{1-x}Sr_xCoO_{3-x/2}$. If compensation is electronic, the vacancies are imagined to be filled by atmospheric oxygen:

$$\tfrac{1}{2}O_2 + V_O^{2\bullet} \rightleftharpoons O_O + 2h^\bullet$$

That is,

$$2SrO + \tfrac{1}{2}O_2(LaCoO_3) \longrightarrow 2Sr'_{La} + 2Co_{Co} + 6O_O + 2h^\bullet$$

The number of holes is thus equal to the dopant concentration. The possibility that the holes are localized on Co atoms is reasonable:

$$Co_{Co} + h^\bullet \longrightarrow Co_{Co}^\bullet$$

That is,

$$2SrO + \tfrac{1}{2}O_2(LaCoO_3) \longrightarrow 2Sr'_{La} + 2Co_{Co}^\bullet + 6O_O$$

The ion Co$^\bullet$ corresponds to Co^{4+} and the number of Co^{4+} ions is then equal to the number of dopant Sr^{2+} ions. The reaction can be formally considered to be the formation of a solid solution between the perovskite structure phases LaCoO$_3$ (Co^{3+}) and SrCoO$_3$ (Co^{4+}). The formula is formally La$_{1-x}$Sr$_x$Co$^{3+}_{1-x}$Co$^{4+}_x$O$_3$.

Note that other equilibria could be written, including the disproportionation:

$$2Co_{Co} \rightleftharpoons Co'_{Co} + Co^\bullet_{Co}$$

That is,

$$Co^{3+} + Co^{3+} \rightleftharpoons Co^{2+} + Co^{4+}$$

Which of these possibilities is preferred can be answered by experiment. A plot of the strontium content, x, in La$_{1-x}$Sr$_x$CoO$_{3-\delta}$, versus the oxygen content, $3-\delta$, for the two possibilities, V$^{2\bullet}_O$ or Co$^\bullet_{Co}$, can be drawn and compared with experimental measurements, which show that both defect types are present with vacancies dominating (Fig. 8.15). The material is a mixed conductor.

Preparations under higher oxygen partial pressure would be expected to increase hole concentrations (Co$_{Co}^\bullet$) and under lower partial pressures to increase the concentration of vacancies (V$^{2\bullet}_O$). The exact defect population in a sample will depend upon temperature, oxygen partial pressure, and the rate at which samples are cooled.

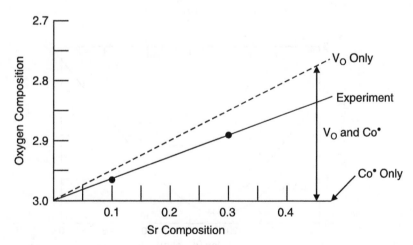

Figure 8.15 Calculated composition versus oxygen stoichiometry curves for La$_{1-x}$Sr$_x$CoO$_{3-\delta}$. [The two experimental points are taken from data in A. N. Petrov, V. A. Cherepanov, and A. Y. Zuev, Thermodynamics, Defect Structure and Charge Transfer in Doped Lanthanum Cobaltites: An Overview, *J. Solid State Electrochem.*, **10**, 517–537 (2006).]

8.7.3.2 B-Site Substitution Acceptor doping of the normally insulating perovskite structure $SrTiO_3$ has been widely explored for the purpose of fabricating mixed conductors. Replacement of part of the Ti^{4+} by a lower valence acceptor cation such as Fe^{3+} leads to enhanced total conductivity with greatly enhanced O^{2-} migration.

The substitution reaction can be written:

$$Fe_2O_3(SrTiO_3) \longrightarrow 2Sr_{Sr} + 2Fe'_{Ti} + 5O_O + V_O^{2\bullet}$$

An idea of how the conductivity will change over the partial pressure range can be gained by writing out the potential defect formation equations. At high oxygen

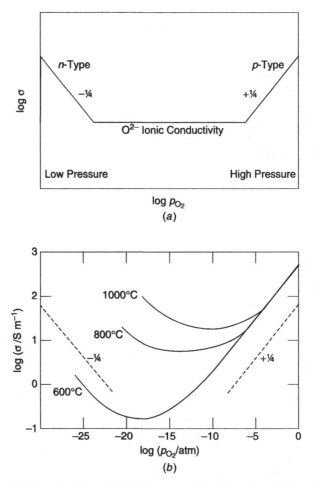

Figure 8.16 Mixed conductivity in $SrTi_xFe_{1-x}O_{3-\delta}$: (a) schematic variation of conductivity and (b) experimental conductivity for $SrTi_{0.5}Fe_{0.5}O_{3-\delta}$. [Data adapted from S. Steinsvik, R. Bugge, J. Gjønnes, J. Taftø, and T. Norby, *J. Phys. Chem. Solids*, **58**, 969–979 (1997).]

pressure the oxygen vacancies will tend to fill:

$$V_O^{2\bullet} + \tfrac{1}{2}O_2 \rightleftharpoons O_O + 2h^\bullet$$

$$K_o = \frac{[h^\bullet]^2}{[V_O^{2\bullet}]p_{O_2}^{1/2}}$$

The conductivity will be proportional to the concentration of holes, so that:

$$\sigma \propto [h^\bullet] \propto p_{O_2}^{1/4}$$

At low oxygen pressure more vacancies will tend to be created:

$$O_O \rightleftharpoons \tfrac{1}{2}O_2 + V_O^{2\bullet} + 2e'$$

$$K_r = [e']^2[V_O^{2\bullet}]p_{O_2}^{1/2}$$

The conductivity will be proportional to the concentration of electrons so that:

$$\sigma \propto [e'] \propto p_{O_2}^{-1/4}$$

The way in which the conductivity will change with oxygen partial pressure can then be estimated (Fig. 8.16a). Experimental data (Fig. 8.16b) are in fair agreement with these ideas.

8.8 MIXED PROTON/ELECTRONIC CONDUCTORS

8.8.1 Proton Mixed Conductors

Acceptor doping in perovskite oxides gives materials with a vacancy population that can act as proton conductors in moist atmospheres (Section 6.9). In addition, the doped materials are generally p-type semiconductors. This means that in moist atmospheres there is the possibility of mixed conductivity involving three charge carriers (H^+, O^{2-}, and $h^\bullet$) or four if electrons, e', are included.

The situation can be illustrated with respect to the acceptor-doped perovskite structure $SrZrO_3$, with Y^{3+} substituted for Zr^{4+} to give compositions $SrZr_{1-x}Y_xO_{3-0.5x}$. The doping reaction can be written:

$$2SrO + Y_2O_3(SrZrO_3) \longrightarrow 2Sr_{Sr} + 2Y'_{Zr} + 5O_O + V_O^{2\bullet}$$

Dopant levels of $x = 0.05{-}0.2$ are common.

The successfully doped materials are generally p-type semiconductors. This arises from a combination of atmospheric oxygen with the vacancies, thus:

$$V_O^{2\bullet} + \tfrac{1}{2}O_2 \rightleftharpoons O_O + 2h^\bullet$$

On exposure to water vapor the oxygen vacancies can also react to form OH^-:

$$O_O + V_O^{2\bullet} + H_2O \longrightarrow 2OH_O^{\bullet}$$

The resulting phases are good proton conductors. The total conductivity is due to contributions from oxygen ions, protons, and holes:

$$\sigma(\text{total}) = \sigma(O^{2-}) + \sigma(H^+) + \sigma(h^{\bullet})$$

The contribution of each of these is strongly temperature dependent.

The transport numbers of the ions can be determined by using a solid-state electrolyte. The cell voltage across an oxygen-conducting electrolyte subjected to an oxygen pressure gradient is given by the Nernst equation (Section 6.8.3):

$$E = -\left(\frac{RT}{4F}\right)\ln\left(\frac{p'_{O_2}}{p''_{O_2}}\right)$$

$$= -\left(\frac{RT}{4F}\right)\ln\left[\frac{\text{anode pressure (low)}}{\text{cathode pressure (high)}}\right]$$

Similarly, the cell voltage across a proton-conducting electrolyte subjected to a hydrogen pressure gradient is given by the Nernst equation:

$$E = -\left(\frac{RT}{2F}\right)\ln\left(\frac{p'_{H_2}}{p''_{H_2}}\right)$$

$$= -\left(\frac{RT}{2F}\right)\ln\left[\frac{\text{anode pressure (low)}}{\text{cathode pressure (high)}}\right]$$

If the electrolyte conducts both ions, the equation becomes

$$E = t_O\left(\frac{-RT}{4F}\right)\ln\left(\frac{p'_{O_2}}{p''_{O_2}}\right) - t_H\left(\frac{-RT}{2F}\right)\ln\left[\frac{(p'_{H_2})}{(p''_{H_2})}\right]$$

where t_O and t_H are the transport numbers for oxygen and hydrogen ions respectively, and the $-$ sign is needed as the ions travel in opposite directions. In practical terms it is more convenient to work with water vapour pressures, using the equilibrium:

$$2H_2O(g) \rightleftharpoons 2H_2 + O_2$$

so that

$$E = t_O\left(\frac{-RT}{4F}\right)\ln\left(\frac{p'_{O_2}}{p''_{O_2}}\right) - t_H\left(\frac{-RT}{2F}\right)\ln\left[\frac{(p'_{H_2O})}{(p''_{H_2O})}\right]$$

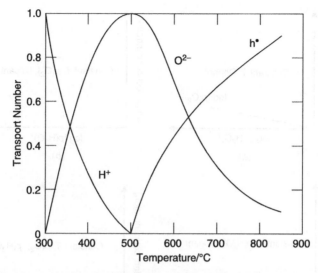

Figure 8.17 Schematic representation of the variation of the transport numbers for H^+, O^{2-}. and $h^\bullet$ in $SrZr_{1-x}Y_xO_{3-\delta}$. [Data adapted from P. Huang and A. Petric, *J. Mater. Chem.*, **5**, 53–56 (1995).]

Thus, measurement of the total conductivity together with the cell voltage allows the transport numbers of the ions to be determined (Fig. 8.17). The results show that at lower temperatures proton conductivity is of greatest importance, at middle temperatures oxygen ion conductivity becomes dominant, and at high temperatures the material is predominantly a hole conductor. Between these temperatures, at approximately 350°C the solid is a mixed H^+ and O^{2-} conductor while at approximately 650°C it is a mixed hole and O^{2-} conductor.

8.8.2 Brouwer Diagram Representation of Mixed Proton Conductivity

A number of factors must be taken into account when the diagrammatic representation of mixed proton conductivity is attempted. The behavior of the solid depends upon the temperature, the dopant concentration, the partial pressure of oxygen, and the partial pressure of hydrogen or water vapor. Schematic representation of defect concentrations in mixed proton conductors on a Brouwer diagram therefore requires a four-dimensional depiction. A three-dimensional plot can be constructed if two variables, often temperature and dopant concentration, are fixed (Fig. 8.18a). It is often clearer to use two-dimensional sections of such a plot, constructed with three variables fixed (Fig. 8.18b–8.18d).

The Brouwer diagram approach can be illustrated with reference to the perovskite structure oxide system $BaYb_xPr_{1-x}O_3$, which has been explored as a potential cathode material for use in solid oxide fuel cells. The parent phase

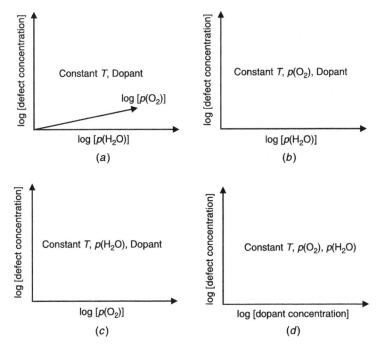

Figure 8.18 Schematic representation of defect concentrations in mixed proton conductors on a Brouwer diagram: (a) three-dimensional plot with two variables fixed; (b)–(d) two-dimensional plots with three variables fixed.

is $Ba^{2+}Pr^{4+}O_3$ and the dopant Y_2O_3 is incorporated to generate a population of vacancies:

$$2BaO + Y_2O_3(BaPrO_3) \longrightarrow 2Ba_{Ba} + 2Y'_{Pr} + 5O_O + V_O^{2\bullet}$$

The vacancies can react with oxygen gas to generate holes:

$$\tfrac{1}{2}O_2 + V_O^{2\bullet} \rightleftharpoons O_O + 2h^\bullet \tag{8.15}$$

or with water vapor to create hydroxyl:

$$O_O + V_O^{2\bullet} + H_2O \longrightarrow 2OH_O^\bullet \tag{8.16}$$

The Pr^{4+} ion is easily reduced to Pr^{3+}, which can be considered in terms of the electronic equilibrium:

$$Pr_{Pr} \rightleftharpoons Pr'_{Pr} + h^\bullet \tag{8.17}$$

Equations (8.15)–(8.17) can be considered to be the basic equations for the diagram. The defect population in the doped solid under moist conditions then

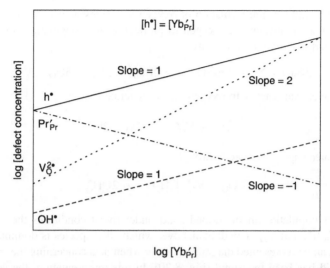

Figure 8.19 Schematic representation of the variation of the defect concentrations in $BaPr_{1-x}Yb_xO_{3-\delta}$ as a function of dopant concentration, assuming fixed temperature, water, and oxygen pressure. The electroneutrality equation used is $[h^\bullet] = [Yb'_{Pr}]$. [Adapted from data in S. Mimuro, S. Shibako, Y. Oyama, K. Kobayashi, T. Higuchi, S. Shin, and S. Yamaguchi, *Solid State Ionics*, **178**, 641–647 (2007).]

consists of $V_O^{2\bullet}$, $OH_O^\bullet$, $h^\bullet$, Y'_{Pr}, and Pr'_{Pr}. The electroneutrality equation is

$$2[V_O^{2\bullet}] + [OH_O^\bullet] + [h^\bullet] = [Y'_{Pr}] + [Pr'_{Pr}] \qquad (8.18)$$

These equations then allow the Brouwer diagrams to be constructed (Fig. 8.19).

8.8.3   Charge Carrier Map Representation of Mixed Conductivity

The Brouwer diagram approach to representing defect concentrations can be augmented by a further description that outlines the domains over which particular defect types dominate the conductivity process. This is analogous to the use of Pourbaix diagrams in the study of corrosion and gives an overview of the conductivity behavior of the solid. In proton mixed conductors the important conductivity domains are those dominated by hole conduction, oxygen ion conduction, and proton conduction. The designation of these domains is to a certain extent arbitrary; a convenient measure is the region over which the transport number for any defect is greater than 0.5. As with the Brouwer diagram, these domains will be a function of the temperature, dopant concentration, and the partial pressures of oxygen and hydrogen or water vapor.

The use of this approach can be illustrated by the perovskite structure proton conductor $BaY_{0.2}Zr_{0.8}O_{3-\delta}$. This material has been investigated for possible use in solid oxide fuel cells, hydrogen sensors and pumps, and as catalysts. It is similar to the $BaPrO_3$ oxide described above. The parent phase is $Ba^{2+}Zr^{4+}O_3$, and doping with trivalent Y^{3+} induces the formation of oxygen vacancies, which can be partly filled by reaction with surrounding oxygen to produce a hole population, or with water vapor to lead to proton conductivity:

$$2BaO + Y_2O_3(BaZrO_3) \longrightarrow 2Ba_{Ba} + 2Y'_{Zr} + 5O_O + V_O^{2\bullet}$$

The vacancies can react with oxygen gas to generate holes:

$$\tfrac{1}{2}O_2 + V_O^{2\bullet} \rightleftharpoons O_O + 2h^\bullet \qquad (8.15)$$

or with water vapor to create hydroxyl:

$$O_O + V_O^{2\bullet} + H_2O \longrightarrow 2OH_O^\bullet \qquad (8.16)$$

The defect population in the doped solid under moist conditions then consists of $V_O^{2\bullet}$, $OH_O^\bullet$, $h^\bullet$, and Y'_{Zr}. The domains over which each species is dominant for conductivity can be represented diagrammatically when data concerning the conductivity of the solid has been measured (Fig. 8.20). In this representation, the conductivity fields are bounded by lines tracing the locus where the transport number for a pair of defects is equal to 0.5. (The diagram could equally well be drawn in terms of domains delineating the defect species that predominate.)

The boundary between the mainly oxygen ion conducting region and the mainly hole conducting region is defined by Eq. (8.15). There is no involvement of water vapor in this equilibrium, so the boundary will be uninfluenced by hydrogen partial pressure and lies parallel to that axis, intersecting the oxygen partial pressure axis at a value appropriate to that at which the defect concentration gives rise to the conductivity relation $t(O^{2-}) = t(h^\bullet) = 0.5$. The boundary between the mainly proton conducting domain and the mainly hole conducting domain will be defined by Eq. (8.16). There is no involvement of oxygen gas in this equilibrium, so the boundary will be uninfluenced by oxygen partial pressure and lies parallel to that axis, intersecting the hydrogen partial pressure axis at a value appropriate to that at which the defect concentration gives rise to the conductivity relation $t(H^+) = t(h^\bullet) = 0.5$. The boundary between the mainly oxygen conducting regime and the proton conducting regime depends upon both Eq. (8.15) and (8.16) and is given by

$$\tfrac{1}{2}O_2 + 2V_O^{2\bullet} + H_2O \longrightarrow 2OH_O^\bullet + 2h^\bullet \qquad (8.19)$$

Because this equation involves both water vapor and oxygen, it will slope at an angle to the axes. The boundary is drawn where the conductivity relation $t(H^+) = t(O^{2-}) = 0.5$ holds.

The diagram then clearly shows the situation for a particular temperature and so can be used to determine the most important conductivity mechanism for a particular operating condition at the temperature of the diagram. Similar diagrams must be constructed for other temperatures, which, when stacked up, will give a three-dimensional representation of the way in which the predominant conduction mechanism changes with temperature. Another set of diagrams will also need to be constructed for other dopant concentrations.

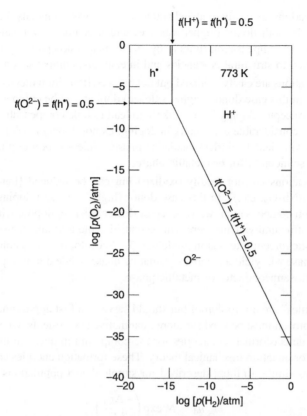

Figure 8.20 Charge carrier map representation of conductivity in the mixed conductor $BaZr_{0.8}Y_{0.2}O_{3-\delta}$ at 773 K. The domains are areas within which conductivity of one species, H^+, O^{2-}, or $h^\bullet$, predominate. Boundaries represent the locus of $t(\text{species}) = 0.5$. [Adapted from data in K. Nomura and H. Kageyama, *Solid State Ionics*, **178**, 661–665 (2007).]

8.9 CHOICE OF COMPENSATION MECHANISM

There are several ways in which a solid doped with an aliovalent impurity can maintain charge balance. It is by no means simple to be sure which compensation mechanism will hold, or even if one mechanism will hold over all of the doping range. However, there are some quantitative guidelines that apply, especially for oxides. The principal mechanism will depend upon how easily the host cation that is being replaced is oxidized or reduced.

1. Host cations are neither easily oxidized or reduced [group 1 metals (Na_2O, etc.), group 2 metals (MgO, etc.), and group 3 metals (Al_2O_3, Sc_2O_3, etc.)]. In this case both donor (higher valence) and acceptor (lower valence) doping will be compensated structurally, that is, by interstitials or vacancies.

2. Host cations are easily oxidized and reduced [transition metals (Fe_2O_3, Mn_2O_3, etc.)]. In both donor (higher valence) and acceptor (lower valence) doping electronic compensation (donors by electrons and acceptors by holes) will be preferred to structural (vacancies and interstitials) compensation.

3. Host cations are easily oxidized but not reduced [transition metals (Cr_2O_3, NiO, etc.)]. In this case donor (higher valence) doping will be compensated structurally. Acceptor doping is most likely to lead to hole compensation. Thus there will be considerable asymmetry in doping. Donor doping a colorless insulating phase will yield a colorless insulating phase, while acceptor doping will yield a black semiconductor or metallic phase.

4. Host cations are not readily oxidized but can be reduced [transition metals (TiO_2, Nb_2O_5, etc.)]. In this case donor (higher valence) doping will lead to electron compensation, while acceptor (lower valence) doping will be compensated structurally. Thus there will be considerable asymmetry in doping in the opposite direction to that in guideline (3). Acceptor doping a colorless insulating phase will yield a colorless insulating phase, while donor doping will yield a black semiconductor or metallic phase.

These "rules" are not foolproof but should serve as a first approximation. A more rigorous approach must be used for more quantitative assessments. One possibility is to compute defect formation energies for the compound in question using atomistic simulations or quantum mechanical theory. These formation energies can be inserted into formulas similar to those described for simple defect populations (Chapter 2):

$$n_d = N \exp\left(\frac{-\Delta g_d}{kT}\right)$$

where n_d is the number of defects, N is the number of atom sites supporting the defect population, and Δg_d is the (Gibbs) energy of formation of the defect. In this way, the relative numbers of the various defects may be ranked.

The variation of the defect species present as a function of dopant concentration, oxygen partial pressure, or temperature can then be determined by solution of the polynomial equations that connect the various defect populations. The Brouwer approximation may be satisfactory in many cases. As always, comparison with experimental data is essential.

8.10 ANSWERS TO INTRODUCTORY QUESTIONS

What are donor and acceptor dopants?

Aliovalent additives are often called donor dopants, when they tend to provide electrons and enhance intrinsic *n*-type semiconducting behavior, or acceptor dopants, when they tend to give a population of mobile holes and enhance *p*-type semiconducting behavior. The process of creating electronic defects in a crystal in this way is called valence induction.

Donor dopants are impurity ions of a *higher* valence than that of the parent ions, as when small amounts of Nb_2O_5 are incorporated into TiO_2, so that Nb^{5+} substitutes for Ti^{4+}. The donor species has an effective *positive* charge, $Nb^{\bullet}_{Ti}$, in this example.

Acceptor dopants are impurity ions of a *lower* valence than that of the parent ions, as when small amounts of Al_2O_3 incorporated into TiO_2, so that the Al^{3+} ion substitutes for Ti^{4+}. The acceptor species has an effective *negative* charge, Al'_{Ti}, in this example.

How do negative temperature coefficient (NTC) thermistors respond to temperature changes?

Negative temperature coefficient thermistors are made from impurity-doped transition-metal oxides. Donor doping, for example, Fe_2O_3 doped with TiO_2, produces *n*-type thermistors. The favored mechanism is the formation of electrons:

$$2TiO_2(Fe_2O_3) \longrightarrow 2Ti^{\bullet}_{Fe} + 2e' + 3O_O + \tfrac{1}{2}O_2$$

Acceptor doping, as in lithium oxide doping of nickel oxide, leads to the production of holes and produces *p*-type thermistors:

$$Li_2O(NiO) + \tfrac{1}{2}O_2 \longrightarrow 2Li'_{Ni} + 2O_O + 2h^{\bullet}$$

These materials are often described as hopping semiconductors in which the mobility is proportional to the exponential of the energy required to liberate the charge carriers so that the resistance decreases exponentially with temperature. Thermistor resistance R is frequently written as

$$R_T = R_{T0} \exp\left[\frac{\beta(T_0 - T)}{T_0 T}\right]$$

where R_T is the resistance at a temperature T (K), β is the thermistor constant, which is the slope of the R_T versus T curve, and R_{T0} is the resistance at a specified reference temperature T_0.

What are mixed conductors?

Mixed conductors are solids in which the total conductivity is made up of both electronic and ionic components, each of which makes an appreciable contribution to the overall conductivity. There is considerable interest in mixed conductors for applications such as fuel cells, batteries, electrochromic devices, and membranes for gas separation or hydrocarbon oxidation.

There are a number of ways in which mixed conductors can be obtained. In one, a material that is a good ionic conductor by virtue of structural features, typified by layer structures, can have the rest of the structure modified to become electronically conducting. In another approach, impurities can be introduced into a matrix to induce both electronic or ionic defects, so as to generate a mixed conducting matrix.

An example of a layer structure mixed conductor is provided by the cathode material Li_xCoO_2 used in lithium batteries. In this solid the ionic conductivity component is due to the migration of Li^+ ions between sheets of electronically conducting CoO_2. The production of a successful mixed conductor by doping can be illustrated by the oxide $Ce_{1-x}Pr_xO_2$. Reduction of this solid produces oxygen vacancies and Pr^{3+} ions. The electronic conductivity mechanism in these oxides is believed to be by way of electron hopping between Pr^{4+} and Pr^{3+}, and the ionic conductivity is essentially vacancy diffusion of O^{2-} ions.

PROBLEMS AND EXERCISES

Quick Quiz

1. Acceptor dopants are NOT compensated by one of:
    (a) Interstitial cations
    (b) Anion vacancies
    (c) Electrons

2. The substitution of Mn^{3+} by Al^{3+} in $MgMn_2O_4$ is:
    (a) Donor doping
    (b) Acceptor doping
    (c) Neither donor or acceptor doping

3. When a donor or acceptor is doped into an oxide, changes must be made to:
    (a) The electroneutrality equation
    (b) The electronic equilibrium equation
    (c) The point defect equilibrium equation

4. Each Sr^{2+} ion incorporated into La_2CuO_4 generates:
    (a) One electron
    (b) One hole
    (c) One Cu^{2+} vacancy

5. To make the metallic oxide $LiTi_2O_4$ insulating, dope with:
    (a) Li^+
    (b) Mg^{2+}
    (c) Mn^{3+}

6. The charge reservoir layer in the cuprate superconductor $HgBa_2CaCu_2O_{6+\delta}$ is:
    (a) BaO Ca BaO
    (b) BaO Hg BaO
    (c) BaO Cu BaO

7. The main defects responsible for mixed conductivity in $Ce_{0.8}Pr_{0.2}O_{2-\delta}$ are:

(a) $Pr_{Ce}' + Ce_{Ce}'$

(b) $V_O^{2\bullet} + Ce_{Ce}'$

(c) $V_O^{2\bullet} + Pr_{Ce}'$

8. The layered material Li_xNiO_2 is of use in a battery as the:

(a) Anode

(b) Cathode

(c) Electrolyte

9. In dry air the proton mixed conductor $SrZr_{1-x}Y_xO_3$ is a:

(a) p-type semiconductor

(b) n-type semiconductor

(c) Insulator

10. Which of the following dopants will change colorless insulating TiO_2 into a blue-black semiconductor?

(a) Nb^{5+}

(b) Hf^{4+}

(c) Li^+

Calculations and Questions

1. Write defect equations for (a) TiO_2 doped with Al_2O_3; (b) TiO_2 doped with Nb_2O_5; (c) Mn_3O_4 doped with NiO; (d) MgO doped with Cr_2O_3; and (e) Nb_2O_5 doped with Fe_2O_3.

2. A plot of the conductivity of the thermistor material $MgGaMnO_4$ is given in the following table. Estimate the value of the thermistor constant β.

Conductivity/S m^{-1}	Temperature/1000 K
8.21	1
0.674	1.35
0.0184	2.0
3.04×10^{-3}	2.23
1.12×10^{-3}	2.5
3.06×10^{-5}	3.25

Data adapted from A. Veres et al., Solid State Ionics, **178**, 423–428 (2007).

3. Write down the explicit polynomial form of Eqs. (8.13) and (8.14) for the case where Schottky defects are more important than electronic defects, $K_S = 10^{40}$ defects m^{-3}, $K_e = 10^{32}$ defects m^{-3}, $D_M = 10^{25}$ donors m^{-3}, $A_M = 10^{25}$ accceptors m^{-3}. (Determine values of K_o and K_r from Section 7.5.2.)

4. The normal spinel $MgTi_2O_4$ is a black semiconductor with a high conductivity. (a) Write defect equations to express the results of doping with Mg^{2+} into Ti sites. (b) What is the limiting composition? (c) Is the final spinel a metal or an insulator?

5. Determine the charge on the charge reservoir and the superconducting layers for (a) $Bi_2Sr_2CaCu_2O_8$, (b) $HgBa_2Ca_2Cu_3O_8$, (c) $TlBa_2Ca_2Cu_3O_9$, (d) $Tl_2Sr_2CaCu_2O_8$, and (e) $Pb_2Sr_2Y_{0.5}Ca_{0.5}Cu_3O_8$.

6. $LaCoO_3$ can be donor doped with CeO_2 to give a mixed conductor. (a) Write defect equations for the reaction. A sample of nominal composition $La_{0.99}Ce_{0.01}CoO_3$ has a Seebeck coefficient of approximately $-300\ \mu VK^{-1}$. (b) What are the charge carriers? (c) What is the ratio of defects to normal Co sites?

7. The dependence of the conductivity of the mixed conductor $La_2Ni_{0.8}Cu_{0.2}O_{4+\delta}$ at 750°C is given in the following table. Explain the oxygen dependence and determine the nature of the charge carriers.

Conductivity/S m^{-1}	Oxygen Pressure/Pa
83.18	3.16×10^{-10}
104.71	1.41×10^{-9}
144.54	7.94×10^{-9}
190.55	2.82×10^{-8}
263.03	1.26×10^{-7}
331.13	0.56×10^{-6}

Data adapted from V. V. Kharton et al., J. Solid State Chem., **177**, 26–37 (2004).

8. Write out the specific equations needed to draw Figure 8.19, using the approximation that $[h^\bullet] = [Yb'_{Pr}]$ (see Section 8.8.2).

9. V_2O_3 is doped with NiO to give $Ni_xV_{2-x}O_3$, which is a hopping semiconductor. Assuming that Ni^{2+} substitutes for V^{3+} and is compensated by V^{4+}; (a) write equations for the doping of V_2O_3 by NiO and the overall equation showing compensation of Ni^{2+} by V^{4+}. (b) Calculate the activation energy for hopping from the data in the following table. (c) At what composition will the conductivity be a maximum, if hopping is between V^{3+} and V^{4+}?

Temperature/K	Conductivity/S m^{-1}
300	70.92
350	94.31
400	116.41
450	134.41
500	149.93

Data adapted from P. Rozier, A. Ratuszna, and J. Galy, Z. Anorg. Allgem. Chem., **628**, 1236–1242 (2002).

10. The figure shows data for the total conductivity of a mixed electronic/proton conductor $La_{0.95}Ca_{0.05}WO_{12}$ as a function of partial pressure in a wet atmosphere. The material is prepared by doping La_6WO_{12} with Ca. [Data adapted from R. Haugsrud, *Solid State Ionics*, **178**, 555–560 (2007).] **(a)** What type of conductivity dominates in each of the three oxygen partial pressure ranges? **(b)** Write a defect equation for the consequence of Ca^{2+} substitution on La^{3+} sites. **(c)** Write a defect equation explaining the origin of proton conductivity in wet atmospheres in the doped phase. **(d)** Explain the slopes of the lines in the oxygen-poor and oxygen-rich regions.

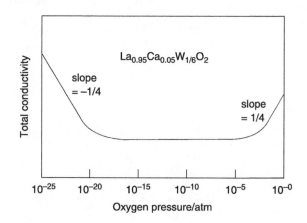

FURTHER READING

Background material on Pourbaix diagrams, electronic properties, and superconductivity is given in:

R. J. D. Tilley, *Understanding Solids*, Wiley, Chichester, 2004.

Much of the defect chemistry is discussed in:

D. M. Smyth, *The Defect Chemistry of Metal Oxides*, Oxford University Press, Oxford, United Kingdom, 2000.

Information on polynomial forms of Brouwer diagrams for Cr_2O_3 is in:

P. Karen, *J. Solid State Chem.*, **179**, 3167–3183 (2006).

The relationship between electronic conductivity, chemical bonding, and structure in oxides is given by:

P. A. Cox, *Transition Metal Oxides*, Oxford University Press, Oxford, United Kingdom, 1992.

P. A. Cox, *The Electronic Structure and Chemistry of Solids*, Oxford University Press, Oxford, United Kingdom, 1987.

Metal-insulator transitions are treated in:

N. F. Mott, *Metal-Insulator Transitions*, Taylor and Francis, London, 1974.

Reviews of superconducting oxides with emphasis on structures are:

E. Takayama-Muromachi, *Chem. Mater.*, **10**, 2686–2698 (1998).

S. Adachi, T. Tatsuki, T. Tamura, and K. Tanabe, *Mater. Chem.*, **10**, 2860–2869 (1998).

P. Majewski, *J. Mater. Res.*, **15**, 854–870 (2000).

Mixed conductivity is discussed by:

J. W. Fergus, *Solid State Ionics*, **177**, 1529–1541 (2006).

M. S. Whittingham, *Mat. Res. Bull.*, **XIV**, September, 31–38 (1989).

Lanthanum colbaltes are reviewed in:

A. N. Petrov, V. A. Cherapanov, and A. Yu. Zuev, *J. Solid State Electrochem.*, **10**, 517–537 (2000).

CHAPTER 9

Magnetic and Optical Defects

What is spintronics?
What is colossal magnetoresistance?
What defects commonly color gemstones?

The relationships between both magnetic and optical properties and defects are complex, and each is deserving of a complete volume. Here a brief survey is given of some aspects of these defects, particularly where they connect with previous discussions. In the case of magnetic defects, it is not possible to view the defects without some brief acquaintance with magnetism and magnetic structures (see Supplementary Material S6). Emphasis is on defects rather than the innumerable current applications, particularly in data storage. In the case of optical defects, most emphasis is placed upon defects that interact with light and produce changes that lie in the approximate wavelength range 400–700 nm (the visible region of the electromagnetic spectrum). As with magnetic materials, space does not allow for extensive descriptions of devices, and discussion will be restricted to the optical defects themselves.

9.1 MAGNETIC DEFECTS

Magnetic properties reside mainly in the electrons surrounding an atom. Each electron has a magnetic moment due to the existence of a magnetic dipole, which can be thought of as a minute bar magnet linked to the electron. Although magnetic effects are associated with some bond configurations in molecules and especially with free radicals or bonds that have been ruptured, most stable magnetic effects are associated with metal atoms or ions. There are two contributions to the magnetic dipole moment of an electron bound to an atomic nucleus, which, in semiclassical models, are attributed to orbital motion and spin. The orbital and spin components of all the electrons on an atom or ion are linked, or coupled, to give an overall magnetic dipole moment for the atom or ion. Atoms or ions with closed shells are found

Defects in Solids, by Richard J. D. Tilley
Copyright © 2008 John Wiley & Sons, Inc.

to have no magnetic moment. The only atoms that display a magnetic moment are those with incompletely filled shells. These are particularly found in the transition metals, with partly filled d shells, and the lanthanides and actinides, which have partly filled f shells. The $3d$ transition-metal ions often have magnetic dipole moments corresponding only to the electron spin contribution, and in these compounds the orbital moment is said to be quenched. In such materials, the magnetic moment of an ion is given by the "spin-only" formula:

$$\mathbf{m} = [n(n + 2)]^{1/2} \mu_B$$

where n is the number of unpaired d electrons on each ion. The magnetic moments are usually expressed in units of μ_B, Bohr magnetons.

Paramagnetic solids are those in which some of the atoms, ions, or molecules making up the solid possess a permanent magnetic dipole moment. These dipoles are isolated from one another. The solid, in effect, contains small, noninteracting atomic magnets. In the absence of a magnetic field, these are arranged at random and the solid shows no net magnetic moment. In a magnetic field, the elementary dipoles will attempt to orient themselves parallel to the magnetic induction in the solid, and this will enhance the internal field within the solid and give rise to the observed paramagnetic effect.

The partial orientation of the elementary dipoles in a paramagnetic solid is counteracted by thermal agitation, and it would be expected that at high temperatures the random motion of the atoms in the solid would cancel the alignment due to the magnetic field. The paramagnetic susceptibility would therefore be expected to vary with temperature. The temperature dependence is given by the Curie law:

$$\chi = \frac{C}{T}$$

where χ is the magnetic susceptibility, T is the absolute temperature, and C is the Curie constant. Curie law dependence in a solid is indicative of the presence of isolated paramagnetic ions or atoms in the material.

Frequently the magnetic behavior of a paramagnetic solid does not follow the Curie law exactly but is rather better fitted by the Curie–Weiss law:

$$\chi = \frac{C}{T - \theta}$$

The Curie–Weiss constant (or Weiss constant), θ, is positive and has the dimensions of temperature. The Curie–Weiss constant is a measure of the departure of the system from the ideal behavior represented by the Curie law.

Magnetic defects can be thought of as defects in the magnetic dipole state of an atom or group of atoms compared to those of the parent structure. For instance, magnetic defects can be considered to form when magnetic ions are introduced into a nonmagnetic structure, either as substituents or as interstitials (Fig. 9.1a and 9.1b). These

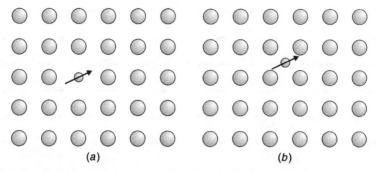

Figure 9.1 Schematic representation of magnetic defects in a nonmagnetic matrix: (*a*) impurity substituent and (*b*) impurity interstitial.

Figure 9.2 Schematic representation of magnetic defects in a magnetic matrix: (*a*) vacancy or nonmagnetic impurity, (*b*) self-interstitial, (*c*) magnetic "Frenkel defect", (*d*) magnetic foreign substituent, and (*e*) magnetic foreign interstitial. The magnetic matrix can have ordered (as drawn) or disordered spins.

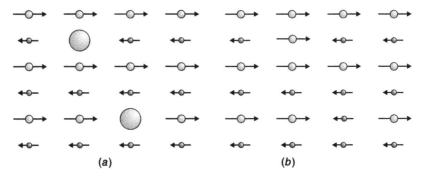

(a)   (b)

Figure 9.3 Schematic representation of magnetic defects in a ferrimagnetic matrix: (a) a pair of vacancies or nonmagnetic impurities similar to a Schottky defect and (b) an antisite defect.

materials would be expected to obey the Curie law. However, magnetic defects also form when there are mistakes in an array of magnetic ions, whether ordered or not, which are then analogous to point defects (Fig. 9.2). Such defects occur when an ion of one spin, say Ni^{3+}, occurs in a matrix of Ni^{2+} or when an ion such as Mn^{3+} can exist in several spin states. Ferrimagnetic materials have two arrays of spins aligned in an antiparallel fashion, and in these materials magnetic defects analogous to Schottky defects and antisite defects can occur (Fig. 9.3) as well as more complex arrangements. In addition, there are extended defects in magnetic structures that can be considered to be equivalent to dislocations, planar faults, or volume defects. The best known of these are the domain boundaries that separate Weiss domains that occur in ferromagnetic solids (Fig. 9.4).

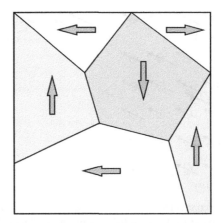

Figure 9.4 Schematic representation of Weiss domains in a ferromagnetic crystal. The arrows represent the direction of alignment of the magnetic moments in each domain.

9.2 MAGNETIC DEFECTS IN SEMICONDUCTORS

Traditionally, semiconductor devices utilize the charge on electrons and ignore the spin component. However, the addition of spin to the observable properties of electrons in semiconductors has excited considerable interest for data storage and computing. The area in which both conductivity and spin are exploited simultaneously is called spintronics. Semiconductors that have a measurable spin or magnetic component added to them are called diluted magnetic semiconductors. The first materials to attempt to exploit this combination were semiconductors such as ZnSe or GaAs, in which some of the nonmagnetic ions are replaced by magnetic ions, such as Mn^{2+}. This ion has a $3d^5$ electron configuration and potentially a magnetic moment due to five unpaired electron spins. The magnetic ions form substitution impurity defects, for example, in the case of Mn^{2+} doping, nominally Mn_{Zn} in ZnSe or Mn'_{Ga} in GaAs. (Antisite defects complicate the real defect structure in some cases.)

The magnetic ions can interact in the same way as they would in any other solid. Thus at low concentrations, the material is often paramagnetic, but higher concentrations lead to ferromagnetic or antiferromagnetic structures. However, the magnetic properties do not reside only in the magnetic impurities. There is considerable evidence that the magnetic properties are closely connected with other defects in the solid. For instance, doping of vanadium into bulk Sb_2Te_3 gives rise to a low-temperature ferromagnetic phase with a Curie temperature of about 20 K, while the Curie temperature of thin films is close to 200 K. The defect structure of the thin films are different than that of the bulk, but the complex relationship between magnetic and nonmagnetic defects has yet to be clarified.

Oxide films are being intensively studied as diluted magnetic semiconductors because doping with magnetic cations can produce materials that combine transparency, semiconductivity, and room temperature ferromagnetism. The first oxide to show ferromagnetic behavior above room temperature was ZnO doped with Mn^{4+} to form $Zn_{1-x}Mn_xO$ with x taking values below 0.02. The dopant substitutes for Zn^{2+}. In a bulk sample these would be written $Mn_{Zn}^{2\bullet}$ defects, but there is evidence that many of the impurity cations may favor surface rather than bulk sites. The properties of these solids also depend upon preparation conditions. When they are prepared at moderate temperatures, the Mn^{4+} ions are distributed at random, but they cluster into antiferromagnetic units if the materials are heated at 700°C or above, and in this case ferromagnetism is lost.

Despite results such as these, there is still much uncertainty about the defects in oxide films and the origin of the ferromagnetic properties. In part this is because the magnetic properties depend upon the nature of the substrate, the relative oxidizing or reducing conditions employed while the films are being prepared, and the degree of crystallinity of the films. The matter is made more complex by the fact that some oxide films, typically TiO_2 and HfO_2, can be prepared in a ferromagnetic state without transition-metal ion impurities.

There are a number of defects that could give rise to magnetic properties alone or interact with magnetic impurities to enhance magnetic behavior. In ZnO-based films such as $Zn_{1-x}Cu_xO$, ferromagnetism is believed to arise with $Zn_i^{2\prime}$ or Zn_i' interstitials.

In films such as SnO_2, oxygen vacancies together with liberated electrons are implicated in the ferromagnetic properties.

The interaction between a charged point defect and neighboring magnetic ions in
magnetically doped thin films has been described in terms of a defect cluster called a
bound magnetic polaron (Fig. 9.5a). The radius of a bound magnetic polaron due to
an electron located on the defect, r, is given by

$$r = \varepsilon_r \left(\frac{m}{m_{eff}} \right)$$

where ε_r is the relative permittivity (dielectric constant) of the solid, m is the electron mass,
m_{eff} the effective mass of the electron and the units are Bohr radius (0.0529 nm). The

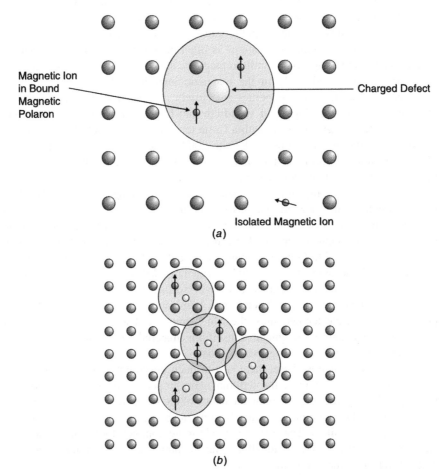

Figure 9.5 Schematic representation of a bound magnetic polaron: (a) one bound magnetic
polaron located on a charged defect; and (b) overlapping bound magnetic polarons leading to
ferromagnetic alignment of magnetic ions.

bound magnetic polaron then has a volume of r^3. The charged point defect acts to align the spins on magnetic ions that lie within the polaron volume. If there is an overlap of bound magnetic polarons, then ferromagnetic order can occur throughout the sample (Fig. 9.5b).

9.3 MAGNETIC DEFECTS IN FERRITES

The magnetically important ferrites are oxides that adopt the spinel structure and have a general formula $A^{2+}Fe_2^{3+}O_4$. The cations are distributed between two types of site, tetrahedral and octahedral (Supplementary Material S1). The ferrite $ZnFe_2O_4$, a normal spinel in which the Zn^{2+} ions occupy the tetrahedral sites and the Fe^{3+} ions the octahedral sites, is written $Zn^{2+}[Fe^{3+}]_2O_4$, where the ions in octahedral sites are enclosed in square brackets. The magnetic moments of the individual Fe^{3+} are ions opposed even though they both occupy the octahedral sites, and the formula can be written $Zn^{2+}[Fe^{3+}\uparrow Fe^{3+}\downarrow]O_4$ where the arrows symbolize the directions of the magnetic moments on the ions. Because the magnetic moments are opposed, the material is effectively nonmagnetic. In reality, the normal spinel structure is not perfect and the cation distribution between the tetrahedral and octahedral sites is temperature dependent, leading to magnetic defects in which some Fe^{3+} replaces the Zn^{2+}.

Many important ferrites crystallize with the inverse spinel structure, written as $Fe^{3+}[A^{2+}Fe^{3+}]O_4$, where the octahedrally coordinated ions are in square brackets. In these materials, the magnetic moment of the Fe^{3+} ions in the tetrahedral sites is opposed to that of the Fe^{3+} ions in the octahedral sites, so that the net magnetic moment due to Fe^{3+} is zero. The first magnetic material known, and for many centuries the only magnetic material known, was the magnetic oxide Fe_3O_4, which is also called *lodestone*[1] or *magnetite*. The oxide Fe_3O_4 is an inverse spinel, although this fact is obscured in the formula by the fact that the A and B cations are both Fe. The magnetic arrangement is given by the formula $Fe^{3+}\uparrow [Fe^{3+}\downarrow Fe^{2+}\downarrow]O_4$, which shows that the overall magnetic moment is solely due to the Fe^{2+} contribution. Similarly, nickel ferrite can be written as $Fe^{3+}\uparrow[Fe^{3+}\downarrow Ni^{2+}\uparrow]O_4$, and, because the magnetic moments on the Fe^{3+} ions cancel, the net magnetic moment is due to the Ni^{2+} ions alone.

By forming solid solutions between two ferrites, it is possible to tune the properties magnetically. An example is provided by the solid solution $Zn_xNi_{1-x}Fe_2O_4$, which is between the normal spinel $ZnFe_2O_4$ and the inverse spinel $NiFe_2O_4$. Reaction to form a solid solution with $NiFe_2O_4$:

$$(1-x)NiFe_2O_4 + xZnFe_2O_4 \longrightarrow Zn_xNi_{1-x}Fe_2O_4$$

introduces substitutional defects. However, although the Ni^{2+} ions substitute for Zn^{2+}, they do not occupy the tetrahedral sites but the octahedral sites. This has

---

[1]The name lodestone is a corruption of the expression *leading stone*, which reflects on the fact that one of its earliest and most important uses was as a compass pointer.

the effect of forcing some Fe^{3+} into tetrahedral sites to give a formula $Zn_x^{2+}Fe_{1-x}^{3+}[Ni_{1-x}^{2+}Fe_{1+x}^{3+}]O_4$. The magnetic properties of the resulting material, which rises to a maximum when x is near to 0.5, depends upon whether the $Ni^{2+}\uparrow$ replaces octahedral $Fe^{3+}\uparrow$ or $Fe^{3+}\downarrow$. The resulting phases can be written $Zn_x^{2+}Fe_{1-x}^{3+}\uparrow[Fe^{3+}\downarrow\ Ni_{1-x}^{2+}\uparrow Fe_x^{3+}\uparrow]O_4$ or $Zn_x^{2+}Fe_{1-x}^{3+}\uparrow[Fe_x^{3+}\downarrow\ Ni_{1-x}^{2+}\uparrow Fe^{3+}\uparrow]O_4$. Experimentally, it is found that the occupation of the tetrahedral and octahedral sites varies with temperature, and a combination of both arrangements prevails, so that careful processing is needed to ensure reproducible magnetic properties. Nickel–zinc ferrites were at one time used in recording heads for audio and video recorders.

9.4 CHARGE AND SPIN STATES IN COBALTITES AND MANGANITES

Cobaltites and manganitites, especially those related to $LaCoO_3$ and $LaMnO_3$, have been studied in detail because of their important electrical and magnetic properties. Both adopt the cubic perovskite structure, with nominal formulas $La^{3+}Co^{3+}O_3$ and $La^{3+}Mn^{3+}O_3$. Under high oxygen pressures both, but especially $LaMnO_3$, can take on more oxygen to give, for example, $LaMnO_{3+\delta}$, while under low oxygen pressures they can lose oxygen to form, for example, $LaMnO_{3-\delta}$. These changes are due to the existence of the ions Co^{2+}, Co^{3+}, and Co^{4+} in the cobaltites and Mn^{2+}, Mn^{3+}, and Mn^{4+} in the manganites. Because these ions have different spin states they can give rise to a variety of magnetic defects.

Undoped $LaCoO_3$ is a nonmagnetic semiconductor at ordinary temperatures. The transition-metal cations occupy octahedral sites in the perovskite structure and the five $3d$-orbital energy levels split into two groups, a higher energy e_g pair and three t_{2g} lower energy levels. At lowest temperatures the Co^{3+} is in the low-spin (LS) state, with an electron configuration t_{2g}^6 (Fig. 9.6a). However, the energy

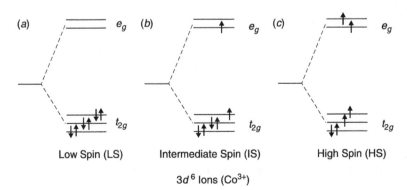

Figure 9.6 Electron configurations possible for Co^{3+} ($3d^6$) cations in an octahedral crystal field: (a) low spin (LS), (b) intermediate spin (IS), and (c) high spin (HS).

difference between the two sets of energy levels is small, and, as the temperature increases, a proportion of electrons from the ground t_{2g} state are promoted into the upper e_g state. The configuration of these ions is $t_{2g}^5 e_g^1$, the intermediate-spin (IS) state (Fig. 9.6b) or $t_{2g}^4 e_g^2$, the high-spin (HS) state (Fig. 9.6c). There is evidence that these two spin states are ordered within the Co atom structure, especially when there are equal numbers of the two states present. In the temperature range from 350 to 650 K the materials smoothly transform from a semi-conductor to a metal. At the same time the Co^{3+} ions transform to the high-spin state, so that at high temperatures in the metallic phase the configuration is 50% HS and 50% IS.

The magnetic structure becomes more complex when doped materials are considered. Acceptor doping of $LaCoO_3$ by incorporation of an alkaline earth cation, Ca, Sr, Ba, in place of La, as in $La_{1-x}Sr_xCoO_3$ now introduces holes into the system. These are generally located on the Co^{3+} ions to form Co^{4+}. The Co^{3+} ions are thought to be mainly in the IS state and Co^{4+} in the HS state, with an electron configuration for this $3d^5$ ion of $t_{2g}^3 e_g^2$. Further studies are needed to completely resolve the spin and charge distribution.

Interest in $LaMnO_3$ intensified when it was discovered that doped forms of this material show colossal magnetoresistance. The magnetoresistance of a solid is the change of resistance when a magnetic field is applied. The magnetoresitive (MR) ratio is defined as the change in resistance when a magnetic field is applied to that in zero field:

$$\text{MR ratio} = \frac{R_H - R_0}{R_0} = \frac{\Delta R}{R_0}$$

where R_H is the resistance in a magnetic field and R_0 is the resistance in the absence of the field. For the doped manganites the magnetoresistive ratio is close to 100% and is negative, meaning that the resistance falls to almost nothing in a magnetic field. In such materials, a better measure of the change is the ratio of R_0/R_H, where values of up to 10^{11} have been recorded. This extreme behavior is called colossal magneto-resistance (CMR).

The CMR effect is found in a variety of doped systems, including hole-doped $La_{1-x}A_xMnO_3$, with A being an alkaline earth cation (Ca, Sr, Ba) and ordered $Ln_{0.5}A_{0.5}MnO_3$, with Ln being a lanthanide, typically Pr. In all of these compounds, doping introduces holes that are located on Mn^{3+} ions to form Mn^{4+}. In addition, disproportion of $2Mn^{3+}$ into Mn^{2+} and Mn^{4+} may also occur, so that several spin states are found in the matrix. It is believed that the Mn^{3+} ions are in the high-spin $t_{2g}^3 e_g^1$ state as are the Mn^{4+} ions, with an electron configuration $t_{2g}^3 e_g^0$. The Mn^{2+} ions can adopt a low-spin $t_{2g}^5 e_g^0$ configuration, an intermediate-spin $t_{2g}^4 e_g^1$ configuration, or a high-spin $t_{2g}^3 e_g^2$ configuration (Fig. 9.7). The distribution between these states is temperature sensitive, and at room temperature the intermediate-spin state appears to dominate. The Mn defects in $La_{1-x}A_xMnO_3$, and $Ln_{0.5}A_{0.5}MnO_3$ can then be written as Mn'_{Mn} (Mn^{2+}) IS, Mn_{Mn} (Mn^{3+}) HS, and $Mn^{\bullet}_{Mn}$ (Mn^{4+}) HS.

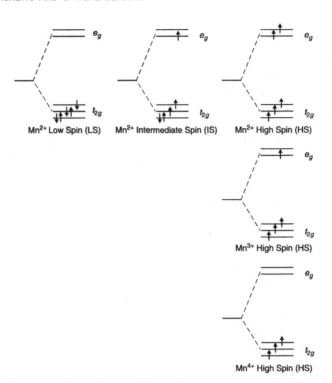

Figure 9.7 Electron configurations possible for Mn cations in an octahedral crystal field: low spin (LS), intermediate spin (IS), and high spin (HS).

Substitution of some of the Mn with both magnetic or nonmagnetic ions, including Ti^{4+}, Sn^{4+}, Fe^{3+}, Cr^{3+}, Al^{3+}, Ga^{3+}, In^{3+}, Mg^{2+}, and Ni^{2+}, increases the CMR effect considerably. For $Pr_{0.7}Ca_{0.2}Sr_{0.1}MnO_3$ the ratio R_0/R_H is approximately 230. This jumps to 4×10^5 for the modestly doped $Pr_{0.7}Ca_{0.2}Sr_{0.1}Mn_{0.98}Mg_{0.02}O_3$.

Neither the magnetic defect structures of these phases nor the mechanism of the CMR effect are yet clear, although Mn^{3+}–Mn^{4+} interactions are believed to be important in inducing CMR.

The differing valence states in cobaltites and manganites can be balanced by oxygen vacancy formation. In some materials, these vacancies are assimilated into the structure, leading to charge and spin-state ordering. This occurs in the phase $Sr_2Mn_2O_5$, derived from the cubic perovskite structure parent compound $SrMn^{4+}O_3$ in which the Mn^{4+} cations occupy octahedra (Fig. 9.8a). In air the Mn^{4+} state is not stable and is replaced by Mn^{3+} to form $Sr_2Mn_2^{3+}O_5$. The structure consists of corner-linked square pyramids (Fig. 9.8b). The magnetic moments on the Mn^{3+} ions are arranged antiferromagnetically in sheets parallel to the **c** axis.

Intermediate phases between these two extremes can be formed. The most stable of these are $Sr_5Mn_5O_{13}$ and $Sr_7Mn_7O_{19}$. In both of these phases the structure consists of both square pyramids, containing Mn^{3+} and octahedra containing Mn^{4+} (Fig. 9.8c and 9.8d). These structures exhibit charge ordering, but the magnetic and spin states of the Mn ions has yet to be resolved.

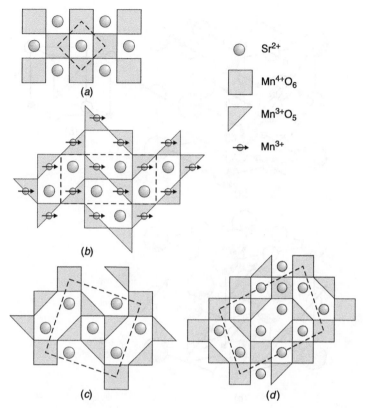

Figure 9.8 Charge-ordered structures (idealized) of $SrMnO_x$: (*a*) cubic perovskite structure $SrMnO_3$, (*b*) $Sr_2Mn_2O_5$, (*c*) $Sr_5Mn_5O_{13}$, and (*d*) $Sr_7Mn_7O_{19}$. The shaded squares represent $[Mn^{4+}O_6]$ octahedra, the shaded triangles $[Mn^{3+}O_5]$ square pyramids. In (b) Mn^{3+} ions have magnetic moments (arrows) aligned antiferromagnetically in layers perpendicular to the page.

9.5 EXTENDED MAGNETIC DEFECTS

Magnetic interactions tend to be long range and extended defects occur frequently in magnetic materials. Nonstoichiometric wüstite, ~FeO, provides an example. This oxide accommodates composition variation by a variable population of defect clusters made up of ordered oxygen vacancies and "interstitial" cations (Fig. 4.6). Like many of the 3*d* transition-metal monoxides, ~FeO has the magnetic moments on the Fe^{2+} ions antiferromagnetically ordered below 183 K with the magnetic moments pointing perpendicular to the (111) planes, that is, along [111] (Fig. 9.9*a*). However, the magnetic moments of the interstitial Fe ions in the defect clusters that allow for composition variation are arranged differently. Although the interstitial cation central to a cluster (Fig. 9.9*b*) is coupled antiferromagnetically to the surrounding cations, the magnetic moments lie in the (111) plane (Fig. 9.9*c*). As the composition of the sample varies, the concentration of the magnetic clusters in the antiferromagnetic matrix varies, giving rise to a change in overall magnetic behavior.

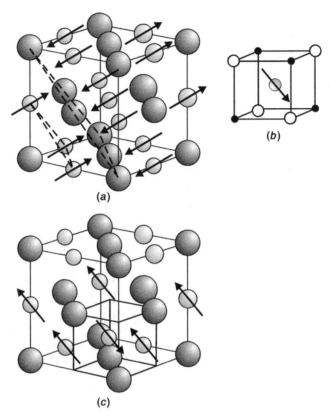

Figure 9.9 Magnetic defects in $\sim$FeO: (*a*) antiferromagnetic alignment of magnetic moments in nominally stoichiometric FeO with the spins perpendicular to [111]; (*b*) the simplest defect cluster in $\sim$FeO, with the spin on the interstitial Fe lying in (111); and (*c*) antiferromagnetic coupling of the surrounding Fe ions with all spins lying in (111).

The potential cathode material Li_xNiO_2 (Section 8.7.2) also shows evidence of magnetic clusters. Under normal preparation conditions, the material is always nickel rich, with a proportion of Ni^{2+} ions occupying the Li^+ layers to give $(Li_{1-y}Ni_y^{2+})(Ni_y^{2+}Ni_{1-y}^{3+})O_2$. Magnetic measurements show that the Ni^{2+} ions located in the Li^+ layers interact to form magnetic clusters. The material is ferrimagnetic at low temperatures. This can be altered by doping. The addition of small amounts of Mg^{2+} results in compositions in which the formation of Ni^{2+} is suppressed and the Mg^{2+} ions preferentially occupy the Li^+ layers, eliminating the ferrimagnetic behavior. For a different reason, doping with Co^{3+} also eliminates cluster formation by stabilizing the Ni^{3+} ions to such an extent that Ni^{2+} ions are not present in sufficient amounts to cluster in the Li^+ layers.

The magnetic coupling between ions in ferromagnetic and antiferromagnetic arrays has important consequences for physical properties. This ordering can be disrupted by the introduction of nonmagnetic species into the array, greatly modifying the physical properties of the solid. For example, in the cuprate La_2CuO_4 the magnetic moments on the Cu^{2+} ions form an antiferromagnetic array (Fig. 9.10a and 9.10b). The presence of holes (or Cu^{3+} ions) due to Sr^{2+} doping in $La_{2-x}Sr_xCuO_4$ together with hole hopping from one Cu to another (Fig. 9.10c) rapidly leads to a destruction of the antiferromagnetic state and the onset of superconductivity.

Extended magnetic defects in magnetic arrays were, at one stage, seriously considered for memory devices. These were based upon thin films of magnetic garnets. The garnet structure is adopted by a number of silicates with composition $A_3B_2Si_3O_{12}$, where A^{2+} are large cations that occupy 8-coordinate (approximately cubic) sites, B^{3+} cations are smaller and occupy octahedral sites, and Si^{4+} occupies tetrahedral sites. There are many combinations of cations that can be introduced into the

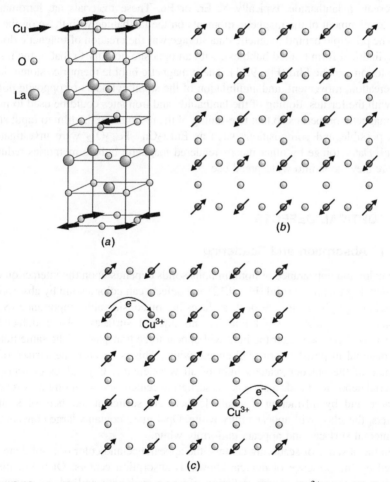

Figure 9.10 Antiferromagnetic ordering of magnetic moments on Cu^{2+} ions: (a) the idealized structure of La_2CuO_4 showing magnetic moments on Cu^{2+}; (b) the ordering in one CuO_2 sheet; and (c) introduction of holes (Cu^{3+}) and electron hopping destroys the antiferromagnetic array.

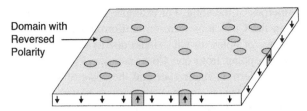

Domain with Reversed Polarity

Figure 9.11   Magnetic bubble memory schematic: domains of reversed polarity (bubbles) are induced in the thin ferromagnetic garnet film.

garnet structure; the one chosen for magnetic memories being $Ln_3Fe_5O_{12}$, where Ln represents a lanthanide, typically Y, Er, or Eu. These materials are ferromagnetic due to alignment of the magnetic moments on the lanthanide with those on the Fe.

The principle behind magnetic data storage was the creation of magnetic domains of cylindrical form (called bubbles), with an opposite polarity to that of the rest of a thin ferromagnetic film (Fig. 9.11). These magnetic bubble memories stored data by the creation, movement, and annihilation of the small cylinders of opposite polarity. As with the ferrites, doping of the lanthanide and iron sites could be used to modify the magnetic properties so that fine-tuning of the response of the film to input signals was possible and compositions such as $EuEr_2Ga_{0.7}Fe_{4.3}O_{12}$ were investigated in detail. Data storage by other means rendered magnetic bubble memories redundant before they came into widespread use.

9.6   OPTICAL DEFECTS

9.6.1   Absorption and Scattering

The color and appearance of nonluminous solids depends upon the interaction of the incident light with the solid (Fig. 9.12) and defects can color a solid by absorption or emission of light. In this respect, surfaces are of considerable importance (Section 3.14.1). An insulating solid that has no internal surfaces (planar defects) that reflect or otherwise scatter the light will appear to be transparent. The same transparent material in powder form will appear white. This is because the surface of each granule of the powder scatters light of all wavelengths. Crystallites of transparent material deposited inside a glass act as scattering centers and scatter light by reflection if large and by diffraction if small. If there are sufficient numbers of scattering centers, the glass will appear milky white. Opal glass contains large concentrations of internal surfaces and appears uniformly white.

In the absence of scattering centers, the appearance and color of a solid are influenced by the presence of defects known as absorption centers. Of these, the best known are those that absorb radiation of one wavelength and release radiation of another wavelength. Fluorescence refers to the instantaneous emission of radiation after excitation by some external radiation. The radiation emitted has a longer wavelength (lower energy) than the exciting radiation. The "missing energy" is lost via nonradiative transitions (also called photon-assisted transitions or, loosely speaking,

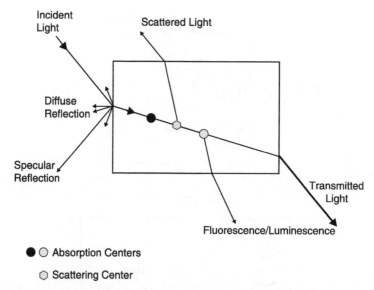

Figure 9.12 Interaction of light with a solid. Some absorption centers can re-emit light as fluorescence or luminescence. Scattering centers can reflect or diffract light depending upon size.

heat), which increase lattice vibrations, resulting in a warming of the fluorescent material. Phosphorescence is the delayed emission of radiation but is otherwise identical to fluorescence.

Phosphors are materials that convert radiation of one wavelength to radiation of another wavelength, the conversion being from a higher energy to a lower energy. The processes taking place in a phosphor frequently involve defect species. The (usually inactive) matrix has a defect incorporated within it termed an *activator, A*, which is able to absorb energy and reemit it as a photon of lower energy (Fig. 9.13*a*). Sometimes the activator *A* cannot absorb the photons directly, in which case a second defect center, a *sensitisor S*, is added as well. In this case, the sensitisor absorbs the exciting radiation and passes the energy to the activator *A* for conversion and emission (Fig. 9.13*b*). Phosphors are widely used in, for example, fluorescent lamps (ultraviolet to visible), cathode ray tube (CRT) displays (electron impact to visible), and scintillators (X rays and γ rays to visible).

Phosphors convert a high-energy input into a lower energy input. The reverse process in which low-energy radiation is converted into higher energy output is known as frequency up-conversion, anti-Stokes fluorescence, or cooperative luminescence. The majority of studies of up-conversion have dealt with the behavior of polycrystalline or glass phosphors containing impurity defects in the form of a pair of lanthanide ions, such as Yb^{3+}/Er^{3+} (Section 9.8.2).

9.6.2 Energy Levels

The interaction of light with an absorption center involves a change in energy. The absorption of energy by an absorption center will cause it to move from a

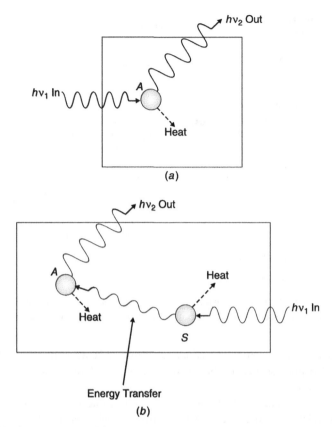

Figure 9.13   Fluorescence and luminescence: (*a*) an activator, *A*, absorbs incident radiation and re-emits it at a lower frequency; (*b*) a sensitisor, *S*, absorbs incident radiation and passes it to an activator, A, which re-emits it at a lower frequency.

low-energy ground state, at energy E_0, to a higher energy excited state at energy E_1, (Fig. 9.14*a*). The relationship between the energy change, ΔE, and the frequency, ν, or the wavelength, λ, of the light absorbed or emitted, is

$$E_1 - E_0 = \Delta E = h\nu = \frac{hc}{\lambda}$$

where h is the Planck constant and c is the speed of light. In up-conversion, two (or more) photons are absorbed, promoting the energy to a higher level E_2 (E_3, etc.) (Fig. 9.14*b*).

When energy is lost from an absorption center, it moves from the excited state back to the ground state. The energy can be released directly by passing from E_1 to E_0 (Fig. 9.14*a*), with an energy output given by

$$E_1 - E_0 = \Delta E = h\nu = \frac{hc}{\lambda}$$

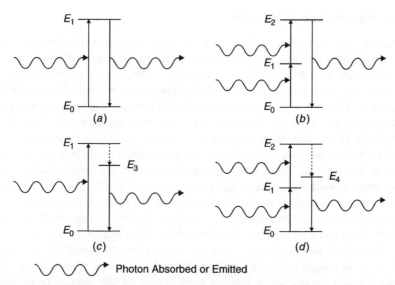

Figure 9.14 Energy transitions in defects: (*a*) simple excitation and release of energy; (*b*) up-conversion of two low-energy photons to one high-energy photon; (*c*) typical fluorescence in which some energy is lost as heat to the solid (dotted arrow) before transition to the ground state; and (*d*) up-conversion of two low-energy photons to a photon of intermediate energy.

which is identical to that of the exciting radiation. A similar process can occur with up-conversion in which case the output photon is twice as energetic as the input photons:

$$E_2 - E_0 = 2(E_1 - E_0) = 2\,\Delta E$$

Commonly some energy is lost to the crystal structure as vibrational energy, via a series of closely spaced energy levels, to a lower stable state E_3. The remaining energy is then lost in a transition to the ground state (Fig. 9.14c). The photon emitted is of a lower energy than that absorbed. A similar process can occur in up-conversion, when energy is lost to arrive at a lower stable state E_4, thereafter emitting a photon of lower energy than the sum of the exciting photons (Fig. 9.14d).

For many optical processes, the number of atoms in each energy state is of critical importance. Under thermal equilibrium, the relative population of two energy levels, E_0 and E_1, is given by the Boltzmann law:

$$\frac{N_1}{N_0} = \exp\left[-\frac{(E_1 - E_0)}{kT}\right]$$

where N_1 is the population of the level with higher energy, E_1 and N_0 is the population of the level with lower energy, E_0. For the energies relevant to optical transitions, only the lowest energy state, E_0, has a significant population at normal temperatures.

9.6.3 Energy Levels in Solids

The energy levels described in the previous section must be viewed in the context of the solid surrounding the defects. The main energy landscape in a solid is the band structure (Supplementary Material S2). In the simplest depictions, the upper energy band (the conduction band) is separated from the lower energy band (the valence band) by a constant band gap. In real structures, the band architecture is more complex.

Important defect energy levels are associated with acceptor and donor dopants (Fig. 9.15). Donor dopants may give energy levels close to the conduction band or far from it. Similarly, acceptor dopants may give energy levels close to the valence band or far from it. Those close to the band edges are called shallow levels, while those toward the center of the band gap are called deep levels. The energy of transitions to and from the conduction and valence bands depends upon whether the levels involved are deep or shallow. Transitions from shallow levels to a nearby band will tend to be associated with infrared energies while transitions involving deep levels will tend to be of high energy, in the ultraviolet. As well as these defect–valence band or defect–conduction band transitions, transitions can occur between defect energy levels without participation of the valence or conduction bands (Fig. 9.15). Transitions can also take place between the valence band and conduction band itself, although these latter transitions do not involve defects.

Processes involving defect energy levels are responsible for coloration of diamonds containing races of nitrogen or boron impurities. Diamond has a band gap of about 8.65×10^{-19} J (5.4 eV), which is too large to absorb visible light and

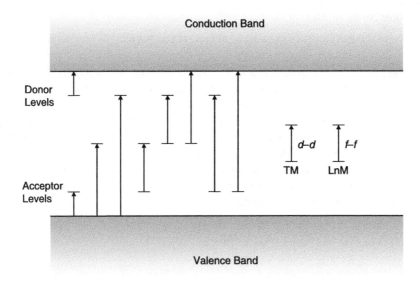

Figure 9.15 Some possible transitions involving defects in a solid. Transitions involving transition-metal (TM) ions are between d orbitals, transitions involving lanthanide (LnM) ions are between f orbitals.

diamonds are therefore clear. Nitrogen impurities occupy normal carbon atom positions to form N'_C defects. Nitrogen has five bonding electrons, one more than carbon. Four of the electrons around each impurity nitrogen atom are used to fulfill the local bonding requirements of the crystal structure and one electron remains unused. The extra electron, one per nitrogen atom impurity, sits in the energy gap to form a donor level at 6.41×10^{-19} J (4 eV) below the conduction band. This energy is in the ultraviolet and corresponds to a wavelength of 310 nm, but the absorption spectrum is broad enough for the low-energy tail to creep into the violet beyond 400 nm. Thus, diamond containing a few atoms of nitrogen absorbs slightly in the violet region, giving the stones a faint yellow aspect. As the nitrogen concentration increases, the color intensifies.

In the case of boron impurities a complementary situation occurs. Boron has only three outer bonding electrons instead of the four found on carbon. Each boron impurity atom occupies a carbon position, forming $B_C^{\bullet}$, which results in the creation of a set of new acceptor energy levels just 0.64×10^{-19} J (0.4 eV) above the valence band. The transition of an electron from the valence band to this acceptor level has an absorption peak in the infrared, but the high-energy tail of the absorption band spills into the red at 700 nm. The boron-doped diamonds therefore absorb some red light and leave the gemstone with an overall blue color.

Optical transitions are of interest for a number of device applications, but the exact mechanism of light production is often hard to pin down. For example, thin films of zinc oxide, ZnO, reduced in an atmosphere of N_2/H_2 show a bright green luminescence after irradiation with ultraviolet light. The active defect is believed to be a singly ionized oxygen vacancy, $V_O^{\bullet}$, which exists as a deep level in the band gap. However, two mechanisms have been proposed for light production, involving either the transition of an electron from the conduction band to form a neutral defect V_O or interaction of the $V_O^{\bullet}$ defect with a hole in the valence band to form a $V_O^{2\bullet}$ defect. Doping with Mg broadens the band gap of the ZnO moving the green luminescence toward blue.

Electron transitions in transition-metal ions usually involve electron movement between the d orbitals ($d–d$ transitions) and in lanthanides between the f orbitals ($f–f$ transitions). The band structure of the solid plays only a small part in the energy of these transitions, and, when these atoms are introduced into crystals, they can be represented as a set of levels within the wide band gap of the oxide (Fig. 9.15).

9.7 PIGMENTS, MINERALS AND GEMSTONES

9.7.1 Transition-Metal and Lanthanide Ion Colors

Pigments, minerals, gemstones, glasses, and many related materials are colored by impurity defects that absorb some of the incident white light, leaving a depleted spectrum to color the solid. Colors in these materials are thus characterized by the absorption spectrum of the solid. Common inorganic colorants are the transition-metal and lanthanide metal ions. The colors are characteristic of the ions themselves and are due to transitions between the partly filled d orbitals of transition-metals (d–d transitions) or the partly filled f orbitals of lanthanides (f–f transitions).

In the $3d$ transition-metal ions, the $3d$ orbitals contain one or more electrons. When these ions are introduced into a solid, the d-orbital energies are split by interactions with the surrounding anions. The commonest coordination for these ions is octahedral, and in this geometry the five d-orbital energy levels split into two groups, a higher energy e_g pair and three t_{2g} lower energy levels (Supplementary Material S6 and Figs. 9.6 and 9.7). The energy gap between the lower and upper set is written Δ(oct) or 10 Dq(oct). This rearrangement in the energies of the d orbitals is called crystal field splitting or ligand field splitting.

Although the way in which the d-orbital energy levels are split depends solely upon the geometry of surrounding atoms, the degree of splitting is dependent upon the interatomic distances between the ion and its neighbors. Short interionic distances lead to a large splitting, that is, large values of Δ, due to a strong crystal field. Long interionic distances produce a smaller splitting, that is, small values of Δ, due to a weak crystal field. The color produced by an ion is due to d electrons moving between the energy levels created by the crystal field splitting. The actual color perceived will therefore be a function of both the geometry of the surrounding coordination polyhedron and its size.

There is an important difference between the interaction of electromagnetic radiation with isolated transition-metal ions and with ions in a solid. In the case of isolated ions, the electrons move between sharp energy levels, E_0, E_1, and so on (Fig. 9.14). This means that light given out corresponds to a sharply defined energy difference and will ideally consist of only one (or a very small range) of frequencies. In solids, the interaction of the d orbitals with the surrounding structure, especially lattice vibrations, mean that the spacing between the energy levels will vary slightly from atom to atom in the crystal. This leads to a spread of energy levels in the crystal, and the absorption or emission spectrum of a solid due to electron transitions will consist of a series of bell-shaped curves rather than sharp lines.

The lanthanides have electrons in partly filled $4f$ orbitals. Many lanthanides show colors due to electron transitions involving the $4f$ orbitals. However, there is a considerable difference between the lanthanides and the $3d$ transition-metal ions. The $4f$ electrons in the lanthanides are well shielded beneath an outer electron configuration, ($5s^2\ 5p^6\ 6s^2$) and are little influenced by the crystal surroundings. Hence the important optical and magnetic properties attributed to the $4f$ electrons on any particular lanthanide ion are rather unvarying and do not depend significantly upon the host structure. Moreover, the energy levels are sharper than those of transition-metal ions and the spectra resemble those of free ions.

9.7.2 Colors and Impurity Defects

Glasses are frequently colored with transition-metal ions such as Mn^{2+}, Ni^{2+}, Co^{2+}, lanthanides, or actinides.[2] The addition of Co^{2+} impurities to silica glass leads to a

---

[2]The major use for uranium before the advent of the nuclear energy industry was as an additive to glass, to produce a yellow-green coloration.

bright blue color, while the pale green color of ordinary window glass is due to the presence of Fe^{2+} impurities. Many pigments for ceramic use contain transition-metal ions as the color producing centers. Similarly, the number of precious and semiprecious gemstones that owe their color and hence value to transition-metal impurities is considerable. Ruby consists of corundum, Al_2O_3, containing traces of Cr^{3+} (Section 1.4).

One simple example must suffice to cover principles underlying all of these systems. The color of Cu^{2+}-containing compounds, such as copper sulfate, $CuSO_4 \cdot 5H_2O$, and the minerals azurite, $Cu_3(OH)_2(CO_3)_2$, and malachite, $Cu_2(OH)_2(CO_3)$, is blue-green. The copper ions are found in octahedral coordination. The octahedral crystal field, due to surrounding oxygen ions, splits the Cu^{2+} d orbitals into two sets, as described above. There are nine d electrons to be allocated to the orbitals so that one vacancy exists in the higher e_g set (Fig. 9.16a). The energy to excite an electron from the lower t_{2g} set to the upper e_g set, equal to Δ(oct), will give rise to a peak in the absorption spectrum (Fig. 9.16b). In the case of $Cu(H_2O)_6^{2+}$, the source of the color in copper sulfate, the value of the octahedral crystal field splitting, Δ(oct), produced by the surrounding oxygen atoms on the water molecules, is 2.58×10^{-19} J (1.61 eV). A transition of this energy produces an absorption peak with a maximum wavelength near to 770 nm in the near infrared. The high-energy tail of this peak encroaches into the red end of the visible spectrum,

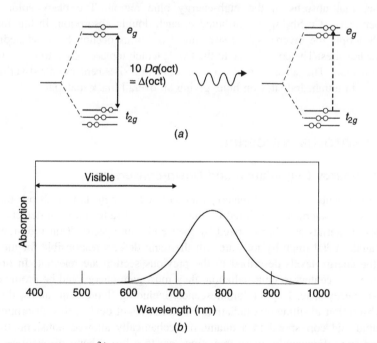

(a)

(b)

Figure 9.16 Color of Cu^{2+} ions in octahedral coordination: (a) the electron distribution between the t_{2g} and e_g split d orbitals before and after absorption of radiation; and (b) the absorption spectrum (schematic) of Cu^{2+} in $CuSO_4 \cdot 5H_2O$.

preferentially absorbing red, and leaving the observer to perceive a blue color. The coordination polyhedra in azurite and malachite are of slightly different sizes. Although this only alters the values of Δ(oct) in these compounds by a small amount, the position of the absorption band is changed enough for the color of these minerals to be easily differentiated from each other and from copper sulfate.

The effect of the strength of the crystal field on transition-metal ion color can be demonstrated by comparing ruby with emerald. Both minerals are colored by Cr^{3+} ions in octahedral sites. Emeralds possess the beryl ($Be_3Al_2Si_6O_{18}$) structure, with a color derived from traces of Cr^{3+} substituted for Al^{3+} in octahedral sites In beryl, the octahedra surrounding the Cr^{3+} ions are slightly larger than in corundum, and so the crystal field experienced by the Cr^{3+} in emerald is weaker than in ruby and Δ(oct) is smaller. Thus, although the energy level diagram remains essentially the same, the shift in the energy levels due to a change in the crystal field changes the perceived color enormously from red to green. Note that the colors of solids doped with lanthanides do not vary so much, as the f-electron energy levels are well shielded from the external surroundings.

In all of these examples, the agent causing the color is a defect in the form of (usually) a substituted transition-metal or lanthanide ion. The degree of coloration can be adjusted by changing the amount of dopant or by adjusting co-dopants to change the dimensions of the surrounding crystal structure. Thus, a colorless crystal can be made to appear black by doping with two substituents, one of which absorbs radiation in the low-energy yellow part of the visible spectrum and another that absorbs in the high-energy blue region. The black color of the pigment $Zn_{7-x}Co_xSb_2O_{12}$ is attributed to such double absorption. In this material, which adopts the inverse spinel structure, Co^{2+} ions occupy both octahedral and tetrahedral sites. The color is due to the Co^{2+}, which replaces Zn^{2+} to form substitutional defects. The absorption of Co^{2+} in octahedral sites centers on red-yellow, and of Co^{2+} in tetrahedral sites on blue, giving an overall black material.

9.8   PHOTOLUMINESCENCE

9.8.1   Energy Degradation and Down-conversion

While pigments and related materials are colored by defects that absorb radiation, luminescent materials are colored by the emission of radiation from defects. Colors in these materials are characterized by their emission spectra. Transition-metal and lanthanide metal impurity ions are still the main defects responsible for the effect, and the energy levels described in the previous section are relevant. In order for the impurity centers to emit radiation, they must gain energy and be promoted to a higher energy state. In photoluminescence, including fluorescent lamps, this is by the absorption of ultraviolet radiation. For an efficient device, this absorption must be rapid and correspond to a quantum mechanically allowed transition from the ground state. Frequently, these transitions are to a broad band of energies formed by the interaction of metal orbitals with nearby anion orbitals to create a wide energy band. In order for the ultraviolet (UV) exciting radiation to be converted into visible light, some of the energy must be lost. This comes about by the transfer of energy into crystal structure vibrations, that is, heat energy, due to transitions through a series of closely spaced energy levels that do not produce externally emitted photons, until a lower level is reached, only after which optical emission is possible. Because of the change from high-energy input to lower energy output, photoluminescence is sometimes called down-conversion.

The general principles of photoluminescence can be illustrated with respect to tri-chromatic fluorescent lamps. These use a phosphor mixture that emits equal amounts of the colors red, blue, and green, all of which generate color emission from impurity atoms in an oxide matrix. The favored red emitter is Eu^{3+} doped into a Y_2O_3 matrix forming Eu_Y defects. Ultraviolet radiation is readily absorbed due to charge transfer between Eu^{3+} and O^{2-} ions. Energy is then lost to the surrounding matrix as non-radiative transitions until the Eu^{3+} ion reaches the upper f-electron energy levels. At this point photons at a wavelength of 612 nm, corresponding to a transition between the energy levels $^5D_0-^7F_2$, are emitted, allowing the return of the ion to the ground state. The green emission is from Tb^{3+} coupled with a sensitisor, usually Ce^{3+}, and occurs at wavelength close to 550 nm, mainly from a $^5D_4-^7F_5$ transition. Typical host matrices are $La(Ce)PO_4$, $LaMg(Ce)Al_{11}O_{19}$, and $La(Ce)MgB_5O_{10}$. In each case the Tb^{3+} and Ce^{3+} ions replace La^{3+} ions to form Tb_{La} and Ce_{La} defects. The blue emission is produced by Eu^{2+} ions. Among the most widely used host lattices for Eu^{2+} ions are $BaMgAl_{11}O_{19}$ and $Sr_4Al_{14}O_{25}$. However, the introduction of Eu^{2+} into β-alumina structures has also been widely investigated as an alternative. In these latter materials, the Eu^{2+} occupies the Beevers–Ross and anti-Beevers–Ross sites normally occupied by Na^+, to form $Eu_{Na}^{\bullet}$ defects (Section 6.6).

The emission spectra of all defect centers depend upon the degree of doping and the host structure. Because of the sensitivity of normal color vision, the color perceived can change appreciably with even small changes of dopant concentration or host structure. Thus, although the principles of color production using defects are understood, great skill is required to produce a satisfactory working phosphor.

9.8.2 Up-conversion

Up-conversion is the opposite of photoluminescence. In this latter process, high-energy ultraviolet radiation is transformed into visible light. In up-conversion low-energy infrared radiation is transformed into visible light. The majority of studies of up-conversion have dealt with the behavior of polycrystalline or glass matrices containing lanthanide ions, especially Er^{3+}, Tm^{3+}, Ho^{3+}, and Yb^{3+}, as defect absorption centers. The Er^{3+} ion has an appeal because the up-conversion can be achieved with low-cost diode lasers operating at wavelengths near to 800 and 900 nm. A number of host structures have been used for these ions, including binary oxides Y_2O_3, Gd_2O_3, and ZrO_2, perovskite structure $BaTiO_3$, $SrTiO_3$, and $PbTiO_3$, fluorides such as $NaYF_4$ and oxide and oxyfluoride glasses. The up-conversion efficiency is strongly dependent upon the nature of the host and dopant levels.

In general, low concentrations of the active ion, of the order of 1%, are used. At these concentrations, the ions form point defects well isolated from each other. At higher concentrations, dopant ions tend to cluster and other energy loss mechanisms interfere with up-conversion.

The energy for up-conversion can be gained by the active ion in two principal ways. In the first, the ion itself can pick up more than one photon. The first photon excites the ion from the ground state to an excited energy level, a process referred to as ground-state absorption (GSA). A subsequent photon is absorbed to excite the ion to a higher energy level again, referred to as excited state absorption (ESA). In the second mechanism, energy is picked up by another ion and transferred to the emitter, a mechanism called energy transfer (ET). In either case, the intensity of the up-converted output, I_{up}, is related to the intensity of the exciting radiation, I_{ex}, by the formula:

$$I_{up} \propto I_{ex}^n$$

where n is the number of photons absorbed per up-converted photon emitted. A graph of log I_{up} versus log I_{ex} will give information on the mechanism of the process.

The oxide CeO_2 doped with approximately 1% Er^{3+} exhibits up-conversion involving only one active ion. The Er^{3+} ions substitute for Ce^{4+} to form a low concentration of Er'_{Ce} defects randomly distributed within the oxide matrix. Irradiation with near-infrared photons with a wavelength of 785 nm excites the Er^{3+} ions from the $^4I_{15/2}$ ground state to the $^4I_{9/2}$ level, that is, a GSA mechanism:

$$^4I_{15/2} \text{ (Er)} + h\nu \text{ (785 nm)} \longrightarrow {}^4I_{9/2} \text{ (Er)}$$

These ions loose energy internally to phonons (lattice vibrations), a process called decay or relaxation, to reach the $^4I_{11/2}$ and $^4I_{13/2}$ energy levels:

$$^4I_{9/2} \text{ (Er)} \longrightarrow {}^4I_{11/2} \text{ (Er)} + {}^4I_{13/2} \text{ (Er)} + \text{phonons}$$

(Fig. 9.17a). Ions in both these levels are further excited by an ESA mechanism to gain energy from the irradiating 785-nm radiation. Those in the $^4I_{11/2}$ energy level are excited to the $^4F_{3/2, 5/2}$ doublet:

$$^4I_{11/2} \text{ (Er)} + h\nu \text{ (785 nm)} \longrightarrow {}^4F_{3/2,5/2} \text{ (Er)}$$

These states subsequently decay by internal energy loss to the $^2H_{11/2}$, $^4S_{3/2}$, and $^4F_{9/2}$ energy levels:

$$^4F_{3/2,5/2} \text{ (Er)} \longrightarrow {}^2H_{11/2} \text{ (Er)} + {}^4S_{3/2} \text{ (Er)} + {}^4F_{9/2} \text{ (Er)} + \text{phonons}$$

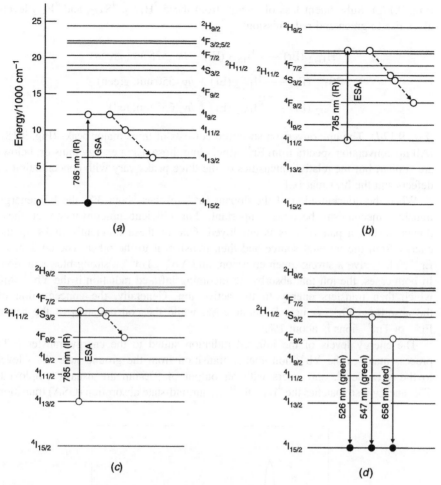

Figure 9.17 Simplified energy level diagram for Er^{3+} showing the important up-conversion transitions: (*a*) GSA and relaxation, (*b*) ESA and relaxation, (*c*) ESA and relaxation, and (*d*) emission of green and red. [GSA = ground-state absorption; ESA = excited-state absorption. Spectroscopist's energy units (cm^{-1}) are converted thus: $25,000\,cm^{-1} = 3.1\,eV = 5.0 \times 10^{-19}\,J$.]

(Fig. 9.17*b*). The ions in the $^4I_{13/2}$ energy level follow a similar path, being excited to the $^2H_{11/2}$ energy level:

$$^4I_{13/2}\,(Er) + h\nu\,(785\,nm) \longrightarrow {}^2H_{11/2}\,(Er)$$

then subsequently relax to the $^4S_{3/2}$ and $^4F_{9/2}$ energy levels:

$$^2H_{11/2}\,(Er) \longrightarrow {}^4S_{3/2}\,(Er) + {}^4F_{9/2}\,(Er) + phonons$$

(Fig. 9.17c). Subsequent loss of energy from these $^2H_{11/2}$, $^4S_{3/2}$, and $^4F_{9/2}$ levels gives rise to green and red emission:

$$^2H_{11/2} \text{ (Er)} \longrightarrow {}^4I_{15/2} \text{ (Er)} + h\nu \text{ (525 nm, green)}$$

$$^4S_{3/2} \text{ (Er)} \longrightarrow {}^4I_{15/2} \text{ (Er)} + h\nu \text{ (550 nm, green)}$$

$$^4F_{9/2} \text{ (Er)} \longrightarrow {}^4I_{15/2} \text{ (Er)} + h\nu \text{ (655 nm, red)}$$

(Fig. 9.17d). The up-conversion spectrum consists of three major peaks (Fig. 9.18). [All up-conversion spectra from Er^{3+} (including those using energy transfer, below) are similar, but the relative intensities of the three peaks vary with concentration of defects and the host matrix.]

When the concentration of the dopant Er^{3+} reaches about 3%, the ET (energy transfer) mechanism becomes important. For efficient up-conversion at these dopant levels, a pair of ions is employed. One of these efficiently picks up the energy from the infrared source and then transfers it to the other. The co-dopants Er^{3+}/Yb^{3+} give a strong green emission, and Yb^{3+}/Tm^{3+} a strong blue emission. In both cases, the ion that absorbs the incoming infrared radiation is the Yb^{3+} ion, which then transfers energy to the active ion. Generally, the concentration of the absorbing Yb^{3+} centers is about 20%, while the concentration of the active Er^{3+} or Tm^{3+} ions is about 1%.

The energy levels of the infrared radiation suited to the energy-transfer (ET) process matches the Yb^{3+} ion energy transition from the ground-state $^2F_{7/2}$ level to the $^2F_{5/2}$ level, and lasers with an output of 975 nm are usually employed. This energy also matches the $^4I_{15/2}$ to $^4I_{11/2}$ ground-state absorption (GSA) transition

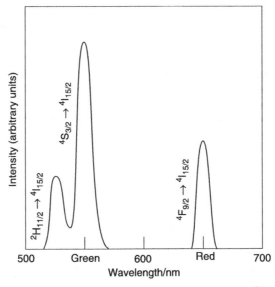

Figure 9.18   Up-conversion emission spectrum from Er^{3+} ions.

of Er^{3+} centers, but energy transfer (ET) from the Yb^{3+} centers dominates the process:

$$\text{GSA} \quad {}^4I_{15/2}\,(Er) + h\nu\,(975\,nm) \longrightarrow {}^4I_{11/2}\,(Er)$$

$$\phantom{\text{GSA}} \quad {}^2F_{7/2}\,(Yb) + h\nu\,(975\,nm) \longrightarrow {}^2F_{5/2}\,(Yb)$$

$$\text{ET} \quad {}^2F_{5/2}\,(Yb) + {}^4I_{15/2}\,(Er) \longrightarrow {}^2F_{7/2}\,(Yb) + {}^4I_{11/2}\,(Er)$$

Some of these excited ${}^4I_{11/2}$ (Er) ions relax to the ${}^4I_{13/2}$ level:

$$ {}^4I_{11/2}\,(Er) \longrightarrow {}^4I_{13/2}\,(Er) + phonons $$

(Fig. 9.19a). The second stage in the excitation process can use any of three mechanisms. Two are similar to those just described, absorption of a photon (ESA) or gain of energy from Yb^{3+} (ET):

$$\text{ESA} \quad {}^4I_{11/2}\,(Er) + h\nu\,(975\,nm) \longrightarrow {}^4F_{7/2}\,(Er)$$

$$\text{ET} \quad {}^2F_{5/2}\,(Yb) + {}^4I_{11/2}\,(Er) \longrightarrow {}^2F_{7/2}\,(Yb) + {}^4F_{7/2}\,(Er)$$

A third mechanism can occur in which two excited Er^{3+} ions exchange energy resulting in further excitation of one and return of the other to the ground state—a process called cross relaxation (CR):

$$\text{CR} \quad {}^4I_{11/2}\,(Er) + {}^4I_{11/2}\,(Er) \longrightarrow {}^4F_{7/2}\,(Er) + {}^4I_{15/2}\,(Er)$$

The populated ${}^4F_{7/2}$ (Er) level is able to lose energy and drop to the ${}^2H_{11/2}$ (Er) and ${}^4S_{3/2}$ (Er) levels:

$$ {}^4F_{7/2}\,(Er) \longrightarrow {}^2H_{11/2}\,(Er) + {}^4S_{3/2}\,(Er) + {}^4F_{9/2}\,(Er) + phonons $$

(Fig. 9.19b).

Although this process populates the ${}^4F_{9/2}$ (Er) level, it is filled in a more efficient way by the following transitions from the ${}^4I_{13/2}$ (Er) level, populated by nonradiative relaxation from the ${}^4I_{11/2}$ (Er) level:

$$\text{ESA} \quad {}^4I_{13/2}\,(Er) + h\nu\,(975\,nm) \longrightarrow {}^4F_{9/2}\,(Er)$$

$$\text{ET} \quad {}^2F_{5/2}\,(Yb) + {}^4I_{13/2}\,(Er) \longrightarrow {}^2F_{7/2}\,(Yb) + {}^4F_{9/2}\,(Er)$$

$$\text{CR} \quad {}^4I_{13/2}\,(Er) + {}^4I_{11/2}\,(Er) \longrightarrow {}^4F_{9/2}\,(Er) + {}^4I_{15/2}\,(Er)$$

(Fig. 9.19c). The emission transitions are the same as detailed above:

$$^2H_{11/2}\ (Er) \longrightarrow\ ^4I_{15/2}\ (Er) + h\nu\ (525\ nm,\ green)$$

$$^4S_{3/2}\ (Er) \longrightarrow\ ^4I_{15/2}\ (Er) + h\nu\ (550\ nm,\ green)$$

$$^4F_{9/2}\ (Er) \longrightarrow\ ^4I_{15/2}\ (Er) + h\nu\ (655nm,\ red)$$

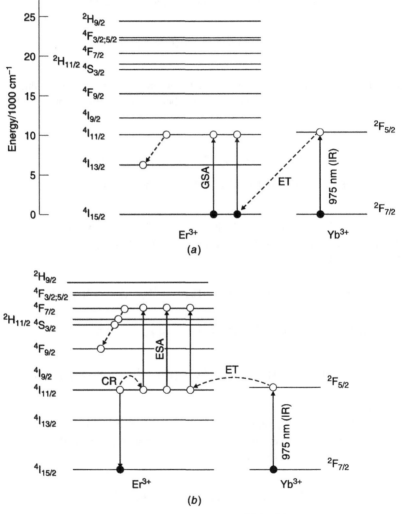

Figure 9.19 Simplified energy level diagram for the Er^{3+}/Yb^{3+} couple showing the important up-conversion and energy transfer transitions: (a) GSA, ET, and relaxation; (b) CR, ESA, ET, and relaxation

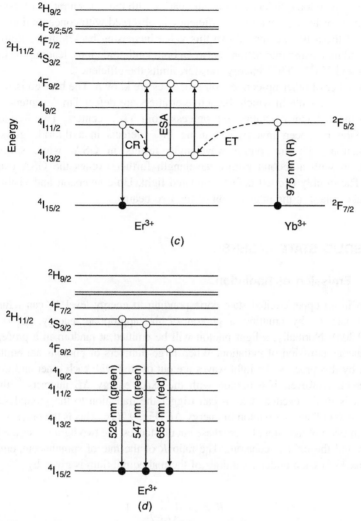

Figure 9.19 (*Continued*). (*c*) CR, ESA, and ET; (*d*) emission of green and red. [GSA = ground-state absorption;   ESA = excited-state   absorption;   CR = cross   relaxation. Spectroscopist's energy units (cm^{-1}) are converted thus: $25{,}000\,\text{cm}^{-1} = 3.1\,\text{eV} = 5.0 \times 10^{-19}\,\text{J}$.]

(Fig. 9.19*d*). Despite the complexity of the excitation processes, the energy transfer steps are the most efficient and responsible for the main energy output.

The up-conversion efficiency is defined as the ratio:

$$\text{Efficiency} = \frac{\text{power emitted at }\lambda}{\text{power absorbed (infrared)}}$$

The up-conversion efficiency is low and varies with the concentration of the activator and sensitisor ions. A maximum efficiency is observed with concentrations of about 1–3% of the active center. Above this value increasing back transfer from Er^{3+} to Yb^{3+} and increasing interactions between both lanthanide ions, leading to cluster formation and $Yb^{3+}-Yb^{3+}$ energy transfer, limits the efficiency.

A number of other up-conversion processes are known. The blue emission from a Yb^{3+}/Tm^{3+} couple in which the active emitters are defect Tm^{3+} centers is mainly due to the efficient excitation ET process from Yb^{3+} centers. Two-frequency up-conversion has been investigated using Pr^{3+} defects in a fluoride glass matrix. Illumination with one pump wavelength results in GSA, while simultaneous irradiation with a second pump wavelength further excites the GSA centers via ESA. The doubly excited defects emit red light. Up-conversion and visible output only takes place at the intersection of the two beams.

9.9 SOLID-STATE LASERS

9.9.1 Emission of Radiation

An ion in an upper excited state corresponding to energy level E_1 can return to the ground-state E_0 by emitting a photon of the appropriate energy, $\Delta E = E_1 - E_0$ (Fig. 9.20a). Normally, a light photon will be emitted at random in a process called spontaneous emission of radiation. When large numbers of photons are emitted from a solid by this process, the light waves are out of step with each other and the light is incoherent. However, if a photon with the exact energy ΔE interacts with the ion while it is in the excited state, it can trigger the transition to the ground state, with the emission of another photon of energy ΔE (Fig. 9.20b). This is the process of stimulated emission of radiation. Under these circumstances, the two light waves are perfectly in step and the light is coherent. The ratio R of the rate of spontaneous emission to stimulated emission under conditions of thermal equilibrium is given by

$$R = \exp\left(\frac{h\nu}{kT}\right) - 1$$

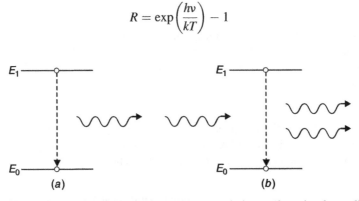

Figure 9.20 Light emission: (a) normal spontaneous emission produces incoherent light, and (b) stimulated emission produces coherent light.

where h is the Planck constant, v the frequency of the radiation, k the Boltzmann constant, and T the temperature (K). For light, R is much greater than 1 and spontaneous emission dominates the release of photons.

Lasers are devices for producing coherent light by way of stimulated emission. (Laser is an acronym for light amplification by stimulated emission of radiation.) In order to impose stimulated emission upon the system, it is necessary to bypass the equilibrium state, characterized by the Boltzmann law (Section 9.6.2), and arrange for more atoms to be in the excited-state E_1 than there are in the ground-state E_0. This state of affairs is called a population inversion and it is a necessary precursor to laser action. In addition, it must be possible to overcome the limitation upon the relative rate of spontaneous emission to stimulated emission, given above. Ways in which this can be achieved are described below, using the ruby laser and the neodymium laser as examples.

9.9.2 Ruby Laser: Three-Level Lasers

The first laser produced was the ruby laser, invented in 1960. Rubies are crystals of aluminum oxide (corundum, Al_2O_3), containing about 0.5% chromium ions Cr^{3+}, as substitution impurities, Cr_{Al}, and laser action, as well as color, is entirely due to these defects (Section 1.4).

The energy levels of importance in laser action are a result of the crystal field splitting of the Cr^{3+} $3d$ orbitals in a (slightly distorted) octahedral crystal field. The three d electrons on Cr^{3+} ions give rise to two sets of energy levels, one set associated with all electron spins parallel ($\uparrow \uparrow \uparrow$), denoted by a spin multiplicity[3] of 4, and one set with one spin-paired couple ($\uparrow \downarrow \uparrow$), denoted by a spin multiplicity of 2. The significance of the spin multiplicity is that electron transitions involved in color (optical transitions) are only allowed between levels in which the total amount of electron spin, that is, the spin multiplicity, does not change. These are known as spin-allowed transitions. As the ground state, $^4A_2$, has electrons with parallel spins, the allowed transitions are also to states with spin multiplicity of 4:

$$^4A_2 \longrightarrow {}^4T_2 \text{ (at 556 nm, absorbs yellow/green)}$$

$$^4A_2 \longrightarrow {}^4T_1 \text{ (at 407 nm, absorbs violet)}$$

(Fig. 9.21a). In ruby, excited Cr^{3+} ions in states $^4T_2$ or $^4T_1$ can lose energy to the crystal structure and drop down to level ^2E. The energy is taken up in lattice vibrations and the ruby crystal warms up. Such a process is a nonradiative or phonon-assisted transition, and because photons are not emitted, is independent of spin multiplicity. Thus, on irradiating the ruby with white light, Cr^{3+} ions will be excited to energy levels $^4T_2$ and $^4T_1$, and a significant number end up in the ^2E state rather than returning to the ground state. The transition from ^2E to the $^4A_2$ ground state is not

---

[3]The multiplicity is written as a superscript to the spectroscopic symbol, representing the energy of the state, for example, the energy level $^4A_2$ has a multiplicity of 4.

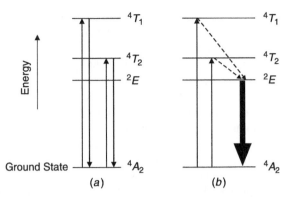

Figure 9.21 Ruby laser: (*a*) main transitions responsible for the color of ruby and (*b*) the main transitions responsible for laser action.

allowed because of the difference in spin multiplicity of the two energy levels, which means that it is possible to build up a population inversion between the ground state and the ^2E state.

Laser operation takes place in the following way. An intense flash of white light is directed onto the crystal, a process called optical pumping. This excites the Cr^{3+} ions into the $^4T_2$ and $^4T_1$ states. These then lose energy by nonradiative transitions and "flow over" into the ^2E state. An intense initial flash will cause a population inversion to form between ^2E and $^4A_2$. About 0.5 ms after the start of the pumping flash, some spontaneous emission will occur from ^2E. These initial photons are reflected back and forth, causing stimulated emission from the other ^2E levels. Once started, the stimulated emission rapidly depopulates these levels in an avalanche, and a burst of coherent red laser light of wavelength 694.3 nm emerges from the ruby (Fig. 9.21*b*).

Following the light burst, the upper levels will be empty. The ruby laser generally operates by emitting energy in short bursts, each of which lasts about 1 ms, a process referred to as pulsed operation. The ruby laser is called a three-level laser because three energy levels are involved in the operation. These are the ground state ($^4A_2$), an excited state reached by optical absorption or pumping ($^4T_2$ or $^4T_1$), and an intermediate state of long lifetime (^2E) reached by radiationless transfer and from which stimulated emission (laser emission) occurs to the ground state.

9.9.3 Neodymium (Nd^{3+}) Laser: Four-Level Lasers

Solid-state lasers using substitutional neodymium (Nd^{3+} ions) as the active defects are widely available. Practical lasers contain about 1% Nd^{3+} dopant. The most common host materials are glass, yttrium aluminum garnet (YAG), $Y_3Al_5O_{12}$, and calcium tungstate, $CaWO_4$. In the crystalline host structures, the defects responsible for amplification are Nd_Y and $Nd^{\bullet}_{Ca}$.

The important transitions taking place in Nd^{3+} ion lasers are between energy levels derived from electrons in the f orbitals. The f-electron levels are rather sharp

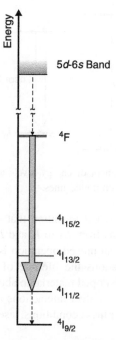

Figure 9.22 Energy levels of most importance in the neodymium laser. The pump transition is from the ground state to the broad $5d$–$6s$ band. The main laser transition is between the ^4F and $^4I_{11/2}$ levels. Internal transitions are marked with dotted lines.

because these levels are shielded from the effects of the surrounding crystal lattice by outer electron orbitals and can be approximated to free ion energy levels. Above the set of f-electron energy levels lie energy bands of considerable width derived from the interaction of the $5d$ and $6s$ orbitals with the surrounding atoms (Fig. 9.22).

Optical pumping excites the ions from the ground state to these wide bands. This process is very efficient, both because broad energy bands allow a wide range of pump wavelengths and because the transition from the ground state is allowed. Following pumping, energy is lost internally until the ^4F pair of f electron energy levels is reached. The principal laser transition is from these ^4F levels to a set of levels labeled $^4I_{11/2}$. The emission is at approximately 1060 nm in the infrared.

This type of laser is called a four-level laser because operation takes place in the following sequence of steps (Fig. 9.23). Atoms in the ground state E_0 are excited to a rather high energy level, E_1, by optical pumping. Subsequently, the atoms in E_1 lose energy internally to an intermediate state I_1. (Once in I_1, atoms should have a long lifetime and not lose energy quickly.) When another intermediate state, I_0, is present and sufficiently high above the ground state to be effectively empty, a small population in I_1 gives a population inversion between I_1 and I_0. Ultimately, a few photons will be released as some atoms drop from I_1 to I_0. These can promote stimulated emission between I_1 and I_0, allowing laser action to take place. Atoms

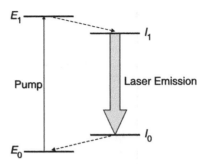

Figure 9.23 Schematic arrangement of energy levels and transitions in a four-level laser. Internal transitions are marked with dotted lines.

return from I_0 to E_0 by an internal transition that needs to be rapid. If the energy corresponding to the transitions from E_1 to I_1 and I_0 to E_0 can be easily dissipated, continuous operation rather than pulsed operation is possible.

At higher Nd^{3+} concentrations the lifetime of the ^4F upper state drops from about 200 μs in a typically 1% doped material to about 5 μs at higher dopant concentrations. This is due to Nd–Nd defect interactions and associated changes in lattice vibration characteristics. Under these conditions, laser operation is no longer possible.

9.10 COLOR CENTERS

Exposure of transparent solids, both glasses and crystals, to high-energy radiation frequently makes them colored. The defects responsible for this are known as color centers. The first of these defects to be characterized was the F center, a term derived from the German *Farbzentrum* (color center).

9.10.1 The F Center

These defects were first produced by exposing alkali halide crystals to high-energy radiation such as X rays. This causes the crystals to become brightly colored with fairly simple bell-shaped absorption spectra. The peak of the absorption curve, λ_{max}, moves to higher wavelengths as both the alkali metal ion size and halide ion size increase (Table 9.1).

Color centers can be introduced in several ways apart from using ionizing radiation. A significant fact is that, regardless of the method used, the color produced in any particular crystal is always the same. Thus, F centers in sodium chloride, NaCl, always give the crystal an orange-brown color. These observations suggest that the centers are defects in the crystal structure that do not involve the chemical nature of the components of the material in a direct fashion. This is so, and it has long been known that the F center is an anion vacancy plus a trapped electron (Fig. 9.24). The trapping is due to the effective positive charge on the vacancy, $V_X^\bullet$, so that the F center is written ($V_X^\bullet$ e′). F centers form deep levels in the band gap of the alkali halide solid. The electron in this location is able to absorb electromagnetic radiation, which gives rise to the color of the solid.

TABLE 9.1   Alkali Metal Halide F Centers

Compound	Absorption Wavelength λ_{max}/nm	Color[a]	Lattice Parameter/nm
LiF	235, UV	Colorless	0.4073
NaF	345, UV	Colorless	0.4620
KF	460, blue	Yellow-brown	0.5347
RbF	510, green	Magenta	0.5640
LiCl	390, UV (just)	Yellow-green	0.5130
NaCl	460, blue	Yellow-brown	0.5641
KCl	565, green	Violet	0.6293
RbCl	620, orange	Blue-green	0.6581
LiBr	460, blue	Yellow-brown	0.5501
NaBr	540, green	Purple	0.5973
KBr	620, orange	Blue-green	0.6600
RbBr	690, red	Blue-green	0.6854

[a]The appearance of the color-center-containing crystal is the complementary color to that removed by the absorption band.

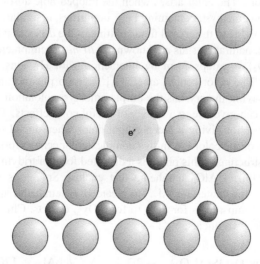

Figure 9.24   Schematic illustration of an F center, an anion vacancy plus a trapped electron, in an alkali halide crystal.

9.10.2 Electron and Hole Centers

Since the original studies of F centers many other color centers have been characterized that may be associated with either trapped electrons or trapped holes. These are called electron excess centers when electrons are trapped and hole excess centers when holes are trapped.

The F center is an electron excess center and arises because the crystal contains a small excess of metal. Similar metal excess F centers exist in compounds other than the alkali halides. An example is provided by the mineral *Blue John*.[4] This is a rare naturally occurring form of fluorite, CaF_2. The coloration is caused by electron excess F centers each consisting of an anion vacancy plus a trapped electron. It is believed that the color centers in Blue John were formed when the fluorite crystals were fortuitously located near to uranium compounds in the rock strata. Radioactive decay of the uranium produced the energetic radiation necessary to form color centers.

One of the best understood hole excess centers gives rise to the color in smoky quartz and some forms of amethyst. These minerals are essentially crystals of silica, SiO_2, which contain small amounts of either Al^{3+} or Fe^{3+} as substitutional impurities, Al'_{Si} or Fe'_{Si}. Charge neutrality is preserved by way of incorporated hydrogen as H^+. The color center giving rise to the smoky color in quartz is formed when an electron is liberated from an $[AlO_4]^{-5}$ group by ionizing radiation and is trapped on one of the H^+ ions present. The reaction can be written as

$$[AlO_4]^{5-} + H^+ \longrightarrow [AlO_4]^{4-} + H$$

The color center is the $[AlO_4]^{4-}$ group, which can be thought of as $[AlO_4]^{5-}$ together with a trapped hole. The color arises when the trapped hole absorbs radiation.

The situation in amethyst, containing Fe^{3+} impurities, is similar. These crystals are a pale yellow color due to the crystal field splitting of the d-electron levels on the Fe^{3+} ions. In this form, natural crystals are known as *citrine*, a semiprecious gemstone. On irradiation, $[FeO_4]^{4-}$ groups form by interaction with H^+ ions, as described for $[AlO_4]^{4-}$ above. The color center, an $[FeO_4]^{5-}$ group plus a trapped hole, is able to absorb light, giving the crystals the purple amethyst coloration.

Color centers can give rise to a variety of useful color effects. The oxide $SrAl_2O_4$ is a long-life phosphor giving a green output color when doped with B, Eu^{2+}, and Dy^{3+}. The origin of the color lies in two complex color centers formed by the impurity cations. The structure of this phase is a distorted form of tridymite, which is composed of corner-linked AlO_4 tetrahedra that enclose Sr^{2+} ions in the cavities so formed. The B^{3+} substitutes for Al^{3+} to create BO_4 tetrahedra and BO_3 triangular groups. The Dy^{3+} substitutes for Sr^{2+} to form $Dy^{\bullet}_{Sr}$ defects. Charge is balanced by the creation of Sr^{2+} vacancies, V''_{Sr}:

$$Dy_2O_3(3SrAl_2O_4) \longrightarrow 2Dy^{\bullet}_{Sr} + V''_{Sr} + 6Al_{Al} + 12O_O$$

Two complex centers form: $(Dy-BO_4 - V'_{Sr} - h^{\bullet})$, which are hole centers formed thermally from $(Dy-BO_4 - V''_{Sr})$, and $(BO_3-V_O^{\bullet} - e')$, which are electron centers formed from $(BO_3-V_O^{2\bullet})$ under violet light. Under normal conditions, the electron and hole centers are metastable and the holes and electrons gradually recombine. The energy liberated is transferred to the Eu^{2+} ions to give a green fluorescence. As there is no radioactivity involved, these materials can be used for luminous

---

[4]The name "Blue John" is a corruption of the French term *bleu-jeune*, which was used to describe the blue form of the normally yellowish fluorite crystals found in nature.

dials on clocks and watches, replacing a historic use involving radioactive materials, or as cold light displays.

9.10.3 Surface Color Centers

The concept of color centers has been extended to surfaces to explain a number of puzzling aspects of surface reactivity. For example, in oxides such as MgO an anion vacancy carries two effective charges, $V_O^{2\bullet}$. These vacancies can trap two electrons to form an F center or one electron to form an F^+ center. When the vacancy is located at a surface, the centers are given a subscript s, that is, F_s^+ represents a single electron trapped at an anion vacancy on an MgO surface. As the trapping energy for the electrons in such centers is weak, they are available to enhance surface reactions.

The concentration of F_s^+ centers can be increased by irradiation with energetic radiation such as X rays or ultraviolet light, as well as by reaction with hydrogen. This latter reaction has lead to the suggestion that several new color centers could form involving hydroxyl. The $F_s^+(OH^-)$ center is imagined to form in the following way. A hydrogen atom reacts on the surface to form a hydroxyl group, OH^-. This leaves the surface to link to a nearby metal cation in the exposed surface, at the same time creating an oxygen vacancy and leaving a trapped electron to create an $F_s^+(OH^-)$ defect (Fig. 9.25).

The properties of defects of this type are difficult to determine experimentally, although absorption spectra do give information about electron or hole binding energies. Much information is obtained by calculation, using density functional or other quantum computational methods. In this way, the relative stabilities of defects on plane faces, steps, terraces, and corners can be explored.

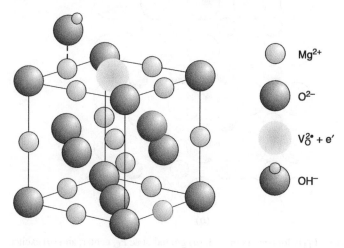

Mg^{2+}

O^{2-}

$V_O^{2\bullet} + e'$

OH^-

Figure 9.25 Schematic diagram of an $F_s^+(OH^-)$ center on an MgO (100) surface.

9.10.4   Complex Color Centers: Laser and Photonic Action

The fabrication of lasers based upon color centers adds a further dimension to the laser wavelengths available. Ordinary F centers do not exhibit laser action, but F centers that have a dopant cation next to the anion vacancy are satisfactory. These are typified by F_{Li} centers, which consist of an F center with a lithium ion neighbor (Fig. 9.26a). Crystals of KCl or RbCl doped with LiCl, containing F_{Li} centers have been found to be good laser materials yielding emission lines with wavelengths between 2.45 and 3.45 μm. A unique property of these crystals is that in the excited state an anion adjacent to the F_{Li} center moves into an interstitial position

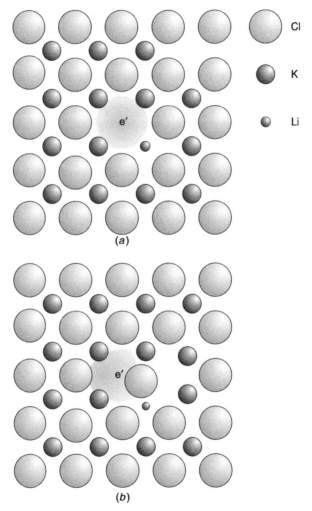

Figure 9.26 F_{Li} color centers in KCl: (a) ground-state F_{Li} center; and (b) excited-state type II F_{Li} center responsible for laser output.

(Fig. 9.26b). This is "type II" laser behavior, and the active centers are called F_{Li} (II) centers.

These complex defects are introduced in a series of steps. Take KCl doped with Li^+ as an example. Initially KCl crystals are grown from a solution containing LiCl as an impurity. The Li^+ cations form substitutional Li_K impurity defects distributed at random throughout the crystal. F centers are introduced by irradiation using X rays. These are not usually located next to a dopant Li^+ cation. To convert the F centers into F_{Li} (II) centers the crystal is subjected to a process called aggregation. In this step, the crystals are cooled to about $-10°C$ and then exposed to white light. This releases the electrons trapped at the F centers, leaving ordinary anion vacancies, which are then able to diffuse through the crystal before recombining with the electrons once more to reform the F center. Ultimately, each vacancy center ends up next to a Li^+ ion. At this position it is strongly trapped and further diffusion is not possible. Recombination with an electron forms the F_{Li} center required. This process of aggregation is permanent if the crystal is kept at $-10°C$ and in this state the crystals are laser active.

The color of diamond due to nitrogen impurities has been described in Section 9.6.3 It has been found that nitrogen impurities that are located next to a carbon vacancy in diamond thin films endow the solid with quite new properties, somewhat similar to the properties of a solid containing F_{Li} centers compared with ordinary F centers. The diamond structure is built up of carbon atoms each surrounded by four

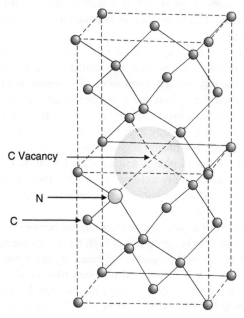

C Vacancy

N

C

Figure 9.27 A (N–V) center in diamond, consisting of a carbon atom vacancy and a neighboring nitrogen atom impurity.

carbon atom neighbors in a tetrahedron. In these centers, called nitrogen–vacancy (N–V) centers, the central carbon atom is missing, forming the vacancy, and a nitrogen atom replaces one of the neighboring carbon atoms, so that the tetrahedron surrounding the vacancy is composed of three carbon atoms and one nitrogen atom (Fig. 9.27). The electron trapped at the color center has an orbital that extends over the cavity, so that it encompasses not only the nitrogen impurity but also the three carbon atoms that also surround the vacancy.

Within its orbit, which has some of the characteristics of a molecular orbital because it is shared with electrons on the surrounding atoms, the electron has two possible spin multiplicity states. These have different energies, and because of the spin-multiplicity rule, when an (N–V) center emits a photon, the transition is allowed from one of these and forbidden from the other. Moreover, the electron can be flipped from one state to another by using low-energy radio-frequency irradiation. Irradiation with an appropriate laser wavelength will excite the electron and as it returns to the ground state will emit fluorescent radiation. The intensity of the emitted photon beam will depend upon the spin state, which can be changed at will by radio-frequency input. These color centers are under active exploration for use as components for the realization of quantum computers.

9.11 ELECTROCHROMIC FILMS

Electrochromic materials, which change color when subjected to an electric field, are widely explored for application as "smart windows" that control the amount of light reflected or transmitted. Color production is linked to defect formation in an otherwise colorless matrix.

The material most widely explored for such devices is tungsten trioxide, WO_3, which ordinarily is pale yellow and an insulator. Thin films are transparent. Electrochromic films rely on the formation of tungsten bronzes, M_xWO_3, as a reduced phase (Section 4.5). There are a number of tungsten bronze structures, but the ones utilized for electrochromic purposes are the perovskite bronzes in which cations such as Li, Na, or K occupy cage sites between the corner-linked WO_6 octahedra of the parent phase. The phase range over which the perovskite bronze structure is stable is greatest for the Li phases and smallest for the K phases. The hydrogen bronzes, H_xWO_3 are also deeply colored, although these are rather different from the alkali metal phases and are probably best regarded as a nonstoichiometric hydroxide, $WO_{3-x}(OH)_x$.

Although the color of the tungsten bronzes has not been explained fully over all of the composition range, it is believed that at high metal M concentrations the color is metallic, due to electrons in the conduction band. At low concentrations, the color induced is so similar to that of reduced tungsten trioxide that it is presumed that charge transfer between two valence states of tungsten is occurring. If so, color may then be attributed to W^{5+}–W^{6+} or W^{4+}–W^{6+} couples, formally equivalent to $W_W^{2\prime}$ or $W_W^\prime$ defects.

The principle of an electrochromic device using tungsten trioxide films is not too difficult to envisage. It is necessary to drive some appropriate metal such as Li into the WO_3 film using an applied voltage. This will make the tungsten trioxide turn into a blue-black tungsten bronze. Reversal of the voltage must remove the interpolated metal and regenerate the colorless state, a process referred to as bleaching. The reaction can be schematically written:

$$WO_3(\text{oxidized state, transparent film}) + xM + xe^-$$
$$\rightleftharpoons M_xWO_3(\text{reduced state, blue-black, } x \approx 0.2)$$

In principle, devices are constructed as a series of thin films on glass. Transparent conducting electrodes, usually indium tin oxide (ITO), serve as electrodes, sandwiching a film of WO_3, an ion-conducting electrolyte, and a source–sink of metal ions (Fig. 9.28). In practice, many designs have been explored. An experimental display has been constructed using the ionic conductor β-alumina as a source and sink for Na, which functions as the interpolated metal M in the bronze phase. In this design, the β-alumina combines the functions of the electrolyte and metal storage films. The power supply can force Na^+ ions to migrate into the WO_3 to form a dark bronze or remove them into the β-alumina reservoir to turn the WO_3 colorless.

A different arrangement has been used to make car mirrors which can be electrically dimmed so as to cut down dazzling reflections from bright lights. This relies upon the decomposition of water vapor to generate H^+ ions, which in turn are used to produce a hydrogen tungsten bronze H_xWO_3.

Decomposition takes place on the outer indium tin oxide electrode:

$$2H_2O \longrightarrow O_2(g) + 4H^+ + 4e^-$$

This electrochemical decomposition requires about 1 V at the electrode surface. To drive the protons into the WO_3 film, a proton-conducting electrolyte, typically

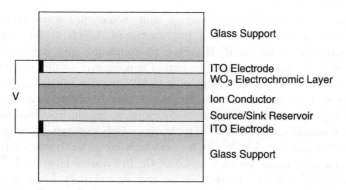

Figure 9.28 Schematic view of an electrochromic film device for the modulation of light intensity (ITO = indium tin oxide).

hydrogen uranyl phosphate, $HUO_2PO_4 \cdot 4H_2O$ (HUP), is utilized (see also Section 8.8). The H^+ produced can pass through the proton-conducting electrolyte to form the bronze, using electrons from the other electrode. If troublesome reflections occur, the unit is switched on and the WO_3 layer is darkened by the formation of H_xWO_3, which cuts down the dazzling reflection. When the problem no longer occurs, the voltage is reversed. The H^+ ions are pulled out from the bronze and the film becomes colorless once more. If a photocell is incorporated into the circuit, the whole device can be automated. In this design, the storage reservoir layer for H^+ is the surrounding atmosphere.

To improve the darkening characteristics of the device, a reservoir film that also darkens can be incorporated into the system. One material examined is hydrated nickel oxide. This is able to store Li reversibly and becomes darker when it is depleted, enhancing the darkening effect of the tungsten bronze layer.

9.12   PHOTOINDUCED MAGNETISM

A number of materials, including molecules, diluted magnetic semiconductors, perovskite structure manganites, and spinel structure ferrites show photoinduced magnetism—modified magnetic behavior when illuminated by laser light. The precise mechanism for the changes observed varies from one material to another and has not been unraveled for everything. However, it is frequently associated with the crystal field splitting of the d-orbital energy levels on transition-metal ions within the structure. The crystal field splitting energy between the t_{2g} and e_g set of orbitals is generally similar to the energy of visible light and is the reason for the color associated with the transition-metal compounds (Section 9.7). It is also responsible for the low-spin versus high-spin magnetic properties of these ions (Section 9.4). This correspondence means that the magnetic properties of an ion can be changed by irradiation of light with a suitable wavelength, which will promote electrons from the lower to the upper energy state, a phenomenon called spin crossover (Fig. 9.29a).

In general, this transition is transitory, with photoexcited states reverting to the ground state rapidly, thus creating just short-lived defects in the magnetic and optical structure of the solid. However, photoinduced spin crossover is being studied in single molecules, with the aim of making ever smaller memory storage devices, and for this a stable change of magnetic structure is required. A necessary precursor is a molecule in which the ground state is neither associated strongly with the high-spin nor low-spin state of the cation, so that the transformation is energetically reasonable. One such set of molecules widely explored for this purpose are related to Prussian blue, $KFe^{II}Fe^{III}(CN)_6$, in which one of the Fe atoms is replaced by another transition-metal ion such as Co or Cr. In Prussian blue, the Fe^{II} and Fe^{III} (nominally Fe^{2+} and Fe^{3+}) ions alternate and are linked by CN groups, and the intense color is due to a transfer of an electron from Fe^{II} to Fe^{III} via the cyanide intermediary. In molecules containing other transition-metal ions, the aim is to induce this transfer optically and then to stabilize the photoinduced state. One such

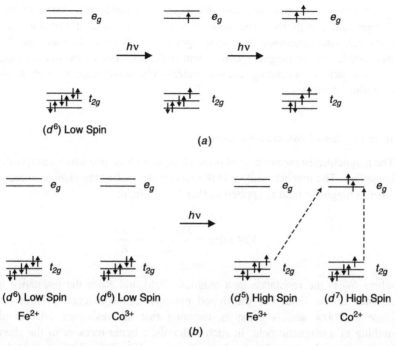

Figure 9.29 Photo-induced spin crossover in transition metal ions: (*a*) incident photons can successively promote electrons from the ground t_{2g} state to the high-energy e_g state; and (*b*) spin crossover involving electron transfer and excitation in KFeCo(CN)$_6$.

molecule that shows promise is KCoFe(CN)$_6$. The nonmagnetic to magnetic state transition is

$$Fe^{II}(t_{2g}{}^6 LS)-CN-Co^{III}(t_{2g}{}^6 LS) \longrightarrow Fe^{III}(t_{2g}{}^5 HS)-CN-Co^{III}(t_{2g}{}^5 e_g{}^2 HS)$$

In the initial state, all transition-metal ions are in the low-spin state, giving an effectively nonmagnetic solid. Irradiation transfers an electron from Fe to Co and promotes electron excitation to create all high-spin ions with a considerable magnetic moment (Fig. 9.29*b*). The magnetic defects in this case are the triple group of metal ion–cyanide–metal ion. As in all materials for device use, variation in dopant concentrations can be used to tune the desirable properties of the solid.

9.13 ANSWERS TO INTRODUCTORY QUESTIONS

What is spintronics?

Traditionally, semiconductor devices utilize the charge on electrons and ignore the spin component. However, the addition of spin to the observable properties of electrons in semiconductors has excited considerable interest for data storage and computing. The area in which both conductivity and spin are exploited simultaneously is called spintronics. Semiconductors that have a measurable spin or magnetic component added to them, called diluted magnetic semiconductors, are being studied extensively with respect to their use in spintronic devices.

What is colossal magnetoresistance?

The magnetoresistance of a solid is the change of resistance when a magnetic field is applied. The magnetoresitive (MR) ratio is defined as the change in resistance when a magnetic field is applied to that in zero field:

$$\text{MR ratio} = \frac{R_H - R_0}{R_0} = \frac{\Delta R}{R_0}$$

where R_H is the resistance in a magnetic field and R_0 is the resistance in the absence of the field. For the doped manganites the magnetoresistive ratio is close to 100% and is negative, meaning that the resistance falls to almost nothing in a magnetic field. In such materials, a better measure of the change is the ratio of R_0/R_H, where values of up to 10^{11} have been recorded. This extreme behavior has been called colossal magnetoresistance (CMR).

The CMR effect is found in a variety of doped systems, including hole doped $La_{1-x}A_xMnO_3$, with A being an alkaline earth cation (Ca, Sr, Ba), ordered $Ln_{0.5}A_{0.5}MnO_3$, with Ln being a lanthanide, typically Pr. In all of these compounds, doping introduces holes that are located on Mn^{3+} ions to form Mn^{4+}. Substitution of some of the Mn with both magnetic or nonmagnetic ions, including Ti^{4+}, Sn^{4+}, Fe^{3+}, Cr^{3+}, Al^{3+}, Ga^{3+}, In^{3+}, Mg^{2+}, and Ni^{2+}, has been found to increase the CMR effect considerably.

What defects commonly color glass and gemstones?

These materials are normally colored by low concentrations of $3d$ transition-metal ions or more rarely by lanthanide ions. The pale green color of ordinary window glass is due to the presence of Fe^{2+} impurities and small amounts of doping of Cr^{3+} into Al_2O_3 (corundum) creates ruby.

The colors are characteristic of the ions themselves and are due to transitions between the partly filled d orbitals of transition metals (d-d transitions) or the partly filled f orbitals of lanthanides (f-f transitions). In the $3d$ transition-metal ions, the $3d$ orbitals contain one or more electrons. When these ions are introduced into a solid, the d-orbital energies are split by interactions with the surrounding anions. The color observed is due to transitions between these split energy levels. The color observed varies considerably as the interactions are dependent upon the closeness of the surrounding anions and their chemical nature. The colors of solids doped with lanthanides do not vary as much as those doped with $3d$ transition-metal ions, because, unlike the d orbitals, the f electron energy levels are well shielded from the external surroundings. Thus, lanthanide ions in both glasses and crystals tend to exhibit very similar colors.

PROBLEMS AND EXERCISES

Quick Quiz

1. Magnetism is associated with partly filled electron:
   (a) s shells
   (b) p shells
   (c) d shells

2. Magnetic semiconductors are doped with:
   (a) Transition-metal ions
   (b) Alkaline earth metal ions
   (c) Donor ions

3. The high-spin (HS) state of the d^6 ion Co^{3+} has the configuration:
   (a) $t_{2g}^6 \, e_g^0$
   (b) $t_{2g}^5 \, e_g^1$
   (c) $t_{2g}^4 \, e_g^2$

4. Colossal magnetoresistance is associated with:
   (a) Ferrites
   (b) Manganites
   (c) Cobaltites

5. Data storage via magnetic bubble memories used thin films of:
   (a) Garnet
   (b) Perovskite
   (c) Spinel

6. The color of a glass containing transition-metal ion impurities is due to:
   (a) Emission of radiation
   (b) Absorption of radiation
   (c) Scattering of radiation

7. Up-conversion is the name of the process in which:
   (a) Ultraviolet absorption gives visible light output
   (b) Visible light absorption gives infrared output
   (c) Infrared absorption gives visible output

8. Laser light is distinguished by being:
   (a) Coherent
   (b) Incoherent
   (c) Partly coherent

9. An F center is an electron trapped at:
   (a) A cation vacancy
   (b) An impurity
   (c) An anion vacancy

10. The F_s^+ color center in the surface of an MgO crystal is an oxygen vacancy plus:
    (a) A trapped hole
    (b) A trapped electron
    (c) A trapped H^+

Calculations and Questions

1. A magnetic semiconductor thin film is made by doping ZnO with the $3d^7$ ion Co^{2+}. The crystal filed splitting of the d orbitals in a tetrahedral site is opposite to that in an octahedral site, with the lower pair of levels labeled e and three upper orbitals labeled t_2. (a) What is the spin state of the Co^{2+} ion in ZnO? (b) What is the expected magnetic moment on the Co^{2+} ions? (c) The spectrum has an absorption peak at 660 nm. What is the crystal field splitting of the Co^{2+} ion in the tetrahedral crystal field of ZnO?

2. A plot of the variation of the magnetic moment of Mn ions in the thermistor material $MgGa_{2-x}Mn_xO_4$ is given in the following table. What can be concluded about the spin state of the Mn ions?

x	μ/μ_B
0.10	5.2
0.25	4.92
0.5	4.61
0.75	4.21
1.0	4.29

*Data adapted from A. Veres et al., Solid State Ionics, **178**, 423–428 (2007). See also Question 8.2.

3. Zinc oxide, ZnO, band gap 3.2 eV, gives green ($\lambda = 498$ nm) photoluminescent emission that changes to blue ($\lambda = 468$ nm) for a ZnO film doped with MgO. The photoluminescence is associated with a transition from the valence band to a deep level due to a $V_O^{\bullet}$ defect. (a) Confirm that the photoluminescence does not correspond to a transition from the valance band to the conduction band. (b) What is the energy of the $V_O^{\bullet}$ level compared to the valence band for

both of these cases? (**c**) If the displacement of the valence band and conduction band is the same, what is the band gap of the MgO-doped film compared to that of pure ZnO? [Data adapted from S. Fujihara, Y. Ogawa, and A. Kasai, *Chem. Mater.*, **16**, 2965–2968 (2004).]

4. An Eu^{3+} containing phosphor is excited by radiation with a wavelength of 400 nm and emits radiation at either 592 or 611 nm with equal intensities. What percentage of the exciting radiation is lost as heat to the solid?

5. The intensity of the upconverted beam as a function of the intensity of the exciting radiation for Er^{3+} doped into CeO_2 is given in the following table. Determine whether the up-conversion is likely to be a two- or three-photon absorption process.

Input Intensity, I_{ex} (arbitrary units)	Upconverted Intensity, I_{up} (arbitrary units)
10.0	50.0
7.50	30.0
5.00	15.0
3.75	9.50
2.50	5.00

*Data adapted from H. Guo, *J. Solid State Chem.*, **180**, 127–131 (2007).

6. The F center absorption maximum for KCl is at 565 nm and that for KF is 460 nm (Table 9.1). (**a**) What is the composition of a natural crystal with color centers showing an absorption peak at 500 nm? (**b**) If the absorption peak for KF corresponds to the promotion of an electron from the F center to the conduction band, determine the energy of the color center with respect to the conduction band. (The band gap in KF is 10.7 eV.) If the relative position of the color center energy level remains the same throughout the KF–KCl solid solution range, estimate (**c**) the band gap of KCl and (**d**) the band gap for the natural crystal.

7. (**a**) Calculate the magnetic moments on the Mn ion states in Figure 9.7. (**b**) If the crystal field splitting for Mn^{2+} is 0.97 eV (1.54×10^{-19} J), estimate the relative number of ions in the LS and IS states. (**c**) What wavelength of light will change the magnetic moment of the Mn^{2+} ions in the ground state?

8. $Mn_{0.01}Ga_{0.99}As$ is a ferromagnetic semiconductor prepared by doping the parent semiconductor GaAs with Mn^{2+}. (**a**) What is the magnetic moment of the Mn^{2+} ions if the spin-only formula applies to this material? (**b**) If ferromagnetism is due to the overlap of bound magnetic polarons, estimate the minimum radius of these defects. The unit cell of GaAs is cubic, $a = 0.56533$ nm, $Z = 4$ GaAs.

9. $Co_{0.04}Zn_{0.96}O$ is a ferromagnetic semiconductor. (**a**) Estimate the radius of a bound magnetic polaron in this material. (**b**) If the Co^{2+} ions are uniformly

distributed, calculate the volume associated with each ion. (**c**) Will bound magnetic polarons located on these ions overlap? The relative permittivity of $ZnO = 9.8$ (average), the effective mass of an electron $= 0.28\,m$, the unit cell is hexagonal, $a = 0.3250\,nm$, $c = 0.5206\,nm$, the volume of the unit cell can be taken as $0.866a^2c$.

10. The band gap of the semiconductor HgTe is $0.06\,eV$. (**a**) What is the ratio of spontaneous to stimulated emission from this semiconductor at 300 K, for transitions from the valence band to the conduction band? (**b**) If the band gap is constant, at what temperature does the rate of spontaneous emission equal the rate of stimulated emission?

FURTHER READING

Introductory material concerning the magnetic and optical properties of solids is given in:

R. J. D. Tilley, *Understanding Solids*, Wiley, Chichester, 2004.

Magnetic materials are covered in:

N. Spaldin, *Magnetic Materials*, Cambridge University Press Cambridge, United Kingdom, 2003.

Specific topics of relevance are in:

Various authors, *Mat. Res. Soc. Bull.*, **28**(10) (2003) (magnetic semiconductors).

J. E. Greedan, Magnetic Oxides, in *Encyclopedia of Inorganic Chemistry*, 2nd ed., Wiley, Hoboken, NJ, 2006.

W. E. Hatfield, Magnetic Transition Metal Ions, in *Encyclopedia of Inorganic Chemistry*, 2nd ed., Wiley, Hoboken, NJ, 2006.

More information on color and optical properties in general is in:

P. Bamfield, *Chromic Phenomena*, The Royal Society of Chemistry Cambridge, United Kingdom, 2001.

K. Nassau, *The Physics and Chemistry of Color*, 2nd ed., Wiley, New York, 2001.

R. J. D. Tilley, *Colour and the Optical Properties of Materials*, Wiley, Chichester, 2000.

Information on specific topics is in:

G. Blasse and B. C. Grabmeier, *Luminescent Materials*, Springer, Berlin, 1994.

C. G. Granquist, *Handbook of Inorganic Electrochromic Materials*, Elsevier, Amsterdam, 1995.

C. Fouassier, Luminescence, in *Encyclopedia of Inorganic Chemistry*, 2nd ed., Wiley, Hoboken, NJ, 2006.

M. Greenblatt, Ionic Conductors, in *Encyclopedia of Inorganic Chemistry*, 2nd ed., Wiley, Hoboken, NJ, 2006.

D. R. Smith, Ligand Field Theory and Spectra, in *Encyclopedia of Inorganic Chemistry*, 2nd ed., Wiley, Hoboken, NJ, 2006.

M. S. Whittingham, *Mat. Res. Bull.*, **XIV**, September, 31–38 (1989).

An introduction to (N–V) centers is in:

D. A. Awschalom, R. Epstein, and R. Hanson, *Sci. Am.*, **297**(4), 58–65 (2007).

Supplementary Material

S1 CRYSTAL STRUCTURES

S1.1 Crystal Systems and Unit Cells

In crystals of any material, the atoms present are always arranged in exactly the same way, over the whole extent of the solid, and exhibit *long-range translational order*. A crystal is conventionally described by its crystal structure, which comprises the unit cell, the symmetry of the unit cell, and a list of the positions of the atoms that lie in the unit cell.

Crystal structures and crystal lattices are *different*, although these terms are frequently (and incorrectly) used as synonyms. A crystal structure is built of atoms. A crystal lattice is an infinite pattern of points, each of which must have the same surroundings in the same orientation. A lattice is a mathematical concept.

The external form of crystals, the internal crystal structures, and the three-dimensional lattices need to be defined unambiguously. For this purpose, a set of vectors, **a**, **b**, and **c** are chosen to define the structure. The vectors **a**, **b**, and **c** can be selected in any number of ways, and crystallographic convention is to choose vectors that are small and reveal the underlying symmetry of the lattice. The parallelepiped formed by the three vectors **a**, **b**, and **c** defines the unit cell of the crystal, with edges of length a, b, and c. The numerical values of the unit cell edges and the angles between them are collectively called the lattice parameters or unit cell parameters. Conventionally, the axes are drawn so that the **a** axis points out from the page, the **b** axis points to the right, and the **c** axis is vertical. The angles between the axes are chosen to be equal to or greater than 90° whenever possible. These are labeled α, β, and γ, where α lies between **b** and **c**, β lies between **a** and **c**, and γ lies between **a** and **b**. Just seven different arrangements of axes, called crystal systems, and hence of unit cells, are needed in order to specify all three-dimensional structures and lattices (Table S1.1).

The unique axis in the monoclinic unit cell is mostly taken as the **b** axis. Rhombohedral unit cells are often specified in terms of a bigger hexagonal unit cell.

Different compounds that crystallize with the same crystal structure, for example, the two alums, $NaAl(SO_4)_2 \cdot 12H_2O$ and $NaFe(SO_4)_2 \cdot 12H_2O$, are said to

Defects in Solids, by Richard J. D. Tilley
Copyright © 2008 John Wiley & Sons, Inc.

TABLE S1.1   Crystal Systems

System	Unit Cell Parameters
Cubic (isometric)	$a = b = c$; $\alpha = 90°$, $\beta = 90°$, $\gamma = 90°$
Tetragonal	$a = b \neq c$; $\alpha = 90°$, $\beta = 90°$, $\gamma = 90°$
Orthorhombic	$a \neq b \neq c$; $\alpha = 90°$, $\beta = 90°$, $\gamma = 90°$
Monoclinic	$a \neq b \neq c$; $\alpha = 90°$, $\beta \neq 90°$, $\gamma = 90°$
Triclinic	$a \neq b \neq c$; $\alpha \neq 90°$, $\beta \neq 90°$, $\gamma \neq 90°$
Hexagonal	$a = b \neq c$; $\alpha = 90°$, $\beta = 90°$, $\gamma = 120°$
Rhombohedral	$a = b = c$; $\alpha = \beta = \gamma \neq 90°$
	$a' = b' \neq c'$; $\alpha' = 90°$, $\beta' = 90°$, $\gamma' = 120°$

be isomorphous[1] or isostructural. Sometimes the crystal structure of a compound will change with temperature and with applied pressure. This is called polymorphism. Polymorphs of elements are known as allotropes. Graphite and diamond are two allotropes of carbon, formed at different temperatures and pressures.

Although the unit cell of a solid with a fixed composition varies with temperature and pressure, at room temperature and atmospheric pressure it is regarded as constant. If the solid has a composition range, as in a solid solution, an alloy, or a nonstoichiometric compound, the unit cell parameters *vary* as the composition changes.

S1.2   Crystal Planes and Miller Indices

The facets of a well-formed crystal or internal planes through a crystal structure are specified in terms of Miller indices. These indices, h, k, and l, written in round brackets (hkl), represent not just one plane but the set of all parallel planes (hkl). The values of h, k, and l are the fractions of a unit cell edge, a, b, and c, respectively, intersected by this set of planes (Fig. S1.1).

A plane that lies parallel to a cell edge, and so never cuts it, is given the index 0 (zero). Thus, the set of planes that pass across the end of each unit cell cutting the **a** axis and parallel to the **b** and **c** axes of the unit cell has Miller indices (100). The indices indicate that the plane cuts the cell edge running along the **a** axis at a position $1a$ and does not cut the cell edges parallel to the **b** or **c** axes at all. The set of planes parallel to this, but separated by half the spacing, so that the cell edge is cut at 0, $a/2$, a, $3a/2$, and so on, has indices (200). Similarly, parallel planes cutting the cell edge at 0, $a/3$, $2a/3$, a, and $4a/3$ would have Miller indices of (300). Any general plane parallel to (100) is written ($h00$).

A general plane parallel to the **a** and **c** axes, and so cutting the **b** axis, has indices ($0k0$), and a general plane parallel to the **a** and **b** axes, and so cutting the **c** axis, has indices ($00l$). The nomenclature is exactly the same as described for ($h00$) planes.

The set of planes that cut two unit cell edges and parallel to a third are described by indices ($hk0$), ($0kl$), or ($h0l$). A plane (110) intersects the cell edges in $1a$, $1b$, and lies

---

[1]This description originally applied to the same external form of the crystals rather than the internal arrangement of the atoms.

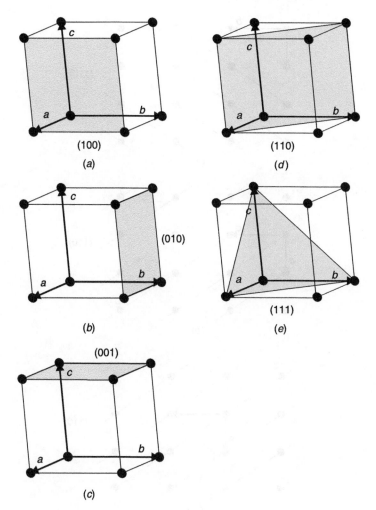

Figure S1.1   Miller indices of crystal or lattice planes.

parallel to the **c** axis. This notation is readily extended to cases where a plane cuts all three unit cell edges, in which case Miller indices (*hkl*) are required. A plane with indices (111) cuts the unit cell edges in 1*a*, 1*b*, and 1*c*.

Negative intersections are written with a negative sign over the index, and pronounced *h* bar, *k* bar, and *l* bar. For example, as well as the (110) plane, a similar plane also cuts the **a** axis in 1*a* and the **b** axis at −*b*, and so has Miller indices (1$\bar{1}$0), pronounced (one, one bar, zero) (Fig. S1.2).

In crystals of high symmetry, there are often several sets of (*hkl*) planes that are identical. For example, in a cubic crystal, the (100), (010), and (001) planes are identical in every way. Similarly, in a tetragonal crystal, (110) and ($\bar{1}$10) planes are identical. Curly brackets, {*hkl*}, designate these related planes. Thus, in the cubic system, the symbol {100} represents the three sets of planes (100), (010), and

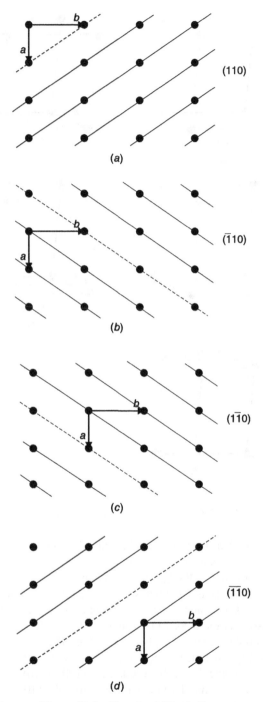

Figure S1.2 Negative Miller indices.

(001), $\{110\}$ represents the six sets of planes (110), (101), (011), ($\bar{1}$10), ($\bar{1}$01), and (0$\bar{1}$1), and $\{111\}$ represents the four sets (111), (11$\bar{1}$), (1$\bar{1}$1), and ($\bar{1}$11).

The Miller indices of planes in crystals with a hexagonal unit cell can be ambiguous. In order to eliminate this ambiguity, four indices, ($hkil$), are often used. These are called Miller–Bravais indices and are only used in the hexagonal system. The index i is given by

$$h + k + i = 0 \quad \text{or} \quad i = -(h + k)$$

In reality this third index is not needed. However, it does help to bring out the relationship between the planes. Because it is a redundant index, the value of i is sometimes replaced by a dot, to give indices ($hk \cdot l$). This nomenclature emphasizes that the hexagonal system is under discussion without actually including a value for i.

S1.3 Directions

The response of a crystal to an external stimulus such as a tensile stress, electric field, and so on is usually dependent upon the direction of the applied stimulus. It is therefore important to be able to specify directions in crystals in an unambiguous fashion. Directions are written generally as [uvw] and are enclosed in square brackets. Note that the symbol [uvw] means all parallel directions or vectors.

The three indices u, v, and w define the coordinates of a point with respect to the crystallographic **a**, **b**, and **c** axes. The index u gives the coordinates in terms of a along the **a** axis, the index v gives the coordinates in terms of b along the **b** axis, and the index w gives the coordinates in terms of c along the **c** axis. The direction [uvw] is simply the vector pointing from the origin to the point with coordinates u, v, w (Fig. S1.3). For example, the direction [100] is parallel to **a**, the direction [010] is parallel to **b**, and [001] is parallel to **c**. Because directions are vectors, [uvw] is not identical to [$\bar{u}\bar{v}\bar{w}$], in the same way that the direction "north" is not

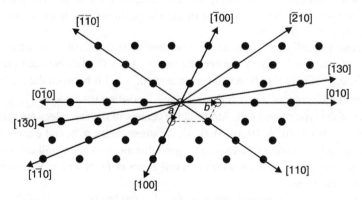

Figure S1.3 Directions in a lattice.

the same as the direction "south." Remember, though, that, any parallel direction shares the symbol [uvw] because the origin of the coordinate system is not fixed and can always be moved to the starting point of the vector. A north wind is always a north wind, regardless of where you stand.

As with Miller indices, it is sometimes convenient to group together all directions that are identical by virtue of the symmetry of the structure. These are represented by the notation $<uvw>$. In a cubic crystal, $<100>$ represents the six directions [100], [$\bar{1}$00], [010], [0$\bar{1}$0], [001], and [00$\bar{1}$].

It is sometimes important to specify a vector with a definite length, perhaps to indicate the displacement of one part of a crystal with respect to another part. In such a case, the direction of the vector is written as above, and a prefix is added to give the length. The prefix is usually expressed in terms of the unit cell dimensions. For example, in a cubic crystal, a displacement of two unit cell lengths parallel to the **b** axis would be written $2a$[010].

The relationship between directions and planes depends upon the symmetry of the crystal. In cubic crystals (and only cubic crystals), the direction [hkl] is normal to the plane (hkl). A zone is a set of planes, all of which are parallel to a single direction, called the zone axis. The zone axis [uvw] is perpendicular to the plane (uvw) in cubic crystals but not in crystals of other symmetry.

S1.4 Crystal Structures

The minimum amount of information needed to specify a crystal structure is the unit cell type, that is, cubic, tetragonal, and so on, the unit cell parameters, and the positions of all of the atoms in the unit cell. The atomic contents of the unit cell are a simple multiple, Z, of the composition of the material. The value of Z is equal to the number of formula units of the solid in the unit cell. Atom positions are expressed in terms of three coordinates, x, y, and z. These are taken as fractions of a, b, and c, the unit cell sides, say $\frac{1}{2}$, $\frac{1}{2}$, and $\frac{1}{4}$. The x, y, and z coordinates are plotted with respect to the unit cell axes, not to a Cartesian set of axes. The space group describes the symmetry of the unit cell, and although it is not mandatory when specifying a structure, its use considerably shortens the list of atomic positions that must be specified in order to built the structure.

An atom at a cell corner is given the coordinates $(0, 0, 0)$. An atom at the center of the face of a unit cell is given the coordinates $\left(\frac{1}{2}, \frac{1}{2}, 0\right)$ if it lies between the **a** and **b** axes, $\left(\frac{1}{2}, 0, \frac{1}{2}\right)$ if between the **a** and **c** axes, and $\left(0, \frac{1}{2}, \frac{1}{2}\right)$ if between the **b** and **c** axes. An atom at the center of a unit cell would have a position specified as $\left(\frac{1}{2}, \frac{1}{2}, \frac{1}{2}\right)$, irrespective of the type of unit cell. Atoms at the centers of the cell edges are specified at positions $\left(\frac{1}{2}, 0, 0\right)$, $\left(0, \frac{1}{2}, 0\right)$, or $\left(0, 0, \frac{1}{2}\right)$, for atoms on the **a**, **b**, and **c** axes. Stacking of the unit cells to build a structure will ensure that an atom at the unit cell origin will appear at every corner, and atoms on unit cell edges or faces will appear on all of the cell edges and faces.

In figures, the conventional origin is placed at the left rear corner of the unit cell. The **a** or x axis is represented as projecting out of the plane of the page, toward the reader, the **b** or y axis points to the right, and the **c** or z axis points toward the top of the page. In projections, the origin is set at the upper left corner of the unit cell projection. A frequently encountered projection is that perpendicular to the **c** axis. In this case, the **a** or x axis is drawn pointing down (from top to bottom of the page) and the **b** or y axis pointing to the right. In projections the x and y coordinates can be determined from the figure. The z position is usually given on the figure as a fraction.

A vast number of structures have been determined, and it is very convenient to group those with topologically identical structures together. On going from one member of the group to another, the atoms in the unit cell differ, reflecting a change in chemical compound, and the atomic coordinates and unit cell dimensions change slightly, reflecting the difference in atomic size, but relative atoms positions are identical or very similar. Frequently, the group name is taken from the name of a mineral, as mineral crystals were the first solids used for structure determination. Thus, all crystals with a structure similar to that of sodium chloride, NaCl (the mineral halite), are said to adopt the sodium chloride, NaCl, rock salt or halite structure. These materials then all have a general formula MX, where M is a metal atom and X a nonmetal atom, for example, MgO. Similarly, crystals with a structure similar to the rutile form of titanium dioxide, TiO_2, are grouped with the rutile structure. These all have a general formula MX_2, for example, FeF_2. Crystallographic details of a number of simple inorganic structures are given below.

Copper (A1, face-centered cubic, fcc)

Structure: cubic; $a = 0.3610$ nm; $Z = 4$; *Space group, $Fm\bar{3}m$* (No. 225);
Atom positions: Cu: $4a$ $0, 0, 0$; $\frac{1}{2}, \frac{1}{2}, 0$; $0, \frac{1}{2}, \frac{1}{2}$; $\frac{1}{2}, 0, \frac{1}{2}$

Tungsten (A2, body-centered cubic, bcc)

Structure: cubic; $a = 0.3160$ nm; $Z = 2$; *Space group, $Im\bar{3}m$* (No. 229);
Atom positions: W: $2a$ $0, 0, 0$; $\frac{1}{2}, \frac{1}{2}, \frac{1}{2}$

Magnesium (A3)

Structure: hexagonal; $a = 0.3200$ nm, $c = 0.5200$ nm; $Z = 2$; *Space group, $P\,6_3/mmc$* (No. 194);
Atom positions: Mg: $2d$ $\frac{2}{3}, \frac{1}{3}, \frac{1}{4}$; $\frac{1}{3}, \frac{2}{3}, \frac{3}{4}$

Diamond (A4)

Structure: cubic; $a = 0.3567$ nm; $Z = 8$; *Space group, $Fd\bar{3}m$* (No. 227);
Atom positions: C: $8a$ $0, 0, 0$; $\frac{1}{2}, \frac{1}{2}, 0$; $0, \frac{1}{2}, \frac{1}{2}$; $\frac{1}{2}, 0, \frac{1}{2}$; $\frac{1}{4}, \frac{1}{4}, \frac{1}{4}$; $\frac{3}{4}, \frac{3}{4}, \frac{1}{4}$; $\frac{3}{4}, \frac{1}{4}, \frac{3}{4}$; $\frac{1}{4}, \frac{3}{4}, \frac{3}{4}$

Each atom is connected to its neighbors by four bonds pointing toward the vertices of a tetrahedron. The structure can also be considered to be made up of carbon tetrahedra, each containing a central carbon atom. Two other members of group 14, Si and Ge, as well as the allotrope of tin stable below $13.2°C$, gray tin or α-Sn, also adopt the diamond structure.

Graphite

Structure: hexagonal, $a = 0.2460$ nm, $c = 0.6701$ nm; $Z = 4$; *Space group, $P6_3mc$* (No. 186);
Atom positions: C1: $2a$ $0, 0, 0$; $0, 0, \frac{1}{2}$;
            C2: $2b$ $\frac{1}{3}, \frac{2}{3}, 0$; $\frac{2}{3}, \frac{1}{3}, \frac{1}{2}$

The carbon sheets form hexagonal nets, each of which is displaced with respect to the one below.

Sodium Chloride (halite, rock salt), NaCl

Structure: cubic; $a = 0.5630$ nm; $Z = 4$; *Space group*, $Fm\bar{3}m$ (No. 225);

Atom positions: Na: $4a$ $0, 0, 0;$ $\frac{1}{2}, \frac{1}{2}, 0;$ $\frac{1}{2}, 0, \frac{1}{2};$ $0, \frac{1}{2}, \frac{1}{2};$

Cl: $4b$ $\frac{1}{2}, 0, 0;$ $0, 0, \frac{1}{2};$ $0, \frac{1}{2}, 0;$ $\frac{1}{2}, \frac{1}{2}, \frac{1}{2};$

(or *vice versa*).

This structure is commonly adopted by oxides, nitrides halides, and sulfides MX, including the nonstoichiometric $3d$ transition-metal oxides $\sim$TiO, $\sim$VO, $\sim$MnO, $\sim$FeO, $\sim$CoO, and NiO.

Zinc Blende (sphalerite), cubic ZnS

Structure: cubic; $a = 0.5420$ nm; $Z = 4$; *Space group*, $F\bar{4}3m$ (No. 216);

Atom positions: Zn: $4a$ $0, 0, 0;$ $\frac{1}{2}, \frac{1}{2}, 0;$ $\frac{1}{2}, 0, \frac{1}{2};$ $0, \frac{1}{2}, \frac{1}{2};$

S: $4c$ $\frac{1}{4}, \frac{1}{4}, \frac{1}{4};$ $\frac{3}{4}, \frac{3}{4}, \frac{1}{4};$ $\frac{3}{4}, \frac{1}{4}, \frac{3}{4};$ $\frac{1}{4}, \frac{3}{4}, \frac{3}{4}$

The cubic zinc blende or sphalerite structure of ZnS is similar to that of diamond but with alternating sheets of Zn and S stacked parallel to the axes, replacing C.

This structure is adopted by a large number of III/V semiconductors, including GaAs, InP, and halides, including AgI, CuF, CuCl, CuBr, and CuI.

Wurtzite, hexagonal ZnS

Structure: hexagonal; $a = 0.3810$ nm, $c = 0.6230$ nm; $Z = 2$; *Space group*, $P6_3mc$ (No. 186);

Atom positions: Zn: $2b$ $\frac{1}{3}, \frac{2}{3}, \frac{1}{2};$ $\frac{2}{3}, \frac{1}{3}, 0;$

S: $2b$ $\frac{1}{3}, \frac{2}{3}, \frac{3}{8};$ $\frac{2}{3}, \frac{1}{3}, \frac{7}{8}$

In the wurtzite form of ZnS the sulfur atoms are arranged in hexagonal close packing, with the metal atoms in one-half of the tetrahedral positions. There are two layers of tetrahedra in the repeat distance, c, and these point in the same direction. This gives the materials a unique axis, the **c** axis, and these compounds show piezoelectricity.

The structure is adopted by a number of other compounds with a formula MX, including SiC, AlN, GaN, InN, ZnO, β-BeO, CdS, CdSe, CdTe, CuCl, CuBr, CuI, and β-AgI.

Corundum, Al₂O₃

Structure: trigonal, hexagonal axes; $a = 0.4763$ nm; $c = 1.3009$ nm; $Z = 6$; *Space group*, $R\bar{3}c$ (No. 167);

Atom positions: each of $(0, 0, 0);$ $\left(\frac{2}{3}, \frac{1}{3}, \frac{1}{3}\right);$ $\left(\frac{1}{3}, \frac{2}{3}, \frac{2}{3}\right);$ plus

Al: $12c$ $0, 0, z;$ $0, 0, \bar{z} + \frac{1}{2};$ $0, 0, \bar{z};$ $0, 0, z + \frac{1}{2};$

O: $18e$ $x, 0, \frac{1}{4};$ $0, x, \frac{1}{4};$ $\bar{x}, \bar{x}, \frac{1}{4};$ $\bar{x}, 0, \frac{3}{4};$ $0, \bar{x}, \frac{3}{4};$ $x, x, \frac{3}{4}$

The x coordinate (O) and the z coordinate (Al) can be approximated to $\frac{1}{3}$. In general, these can be written $x = \frac{1}{3} + u$ and $z = \frac{1}{3} + w$, where u and w are both small. Taking typical values of $x = 0.306$, and $z = 0.352$, the positions are:

Al: $12c$ $0, 0, 0.352$; $0, 0, 0.148$; $0, 0, 0.648$; $0, 0, 0.852$

O: $18e$ $0.306, 0, \frac{1}{4}$; $0, 0.306, \frac{1}{4}$; $0.694, 0.694, \frac{1}{4}$; $0.694, 0, \frac{3}{4}$;
$0, 0.694, \frac{3}{4}$; $0.306, 0.306, \frac{3}{4}$

The unit cell of corundum is rhombohedral, but the structure is usually described with respect to hexagonal axes. Each of the cations is surrounded by six oxygen ions in a slightly distorted octahedral coordination. The anions are close to a hexagonal close-packed array and the cations occupy two-thirds of the available octahedral positions in this array in an ordered fashion. The structure is also adopted by α-Fe_2O_3 and Cr_2O_3.

Fluorite, CaF$_2$

Structure: cubic; $a = 0.5450$ nm; $Z = 4$; *Space group*, $Fm\bar{3}m$ (No. 225);
Atom positions: Ca: $4a$ $0, 0, 0$; $\frac{1}{2}, \frac{1}{2}, 0$; $\frac{1}{2}, 0, \frac{1}{2}$; $0, \frac{1}{2}, \frac{1}{2}$;
F: $8c$ $\frac{1}{4}, \frac{1}{4}, \frac{1}{4}$; $\frac{1}{4}, \frac{3}{4}, \frac{3}{4}$; $\frac{3}{4}, \frac{1}{4}, \frac{3}{4}$; $\frac{3}{4}, \frac{3}{4}, \frac{1}{4}$; $\frac{1}{4}, \frac{1}{4}, \frac{3}{4}$; $\frac{1}{4}, \frac{3}{4}, \frac{1}{4}$;
$\frac{3}{4}, \frac{1}{4}, \frac{1}{4}$; $\frac{3}{4}, \frac{3}{4}, \frac{3}{4}$

Rutile, TiO$_2$

Structure: tetragonal; $a = 0.4594$ nm, $c = 0.2959$ nm; $Z = 2$; *Space group*, $P4_2/mnm$ (No. 136);
Atom positions: Ti: $2a$ $0, 0, 0$; $\frac{1}{2}, \frac{1}{2}, \frac{1}{2}$;
O: $4f$ $x, x, 0$; $\bar{x}, \bar{x}, 0$; $\bar{x} + \frac{1}{2}, x + \frac{1}{2}, \frac{1}{2}$; $x + \frac{1}{2}, \bar{x} + \frac{1}{2}, \frac{1}{2}$

The x coordinate for O is close to $\frac{1}{3}$. Taking a typical value of $x = 0.305$, the positions are:

O: $4f$ $0.305, 0.305, 0$; $0.695, 0.695, 0$; $0.195, 0.805, \frac{1}{2}$; $0.805, 0.195, \frac{1}{2}$

Titanium dioxide crystallizes in several forms. The most important is the rutile form. This structure is also adopted by SnO_2, MgF_2, and ZnF_2. A number of oxides that show metallic or metal–insulator transitions, for example, VO_2, NbO_2, and CrO_2, have a slightly distorted form of the structure.

Cadmium Iodide (ideal), CdI$_2$, Cadmium Hydroxide, Cd(OH)$_2$

Structure: hexagonal; $a = 0.424$ nm, $c = 0.684$ nm; $Z = 2$; *Space group*, $P\bar{3}m1$ (No. 164);
Atom positions: Cd: $1a$ $0, 0, 0$;
I: $2d$ $\frac{1}{3}, \frac{2}{3}, \frac{1}{4}$; $\frac{2}{3}, \frac{1}{3}, -\frac{1}{4}$

Because CdI_2 has a large number of polymorphic forms, the structure type is now referred to as the $Cd(OH)_2$ structure, which is unambiguous. The idealized structure given above is sufficient for the descriptions in this book.

The CdI_2 structure made up of layers of iodine atoms stacked up in a hexagonal ... ABAB ... fashion. Every other sheet of octahedral sites is filled by cadmium to generate layers of composition CdI_2. The stacking sequence is:

$$...A\gamma B \quad A\gamma B \quad A\gamma B \quad A\gamma B ...$$

where γ represents the octahedral site occupied, following the convention $\alpha = A$, $\beta = B$, $\gamma = C$ where A and so forth is used for anions and α etc for cations. These layers are only weakly held together by secondary bonding. CdI_2 is easily cleaved into sheets parallel to the layers because of this.

This structure is adopted by many halides and hydroxides, including $Mg(OH)_2$, $Cd(OH)_2$, and many sulfides including TiS_2, ZrS_2, HfS_2, α-TaS_2, PtS_2, and SnS_2.

Cadmium Chloride (ideal), CdCl₂

Structure: rhombohedral, hexagonal axes; $a = 0.385$ nm, $c = 1.746$ nm; $Z = 3$; space group, $R\bar{3}m$ (No. 166);

Atom positions: Cd: $3a$ $\quad 0, 0, 0;$ $\quad \frac{1}{3}, \frac{2}{3}, \frac{2}{3};$ $\quad \frac{2}{3}, \frac{1}{3}, \frac{1}{3};$

Cl: $6c$ $\quad \pm(u, u, u;$ $\frac{1}{3}+u, \frac{2}{3}+u, \frac{2}{3}+u;$ $\frac{2}{3}+u, \frac{1}{3}+u, \frac{1}{3}+u);$ $\quad u = \frac{1}{4}$

This structure has a rhombohedral symmetry but is usually described in terms of a triple-volume hexagonal cell, which makes comparison with the idealized CdI_2 structure simpler. The idealized unit cell is adequate for the purposes of this book. In this representation the anion layers are in cubic closest packing ... ABCABCABC...

The metal and nonmetal stacking sequence is

$$...A\gamma B \quad C\beta A \quad B\alpha C ...$$

Molybdenite (β-MoS₂) and Related Sulfides

Structure: hexagonal; $a = 0.316$ nm, $c = 1.230$ nm; $Z = 2$; *Space group*, $P6_3/mmc$ (No. 194);

Atom positions: Mo: $2c$ $\quad \pm(\frac{1}{3}, \frac{2}{3}, \frac{1}{4})$

S: $4f$ $\quad \pm(\frac{1}{3}, \frac{2}{3}, 0.63;$ $\frac{2}{3}, \frac{1}{3}, 0.13)$

This is a layer structure similar to the CdI_2 structure except that the pairs of S layers are directly superimposed upon each other, so that the central cation is in trigonal prismatic coordination, not octahedral or tetrahedral. In the structures derived from this S–Mo–S unit, these layers can be stacked like the simple layers in atomic close-packed sequences. The molybdenite (β-MoS_2) structure has these layers in hexagonal packing, so that the sequence of layers is ... BαB AβA ....

This is also known as the C7 structure or given the symbol $2H_1$ where the initial number gives the number of S–Mo–S layers in the repeat. Other compounds with this structure are

$$2H_1 \text{ β-}MoS_2, (MoS_2 \text{ hex}), \text{ and } WS_2.$$

An alternative two-layer structure is adopted by a hexagonal form of NbS_2 (NbS_2 hex), with the sequence ... AγA BγB ....

The cubic stacking, with a sequence ... AβA BγB CαC .... is called the 3R form (with three composite S–M–S layers) and is adopted by 3R-MoS$_2$ (Rh), NbS$_2$ (Rh) TaS$_2$, WS$_2$, ReS$_2$.

Rhenium Trioxide, ReO$_3$

Structure: cubic; $a = 0.3750$ nm; $Z = 1$; *Space group, Pm$\bar{3}$m* (No. 221);
Atom positions: Re: 1*a*   0, 0, 0;
O: 3*d* $\frac{1}{2}$, 0, 0;   0, $\frac{1}{2}$, 0;   0, 0, $\frac{1}{2}$

This structure is often used to represent an idealized cubic form of the monoclinic oxide WO$_3$.

Strontium Titanate, SrTiO$_3$ and Perovskites

Structure: cubic; $a = 0.3905$ nm; $Z = 1$; *Space group, Pm$\bar{3}$m* (No. 221);
Atom positions: Ti: 1*a*   0, 0, 0;
Sr: 1*b* $\frac{1}{2}$, $\frac{1}{2}$, $\frac{1}{2}$;
O: 3*c*   0, $\frac{1}{2}$, $\frac{1}{2}$; $\frac{1}{2}$, 0, $\frac{1}{2}$; $\frac{1}{2}$, $\frac{1}{2}$, 0

The perovskite structure is adopted by number of compounds with a formula ABX$_3$, where A is a large cation, B a medium-sized cation, and X an anion. Perovskite is a mineral of composition CaTiO$_3$. The structure of this compound, initially thought to be cubic, was later shown to be of lower symmetry than the simple cubic structure first assigned to it. The high-symmetry version, called the ideal or cubic perovskite structure, is adopted by the oxide SrTiO$_3$. It is often convenient to think of the structure as built up from an array of corner sharing TiO$_6$ octahedra with the large Sr^{2+} ion located at the cell center. The TiO$_6$ framework is then similar to that in ReO$_3$, described above.

The framework representation of the structure indicates that the edge of the cubic unit cell, a, is equal to twice the Ti–O bond length:

$$a = 2(r_{Ti} + r_O)$$

where r_{Ti} and r_O are the radii of the Ti and O ions. This relationship fixes the size of the internal cage site. If it is assumed that for a stable structure to form the central cation must just touch the surrounding oxygen anions (Goldschmidt's rule), then

$$(r_{Sr} + r_O) = \left(\frac{\sqrt{2}}{2}\right) a = \sqrt{2}(r_{Ti} + r_O)$$

This relationship can be generalized for any cubic perovskite ABX$_3$ and provides a useful guide to the formation of these phases if it is written in the form:

$$(r_A + r_X) = \left(\frac{\sqrt{2}}{2}\right) a = t\sqrt{2}(r_B + r_X)$$

or

$$t = \frac{r_A + r_X}{\sqrt{2}(r_B + r_X)}$$

In this equation r_A is the radius of the cage site cation, r_B is the radius of the octahedrally coordinated cation, and r_X is the radius of the anion. The factor t is called the *tolerance factor*. Ideally, t should be equal to 1.0, and it has been found empirically that if t lies in the approximate range 0.9–1.0, a cubic perovskite structure is stable. However, some care must be exercised when using this simple concept. It is necessary to use ionic radii appropriate to the coordination geometry of the ions. Thus, r_A should be appropriate to 12 coordination, r_B to octahedral coordination, and r_X to linear coordination. Within this limitation the tolerance factor has good predictive power.

The structure must be electrically neutral. If the charges on the ions are written as q_A, q_B, and q_X, then

$$q_A + q_B = -3q_X$$

where the anions are taken to show negative charge. The most likely combinations are

$$X = -1 \quad (\text{i.e., } F^-, \ Cl^-, \ Br^-), \ (A, B) = (1, 2)$$
$$= -2 \quad (\text{i.e., } O^{2-}), \ (A, B) = (1, 5), (2, 4), (3, 3)$$
$$= -3 \quad (\text{i.e., } N^{3-}), \ (A, B) = (3, 6)$$

The largest number of ideal perovskites synthesized to date are oxides of general formula ABO_3. Some examples are:

$A^+B^{5+}O_3$: $RbUO_3$, $KTaO_3$
$A^{2+}B^{4+}O_3$: $BaSnO_3$, $BaZrO_3$, $SrMnO_3$
$A^+B^{2+}X_3$: $CsCdF_3$, $CsCaF_3$, $KMgF_3$, $KNiF_3$, $CsCaCl_3$, $CsCdCl_3$, $CsHgBr_3$

In many compounds of formula ABX_3, slight distortions of the BX_6 octahedra lower the symmetry away from cubic to tetragonal, orthorhombic, and so on. This enormous family of compounds includes many technologically important materials, including $BaTiO_3$, $NaNbO_3$, and $LiNbO_3$.

Spinel, MgAl₂O₄ and AB₂O₄ Spinels

Structure: cubic; $a = 0.8090$ nm; $Z = 8$; *Space group*, $Fd\bar{3}m$ (No. 227);
Atom positions: each of $(0, 0, 0)$; $(0, \frac{1}{2}, \frac{1}{2})$; $(\frac{1}{2}, 0, \frac{1}{2})$; $(\frac{1}{2}, \frac{1}{2}, 0)$; plus

Mg: 8a $0, 0, 0$; $\frac{3}{4}, \frac{1}{4}, \frac{3}{4}$;

Al: 16d $\frac{5}{8}, \frac{5}{8}, \frac{5}{8}$; $\frac{3}{8}, \frac{7}{8}, \frac{1}{8}$; $\frac{7}{8}, \frac{1}{8}, \frac{3}{8}$; $\frac{1}{8}, \frac{3}{8}, \frac{7}{8}$;

O: 32e x, x, x; $\bar{x}, \bar{x}+\frac{1}{2}, x+\frac{1}{2}$; $\bar{x}+\frac{1}{2}, x+\frac{1}{2}, \bar{x}$; $x+\frac{1}{2}, \bar{x}, \bar{x}+\frac{1}{2}$; $x+\frac{3}{4}, x+\frac{1}{4}$,
$\bar{x}+\frac{1}{4}$; $x+\frac{1}{4}, \bar{x}+\frac{3}{4}, x+\frac{3}{4}$; $\bar{x}+\frac{3}{4}, x+\frac{3}{4}, x+\frac{1}{4}$

The x coordinate for O is approximately equal to $\frac{3}{8}$ and is often given in the form $\frac{3}{8} + u$, where u is of the order of 0.01. Taking a typical value of $x = 0.388$, in which

$u = 0.013$, the positions are

O: 32e   0.388, 0.388, 0.388;   0.612, 0.112, 0.888;   0.112, 0.888, 0.612;

0.888, 0.612, 0.112;   0.138, 0.638, 0.862;   0.862, 0.862, 0.862;

0.368, 0.362, 0.138;   0.362, 0.138, 0.638

There are four lattice points in the face-centered unit cell, and the motif is two $MgAl_2O_4$ complexes. This structure is named after the mineral spinel, $MgAl_2O_4$. The oxygen atoms in the crystal structure are in the same relative positions as the chlorine atoms in 8 unit cells of NaCl, stacked together to form a $2 \times 2 \times 2$ cube. Thus, in the cubic unit cell of spinel there are 32 oxygen atoms, to give a unit cell contents of $Mg_8Al_{16}O_{32}$. The Mg and Al atoms are inserted into this array in an ordered fashion. To a good approximation, all of the magnesium atoms are surrounded by 4 oxygen atoms in the form of a tetrahedron and are said to occupy tetrahedral positions, or sites, in the structure. Similarly, to a good approximation, the aluminum atoms are surrounded by 6 oxygen atoms and are said to occupy octahedral positions or sites. In each unit cell there are the same number of octahedral sites as there are oxygen ions, that is, 32, and twice as many tetrahedral sites as oxygen ions, that is 64. The Mg^{2+} and Al^{3+} cations fill one-half of the available octahedral positions and one-eighth of the available tetrahedral positions. This means that there are 8 occupied tetrahedral sites and 16 occupied octahedral sites in a unit cell.

When the structure is viewed down [111], the oxygen atoms can be seen to form close cubic packed layers, emphasizing the relationship with the NaCl structure.

The mineral spinel, $MgAl_2O_4$, has given its name to an important family of compounds with the same structure, collectively known as *spinels*, which includes halides, sulfides, and nitrides as well as oxides. These are often regarded as ionic compounds. The formula of the oxide spinels, AB_2O_4, is satisfied by a number of combinations of cations, the commonest of which is A^{2+} and B^{3+}, typified by Mg^{2+} and Al^{3+} in spinel itself. There are two principal arrangements of cations found. If the 8 A^{2+} ions per unit cell are confined to the available tetrahedral sites, these are filled completely. The 16 B^{3+} ions are then confined to the octahedral sites. This cation distribution is often depicted as $(A)[B_2]O_4$, with the tetrahedral cations in ( ) and the octahedral cations in [ ], or just $A[B_2]O_4$ where only the octahedral sites are distinguished, it being taken for granted that the other ions are in tetrahedral sites. This is called the *normal spinel structure*, and compounds with this arrangement of cations are said to be *normal spinels*. If the 8 A^{2+} ions are placed in half of the available 16 octahedral sites, half of the B^{3+} ions must be placed in the remaining octahedral sites and the other half in the tetrahedral sites. This can be written as $(B)[AB]O_4$ or $B[AB]O_4$. This arrangement is called the *inverse spinel structure* and compounds with this cation arrangement are said to be *inverse spinels*.

In reality, very few spinels have exactly the normal or inverse structure, and these are called *mixed spinels*. The cation distribution between the two sites is a function of a number of parameters, including temperature. This variability is described by an occupation factor, λ, which gives the fraction of B^{3+} cations in tetrahedral positions. The spinel formula is written $(A_{1-2\lambda}B_{2\lambda})[A_{2\lambda}B_{2-2\lambda}]O_4$. A normal spinel is characterized by a value of λ of 0, and an inverse spinel by a value of λ of 0.5. An alternative description uses an inversion parameter, i, which is the total amount of B^{3+} cations in tetrahedral positions. The spinel formula is written $(A_{1-i}B_i)[A_iB_{2-i}]O_4$. The relationship between the two is $i = 2\lambda$. The data for spinel, $MgAl_2O_4$, given above has been taken from a sample with a value of $\lambda = 0.05$ ($i = 0.1$) and so is quite a good approximation to a normal spinel. Many samples of $MgAl_2O_4$, including natural samples, are mixed spinels with recorded values of λ up to 0.1 (i up to 0.2).

Cubic $A^{2+}Fe^{3+}O_4$, ferrites form an important group of magnetic oxides that have (in idealized circumstances) an inverse spinel structure $(Fe^{3+})[A^{2+}Fe^{3+}]O_4$. Lodestone, or magnetite, Fe_3O_4, is an inverse spinel with a formula of $(Fe^{3+})[Fe^{2+}Fe^{3+}]O_4$. The tetrahedral and octahedral sites in cubic ferrites provide two magnetic substructures, which give these oxides very flexible magnetic properties that can be tailored by varying the cations and the distribution between the octahedral and tetrahedral sites. In reality, the distribution of the cations in cubic ferrites is rarely perfectly normal or inverse, and the distribution tends to vary with temperature. Thus processing conditions are important if the desired magnetic properties are to be obtained.

Potassium Nickel Fluoride, K_2NiF_4

Structure: tetragonal; $a = 0.4006$ nm, $c = 1.3706$ nm; $Z = 2$; *Space group*, $I4/mmm$ (No. 139);

Atom positions: Ni: $2a$ $0, 0, 0$; $\frac{1}{2}, \frac{1}{2}, \frac{1}{2}$;

K: $4e$ $\pm (0, 0, z;$ $\frac{1}{2}, \frac{1}{2}, z + \frac{1}{2})$; $z = 0.35$;

F(1): $4c$ $0, \frac{1}{2}, 0$; $\frac{1}{2}, 0, 0$; $\frac{1}{2}, 0, \frac{1}{2}$; $0, \frac{1}{2}, \frac{1}{2}$;

F(2): $4e$ $\pm (0, 0, z;$ $\frac{1}{2}, \frac{1}{2}, z + \frac{1}{2})$; $z = 0.15$

The K_2NiF_4 structure, also known as the K_2MgF_4 structure, is made up of sheets of the perovskite type one octahedron in thickness stacked up one on top of the other. The octahedra contain the small Ni^{2+} cations; the large K^+ ions lie between the sheets of octahedra. The positions of the K^+ ions and the neighboring F^- ions are similar to the cation and anion positions in the NaCl structure, and so the K_2NiF_4 structure is often said to be made up of an *intergrowth* of perovskite and sodium-chloride-like slabs. As well as fluorides and a few chlorides, a large number of oxides crystallize with this structure, including La_2NiO_4, Sr_2TiO_4, and Sr_2SnO_4.

The structure of La_2CuO_4 is closely related to that of K_2NiF_4, but the CuO_6 octahedra are lengthened along the **c** axis, compared to the regular octahedra in the above compounds. At room temperature the cell is orthorhombic with a and b approximately equal to the cell diagonal of the tetragonal ideal K_2NiF_4 cell, $a \sim b \sim \sqrt{2}\, a_{tet.}$. Above 260°C it becomes tetragonal. In the superconductor literature this structure is often called the T or T/O structure.

FURTHER READING

R. J. D. Tilley, *Crystals and Crystal Structures*, Wiley Chichester, 2006, and references therein.

S2   BAND THEORY

S2.1   Energy Bands

At the beginning of the twentieth century explanations of the electrical properties of solids were hazy. The best that could be said was that metals could be supposed to contain a gas of free electrons, with properties derived from an approach that mirrored the kinetic theory of gases. The theory of the electron pair bond and the notion of valence were not as obviously applicable to solids as to molecules. In fact progress in understanding the nature of the electrical properties in solids came from the direct application of the wave equation to the free electron theory. This was first attempted by Sommerfeld in 1928, rapidly followed by Bloch in 1929, who took into account the periodic nature of the atom array in a crystal, and Wilson, in 1931, who distinguished between metals, semiconductors, and insulators on the basis of Bloch's calculations. It was soon realized that the breakdown into these three classes was incomplete. Some solids, especially transition-metal oxides such as CoO or NiO should be metals in terms of the simple theory but were, in fact, insulators. The dilemma was resolved by Mott in the years between 1937 and 1949, and the possibility of transforming insulating solids into metals was appreciated. In the years between 1946 and 1950 the semiconductor industry was founded on the practical and theoretical studies of Bardeen, Brattain, and Shockley into the construction and operation of the transistor. By the middle of the twentieth century, the mystery surrounding the electronic behavior of solids was erased. The theory as a whole is generally called band theory.

In band theory the electrons responsible for conduction are not linked to any particular atom. They can move easily throughout the crystal and are said to be free or very nearly so. The wave functions of these electrons are considered to extend throughout the whole of the crystal and are delocalized. The outer electrons in a solid, that is, the electrons that are of greatest importance from the point of view of both chemical and electronic properties, occupy *bands of allowed energies*. Between these bands are regions that cannot be occupied, called band gaps.

The energy bands are derived from the overlap of the outer electron energy levels on the atoms making up the solid. The respective bands are then given names describing the principal contributing orbitals. Thus bands may be described as mainly of s, p, or d character. The band structure of a crystal is quite complex, and energy bands can overlap in certain crystallographic directions while being well separated in others. Fortunately, the complex three-dimensional nature of the bands in a solid can be ignored for many purposes and a "flat-band" picture is then adequate.

S2.2   Insulators, Semiconductors and Metals

The fundamental division of materials when electrical properties are considered is into metals, insulators, and semiconductors. An *insulator* is a material that normally shows no electrical conductivity. Metals and semiconductors were originally classified more or less in terms of the magnitude of the measured electrical conductivity. However, a better definition is to include in *metals* those materials for which the *electrical conductivity falls* as the temperature *increases*. *Semiconductors* show an *increase in electrical* conductivity as the temperature *increases*.

Metals are defined as materials in which the uppermost energy band is only partly filled. The uppermost energy level filled is called the *Fermi energy* or the *Fermi level*. Conduction can take place because of the easy availability of empty energy levels just above the Fermi energy. In a crystalline metal the Fermi level possesses a complex shape and is called the *Fermi surface*. Traditionally, typical metals are those of the alkali metals, Li, Na, K, and the like. However, the criterion is not restricted to elements, but some oxides, and many sulfides, are metallic in their electronic properties.

Insulators have the upper band completely empty and the lower band completely filled by electrons. The energy gap between the top of the filled band and the bottom of empty band is quite large. The *filled* band of energies is called the *valence band*, and the empty band is called the *conduction band*. The energy difference between the top of the valence band and the bottom of the conduction band is called the band gap. When a material conducts electricity the electrons pick up some energy and are transferred to slightly higher energy levels. If the electron-containing band is full, conductivity cannot occur because there are no slightly higher energy levels available.[2]

Intrinsic semiconductors have a similar band picture to insulators except that the separation of the empty and filled bands is small. The band gap must be such that some electrons will be transferred from the top of the valence band to the bottom of the conduction band at room temperature. The electrons in the conduction band will have plenty of slightly higher energy levels available, and, when a voltage is imposed upon the material, they can take up some energy, and this leads to some degree of conductivity. At 0 K these materials will be insulators because no electrons will cross from the valence band to the conduction band. The magnitude of their electrical conductivity will increase as the temperature increases because more electrons will gain sufficient energy to cross the band gap. The Fermi level for an intrinsic semiconductor lies at the midpoint of the band gap.

Rather surprisingly, an equal contribution to the conductivity will come from positive charge carriers equal in number to the electrons promoted into the conduction band. These are "vacancies" in the valence band and are called *positive holes* or more generally holes. Each time an electron is removed from the full valence band to the conduction band, *two* mobile charge carriers are therefore created, an *electron* and a *hole*.

The magnitude of the band gap can be estimated by measuring the variation of the conductivity of a semiconductor with temperature, to give the *thermal band gap*. Alternatively, the energy of light photons that are just sufficient to excite an electron across the band gap can be determined to give an estimate of the *optical band gap*. Optical band gap measurements show more transitions than just a single energy peak corresponding to the valence band – conduction band energy gap. These additional peaks in the absorption spectra of intrinsic semiconductors were

---

[2]Naturally, if an enormous voltage is applied, as in a thunderstorm, the electrons are given so much energy that they are ripped from the valence band and can transfer to the conduction band. In these conditions the insulator is said to break down.

interpreted in terms of bound hole–electron pairs, called excitons, during the years between 1938 and 1948 by Peierls, Frenkel, and Davydov.

Extrinsic semiconductors contain an appreciable number of foreign atoms that may have been added intentionally as *dopants* or occur unintentionally as *impurities*. At very low temperatures these may have no effect on the electronic properties of the material, but, as the temperature rises, they can influence the behavior in two very different ways. They can act as *donors*, donating electrons to the conduction band, or as *acceptors*, accepting electrons from the valence band, which is equivalent to donating holes to the valence band. Donors and acceptors are often represented by energy levels within the band gap. Large number of impurities can give rise to narrow energy bands in place of localized energy levels. When donors are the main impurities present in the crystals, the conduction is mainly by way of electrons and the material is called an *n-type semiconductor*. Similarly, if acceptors are the major impurities present, conduction is mainly by way of holes and the material is called a *p-type semiconductor*. In intrinsic semiconductors electrons and holes are present in roughly equal numbers. In extrinsic conduction the conductivity is dominated by either one or the other.

The position of the Fermi level in an extrinsic semiconductor depends upon the dopant concentrations and the temperature. As a rough guide the Fermi level can be taken as half way between the donor levels and the bottom of the valence band for *n*-type materials or half way between the top of the valence band and the acceptor levels for *p*-type semiconductor, both referred to 0 K. As the temperature rises the Fermi level in both cases moves toward the center of the band gap.

If the donors and acceptors are present in equal numbers, the material is said to be a *compensated semiconductor*. At 0 K these materials are insulators, and it is difficult in practice to distinguish between compensated and intrinsic semiconductors. When all of the impurities are *fully ionized* so that either all the donor levels have lost an electron or all the acceptor levels have gained an electron, the *exhaustion range* has been reached.

Degenerate semiconductors can be intrinsic or extrinsic semiconductors, but in these materials the band gap is similar to or less than the thermal energy. In such cases the number of charge carriers in each band becomes very high, as does the electronic conductivity. The compounds are said to show quasi-metallic behavior.

Semimetals show metallic conductivity due to the overlap of a filled and an empty band. In this case electrons spill over from the filled band into the bottom of the empty band until the Fermi surface intersects both sets of bands. In semimetals holes and electrons coexist even at 0 K.

S2.3 Point Defects and Energy Bands in Semiconductors and Insulators

A point defect in an insulator or semiconductor is represented on band diagrams as an energy level. These energy levels can lie within the conduction or valence bands, but those of most consequence for electronic and optical properties are those that lie in the band gap. The effects of these impurities on the electronic properties of the solid will depend upon how close the energy levels lie to either the bottom of the conduction band or the top of the valence band. Those that are close to these edges are called *shallow levels*, while those nearer to the center of the band gap are called *deep levels*.

The most fundamental transition that can take place is the transfer of an electron from the valence band to the conduction band. This creates a mobile electron and a mobile hole, both of which can often be treated as defects. Transitions of this type, and the reverse, when an electron in the conduction band drops to the valence band, eliminating a hole in the process and liberating energy, are called *interband transitions*. Apart from the electrons and holes themselves, interband transitions do not involve defects. All other transitions do.

Shallow levels play an important part in electronic conductivity. Shallow donor levels lie close to the conduction band in energy and liberate electrons to it to produce *n*-type semiconductors. Interstitial metal atoms added to an insulating ionic oxide often act in this way because metal atoms tend to ionize by losing electrons. When a donor level looses one or more electrons to the conduction band, it is said to be ionized. The energy level representing an ionized donor will be lower than that of the un-ionized (neutral) donor by the same amount as required to move the electron into the conduction band. The presence of shallow donor levels causes the material to become an *n*-type semiconductor.

Shallow acceptor levels lie close to the valence band and take up electrons from it to create holes in the valence band and produce *p*-type semiconductors. Interstitial nonmetal atoms often generate shallow acceptor levels because anion formation involves taking up extra electrons. Acceptor levels are said to be ionized when they take electrons from the valence band, creating holes in the process. The energy of a neutral acceptor atom is different to that of an ionized acceptor. The electrons on the ionized anions are often "trapped" and do not contribute to the conductivity.

The same sort of considerations will apply to vacancies. For instance, an anion vacancy may give rise to a set of shallow donor levels just below the lower edge of the conduction band. If the vacancy is created by removing a neutral nonmetal atom from the crystal, the electrons that were on the anion are transferred to the conduction band to produce an *n*-type semiconductor. The energies of neutral and ionized vacancies are slightly different.

A cation vacancy will be opposite to this in behavior. Removal of a neutral metal atom from a material will involve removal of a cation plus the correct number of electrons, which are taken from the valence band. Cation vacancies will therefore be represented as acceptor levels situated near to the valence band together with an equivalent number of holes in the band. These materials are *p*-type semiconductors.

Deep levels generally have less effect on electronic conductivity but are important in other ways, and especially influence optical properties.

S2.4 Transition-Metal Oxides

Shallow levels are often found in transition-metal compounds in which the cations can exist in several valence states. The most familiar of these are the transition metal oxides. In these compounds, the valence band is mainly derived from oxygen p orbitals. The conduction band is made up mainly from the metal atom d orbitals but can also contain considerable mixing in of metal s orbitals as well. The situation can be illustrated with respect to two transition-metal oxides, titanium dioxide, TiO_2, and nickel oxide, NiO.

Titanium dioxide, nominally Ti^{4+} and O^{2-}, has a band structure with a filled valence band mainly derived from oxygen $2p$ orbitals and an empty conduction band derived from the titanium d orbitals. The oxide is an insulator. The oxide Ti_2O_3, nominally Ti^{3+} and O^{2-}, has an extra electron on each cation, located in the $3d$ orbitals. In the band picture, these electrons are distributed in the d band, which is only partly filled so that the material behaves as a metal. The situation that is found when TiO_2 is slightly reduced to TiO_{2-x} is a step toward this latter situation. The charge imbalance resulting from oxygen loss is restored by the creation of Ti^{3+} ions; nominally two Ti^{3+} form for each O^{2-} lost. Each Ti^{3+} can be represented as an impurity level close to the bottom of the conduction band. Thermal energy is generally sufficient to ionize these centers, and the electrons are liberated into the conduction band. Slightly reduced TiO_2 would then be expected to show n-type semiconductivity.

In nickel oxide the d band should be partly full, as Ni^{2+} has $8d$ electrons and the band can hold 10. However, in this material the d band is split into a completely filled part and an empty part, this latter probably overlapping with the $4s$ band. This means that NiO will be an insulator. Nickel oxide can be made nonstoichiometric, and the resulting material has an excess of oxygen, due to Ni^{2+} vacancies, so that the formula is $Ni_{1-x}O$. The charge imbalance is corrected by transforming some Ni^{2+} ions into Ni^{3+} ions, which are equivalent to a normal Ni^{2+} ion together with a hole. Each Ni^{3+} ion can be represented as a shallow acceptor level just above the filled portion of the d band. Thermal energy is generally sufficient to ionize these centers. In this case the holes are transferred from the Ni^{3+} ions into the valence band. Slightly metal-deficient $Ni_{1-x}O$ would then be expected to show p-type semiconductivity.

S3 SEEBECK COEFFICIENT

S3.1 Seebeck Coefficient and Entropy

The origin of the Seebeck effect can be qualitatively understood by assuming that the electrons or holes in the material are defects that behave rather like gas atoms. The charge carriers in the hot region will have a higher kinetic energy, and hence a higher velocity, than those in the cold region. This means that the net velocity of the charge carriers at the hot end moving toward the cold end will be higher than the net velocity of the charge carriers at the cold end moving toward the hot end. In this situation more carriers will flow from the hot end toward the cold end than *vice versa*, which will cause a voltage to build up between the hot and cold ends of the sample. Eventually equilibrium will be established and the potential so set up is the Seebeck voltage. This simple model, the classical or Drude model, gives a formula for the Seebeck coefficient:

$$\alpha = -\frac{k}{2e} \quad \text{for electrons}$$

$$= +\frac{k}{2e} \quad \text{for holes}$$

where k is the Boltzmann constant and e the charge on the electron.

Electrons, holes, or other mobile charge carriers can be considered as chemically reactive species that can be allotted a thermodynamic chemical potential—the electrochemical potential, $\bar{\mu}$, defined as

$$\bar{\mu} = \mu + Ze\phi$$

where μ is the chemical potential of a mobile charge carrier in the absence of an electrical potential, Ze is the charge on the mobile species, and ϕ is the electric potential in the neighborhood of the carrier. In the case of electrons or holes, Z will be equal to -1 or $+1$; hence:

$$\bar{\mu} = \mu + e\phi \quad \text{(holes)}$$

$$= \mu - e\phi \quad \text{(electrons)}$$

In the case of a p-type material subjected to a temperature gradient, at equilibrium the electrochemical potential of the holes at the hot end of the rod must be equal to that of the holes at the cold end of the rod. Hence:

$$\bar{\mu}_H = \bar{\mu}_C$$
$$\bar{\mu}_H + e\phi_H = \bar{\mu}_C + e\phi_C$$
$$e(\phi_H - \phi_C) = \bar{\mu}_C - \bar{\mu}_H$$

The chemical potential of a substance can be equated to the Gibbs energy, and in this case we can write this as g per hole[3] so that:

$$\bar{\mu} = g = h - Ts$$

where h and s represent the enthalpy and entropy of a mobile hole and T is the temperature (K). Using this equation to substitute for $\bar{\mu}$:

$$e(\phi_H - \phi_C) = (h_C - T_C s_C) - (h_H - T_H s_H)$$
$$= h_C - h_H + T_H s_H - T_C s_C$$

If the enthalpy and entropy of the holes at the hot and cold ends of the material can, to a reasonable approximation, be taken as equal over the small temperature ranges

---

[3]The use of lower case letters, g, h, and s, indicates that the free energy *per particle* is being discussed.

that are normal in experiments:

$$h_C = h_H = h \qquad s_C = s_H = s$$

thus:

$$e(\phi_H - \phi_C) = (T_H - T_C)s$$
$$\frac{\phi_H - \phi_C}{T_H - T_C} = \frac{s}{e}$$

The Seebeck coefficient, α, is defined as

$$\alpha = \frac{\phi_H - \phi_C}{T_H - T_C}$$

so that, for holes:

$$\alpha = \frac{s}{e}$$

The analysis can be repeated for electrons, using:

$$\bar{\mu} = \mu - e\phi$$

to give

$$\alpha = -\frac{s}{e}$$

This result is frequently generalized as the entropy per carrier:

$$\alpha = -\left(\frac{1}{e}\right)\left(\frac{\partial S}{\partial N}\right)$$

where N is the number of particles and S the total entropy.

The units of Seebeck coefficient are given by:

$$\frac{\text{Entropy (J K}^{-1}) \text{ per particle}}{\text{Electron charge (C) per particle}}$$

which is (J K^{-1} C^{-1}) per particle. Using the conversion J C^{-1} = V, this becomes volts per degree (V K^{-1}).

S3.2   Seebeck Coefficient and Defect Populations

The Seebeck coefficient can be related to the number of defects present in a crystal for a material in which the mobile charges are located on a particular sublattice and migrate by jumping from one ion to another. The defects are the ions together with the (momentarily localized) charge carrier. For example, slightly reduced titanium dioxide can be written as $Ti^{4+}_{1-2x}Ti^{3+}_{2x}O_{2-x}$. The defects present are Ti^{3+} ions, which, in this context, can be thought of as point defects consisting of a Ti^{4+} ion plus an electron. The electronic conductivity and thermoelectric effects arise because of the ability of electrons to hop from one Ti^{3+} ion to an adjacent Ti^{4+} ion.

The entropy of the defects, which can be related to the Seebeck coefficient, can be broken down into several components. The arrangement of the defects over the available sites gives rise to the configurational entropy. Displacements of the charge carriers due to thermal energy gives rise to the vibrational entropy. Finally, it is necessary to consider the possible arrangements of spin states on the ions, the spin degeneracy.

It is easiest to start with the configurational entropy, S_C. Suppose that the number of defects, which is equal to the number of mobile (localized) holes or electrons, is n_d and moreover that only one type of mobile carrier, either holes or electrons, is present. The configurational entropy S_C is given by using the Boltzmann formula:

$$S_C = k \ln \left[\frac{n_0!}{(n_0 - n_d)! \, n_d!} \right]$$

where S_C is the configurational entropy of n_d electrons or holes arranged on n_0 available sites, k is the Boltzmann constant, and ! represents the factorial of the number. This can be simplified using Stirling's approximation:

$$\ln n! = n \ln(n - n)$$

(see also Supplementary Material Section S4). The configurational entropy can then be written:

$$S_C = k[n_0 \ln n_0 - n_d \ln n_d - (n_0 - n_d) \ln (n_0 - n_d)]$$

The configurational entropy per particle, s_C, is given by dS_C/dn_d, which is

$$s_C = k \ln \left(\frac{n_0 - n_d}{n_d} \right) \tag{S5.1}$$

$$= k \ln \left[\left(\frac{n_0}{n_d} \right) - 1 \right]$$

In general, n_0 is of the order of 10^{28} cation sites m^{-3} and n_d is of the order of 10^{20} defects m^{-3}, so that $n_0/n_d \gg 1$, hence:

$$\alpha = -\left(\frac{k}{e}\right) \ln\left[\left(\frac{n_0}{n_d}\right)\right] \quad \text{for electrons} \qquad \text{(S5.2a)}$$

$$= +\left(\frac{k}{e}\right) \ln\left[\left(\frac{n_0}{n_d}\right)\right] \quad \text{for holes} \qquad \text{(S5.2b)}$$

The positive version applies to p-type materials and the negative expression to n-type materials. Note that n_0/n_d increases as the number of defects falls, so that largest values of α are to be found in materials with the smallest defect populations.

Equation (S5.1) can be written in an alternative form. Write:

Number of sites occupied by a charge carrier $= cn_0 = n_d$

Number of available sites $= (1 - c)n_0 = (n_0 - n_d)$

Substitution into Eq. (S5.1) gives

$$s_C = k \ln\left(\frac{n_0 - n_d}{n_d}\right) = k \ln\left[\frac{(1 - c)n_0}{cn_0}\right] = k \ln\left(\frac{1 - c}{c}\right)$$

where c is the fraction of defects present. This leads directly to the *Heikes equation*:

$$\alpha = -\left(\frac{k}{e}\right) \ln\left(\frac{1 - c}{c}\right) \quad \text{for electrons}$$

$$= +\left(\frac{k}{e}\right) \ln\left(\frac{1 - c}{c}\right) \quad \text{for holes}$$

This form is useful because the value of c is related in a simple way to the composition of the sample.

A vibrational entropy term, S_V, can be added to Eq. (S5.2a) or (S5.2b) to give

$$\alpha = -\left(\frac{k}{e}\right)\left\{ \ln\left(\frac{n_0}{n_d}\right) + \frac{S_V}{k} \right\} \quad \text{for electrons}$$

$$= +\left(\frac{k}{e}\right)\left\{ \ln\left(\frac{n_0}{n_d}\right) + \frac{S_V}{k} \right\} \quad \text{for holes}$$

In general S_V/k is much smaller than $\ln[(n_0/n_d)]$ and as the only temperature variation will come into this equation from the S_V term, α will be approximately independent of temperature in this approximation. These equations are of the form:

$$\alpha = \pm\left(\frac{k}{e}\right)\left[\ln\left(\frac{n_0}{n_d}\right) + A \right]$$

where A is a constant, n_0 is the number of cation sites, and n_d is the number of defects giving rise to mobile electrons or holes. In the Heikes equation the term A is omitted.

The configurational entropy evaluated above omits all mention of electron spin. Because there are two spin directions, a spin degeneracy of 2, these must be included in the possible configurations. When this term is included, the Heikes equation becomes the *Chaikin–Beni equation*:

$$\alpha = -\left(\frac{k}{e}\right) \ln\left[2\left(\frac{1-c}{c}\right)\right]$$

In general, the spin degeneracy can be specified by a parameter β, to give

$$\alpha = -\left(\frac{k}{e}\right) \ln\left[\beta\left(\frac{1-c}{c}\right)\right]$$

This equation is sometimes called the *extended Heikes equation*.

If cations in the material can take two valence states, as occurs in many transition-metal compounds, Ti^{3+}/Ti^{4+} or Co^{3+}/Co^{4+}, for example, then the value of the spin degeneracy, β, must take into account these configurations and can take values greater than 2.

Note that the above model is for a simple system in which there is only one defect and one type of mobile charge carrier. In semiconductors both holes and electrons contribute to the conductivity. In materials where this analysis applies, both holes and electrons contribute to the value of the Seebeck coefficient. If there are equal numbers of mobile electrons and holes, the value of the Seebeck coefficient will be zero (or close to it). Derivation of formulas for the Seebeck coefficient for band theory semiconductors such as Si and Ge, or metals, takes us beyond the scope of this book.

S4 SCHOTTKY AND FRENKEL DEFECTS

S4.1 Equilibrium Concentration of Schottky Defects Derived from Configurational Entropy

The introduction of Schottky defects causes the Gibbs energy of the crystal to change by an amount ΔG_S:

$$\Delta G_S = \Delta H_S - T\,\Delta S_S$$

where ΔH_S is the associated change in enthalpy, and ΔS_S is the change in the entropy of the crystal. In general, the energy increase due to the ΔH_S term will be offset by the energy decrease due to the $T\,\Delta S_S$ term.

In a crystal of overall composition MX, suppose that n_S is the number of Schottky defects in the crystal at a temperature T (K), that is, there are n_S vacant cation sites and n_S vacant anion sites present, distributed over N possible cation sites and N possible anion sites. The configurational entropy change, ΔS_S, due to the distribution of the vacancies over the available cation and anion positions can be determined by using the Boltzmann equation:

$$S = k \ln W$$

where S is the entropy of a system in which W is the number of ways of distributing n defects over N sites at random and k is the Boltzmann constant. The value of W is given by the formula:

$$W = \frac{N!}{(N - n)! \, n!}$$

where the symbol ! represents the factorial of the number. The number of ways that the n_S cation and anion vacancies can be distributed over the N available sites in the crystal is given by

$$w_c = \frac{N!}{(N - n_S)! \, n_S!}$$

for vacancies on cation sites, and

$$w_a = \frac{N!}{(N - n_S)! \, n_S!}$$

for vacancies on anion sites. For a crystal of stoichiometry MX,

$$w_c = w_a$$

The total number of ways of distributing these defects, W, is given by the product of w_c and w_a, hence:

$$W = w_c \cdot w_a = w^2$$

Therefore, the change in configurational entropy caused by introducing these defects is

$$\Delta S_S = k \ln (w^2) = 2k \ln w$$

that is,

$$\Delta S_S = 2k \ln \left[\frac{N!}{(N - n_S)! \, n_S!} \right]$$

This expression is simplified by employing the approximation:

$$\ln N! \approx N \ln N - N$$

(known as Stirling's approximation, but see Section S.4.2). Substituting and simplifying yields

$$\Delta S_S = 2k \{N \ln N - (N - n_S) \ln(N - n_S) - n_S \ln n_S\}$$

It is assumed that the entropy change is solely made up of this contribution.

The enthalpy change involved, ΔH_S, is not explicitly calculated. It is assumed that the enthalpy to form one defect, Δh_S, is constant over the temperature range of interest, so that the total enthalpy change, ΔH_S, is given by

$$\Delta H_S = n_S \, \Delta h_S$$

The values for ΔS_S and ΔH_S are substituted into the Gibbs equation to give

$$\Delta G_S = n_S \, \Delta h_S - 2kT\{N \ln N - (N - n_S) \ln(N - n_S) - n_S \ln n_S\}$$

At equilibrium, ΔG will be equal to zero and the minimum in the ΔG versus n_S curve occurs when

$$\left(\frac{d \, \Delta G}{dn_S}\right)_T = 0$$

that is,

$$\left(\frac{d\Delta G}{dn_S}\right)_T = \frac{d}{dn_S}\{n_S \, \Delta h_S - 2kT[N \ln N - (N - n_S) \ln(N - n_S) - n_S \ln n_S]\} = 0$$

Remembering that $(N \ln N)$ is constant, so its differential is zero, the differential of $(\ln x)$ is $(1/x)$ and of $(x \ln x)$ is $(1 + \ln x)$, on differentiating:

$$\Delta h_S - 2kT\left\{\frac{d}{dn_S}[N \ln N - (N - n_S) \ln(N - n_S) - n_S \ln n_S]\right\} = 0$$

that is,

$$\Delta h_S - 2kT\left[\ln(N - n_S) + \frac{N - n_S}{N - n_S} - \ln n_S - \frac{n_S}{n_S}\right] = 0$$

hence:

$$\Delta h_S = 2kT \ln\left(\frac{N - n_S}{n_S}\right)$$

Rearranging:

$$n_S = (N - n_S)e^{-\Delta h_S/2kT}$$

This expression can be further simplified if we make yet another approximation and suppose the number of Schottky defects, n_S, is a lot less than the number of normally occupied sites N. In this case

$$n_S \approx Ne^{-\Delta h_S/2kT}$$

S4.2 Stirling's Approximation

The approximation described as Stirling's approximation above and in Section S3.2 is not very accurate. It is several percent in error even for values of N as large as 10^{10}. The correct expression for Stirling's approximation is

$$\ln N! \approx N \ln N - N + \tfrac{1}{2} \ln (2\pi N) \qquad (S4.1)$$

which is accurate even for very low values of N.

It is of interest to derive the relationship between defect numbers and configurational entropy using the correct form of Stirling's approximation. The principle can best be illustrated with respect to the population of vacancies in a monatomic crystal (Section 2.1). Substituting from the more accurate Eq. (S4.1):

$$\Delta S_V = k[\{N \ln N - (N - n_V)\ln(N - n_V) - n_V \ln n_V\}$$
$$- \tfrac{1}{2}\ln 2\pi N + \tfrac{1}{2}\ln 2\pi(N - n_V) + \tfrac{1}{2}\ln 2\pi n_V]$$

Proceeding as before:

$$\Delta G_V = n_V\Delta h_V - kT[\{N \ln N - (N - n_V) \ln (N - n_V) - n_V \ln n_V\}$$
$$- \tfrac{1}{2}\ln 2\pi N + \tfrac{1}{2}\ln 2\pi(N - n_V) + \tfrac{1}{2}\ln 2\pi n_V]$$

and setting $(d\ \Delta G_V/dn_V)_T = 0$:

$$\Delta h_V - kT\left\{\frac{d}{dn_V}[\{N \ln N - (N - n_V) \ln (N - n_V) - n_V \ln n_V\}\right.$$
$$\left. - \tfrac{1}{2}\ln 2\pi N + \tfrac{1}{2}\ln 2\pi(N - n_V) + \tfrac{1}{2}\ln 2\pi n_V]\right\} = 0$$

On differentiating:

$$\Delta h_V - kT\{[\ln(N - n_V) - \ln n_V] + \tfrac{1}{2}[\ln\ 2\pi(N - n_V) + \ln 2\pi n_V]\} = 0$$

Hence:

$$\Delta h_V = kT\ \ln\left(\frac{N - n_V}{n_V}\right) + \tfrac{1}{2}\ln [4\pi^2(N - n_V^2)]$$

Assuming that N is much greater than n_V, this can be solved to give

$$n_V = e^{-\Delta h_V/kT}N^{1+1/2kT}(2\pi)^{1/kT}$$

S4.3 Equilibrium Concentration of Frenkel Defects Derived from Configurational Entropy

The calculation of the number of Frenkel defects in a crystal proceeds along lines parallel to those above. The introduction of Frenkel defects causes the Gibbs energy of the crystal to change by an amount ΔG_F:

$$\Delta G_F = \Delta H_F - T\,\Delta S_F$$

where ΔH_F is the associated change in enthalpy and ΔS_F the change in the entropy of the crystal. In general, the energy increase due to the ΔH_F term will be offset by the energy decrease due to the $T\,\Delta S_F$ term.

In a crystal of overall composition MX, suppose that n_F is the number of Frenkel defects in the crystal at a temperature T (K), that is, there are n_F interstitials distributed over N_i interstitial sites and n_F vacant atom sites present, distributed over N possible atom sites.

The configurational entropy change, ΔS_F, due to the distribution of the defects over the available positions, can be determined by using the Boltzmann equation. The number of ways of distributing the n_F vacancies that have been created over the N available positions in the atom array affected by Frenkel defects is

$$w_v = \frac{N!}{(N - n_F)!\, n_F!}$$

Similarly, the number of ways of distributing the n_F interstitial atoms over the N_i available interstitial positions in the crystal is

$$w_i = \frac{N_i!}{(N_i - n_F)!\, n_F!}$$

The total number of ways of distributing these defects, W, is given by the product of w_v and w_i, hence:

$$W = w_v w_i$$

The change in configurational entropy, ΔS_F, due to this distribution is

$$\Delta S_F = k \ln W = k \ln w_v w_i$$

Hence:

$$\Delta S_F = k \left\{ \ln \left[\frac{N!}{(N - n_F)!\, n_F!} \right] + \ln \left[\frac{N_i!}{(N_i - n_F)!\, n_F!} \right] \right\}$$

This expression is simplified by employing Stirling's approximation:

$$\ln N! \approx N \ln N - N$$
$$\ln N_i! \approx N_i \ln N_i - N_i$$

Substituting and simplifying yields the cumbersome expression:

$$\Delta S_F = k[N \ln N + N_i \ln N_i$$
$$- (N - n_F) \ln (N - n_F) - (N_i - n_F) \ln (N_i - n_F) - 2n_F \ln n_F]$$

(Note that if we make N_i and N equal to each other we arrive at the expression for Schottky defects.)

The enthalpy change involved, ΔH_F, is not explicitly calculated. It is assumed that the enthalpy to form one defect, Δh_F, is constant over the temperature range of interest, so that the total enthalpy change, ΔH_F, is given by

$$\Delta H_F = n_F \, \Delta h_F$$

The values for ΔS_F and ΔH_F are substituted into the Gibbs equation to give:

$$\Delta G_F = n_F \, \Delta h_F - kT[N \ln N - N_i \ln N_i - (N - n_F)$$
$$\ln(N - n_F) - (N_i - n_F) \ln(N_i - n_F) - 2n_F \ln n_F]$$

Setting $(d\Delta G_F/dn_F)_T = 0$ at equilibrium and differentiating, remembering that $(N \ln N)$ and $(N_i \ln N_i)$ are constant, so their differentials are zero, and the differential of $(\ln x)$ is $(1/x)$ and of $(x \ln x)$ is $(1 + \ln x)$, yields

$$\Delta h_F = kT \, \ln\left[\frac{(N - n_F)(N_i - n_F)}{n_F^2}\right]$$

that is,

$$n_F^2 = (N - n_F)(N_i - n_F)e^{-\Delta h_F/kT}$$

This expression can be further simplified if we make yet another approximation, and suppose the number of Frenkel defects, n_F, is a lot less than either the number of normal positions N or the number of interstitial positions available N_i. In this case:

$$n_F \approx (NN_i)^{1/2} e^{-\Delta h_F/2kT}$$

A more extended expression, using the correct formulation of Stirling's approximation, can be derived as in Section S4.2.

S5   DIFFUSION

S5.1   Diffusion Equations

Measurement of the diffusion of an atom through a perfect crystal will give a concentration profile. The diffusion coefficient is extracted from the concentration profile by solution of one of two diffusion equations. For one-dimensional diffusion, along x, they are

$$J = -D\frac{dc}{dx} \tag{S5.1}$$

Fick's first law of diffusion, and

$$\frac{dc}{dt} = D\frac{d^2c}{dx^2} \tag{S5.2}$$

This is known as Fick's second law of diffusion or more commonly as the diffusion equation. In these equations, J is called the flux of the diffusing species, with units of [amount of substance (atoms or equivalent units) m^2 s^{-1}], c is the concentration of the diffusing species, with units of [amount of substance (atoms or equivalent units) m^{-3}] at position x (m) after time t (s); D is the diffusion coefficient, units (m^2 s^{-1}).

S5.2   Non-Steady-State Diffusion

The normal state of affairs during a diffusion experiment is one in which the concentration at any point in the solid changes over time. This situation is called *non-steady-state diffusion*, and diffusion coefficients are found by solving the diffusion equation [Eq. (S5.2)]:

$$\frac{dc_x}{dt} = D\left(\frac{d^2c_x}{dx^2}\right)$$

Provided the diffusion coefficient, D, is not dependent upon composition and position, analytical solutions can be found to give an expression for c in terms of x. (The symbol D generally stands for D_{self}, D^*, or D_A^*, depending upon the experimental details.)

For the diffusion couple experimental arrangement (Fig. 5.2), (ignoring all short-circuit diffusion), the solution is

$$c_x = \left[\frac{c_0}{2(\pi Dt)^{1/2}}\right]\exp\left(\frac{-x^2}{4Dt}\right)$$

where c_0 is the initial concentration on the surface, usually measured in moles per square meter. A value for the diffusion coefficient is obtained by taking logarithms

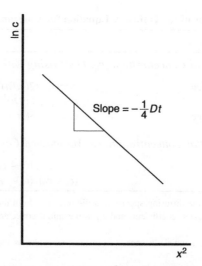

Figure S5.1   Straight-line graph of ln c versus x^2 from a diffusion experiment: The slope of the graph allows a value for the diffusion coefficient, D, to be determined.

of both sides this equation to give:

$$\ln c_x = \ln\left[\frac{c_0}{2(\pi Dt)^{1/2}}\right] - \left(\frac{x^2}{4Dt}\right)$$

This has the form

$$\ln c_x = \text{constant} - \left(\frac{x^2}{4Dt}\right)$$

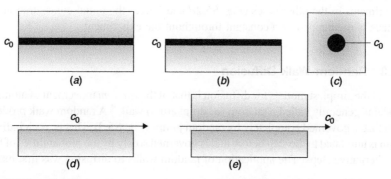

Figure S5.2   Common geometries for non-steady-state diffusion: (a) thin-film planar sandwich, (b) open planar thin film, (c) small spherical precipitate, (d) open plate, and (e) sandwich plate. In parts (a)–(c) the concentration of the diffusant is unreplenished; in parts (d) and (e), the concentration of the diffusant is maintained at a constant value, c_0, by gas or liquid flow.

TABLE S5.1   Solutions of the Diffusion Equation for Constant D

Experimental Arrangement	Solution

Initial Concentration, c_0, Decreasing with Time

Thin-film planar sandwich	$c_x = [c_0/2(\pi Dt)^{1/2}] \exp[-x^2/4Dt]^a$
Open planar thin film	$c_x = [c_0/(\pi Dt)^{1/2}] \exp[-x^2/4Dt]^a$
Small spherical precipitate	$c_r = [c_0/8(\pi Dt)^{3/2}] \exp[-r^2/4Dt]^b$

Initial Concentration, c_0, Maintained Constant

Open plate	$(c_x - c_0)/(c_s - c_0) = 1 - \mathrm{erf}[x/2(Dt)^{1/2}]^c$
Sandwich plate	$(c_x - c_0)/(c_s - c_0) = \frac{1}{2}\{1 - \mathrm{erf}[x/2(Dt)^{1/2}]\}$

$^a c_x$ is the concentration of the diffusing species at a distance of x from the original surface after time t has elapsed, D is the diffusion coefficient, and c_0 is the initial concentration on the surface, usually measured in mol m^{-2}.

$^b c_r$ is the concentration at a radial distance r from the precipitate as the precipitate dissolves in the surroundings.

$^c c_s$ is the constant concentration of the diffusing species at the surface, c_0 is the uniform concentration of the diffusing species already present in the solid before the experiment and c_x is the concentration of the diffusing species at a position x from the surface after time t has elapsed and D is the (constant) diffusion coefficient of the diffusing species. The function erf $[x/2(Dt)^{1/2}]$ is called the error function. The error function is closely related to the area under the normal distribution curve and differs from it only by scaling. It can be expressed as an integral or by the infinite series $\mathrm{erf}(x) = 2/\sqrt{\pi}[x - \frac{1}{1!}\frac{x^3}{3} + \frac{1}{2!}\frac{x^5}{5} - \frac{1}{3!}\frac{x^7}{7}\cdots]$. The complementary error function, $\mathrm{erfc}(z)$ is defined as $1 - \mathrm{erf}(z)$.

and a plot of $\ln c_x$ versus x^2 will have a gradient of $[-1/(4Dt)]$ (Fig. S5.1). A measurement of the gradient gives a value for the tracer diffusion coefficient at the temperature at which the diffusion couple was heated. To obtain the diffusion coefficient over a variety of temperatures, the experiments must be repeated.

Solutions for some diffusion experiment geometries (Fig. S5.2) are summarized in Table S5.1. In Figure S5.2a–S5.2c the initial concentration of the tracer is fixed, and the amount remaining as the diffusion progresses will diminish over the course of the experiment. In the other cases (Fig. S5.2d and S5.2e) the initial concentration at the surface is maintained as a constant throughout the experiment.

S5.3   Random-Walk Diffusion

One of the simplest models for diffusion is that of the *random* movement of atoms. The model is generally called a random (or drunkard's) walk.[4] A random walk produces a path that is governed completely by random jumps (Fig. S5.3). That is, each individual jump is unrelated to the step before and is governed solely by the probabilities of taking the alternative steps. The application of random walks to diffusion was first made by

[4]Random walks are often called *Markov random walks*. A Markov chain is a sequence of random events described in terms of a probability that the event under scrutiny evolved from a defined predecessor. In effect there is no memory of any preceding step in a Markov chain. "Hidden" Markov processes involve some short-term memory of preceding steps.

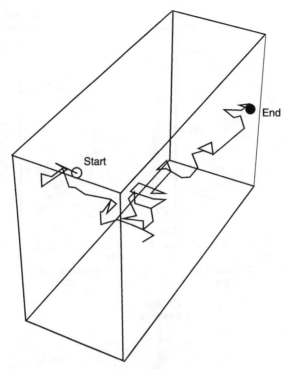

Figure S5.3 Typical three-dimensional random walk, redrawn from a Mathematica®
calculation (see Further Reading, Chapter 5, S. Wolfram).

Einstein to describe Brownian motion.[5] The model can be used to derive the diffusion
equations and to relate the diffusion coefficient to atomic movements.

For convenience, only one-dimensional random movement will be considered. In
this case, an atom is constrained to jump from one stable site to the next in the x direction, the choice of $+x$ or $-x$ being selected in a random way.[6] For example, imagine a
diffusion experiment starting with a thin layer of N atoms on the surface of a crystal.

---

[5]Brownian motion is the apparently random path taken by a minute grain immersed in a fluid. It was first
reported by Brown in 1828, with respect to the endless joggling motion of pollen grains in liquid.
Brownian motion was taken to be evidence of the existence of atoms, a fact still disputed at the outset of
the twentieth century. Einstein's analysis of Brownian motion was a statistical treatment, published in
1905. It made the assumption that the movement of the particle was due to random jostling by atoms in
the fluid, and the agreement between theory and observation was taken as strong confirmatory evidence
for the existence of atoms. [A. Einstein, *Ann. Physik* (4th Series), **17**, 549–560 (1905)]. A better source
for English speakers is: "Investigation on the Theory of the Brownian Movement." This is an English translation of Einstein's study, together with notes by R. Fürth, made by A. D. Couper, published by Melthuen,
London, 1926.
[6]The statistics of this process is identical to those pertaining to the tossing of a coin. The mathematics was
first worked out with respect to games of chance by de Moivre, in 1733. It is formally described by the
binomial distribution.

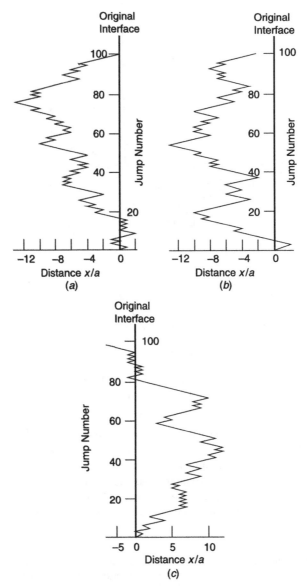

Figure S5.4   Three one-dimensional random walks: each individual walk, (*a*)–(*c*), gives the position of the diffusing atom, *x* after the *N*th jump. The final positions reached are: (*a*) $x = 0$, (*b*), $x = -2a$, and (*c*) $x = -8a$.

After a time t, each atom will have carried out a random walk of n steps, to yield a concentration profile. Three such walks, each of 100 jumps, are shown in Figure S5.4. The net displacement of a diffusing atom after n jumps will be the algebraic sum of the individual jumps. If x_i is the distance moved along the x axis in the i th jump, the distance moved after a total of n jumps, x, will simply be the sum of all the individual steps, that is,

$$x = x_1 + x_2 + x_3 + \cdots = \sum_{i=1}^{n} x_i$$

In a linear "crystal" in which the sites are separated by a unit distance $\pm a$, each individual value of x_i can be $+a$ or $-a$. As the jumps take place with an equal probability in both directions, after n jumps the total displacement in any one walk may have any value between zero and $\pm na$.

[Note that during the random walk a number of diffusing atoms will return to the starting layer. The probability of returning to the starting point after a walk of n steps can readily be calculated. The first step can be made in one of two ways, as can the second, third, and so on. This means that in a walk of n steps, there are 2^n different ways of taking the steps, which is the same as saying that there are 2^n different random walks possible for a set of n steps.

To return to the starting point, the walk must contain an equal number $(n/2)$ of steps $+a$ and $-a$, but these can occur in any order. The number of walks with $(n/2) + a$ steps and $(n/2) - a$ steps is given by $N(0)$:

$$N(0) = \frac{n!}{(n/2)! \, (n/2)!}$$

The probability of returning to the starting point, $P(0)$, is then the number of walks with equal numbers of $+$ and $-$ steps divided by the total number of ways of walks, 2^n, so that:

$$P(0) = \frac{N(0)}{2^n}$$

For

$$n = 2: \quad P(0) = 0.5$$
$$n = 4: \quad P(0) = 0.375$$
$$n = 10: \quad P(0) = 0.246$$
$$n = 100: \quad P(0) = 0.080$$

Surprisingly, in a random-walk diffusion experiment, although the concentration of atoms in the initial layer always remains greater than in the bulk, it does not mean that these atoms have not diffused. They have taken just as many steps as those furthest from the starting layer!]

To relate the average distance traveled by the diffusing atoms to the diffusion coefficient, it is necessary to consider the outcome of the N walks occurring. The average distance traveled for one walk is

$$x = x_1 + x_2 + x_3, \ldots$$

When more and more walks are added together, the number of $+a$ steps will become more and more equal to the number of $-a$ steps, and the average distance traveled will

TABLE S5.2 Statistics of Random Walks

Number of Steps, n	Total Number of Different Walks, 2^n	Walk[a]	Total Distance Traveled, x[b]	$\sum x$	$\langle x \rangle = \sum x/N$	x^2	$\sum x^2$	$\langle x^2 \rangle = \sum x^2/2^n$
2	4	+ +	2	0	0	4	8	2
		+ −	0			0		
		− +	0			0		
		− −	−2			4		
3	8	+ + +	3	0	0	9	24	3
		+ + −	1			1		
		+ − +	1			1		
		− + +	1			1		
		+ − −	−1			1		
		− + −	−1			1		
		− − +	−1			1		
		− − −	−3			9		
4	16	+ + + +	4	0	0	16	64	4
		+ + + −	2			4		
		+ + − +	2			4		
		+ − + +	2			4		
		− + + +	2			4		
		+ + − −	0			0		
		+ − + −	0			0		
		+ − − +	0			0		
		− + + −	0			0		
		− + − +	0			0		
		− − + +	0			0		
		+ − − −	−2			4		
		− + − −	−2			4		
		− − + −	−2			4		
		− + − −	−2			4		
		− − − −	−4			16		

[a] For simplicity the walk is represented as a series of + and −, standing for $+a$ and $-a$.
[b] The walk displacement is represented by the number of steps only. The real distance traversed must be multiplied by the step length a.

approach zero. If all different 2^n walks are included, then the average works out to be exactly zero (Table S5.2).

To get an idea of the probable displacement distribution of the atoms, the jump distances are squared, and the mean of these squares, over the n steps, is computed. Thus, for a single walk:

$$
\begin{aligned}
x^2 &= (x_1 + x_2 + x_3, \ldots, x_n)(x_1 + x_2 + x_3, \ldots, x_n) \\
&= (x_1 x_1 + x_1 x_2 + x_1 x_3 + \cdots + x_1 x_n) \\
&\quad + (x_2 x_1 + x_2 x_2 + x_2 x_3 + \cdots + x_2 x_n) \\
&\quad + \cdots \\
&\quad + x_n x_1 + x_n x_2 + x_n x_3 + \cdots + x_n x_n)
\end{aligned}
$$

This can be written:

$$
\begin{aligned}
x^2 &= \sum_{i=1}^{n} x_i{}^2 + 2 \sum_{i=1}^{n} x_i x_{i+1} + 2 \sum_{i=1}^{n} x_i x_{i+2} + \cdots \\
&= \sum_{i=1}^{n} x_i{}^2 + 2 \sum_{i=1}^{n} \sum_{j=i+1}^{n} x_i x_{i+j}
\end{aligned}
$$

In the limit of a large number of jumps, knowing that each jump may be either positive or negative, all terms except the first average to zero. As each jump, x_i, can be equal to $+a$ or $-a$,

$$
\begin{aligned}
x^2 &= x_1{}^2 + x_2{}^2 + x_3{}^2 + \cdots + x_n{}^2 \\
&= a^2 + a^2 + a^2 + \cdots + a^2 \\
&= na^2
\end{aligned}
$$

where n is the number of steps in the walk. Adding the x^2 terms for all of the N different walks and dividing by the number of walks, N, gives the mean square displacement $<x^2>$:

$$
<x^2> = \frac{N(na^2)}{N} = na^2
$$

Statistics for three simple walks, with $n = 2$, 3, and 4 are set out in Table S5.2.

If the time that has elapsed from the beginning of the random walk is t, and the time for each step is τ, the number of jumps, n is given by

$$
n = \frac{t}{\tau}
$$

and writing Γ as the jump frequency, $1/\tau$:

$$
<x^2> = \Gamma t a^2
$$

S5.4 Concentration Profile

The random-walk model of diffusion can also be applied to derive the shape of the bell-shaped concentration profile characteristic of bulk diffusion. As in the previous section, a planar layer of N tracer atoms is the starting point. Each atom diffuses from the interface by a random walk of n steps in a direction perpendicular to the interface. As mentioned (see footnote 5) the statistics are well known and described by the binomial distribution (Fig. S5.5a–S5.5c). At large values of N, this discrete distribution can be approximated by a *continuous* function, the Gaussian distribution curve[7] with a form:

$$N(x) = \frac{2N}{\sqrt{2\pi n}}\exp\left(\frac{-x^2}{2na^2}\right)$$

where the function $N(x)$ is the number of atoms that reach a position x, a is the step length, N is the number of atoms taking part in the random walk, and n is the number of steps in each walk (Fig. 5.6a). The Gaussian distribution curve is an important probability function encountered in the statistical analysis of random processes and widely used for the estimation of errors. The experimentally determined concentration profile should then be a bell-shaped curve of the same shape as the Gaussian distribution if random-walk diffusion is important. As pointed out above, those atoms that are still located at the initial interface after the diffusion experiment will still have undergone the same random walk as those far removed from the interface. These are the atoms for which the random walk returns to zero.

The statistics of the normal distribution can now be applied to give more information about the statistics of random-walk diffusion. It is then found that the mean of the distribution is zero and the variance (the square of the standard deviation) is $\sqrt{(na^2)}$, equal to the mean-square displacement, $<x^2>$. The standard deviation of the distribution is then the square root of the mean-square displacement, the root-mean-square displacement, $\pm\sqrt{<x^2>}$. The area under the normal distribution curve represents a probability. In the present case, the probability that any particular atom will be found in the region between the starting point of the diffusion and a distance of $\pm\sqrt{<x^2>}$ (the root-mean-square displacement) on either side of it, is approximately 68% (Fig. 5.6b). The probability that any particular atom has diffused *further* than this distance is given by the total area under the curve minus the shaded area, which is approximately 32%. The probability that the atoms have diffused further than $2\sqrt{<x^2>}$ is equal to the total area under the curve minus the area under the curve up to $2\sqrt{<x^2>}$. This is found to be equal to about 5%. Some atoms will have gone further than this distance, but the probability that any one particular atom will have done so is very small.

[7]The Gaussian or normal distribution is generally written as

$$P(x) = \frac{1}{\sqrt{2\pi}\sigma}\exp\left(\frac{-(x-\mu)^2}{2\sigma^2}\right)$$

where μ is the mean and σ the standard deviation of the distribution, the whole being normalized so that the area under the curve is equal to 1.0.

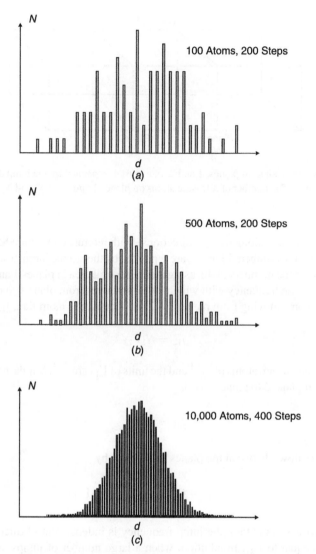

Figure S5.5 Random-walk statistics: Each plot shows the number of atoms, N, reaching a distance d in a random walk, for walks: (a) 100 atoms and 200 steps, (b) 500 atoms and 200 steps, and (c) 10,000 atoms and 400 steps. The curve approximates to the binomial distribution as the number of atoms and steps increases.

S5.5 Fick's Laws and the Diffusion Equations

Fick's (continuum) laws of diffusion can be related to the discrete atomic processes of the random walk, and the diffusion coefficient defined in terms of Fick's law can be equated to the random-walk displacement of the atoms. Again it is easiest to use a one-dimensional random walk in which an atom is constrained to jump from one

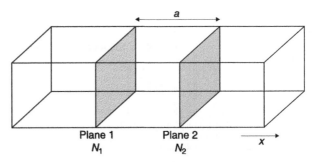

Figure S5.6 Two adjacent planes, 1 and 2, in a crystal, separated by the jump distance of the diffusing atom, a. The number of diffusing atoms on planes 1 and 2 are N_1 and N_2, respectively.

stable site to the next along the $\pm x$ directions. With reference to Figure S5.6, adjacent planes in a crystal, numbered 1 and 2, are separated by the atomic jump distance, a. Take N_1 and N_2 to be the numbers of diffusing atoms per unit area in planes 1 and 2, respectively. If Γ_{12} is the frequency with which an atom moves from plane 1 to plane 2, then the flow of atoms moving from plane 1 to 2 per second, the atom flux, j_{12}, is

$$j_{12} = N_1 \Gamma_{12}$$

where the units of j are atoms m^2 s^{-1} and the units of Γ_{12} are s^{-1}. Similarly, the number moving from plane 2 to plane 1, j_{21} is

$$j_{21} = N_2 \Gamma_{21}$$

The net flux (flow), between the planes, J, is given by

$$J = j_{12} - j_{21} = (N_1 \Gamma_{12} - N_2 \Gamma_{21})$$

If the process is random, the jump frequency is independent of direction and we can set Γ_{12} equal to Γ_{21}. In addition, when a large number of jumps are taken into account, the number of jumps in either direction will be equal, so:

$$\Gamma_{12} = \Gamma_{21} = \tfrac{1}{2}\Gamma$$

where Γ represents the overall jump frequency of the diffusion atoms. Thus:

$$J = \tfrac{1}{2}(N_1 - N_2)\Gamma$$

The number of mobile atoms on plane 1 is N_1 per unit area, so that the concentration per unit volume at plane 1 is N_1/a, or c_1. Similarly, the number of mobile atoms per unit area on plane 2 is N_2, so that the concentration per unit volume at plane 2 is N_2/a or c_2. Thus:

$$(N_1 - N_2) = a(c_1 - c_2)$$

Hence:

$$J = \tfrac{1}{2}a(c_1 - c_2)\Gamma$$

The concentration gradient, dc/dx, is given by the change in concentration between planes 1 and 2 divided by the distance between planes 1 and 2, that is

$$\frac{-dc}{dx} = \frac{c_1 - c_2}{a}$$

where a minus sign is introduced as the concentration falls as we move from plane 1 to plane 2. Hence:

$$c_1 - c_2 = -a\frac{dc}{dx}$$

so that:

$$J = -\tfrac{1}{2}\Gamma\, a^2 \frac{dc}{dx}$$

The units of the term $\tfrac{1}{2}\Gamma a^2$ are $m^2\ s^{-1}$, the units of the diffusion coefficient, and writing:

$$\tfrac{1}{2}\Gamma\, a^2 = D$$

where D is defined as the diffusion coefficient, one obtains Fick's first law of diffusion:

$$J = -D\frac{dc}{dx}$$

In addition, the random-walk statistics show that

$$<x^2> = \tfrac{1}{2}\Gamma\, a^2$$

so that

$$<x^2> = 2Dt \qquad D = \frac{<x^2>}{2t}$$

where D is the diffusion coefficient, t is the time over which diffusion occurs, and $<x^2>$ is the mean-square displacement of the diffusing atoms. This is the Einstein diffusion equation.

The overall flow of atoms into the volume between the two planes 1 and 2 from plane 1 is the flux, j_1, multiplied by the total area of plane 1, A, and so the increase in concentration of atoms in the volume between planes 1 and 2, $dc(+)$, is:

$$\frac{dc(+)}{dt} = \frac{j_1 A}{(Aa)} = \frac{j_1}{a}$$

where a is the separation of planes 1 and 2 and the volume of the region between planes 1 and 2 is Aa. Similarly, the decrease in concentration of the atoms in this region due to the flow of atoms out of the volume at plane 2, $dc(-)$ is

$$\frac{dc(-)}{dt} = \frac{j_2 A}{(Aa)} = \frac{j_2}{a}$$

The net change in concentration, dc/dt is then

$$\frac{dc}{dt} = \left[\frac{dc(+)}{dt}\right] - \left[\frac{dc(-)}{dt}\right] = \frac{j_1 - j_2}{a} = \frac{J}{a}$$

Now this last term is the flux gradient, that is,

$$\frac{j_1 - j_2}{a} = \frac{J}{a} = \frac{-dJ}{dx}$$

Hence:

$$\frac{dc}{dt} = \frac{-dJ}{dx}$$

$$= \frac{-d}{dx}\left(-\frac{1}{2}\Gamma a^2 \frac{dc}{dx}\right)$$

$$= \frac{-d}{dx}\left(-D\frac{dc}{dx}\right)$$

that is,

$$\frac{dc}{dt} = D\frac{d^2c}{dx^2}$$

This is Fick's second law of diffusion, the diffusion equation.

S5.6 Penetration Depth

The distribution of diffusing atoms represented by the Gaussian distribution curve:

$$N(x) = \frac{2N}{\sqrt{2\pi n}} \exp\left(\frac{-x^2}{2na^2}\right)$$

can now be written in terms of the diffusion coefficient because na^2 can be replaced by $2Dt$. Hence:

$$N(x) = \frac{2N}{\sqrt{2\pi n}} \exp\left(\frac{-x^2}{4Dt}\right)$$

It is of considerable practical importance to have some idea of how far an atom or ion will diffuse into a solid during a diffusion experiment. For example, the electronic properties of integrated circuits are created by the careful diffusion of selected dopants into single crystals of very pure silicon, and metallic machine components are hardened by the diffusion of carbon or nitrogen from the surface into the bulk. An approximate estimate of the depth to which diffusion is significant is given by the penetration depth or diffusion length, x_p, which is the depth where an appreciable change in the concentration of the tracer can be said to have occurred after a diffusion time t. The statistics of the Gaussian function indicates that the probability of a diffusing atom being found within the region between zero and $\pm\sqrt{<x^2>}$ from original interface is 68%, and within a of region $\pm 2\sqrt{<x^2>}$ is 95%. Using the information

$$<x^2> = na^2 = 2Dt$$

the penetration depth can be expressed as

$$x_p \approx \sqrt{(2Dt)} \quad \text{or} \quad x_p \approx 2\sqrt{(2Dt)}$$

depending upon the precision required.

S6 MAGNETIC PROPERTIES

S6.1 Atomic Magnetism

Magnetic properties reside in the subatomic particles that make up atoms. Of these, electrons make the biggest contribution, and only these will be considered here. Each electron has a magnetic moment due to the existence of a magnetic dipole, which can be thought of as a minute bar magnet linked to the electron.

There are two contributions to the magnetic dipole moment of an electron bound to an atomic nucleus, which, in semiclassical models, are attributed to orbital motion, represented by quantum number l, and spin, represented by quantum number s. The orbital and spin components are linked, or coupled, on isolated atoms or ions to give an overall magnetic dipole moment for the atom. The total magnetic dipole moment of the atom is given by

$$\mathbf{m}_{atom} = g_J\,\mu_\mathrm{B}[J(J+1)]^{1/2} \tag{S6.1}$$

where g_J is the Landé g-factor, and J is the total quantum angular momentum.

Atoms or ions with completely filled orbitals have J equal to zero, which means that atoms with closed shells have no magnetic moment. The only atoms that display a magnetic moment are those with incompletely filled shells. These are particularly found in the transition metals, with incompletely filled d shells, and the lanthanides and actinides, which have incompletely filled f shells.

Equation (S6.1) is applicable to the salts of lanthanide ions. These have a partly filled 4f shell, and the 4f orbitals are well shielded from any interaction with the surrounding atoms by filled 5s, 5p, and 6s orbitals, so that, with the notable exceptions, Eu^{3+} and Sm^{3+}, they behave like isolated ions. For the transition metals, especially those of the 3d series, interaction with the surroundings is considerable. Because of this, the 3d transition-metal ions often have magnetic dipole moments corresponding only to the electron spin contribution. The orbital moment is said to be quenched. In such materials Eq. (S6.1) can then be replaced by a "spin-only" formula:

$$\mathbf{m} = [n(n+2)]^{1/2}\mu_\mathrm{B}$$

where n is the number of unpaired d electrons on each ion.

S6.2   Types of Magnetic Material

Magnetic materials can be classified in terms of the arrangements of magnetic dipoles in the solid. These dipoles can be thought of, a little imprecisely, as microscopic bar magnets attached to the various atoms present. Materials with no elementary magnetic dipoles at all are *diamagnetic*.

Paramagnetic solids are those in which some of the atoms, ions, or molecules making up the solid possess a permanent magnetic dipole moment. These dipoles are isolated from one another. The solid, in effect, contains small, noninteracting atomic magnets. In the absence of a magnetic field, these are arranged at random and the solid shows no net magnetic moment. In a magnetic field, the elementary dipoles will attempt to orient themselves parallel to the magnetic induction in the solid, and this will enhance the internal field within the solid and give rise to the observed paramagnetic effect (Fig. S6.1a).

The partial orientation of the elementary dipoles in a paramagnetic solid is counteracted by thermal agitation, and it would be expected that at high temperatures

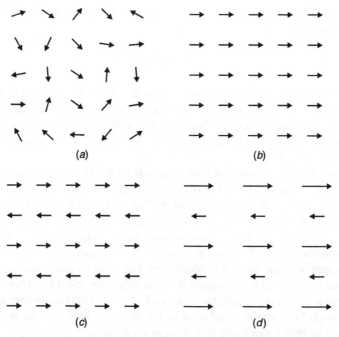

Figure S6.1 Magnetic structures: (*a*) paramagnetic solid, (*b*) ferromagnetic ordering, (*c*) antiferromagnetic ordering, and (*d*) ferrimagnetic ordering. The magnetic dipoles on atoms are represented by arrows.

the random motion of the atoms in the solid would cancel the alignment due to the magnetic field. The paramagnetic susceptibility would therefore be expected to vary with temperature. The temperature dependence is given by the Curie law:

$$\chi = \frac{C}{T}$$

where χ is the magnetic susceptibility, T is the absolute temperature, and C is the Curie constant. Curie law dependence in a solid is indicative of the presence of isolated paramagnetic ions or atoms in the material.

Interacting magnetic dipoles can produce a variety of magnetic properties in a solid. *Ferromagnetic* materials are those in which the magnetic dipoles align parallel to each other over considerable distances in the solid (Fig. S6.1*b*). An intense external magnetic field is produced by this alignment. Ferromagnetism is associated with the transition elements, with unpaired *d* electrons, and the lanthanides and actinides, with unpaired *f* electrons.

Above a temperature called the Curie temperature, T_C, all ferromagnetic materials become paramagnetic. The transition to a paramagnetic state comes about when thermal energy is greater than the magnetic interactions and causes the dipoles to disorder. Well above the Curie temperature, ferromagnetic materials obey the Curie–Weiss law:

$$\chi = \frac{C}{T - \theta}$$

The Curie–Weiss constant, θ, is positive, has the dimensions of temperature, and a value usually close to, but not quite identical to, the Curie temperature, T_C. The transition is reversible, and on cooling, ferromagnetism returns when the magnetic dipoles align parallel to one another as the temperature drops through the Curie temperature.

It is energetically favorable in some materials for the elementary magnetic dipoles to align in an antiparallel fashion (Fig. S6.1c). These are called *antiferromagnetic* compounds. Above a temperature called the Néel temperature, T_N, this arrangement disorders and the materials revert to paramagnetic behavior. Cooling the sample through the Néel temperature causes the antiferromagnetic ordering to reappear.

An important group of solids have two different magnetic dipoles present, one of greater magnitude than the other. When these line up in an antiparallel arrangement, they behave rather like ferromagnetic materials (Fig. S6.1d). They are called ferrimagnetic materials. A ferrimagnetic solid shows a complex temperature dependence. This is because the distribution of the magnetic ions over the available sites is sensitive to both temperature and the spin interactions.

S6.3   Crystal Field Splitting

A feature of the transition-metal ions, especially noticeable in the $3d$ series, is that many ions have two different magnetic states, which is considered to be due to

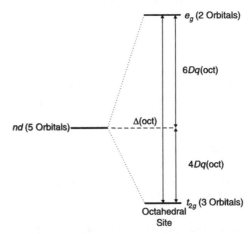

Figure S6.2   Crystal field splitting of the energy of five d orbitals when the ion is placed in a site with octahedral symmetry. The magnitude of the splitting, Δ or $10D_q$, depends upon the size of the octahedral site and the charges on the surrounding ions.

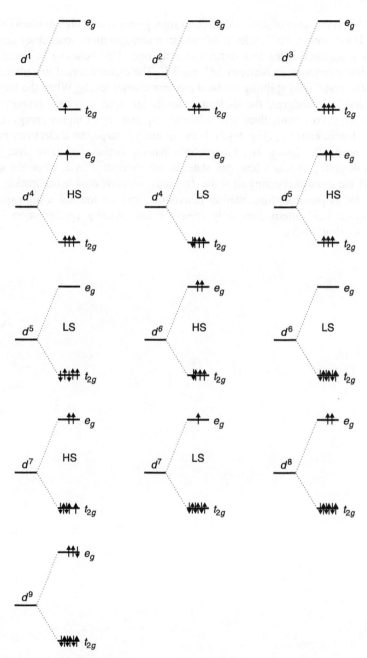

Figure S6.3 Electron configurations possible for d^n cations in an octahedral crystal field. For the ions d^4 to d^7, two configurations are possible. When the crystal field splitting is small, electrons avoid each other and produce a high-spin (HS) configuration. When the crystal field splitting is large, the electrons pair and produce a low-spin (LS) state.

two different amounts of unpaired electron spin, giving rise to high-spin and low-spin states. For example, Fe^{3+}, with a $3d^5$ electron configuration, sometimes appears to have five unpaired spins and sometimes only one. This behavior is found for $3d$ ions with configurations between $3d^4$ and $3d^7$. The explanation of this phenomenon lies with crystal field splitting of the d-electron energy levels. When the ions are in octahedral surroundings, the d-electron levels are split into two groups, one of lower energy, containing three levels, labeled t_{2g}, and one of higher energy, containing two levels, called e_g (Fig. S6.2). If the splitting is large, the d electrons preferentially occupy the lower, t_{2g}, group. Spin pairing occurs for more than three d electrons, giving rise to a low-spin state. In the alternative case, when the splitting is small, the electrons occupy all of the d orbitals, avoiding double occupation of orbitals as far as possible. Spin pairing is avoided, and the ion has a high-spin state. The crystal field alternatives only arise for ions with a configuration between d^4 and d^7 (Fig. S6.3).

Answers to Problems and Exercises

Chapter 1

Quick Quiz

1. (b)   **6. (c)**

2. (a)   **7. (a)**

3. (b)   **8. (b)**

4. (c)   **9. (b)**

5. (a)   **10. (a)**

Calculations and Questions

1. (a) $4.21 \times 10^{20}\,\text{cm}^{-3}$; (b) $MgGa_{1.75}Mn_{0.25}O_4$

2. (a) V_{Zr} most likely, $Zr_{0.796}\,S$; (b) $Ca_{0.1}Zr_{0.9}O_{1.9}$

3. (a) Holes; (b) concentration, 0.00095; $LaCoO_{3.00095}$

4. $182.7\,\mu\text{V K}^{-1}$

5. Representative points are: $La_{0.9}Sr_{0.1}MnO_3$, $\alpha = 189\,\mu\text{V K}^{-1}$; $La_{0.8}Sr_{0.2}MnO_3$, $\alpha = 119\,\mu\text{V K}^{-1}$

6. 3.26

7. (a) $Li_2O\ (MgO) \quad\longrightarrow\quad 2Li'_{Mg} + O_O + V_O^{2\bullet}$
   (b) $PrO_2 - \tfrac{1}{2}O_2 \quad\longrightarrow\quad V_O^{2\bullet} + 2e'$
   (c) $Ga_2O_3\ (TiO_2) \quad\longrightarrow\quad 2Ga'_{Ti} + 3O_O + V_O^{2\bullet}$
   (d) $2CaO\ (La_2O_3) \quad\longrightarrow\quad 2Ca'_{La} + 2O_O + V_O^{2\bullet}$
     or $2CaO\ (LaCoO_3) \quad\longrightarrow\quad 2Ca'_{La} + 2Co_{Co} + 5O_O + V_O^{2\bullet}$
   (e) $V_2O_3\ (Al_2O_3) \quad\longrightarrow\quad 2V_{Al} + 3O_O$
     or $V_2O_3\ (MgAl_2O_4) \quad\longrightarrow\quad Mg_{Mg} + 2V_{Al} + 4O_O$

8. (a) $\alpha = +k/e \ln\left[(1 - 2x)/x\right]$; (b) holes; (c) $0.0884\,\text{mV K}^{-1}$

9. $M_2^+O\ (MgO) \quad\longrightarrow\quad 2M'_{Mg} + O_O + V_O^{2\bullet}$

Defects in Solids, by Richard J. D. Tilley
Copyright © 2008 John Wiley & Sons, Inc.

$$MO\ (MgO) \longrightarrow M_{Mg} + O_O$$
$$M_2O_3\ (MgO) \longrightarrow 2M_{Mg}^\bullet + V_{Mg}^{2\prime} + 3O_O$$
$$MO_2\ (MgO) \longrightarrow M_{Mg}^{2\bullet} + V_{Mg}^{2\prime} + 2O_O$$
$$M_2O_5\ (MgO) \longrightarrow 2M_{Mg}^{3\bullet} + 3V_{Mg}^{2\prime} + 5O_O$$

10. (a) 3565 kg m^{-3}; (b) 4502 kg m^{-3}; (c) 3840 kg m^{-3}; positive deviation from Vegard's law

Chapter 2

Quick Quiz

1. (b)   6. (a)

2. (a)   7. (a)

3. (b)   8. (b)

4. (c)   9. (a)

5. (c)   10. (c)

Calculations and Questions

1.

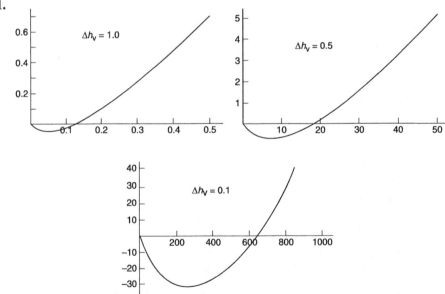

2.

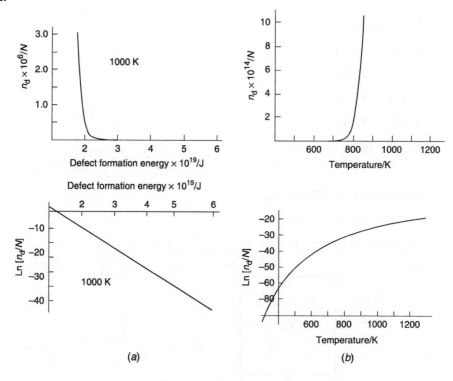

(a)   (b)

3. (a) 3.2×10^{-5}; (b) $1.88 \times 10^{24} \, \text{m}^{-3}$; (c) $1.88 \times 10^{24} \, \text{m}^{-3}$

4. $n_S \approx N \exp(-\Delta H_S/3RT)$

5. (a) anion Frenkel: $V_F^\bullet + F_i'$; cation Frenkel, $V_{Pb}^{2\prime} + Pb_i^{2\bullet}$; Schottky, $V_{Pb}^{2\prime} + 2F_i'$
 (b) $aF : S : cF \approx 1 : 2.6 \times 10^{-6} : 1.3 \times 10^{-17}$

6. (a) $3.7 \times 10^{-19} \, \text{J} = 2.3 \, \text{eV}$; (b) $1.95 \times 10^{22} \, \text{m}^{-3}$

7. $1.04 \times 10^{-25} \, \text{m}^{-3}$ (for both)

8. $E = (2/4 \, \pi \, \varepsilon_0) \{[q_1 \, q_2/(r_1 - r_2)] + [q_1 \, q_3/(r_1 - r_3)] + [q_2 \, q_3/(r_2 - r_3)]\}$

9. (a) $4.78 \times 10^{24} \, \text{m}^{-3}$; (b) $9.56 \times 10^{24} \, \text{m}^{-3}$; $6.99 \times 10^{17} \, \text{g}^{-1}$

10. (a) 7.90×10^{-15}; (b) $2.61 \times 10^{14} \, \text{m}^{-3}$

11. $200 \, \text{kJ mol}^{-1}$

Chapter 3

Quick Quiz

1. (b)   6. (b)
2. (c)   7. (b)
3. (a)   8. (a)
4. (b)   9. (c)
5. (a)   10. (c)

Calculations and Questions

1. (a) 0.255 nm; (b) 0.147 nm
2. (a) $\mathbf{a}/2[110]$; (b) $[111]$
3.

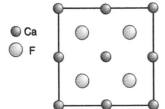

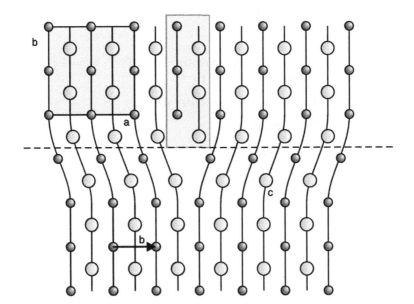

4. $\mathbf{b} = [010]$; $\mathbf{b}1 = \frac{1}{3}[100]$; $\mathbf{b}2 = \frac{1}{3}[010]$; $\mathbf{b}3 = \frac{1}{3}[110]$; $\mathbf{b}4 = \frac{1}{3}[010]$

5. $\mathbf{b} = [110]$; $\mathbf{b}1 = \frac{1}{3}[120]$; $\mathbf{b}2 = \frac{1}{3}[210]$

6. 0.49 nm

7. (a) Antiphase boundary; (b) twin; (c) twin; (d) no boundary; (e) antiphase boundary; (f) twin

8. (100), Ca; (200) Ca; (400), F; (110), CaF; (220), CaF

9. (a) VO_2 O_3 VO_2 O_3 VO_2 O_3 **VO_2** O_3 VO_2 O_3 VO_2 O_3 VO_2 (twin plane bold); (b) no composition change

10. (a) $Y_2O_3 \, (BaO) \rightarrow 2Y_{Ba}^\bullet + 2O_O + O_i^{2\prime}$
    $[Ti_2O_3 \, (TiO_2) \rightarrow 2Ti_{Ti4}^\prime + 3O_O + V_O^{2\bullet}]$
    Adding: $Y_2O_3 \, (BaTiO_3) \rightarrow 2Y_{Ba}^\bullet + 2Ti_{Ti4}^\prime + 6O_O$
    (b) $Nb_2O_5 \, (TiO_2) \rightarrow 2Nb_{Ti4}^\bullet + 4O_O + O_i^{2\prime}$
    $[Ti_2O_3 \, (TiO_2) \rightarrow 2Ti_{Ti4}^\prime + 3O_O + V_O^{2\bullet}]$
    Adding: $Nb_2O_5 \, (TiO_2) \rightarrow 2Nb_{Ti4}^\bullet + 2Ti_{Ti4}^\prime + 8O_O$

Chapter 4

Quick Quiz

1. (b)   6. (c)

2. (c)   7. (c)

3. (b)   8. (b)

4. (a)   9. (a)

5. (a)   10. (b)

Calculations and Questions

1. $V_{Ti}^{4\prime} + V_O^{2\bullet}$; $Lu_{Ti}^\prime + V_O^{2\bullet}$; $Lu_{Ti}^\prime$, $Ti_i^{4\bullet}$

2. (a) $0.4CaTiO_3 + 0.6La_{2/3}TiO_3$; (b) $0.5CaTiO_3 + 0.5\,La_{2/3}TiO_3$; 1339.7 kg m^{-3}

3. (a) $Mg_{0.3}Al_{1.4}O_{2.4}$, $Mg_{0.5}Al_{2.333}O_4$, $Mg_{0.429}Al_2O_{3.43}$
   (b) $4Al_2O_3 \, (MgAl_2O_4) \rightarrow V_{Mg}^{2\prime} + 2Al_{Mg}^\bullet + 6Al_{Al} + 12O_O$
   (c) $Al_2O_3 \, (MgAl_2O_4) \rightarrow V_{Mg}^{2\prime} + 2Al_{Al} + 3O_O + V_O^{2\bullet}$

4. (a) $BaTi_2Al_2O_8$;   (b) $Ba_{0.9}Cs_{0.1}Ti_{2.2}Al_{1.8}O_8$;   (c) $Ba_{0.65}Sr_{0.35}Ti_2Al_2O_8$;
   (d) $Ba_{0.83}La_{0.17}Ti_{1.83}Al_{2.17}O_8$

5. $Bi_8Pb_{4n}S_{4n+12}$ or $(Bi_2S_3)_4(PbS)_N$, $N = 4n$

6. (a) $(Ca, Na)_5Nb_5O_{17}$;   **(b)** $(Ca, Na)_{13}Nb_{13}O_{45}$;   **(c)** $(Ca, Na)_{14}Nb_{14}O_{48}$;
    (d) $(Ca, Na)_{11}Nb_{11}O_{37}$

7. (a) $Sr_{1.1}TiS_3$; 3.0 nm; **(c)** 3024 nm

8. (a) $Ca_{0.847}U_{0.153}F_{2.306}$ $(MF_{2.306})$; **(b)** $Ca_{0.644}U_{0.116}F_2$ $(M_{0.76}F_2)$

9. (a) $Na_{2.2}Al_{11}O_{17.6}$, Na and O interstitials; **(b)** $Na_{2.125}Al_{10.625}O_{17}$, Na interstitials
    and Al vacancies

10. (a) $Ti_{2.57}N$, Ti interstitials; **(b)** $TiN_{0.36}$, N vacancies; **(c)** ρ (vacancies)/ρ
    (perfect) $= p$; **(d)** $Ti_{0.95}N_{0.95}$

Chapter 5

Quick Quiz

1. (b)   **6. (a)**

2. (a)   **7. (c)**

3. (b)   **8. (c)**

4. (c)   **9. (b)**

5. (a)   **10. (c)**

Calculations and Questions

1. $3.6 \times 10^{-14}\, m^2\, s^{-1}$

2. (a) $386.4\, a^2\, s^{-1}$; $3.48 \times 10^{-17}\, m^2\, s^{-1}$

3. $232\, kJ\, mol^{-1} = 2.41\, eV$

4. $87.9\, kJ\, mol^{-1}$

5. $99.0\, kJ\, mol^{-1}$

6. $322°C$

7. $D = \gamma a^2 v\, \exp(-\Delta H_i/RT)\, \exp(-\Delta H_m/RT)$

8. $1.854 \times 10^{-17}\, m^2\, s^{-1}$

9. Bell-shaped curve, representative points: $x = 0, c_x = 1.66 \times 10^7$; $x = 2.0 \times 10^{-7}$ m,
    $c_x = 1.17 \times 10^7$; $x = 4.0 \times 10^{-7}$ m, $c_x = 4.35 \times 10^6$

10. Almost linear, representative points: $A_r(0) = 10000$; $x = 2.3 \times 10^{-4}$ m, $A_r = 7 \times 10^3$; $x = 5.6 \times 10^{-4}$ m, $A_r = 4.6 \times 10^3$

11. Bell-shaped curve, representative points: $x = 0, c_x = 7.0 \times 10^5$; $x = 2.0 \times 10^{-4}$ m,
    $c_x = 5.84 \times 10^5$; $x = 4.8 \times 10^{-4}$ m, $c_x = 2.72 \times 10^5$

12. Almost linear, representative points: $A_r(0) = 100$; $x = 1.95 \times 10^{-6}$ m, $A_r = 80$;
    $x = 6.0 \times 10^{-6}$ m, $A_r = 42.6$

Chapter 6

Quick Quiz

1. (c)	**6.** (c)
2. (a)	**7.** (c)
3. (b)	**8.** (b)
4. (b)	**9.** (a)
5. (a)	**10.** (b)

Calculations and Questions

1. $181 \text{ kJ mol}^{-1} = 1.88 \text{ eV}$

2. (a) $52.1 \text{ kJ mol}^{-1} = 0.54 \text{ eV}$; (b) $15.4 \text{ kJ mol}^{-1} = 0.16 \text{ eV}$

3. (a) Nb_2O_5 (EuNbO$_4$) $\rightarrow 2\text{V}_{\text{Eu}}^{3/} + 2\text{Nb}_{\text{Nb}} + 5\text{O}_\text{O} + 3\text{V}_\text{O}^{2\bullet}$; (b) an increase in $[\text{V}_\text{O}^{2\bullet}]$ increases the O ion conductivity; (c) 72.6 kJ mol^{-1}

4. (a) 87.9 kJ mol^{-1}; (b) $1.63 \times 10^{-13} \text{ m}^2 \text{ s}^{-1}$; (c) 0.20

5. This is due to charge neutrality, $1 \text{ Na}^+ = \frac{1}{2}\text{M}^{2+} = \frac{1}{3}\text{M}^{3+}$. If more is added, there must either be more $\text{O}_i^{2/}$ in the interlayers or $\text{V}_{\text{Al}}^{3/}$ in the spinel blocks. Interstitial O in the spinel blocks is unlikely.

6. 3.66 V

7. (a) $+414 \text{ mV}$; 0.0954 atm

8. (a) Pt, p_{O_2}', BaTiO$_3\|$ZrO$_2\|p_{\text{O}_2}''$, Pt; (b) BaTiO$_{2.99997}$

9. (a) $400°C$: R(bulk) $5 \text{ k}\Omega$, R(grain boundary) $29 \text{ k}\Omega$; $500°C$: R(bulk) $2 \text{ k}\Omega$, R(grain boundary) $16 \text{ k}\Omega$; $600°C$: R(bulk) $1 \text{ k}\Omega$, R(grain boundary) $9 \text{ k}\Omega$; (b) 46 kJ mol^{-1}; (c) 35 kJ mol^{-1}

10. (a) $2.53 \times 10^{-12} \text{ m}^2 \text{ s}^{-1}$; (b) ≈ 1

11. (a) $6.20 \times 10^{-4} \text{ S m}^{-1}$; (b) the values are similar so that both diffusion and conductivity are likely to take place by the same mechanism.

Chapter 7

Quick Quiz

1. (b)	**6.** (b)
2. (b)	**7.** (c)
3. (c)	**8.** (a)

4. (a)   **9. (c)**

5. (a)   **10. (a)**

Calculations and Questions

1. (a) $\frac{x}{2} O_2$ (CoO) $\rightarrow x\, O_O + x\, V''_{Co} + 2\, x\, Co^\bullet_{Co}$ or $\frac{x}{2} O_2$ (CoO) $\rightarrow x\, O_O + x\, V''_{Co} + 2x\, h^\bullet$; **(b)** 0.002/0.999

2. A oxide + B oxide + B metal

3. The slope of the log σ versus log p_{O_2} graph is $1/5.2$. A slope of $\frac{1}{6}$ corresponds to Co^{3+} in $Co_{1-x}O$, a slope of $\frac{1}{4}$ corresponds to Co^{4+}, so the defects are almost all Co^{3+}, but there may be some Co^{4+} present.

4. The slope of the log σ versus log p_{O_2} graph is $\frac{1}{6}$, corresponding to Ni^{3+} in $Ni_{1-x}O$

5. Doubly charged vacancies fit best.

6.

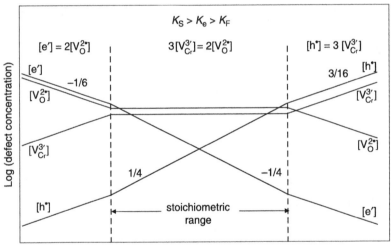

7. At high pressure: $\sigma \propto [h^\bullet] \propto p_{O_2}{}^{1/4}$; at low pressure: $\sigma \propto [e'] \propto p_{O_2}{}^{-1/4}$

8. $[e']$: $2K_o p[e']^4 + K_e^2 p^{1/2} [e']^3 - 2K_r K_e^2 - K_e^2 p^{1/2} [e'] = 0$

   $[e']$: $2 \times 10^{56} p[e']^4 + 1 \times 10^{80} p^{1/2} [e']^3 - 2 \times 10^{136} - 1 \times 10^{120} p^{1/2} [e'] = 0$

   $[V''_M]$: $2K_S^{1/2} p^{1/4} [V''_M]^2 + K_r^{1/2} [V''_M]^{3/2} - 2K_S^{3/2} p^{1/4} - K_o^{1/2} K_S^{1/2} p^{1/2} [V''_M]^{1/2} = 0$

   $[V''_M]$: $2 \times 10^{16} p^{1/4} [V''_M]^2 + 1 \times 10^{28} [V''_M]^{3/2} - 2 \times 10^{48} p^{1/4} -$
   $\qquad 1 \times 10^{44} p^{1/2} [V''_M]^{1/2} = 0$

   $[V_X^{\bullet\bullet}]$: $2K_S^{3/2} p^{1/4} + K_r^{1/2} K_S^{1/2} [V_X^{\bullet\bullet}]^{1/2} - 2K_r^{1/2} p^{1/4} [V_X^{\bullet\bullet}]^2 + K_o^{1/2} p^{1/2} [V_X^{\bullet\bullet}]^{3/2} = 0$

$[V_X^{2\bullet}]$: $2 \times 10^{48} p^{1/4} + 1 \times 10^{44} [V_X^{2\bullet}]^{1/2} - 2 \times 10^{16} p^{1/4} [V_X^{2\bullet}]^2 +$
$1 \times 10^{28} p^{1/2} [V_X^{2\bullet}]^{3/2} = 0$

9. (a) Electrons; (b) Co^{2+}, Co^{3+}; (c) Co^{3+}; (d) $Bi_2^{3+} Sr_2^{2+} Co_{0.320}^{2+} Co_{0.680}^{3+} O_{6+\delta}$; (e) 0.340

10. (a) Fe, $\sim$FeO, Fe_3O_4, Fe_2O_3; (b) $\sim$FeO is nonstoichiometric; (c) $Fe + \frac{1}{2}O_2 \rightarrow$ FeO, $3FeO + \frac{1}{2}O_2 \rightarrow Fe_3O_4$, $\frac{2}{3}Fe_3O_4 + \frac{1}{6}O_2 \rightarrow Fe_2O_3$; (d) -174.2 kJ mol^{-1}, -142.8 kJ mol^{-1}, -14.0 kJ mol^{-1}; (e) 4.37×10^{-13} atm

Chapter 8

Quick Quiz

1. (c)   6. (b)

2. (c)   7. (c)

3. (a)   8. (b)

4. (b)   9. (a)

5. (a)   10. (c)

Calculations and Questions

1. (a) $Al_2O_3 (TiO_2) \rightarrow 2Al'_{Ti} + 3O_O + V_O^{2\bullet}$
   $Al_2O_3 (TiO_2) + \frac{1}{2}O_2 \rightarrow 2Al'_{Ti} + 4O_O + 2h^\bullet$
   $2Al_2O_3 (TiO_2) \rightarrow 3Al'_{Ti} + 6O_O + Al_i^{3\bullet}$

   (b) $2Nb_2O_5 (TiO_2) \rightarrow 4Nb_{Ti}^\bullet + 10O_O + V_{Ti}^{4'}$
   $Nb_2O_5 (TiO_2) \rightarrow 2Nb_{Ti}^\bullet + 4O_O + O_i^{2'}$
   $Nb_2O_5 (TiO_2) \rightarrow 2Nb_{Ti}^\bullet + 4O_O + \frac{1}{2}O_2 + 2e'$

   (c) $NiO (Mn_3O_4) \rightarrow Ni_{Mn2} + 2Mn_{Mn3} + 4O_O$
   $2NiO (Mn_3O_4) \rightarrow 2Ni'_{Mn3} + Mn_{Mn2} + 3O_O + V_O^{2\bullet}$

   (d) $Cr_2O_3 (MgO) \rightarrow 2Cr_{Mg}^\bullet + 2O_O + O_i^{2'}$
   $Cr_2O_3 (MgO) \rightarrow 2Cr_{Mg}^\bullet + 3O_O + V_{Mg}^{2'}$
   $Cr_2O_3 (MgO) \rightarrow 2Cr_{Mg}^\bullet + 2O_O + \frac{1}{2}O_2 + 2e'$

   (e) $5Fe_2O_3 (Nb_2O_5) \rightarrow 6Fe_{Nb}^{2'} + 15O_O + 4Fe_i^{3\bullet}$
   $Fe_2O_3 (Nb_2O_5) \rightarrow 2Fe_{Nb}^{2'} + 3O_O + 2V_O^{2\bullet}$
   $Fe_2O_3 (Nb_2O_5) \rightarrow 2Fe_{Nb}^{2'} + 3O_O + 4h^\bullet$

2. 5560 K

3. Donors: $2 \times 10^{116} p + 1 \times 10^{96} [h^\bullet] p^{1/2} - 2 \times 10^{52} [h^\bullet]^4 - 1 \times 10^{64} [h^\bullet]^3 p^{1/2} - 1 \times 10^{89} [h^\bullet]^2 p^{1/2} = 0$

Acceptors: $2 \times 10^{116}p + 1 \times 10^{96}[h^\bullet]p^{1/2} + 1 \times 10^{89}[h^\bullet]^2p^{1/2} - 2 \times 10^{52}[h^\bullet]^4 - 1 \times 10^{64}[h^\bullet]^3p^{1/2} = 0$

4. (a) $2MgO\ (Ti_2O_3) + \frac{1}{2}O_2 \to 2Mg'_{Ti} + 3O_O + 2h^\bullet$

   or $2MgO\ (MgTi_2O_3) + \frac{1}{2}O_2 \to Mg_{Mg} + 2Mg'_{Ti} + 4O_O + 2h^\bullet$

   or $2MgO\ (MgTi_2O_3) + \frac{1}{2}O_2 \to 2Mg_{Mg} + 2Mg'_{Ti} + 2Ti^\bullet_{Ti} + 4O_O$

   (b) Mg_2TiO_4, that is, $(Mg^{2+})[Mg^{2+}Ti^{4+}]O_4$

   (c) Insulator

5. (a) Charge reservoir: $+2$; superconducting layer: -2

   (b) Charge reservoir: $+2$; superconducting layer: -2

   (c) Charge reservoir: $+1$; superconducting layer: -2

   (d) Charge reservoir: $+2$; superconducting layer: -2

   (e) Charge reservoir: $+1$; superconducting layer: -1.5

6. (a) $3CeO_2\ (LaCoO_3) \to 3Ce^\bullet_{La} + V'''_{La} + 4Co_{Co} + 12O_O$

   or $2CeO_2\ (LaCuO_3) \to 2Ce^\bullet_{La} + 2Co_{Co} + 6O_O + O''_i$

   or $2CeO_2\ (LaCoO_3) \to 2Ce^\bullet_{La} + 2Co_{Co} + 6O_O + \frac{1}{2}O_2 + 2e'$

   or $2CeO_2\ (LaCoO_3) \to 2Ce^\bullet_{La} + 2Co'_{Co} + 6O_O + \frac{1}{2}O_2$

   (b) Electrons

   (c) 0.03

7. Slope $= 1/5.5$. The equation is in agreement with holes plus oxygen interstitials:
   $\frac{1}{2}O_2 \to O''_i + 2h^\bullet$

8. $[Pr'] = K_e/[h^\bullet] = K_e/[Yb'_{Pr}];$ slope $= -1$

   $[V^{2\bullet}_O] = [h^\bullet]^2/K_op^{1/2}_{O_2} = [Yb'_{Pr}]^2/K_op^{1/2}_{O_2};$ slope $= +2$

   $[OH^\bullet_O] = K^{1/2}_w[Yb'_{Pr}]p^{1/2}_{H_2O}/K^{1/2}_op^{1/2}_{O_2};$ slope $= +1$

9. (a) $2NiO\ (V_2O_3) \to 2Ni'_V + 2O_O + V^{2\bullet}_O$

   $V^{2\bullet}_O + \frac{1}{2}O_2 \to O_O + 2h^\bullet$

   Adding: $2NiO\ (V_2O_3) \to 2Ni'_V + 2O_O + 2h^\bullet$

   (b) $7.85\ kJ\ mol^{-1}$; (c) $Ni_{2/3}V^{3+}_{2/3}V^{4+}_{2/3}O_3$

10. (a) Low O_2 pressure: negative slope, n-type electronic conductivity; high O_2 pressure: positive slope, p-type electronic conductivity; middle O_2 pressure: no pressure dependence, proton conductivity

    (b) Taking WO_3 as the "sleeping partner": $2CaO\ (La_2O_3) \to 2Ca'_{La} + 2O_O + V^{2\bullet}_O$ or: $6CaO\ (La_6WO_{12}) \to 6Ca'_{La} + W_W + 9O_O + 3V^{2\bullet}_O$

    (c) $H_2O + O_O + V^{2\bullet}_O \to 2OH^\bullet_O$

    (d) At low O_2 pressure reduction occurs: $O_O \to V^{2\bullet}_O + 2e' + \frac{1}{2}O_2$ This leads to $\sigma \propto [e'] \propto p^{-1/4}_{O_2}$; at high O_2 pressure oxidation occurs: $V^{2\bullet}_O + \frac{1}{2}O_2 \to O_O + 2h^\bullet$. This leads to $\sigma \propto [h^\bullet] \propto p^{1/4}_{O_2}$.

Chapter 9

Quick Quiz

1. (c)   **6.** (b)

2. (a)   **7.** (c)

3. (c)   **8.** (a)

4. (b)   **9.** (c)

5. (a)   **10.** (b)

Calculations and Questions

1. (a) $e^4 t^3$: ↓↑ ↓↑   ↑↑↑
   (b) 3.87 μ_B
   (c) 1.88 eV = 3.01×10^{-19} J

2. The material is mainly Mn^{3+}. At $x = 0.1$ there might be some Mn^{2+} present, and at $x = 1.0$ there might be some Mn^{4+} present.

3. (a) The band gap transition is at a wavelength of 387.5 nm (ultraviolet).
   (b) 2.49 eV = 3.99×10^{-19} J; 2.65 eV = 4.24×10^{-19} J
   (c) 3.52 eV

4. 33.5%

5. The slope is approximately 1.7, so the process is likely to be a two-photon excitation.

6. (a) $KF_{0.63}Cl_{0.37}$
   (b) 2.7 eV = 4.3×10^{-19} J
   (c) 8.8 eV = 14.1×10^{-19} J
   (d) 9.9 eV = 15.9×10^{-19} J

7. (a) Mn^{2+} LS, 1.73 μ_B; Mn^{2+} IS, 3.87 μ_B; Mn^{2+} HS, 5.92 μ_B; Mn^{3+} HS, 4.90 μ_B; Mn^{4+} HS, 3.87 μ_B
   (b) Ratio is 5.08×10^{-17}; none in the upper states
   (c) 1278 nm (infrared)

8. (a) 5.92 μ_B; (b) 1.03 nm

9. (a) 1.85 nm; (b) 0.52 nm; (c) yes, they would overlap

10. (a) 9.18; (b) 1004.5 K

Nonstoichiometric compounds such as $\sim$FeO, $Ni_{1-x}O$, $LaCoO_{3+\delta}$, $LaCoO_{3+x/2}$ etc. are indexed under the notionally stoichiometric phase, i.e. FeO, NiO, $LaCoO_3$.

Doped phases such as $Al_{1.995}Cr_{0.005}O_3$, $LaNi_xCo_{1-x}O_3$ etc. are classed as subset of the parent compound, i.e. Al_2O_3, $LaCoO_3$.

Phases given ionic charges, such as $Ba^{2+}Zr^{4+}O_3$, are mostly indexed under the parent phase without ionic charges, i.e. $BaZrO_3$.

Defects in Solids, by Richard J. D. Tilley
Copyright © 2008 John Wiley & Sons, Inc.

Defects in Solids, by Richard J. D. Tilley
Copyright © 2008 John Wiley & Sons, Inc.

Printed and bound by CPI Group (UK) Ltd, Croydon, CR0 4YY

16/04/2025

14658344-0005